NUCLEAR SYSTEMS II

NUCLEAR SYSTEMS II

Elements of Thermal Hydraulic Design

Neil E. Todreas
Mujid S. Kazimi
Massachusetts Institute of Technology

NUCLEAR SYSTEMS II Elements of Thermal Hydraulic Design

Copyright © 2001 by Taylor and Francis. All rights reserved.
Printed in the United States of America. Except as permitted under the United States Copyright Act of 1976, no part of this publication may be reproduced or distributed in any form or by any means, or stored in a data base or retrieval system, without the prior written permission of the publisher.

Library of Congress Cataloging-in-Publication Data

Todreas, Neil E.
  Nuclear systems.

  Includes bibliographical references and index.
  Contents: 1. Thermal hydraulic fundamentals — 2. Elements of thermal hydraulic design.
  1. Nuclear reactors—Fluid dynamics. 2. Heat—Transmission. 3. Nuclear power plants. I. Kazimi, Mujid S.
TK9202.T59 1990   621.48′3   89-20023

ISBN 0-89116-936-9 (case)
ISBN 1-56032-079-6 (paper)

To our families for their support in this endeavor
Carol, Tim and Ian
Nazik, Yasmeen, Marwan and Omar

CONTENTS

PREFACE

This book can serve as a textbook for two to three courses at the advanced undergraduate and the graduate student level. It is also suitable as a basis for continuing education of engineers in the nuclear power industry, who wish to expand their knowledge of the principles of thermal analysis of nuclear systems. The book, in fact, was an outgrowth of the course notes used for teaching several classes at MIT over a period of nearly 15 years.

The book is meant to cover more than thermal hydraulic design and analysis of the core of a nuclear reactor. Thus, in several parts and examples, other components of the nuclear power plant such as the pressurizer, the containment and the entire primary coolant system are addressed. In this respect the book reflects the importance of such considerations in thermal engineering of a modern nuclear power plant. The traditional concentration on the fuel element design in earlier textbooks was appropriate when the fuel performance had a higher share of the cost of electricity than in modern plants. The cost of the nuclear electricity proved to be more influenced by the steam supply system and the containment building than previously anticipated.

The desirability of providing in one book the basic concepts as well as the complex formulations for advanced applications has resulted in a more comprehensive textbook than any previously authored in the field. The basic ideas of both fluid flow and heat transfer as applicable to nuclear reactors are discussed in Volume I. No assumption is made about the degree to which the reader is already familiar with the subject. Therefore, various reactor types, energy source distribution, and fundamental laws of conservation of mass, momentum, and energy are presented in early chapters. Engineering methods for analysis of flow hydraulics and heat transfer in single-phase as well as two-phase coolants are presented in later chapters. In Volume II, applications of the fundamental ideas to the multichannel flow conditions in the reactor are described as well as specific design considerations such as natural convection and core reliability. They are presented in a way that renders it possible to use the analytical development in simple exercises and as the bases for numerical computations similar to those commonly practiced in the industry.

A consistent nomenclature is used throughout the text and a table of the nomenclature is included in the Appendices. Each chapter includes problems identified as to their topic and the section from which they are drawn. While the SI unit system is principally used, British Engineering Units are given in brackets for those results commonly still reported in the United States in this system.

ACKNOWLEDGMENTS

Much material in Volume I of this book originated from lectures developed at MIT by Professor Manson Benedict with Professor Thomas Pigford for a subject in nuclear reactor engineering and by Professors Warren Rohsenow and Peter Griffith for a subject in boiling heat transfer and two-phase flow. We have had many years of pleasant association with these men as their students and colleagues and owe a great deal of gratitude to them for introducing us to the subject material. The development of the book has benefited from the discussion and comments provided by many of our colleagues and students. In particular Professor George Yadigaroglu participated in the early stage of this work in defining the scope and depth of topics to be covered.

We are at a loss to remember all the other people who influenced us. However, we want to thank particularly those who were kind enough to review nearly completed chapters, while stressing that they are not to be blamed for any weaknesses that may still remain. These reviewers include John Bartzis, Manson Benedict, Greg Branan, Dae Cho, Michael Corradini, Hugo DaSilva, Michael Driscoll, Don Dube, Tolis Efthimiadis, Gang Fu, Elias Gyftopoulos, Pavel Hejzlar, Steve Herring, Dong Wook Jerng, John Kelly, Min Lee, Alan Levin, Joy Maneke, Mahmoud Massoud, John Meyer, Hee Cheon No, Klaus Rehme, Tae Sun Ro, Donald Rowe, Gilberto Russo, Robert Sawdye, Andre Schor, Nathan Siu, Kune Y. Suh, and Robert Witt. Finally, we want to express our appreciation to all students at MIT who proof-tested the material at its various stages of development and provided us with numerous suggestions and corrections that have made their way into the final text.

Most of the figures in this book were prepared by a number of students using a microcomputer under the able direction of Alex Sich. Many others have participated in the typing of the manuscript. We offer our warmest thanks to

Gail Jacobson, Paula Cornelio, and Elizabeth Parmelee for overseeing preparation of major portions of the final text.

A generous grant from the Bernard M. Gordon (1948) Engineering Curriculum Development Fund at MIT was provided for the text preparation and for this we are most grateful.

Mujid S. Kazimi
Neil E. Todreas

FORMULATION OF THE REACTOR THERMAL HYDRAULIC DESIGN PROBLEM

I INTRODUCTION

This chapter presents the mathematical definition of the reactor thermal analysis problem. The various sets of applicable boundary conditions are stressed. The solution procedures that are applicable to the various reactor types are introduced here and presented in detail in subsequent chapters.

II POWER REACTOR HYDRAULIC CONFIGURATIONS

The power reactor we wish to analyze is similar to that sketched in Figure 1-1. The typical core under consideration consists of a heterogeneous arrangement of fuel and coolant. The coolant channels are connected to common plena at both the inlet and outlet of the core.

In pressurized water reactors (PWRs) the coolant channels are in communication with each other over their entire length. In this case the core can be considered as a heterogeneous arrangement of continuously interacting coolant channels in parallel flow.

Prismatic graphite-moderated cores are a variant of this arrangement in that the coolant channels are transversely connected only at distinct axial planes as shown in Figure 1-2. The planes at which transverse coolant flow occurs correspond to the horizontal faces of the stacked graphite moderator

1

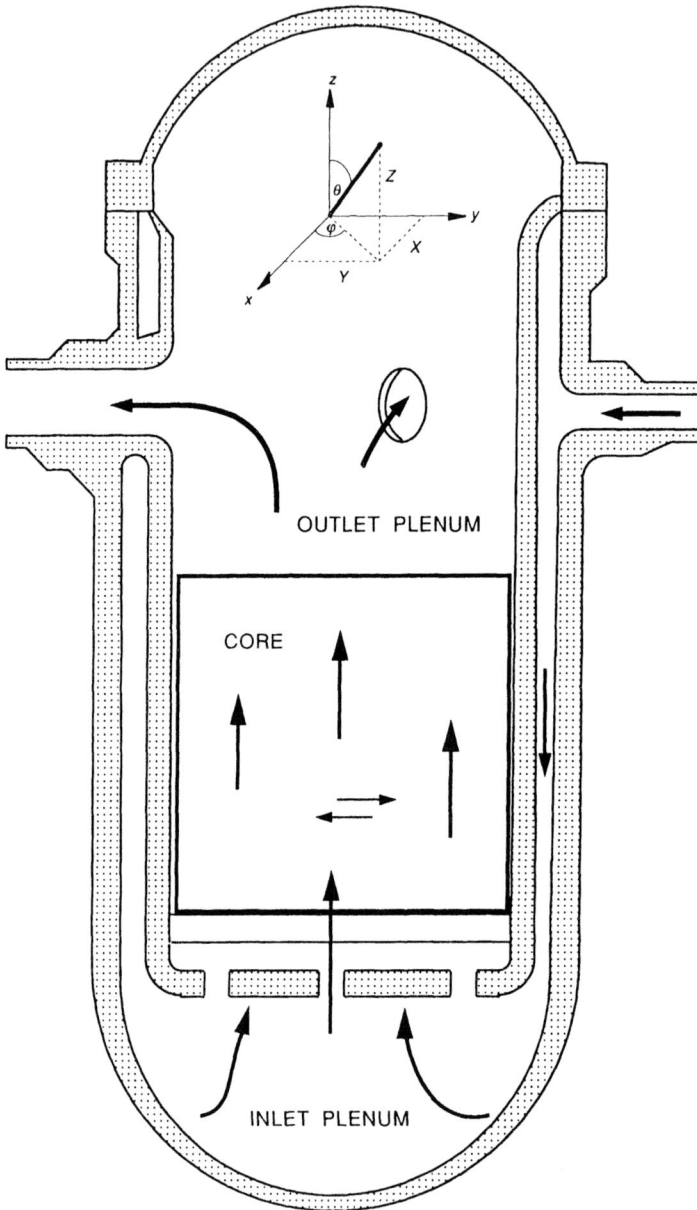

Figure 1-1 Reactor assembly schematic.

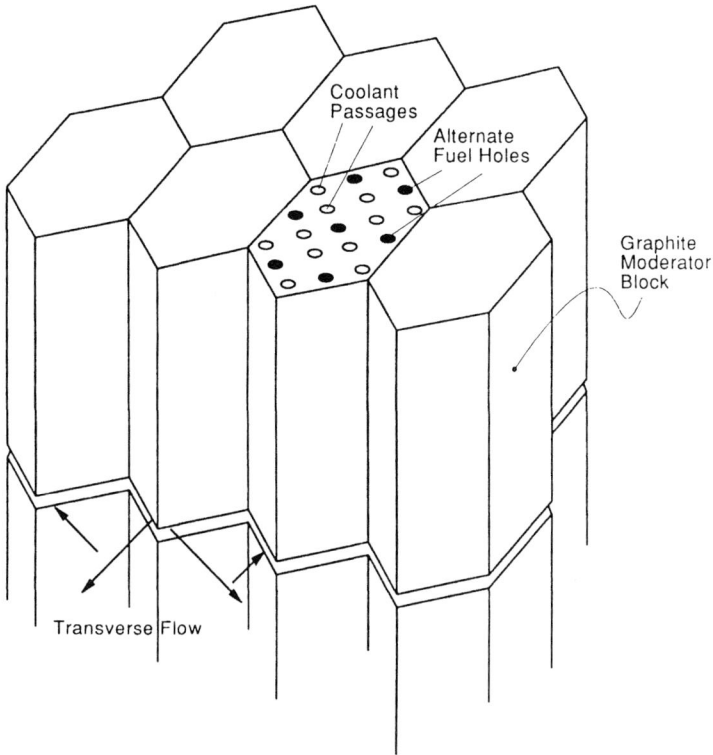

Figure 1-2 Coolant flow in graphite-moderated core of high temperature gas reactor (HTGR).

blocks. The transverse flow paths arise from the nonuniform dimensional changes between blocks which are caused by radiation damage.

In boiling water reactors (BWRs) and liquid metal cooled reactors (LMRs), on the other hand, groups of fuel pins are enclosed by flow boundaries called cans, ducts, or channels to form assemblies. This yields a heterogeneous, parallel flow arrangement with continuous mass, momentum, and energy exchanges taking place only between coolant channels within an assembly. However in LMRs, significant energy exchange can also occur between adjacent assemblies by conduction through the duct walls and the sodium-filled spaces between assemblies.

Generally in all reactor designs a fraction of the inlet coolant is bypassed around the core to maintain the core support structure and the thermal shields near inlet temperature conditions. The bypass flow is mixed with the core flow at the core outlet.

In practice the core neutronic and thermal/hydraulic behavior can be strongly coupled so that the energy generation rate cannot be prescribed independently of the coolant density or fuel temperature distributions. The need for consideration of this coupling exists principally for BWRs under steady-state as well as transient operation and for PWRs and LMRs under transients that may cause large coolant density changes.

Actual core boundary conditions reflect the physical arrangement of the reactor assembly as a component within a flow loop. Although outlet plenum designs generally yield a uniform pressure distribution at the core outlet, inlet plenum configurations often produce a situation in which the inlet mass flow and pressure can have radial variations. One must then consider the details of the flow in the inlet plenum in order to determine the pressure and velocity distributions at the inlet of the core. In some cases such calculations can be made using two- or three-dimensional flow formulations. When the inlet plenum is included in the region under study, the boundary at which the appropriate boundary conditions are applied must be located further upstream. It might be set, for example, at the inlet flow nozzle if the flow conditions are sufficiently well known there. Generally, however, the core inlet plane should be considered as the interface at which the solutions for the inlet plenum region and for the core region must be matched.

Since the core inlet pressure and velocity distributions are interrelated, any analysis of the core alone as a boundary value problem must be set up to ensure that the applied boundary conditions do not overspecify the problem.

III BOUNDARY CONDITIONS FOR THE HYDRAULIC PROBLEM

Historically because of the limited detail with which components could be represented in loop codes, reactor design and analysis have been performed in an iterative fashion using simple overall loop representations and separate detailed component representations. Consequently, reactor cores have been separately analyzed subject to imposed boundary conditions. Two fundamental types of boundary conditions exist—the pressure and the mass flow rate (or velocity) boundary conditions. The application of these boundary conditions depends on the problem definition in terms of:

• Subsonic or supersonic flow, i.e., the Mach number (M)
• Compressible or incompressible flow
• The dimensionality, i.e. single or multiple dimensions

In this chapter only subsonic flow ($M < 1$) will be considered. The general case will be compressible flow, with incompressibility considered as a limiting case. The dimensionality factor will be used to distinguish among the flow arrangements possible between fixed inlet and outlet plena. The flow arrangements useful for nuclear reactor technology are illustrated in Figure 1-3.

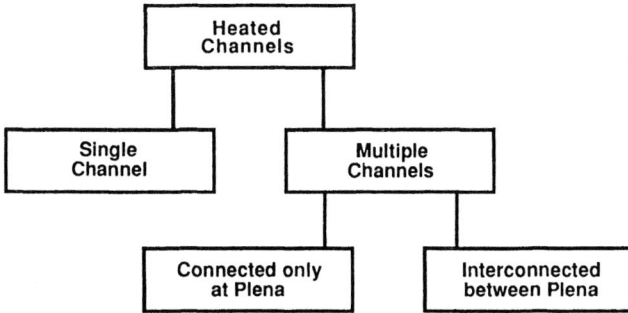

Figure 1-3 Flow arrangements in nuclear reactor technology.

IV PROBLEMS TREATED IN THIS BOOK

The problem statements presented in this chapter identify the spectrum of cases useful in nuclear reactor assembly analysis. Table 1-1 lists these statements and identifies the section in which the method of solution is presented. These problem statements range from the simplest case of a single channel between inlet and outlet plena to multiple channels between common inlet and outlet plena which are interconnected along their length. Additionally, Chapter 3 of this text treats flow loops. A flow loop can be viewed as a single channel whose exit closes upon its inlet. Chapter 3 treats single flow loops under both single- and two-phase conditions.

Table 1-1 Synopsis of reactor assembly problem statements and solutions

Problem	Problem definition	Problem solution
Single channel between plena (decoupled mass, momentum, energy coolant region solution under specified flow or pressure boundary conditions)	Table 1-3	Chapter 13 Volume I
Single channel between plena (coupled mass, momentum, energy coolant region solution under specified flow or pressure boundary conditions)	Table 1-3	Chapter 2
Multiple channels in parallel flow connected only at plena (decoupled mass, momentum, energy coolant region solution)	Table 1-4	Chapter 4
Multiple channels in parallel flow connected only at plena (coupled mass, momentum, energy coolant region solution)	Table 1-4	Chapter 4
Multiple interconnected channels in parallel flow	Table 1-6	Chapters 5, 6, 7

V FLOW IN SINGLE CHANNELS

The principal sets of boundary conditions for flow in a single channel will be described. Flow conditions are usually categorized by the assumption made regarding the dependence of density on pressure and enthalpy. Four cases will be considered involving possible assumptions of pressure and enthalpy influence. These cases are summarized in Table 1-2. For example, for a typical unheated, compressible flow, the density can be taken as only a function of pressure, i.e., $\rho(p)$ and a reference enthalpy, h^*, must be prescribed. The heated, incompressible case in which density is a function of enthalpy only, i.e., $p(h)$, is commonly called a thermally expandable flow.

A Unheated Channel

Consider a single, unheated channel bounded by inlet and outlet plena under compressible flow for which the density is taken as a function of pressure only and a reference enthalpy is prescribed. The principal boundary condition sets for subsonic flow in this channel are listed below. They are also illustrated in Figure 1-4.

Boundary condition set numbers	Imposed conditions
(1)	Pressure at inlet and outlet boundaries
(2)	Mass flow rate or velocity at inlet boundary; pressure at outlet boundary
(3)	Pressure at inlet boundary; mass flow rate or velocity at outlet boundary
(4)	Mass flow rate or velocity at inlet and outlet boundaries

Notice that for compressible, subsonic flow a single boundary condition is applied at each end of the flow channel since conditions at each end do affect flow behavior within the channel. In contrast, in supersonic flow, two boundary conditions must be applied at the boundary across which flow enters the channel. Further, the case of both boundary conditions applied at the outlet boundary does not exist.

For unheated, incompressible flow the boundary condition set is truncated because the density is no longer a function of local pressure. Rather, a single value of density is assumed which is usually prescribed by adoption of a reference pressure from which density as well as other necessary properties can be obtained. Further, since the density is constant, the mass flow rate is spatially uniform at each time, and the sonic propagation velocity is infinite.

Now, consider the boundary sets for unheated, incompressible flow. Since the density of the fluid has been established by the selected reference pressure and enthalpy, the flow conditions in the channel are not influenced in an inde-

Table 1-2 Four treatments of density dependence on pressure and enthalpy

	ρ Dependence	
	Required auxiliary information	
Channel type	Compressible	Incompressible
Unheated†	$\dfrac{\rho(p)}{h^*}$	$\dfrac{\rho = \text{Constant}}{p^*,h^*}$
Heated	$\dfrac{\rho(p,h)}{-}$	$\dfrac{\rho(h)}{p^*}$

† There may be unheated problems in which enthalpy variations are important (e.g., cases with time variation of enthalpy at
the inlet or with rapid pressure changes); in those cases, consider
the categories to be heated.

The error incurred in the computation of pressure variations by assuming fluid incompressibility can be found in terms of
stagnation parameters p_0, ρ_0, and c_0 as:

$$\frac{p_0 - p}{\frac{1}{2}\rho_0 V^2} = 1 - \frac{1}{4}\left(\frac{V}{c_0}\right)^2 + \cdots \text{ (isentropic flow, perfect gas)}$$

$$\text{for } \frac{V}{c_0} = 0.2, \frac{p_0 - p}{\frac{1}{2}\rho_0 V^2} - 1 = -0.01.$$

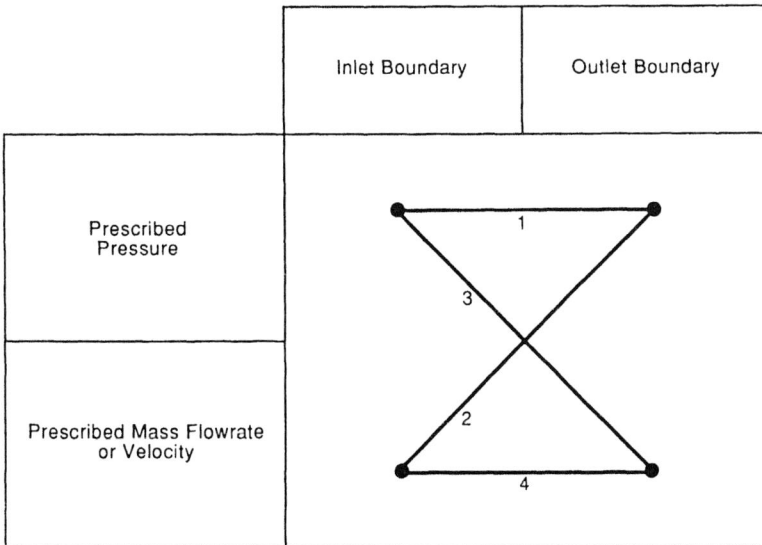

Figure 1-4 Boundary condition sets for subsonic, compressible flow in a single, unheated channel.

Table 1-3 Problem definition for single, heated channel in subsonic, compressible flow

	Specified flow condition	Specified pressure condition
Goals	Enthalpy, $h_{coolant}(z,t)$ Void fraction, $\alpha(z,t)$ Pressure drop, $p_{in}(t) - p_{out}(t)$	Enthalpy, $h_{coolant}(z,t)$ Void fraction, $\alpha(z,t)$ Flow rate, $\dot{m}(z,t)$
Constraints	Design limits (Section 2.4, Volume I) Geometry Materials	Design limits (Section 2.4, Volume I) Geometry Materials
Input conditions	Energy generation rate, $q'''_{coolant}(z,t)$ and/or surface heat flux, $q''(z,t)$ Friction factor, $f(h,p,G,q'')$	Energy generation rate, $q'''_{coolant}(z,t)$ and/or surface heat flux, $q''(z,t)$ Friction factor, $f(h,p,G,q'')$
Boundary conditions†	(2) $\dot{m}_{in}(t), p_{out}(t)$ Enthalpy, $h_{in}(t)$ (3) $p_{in}(t), \dot{m}_{out}(t)$ Enthalpy, $h_{in}(t)$ (4) $\dot{m}_{in}(t), \dot{m}_{out}(t)$ Enthalpy, $h_{in}(t)$	(1) $p_{in}(t), p_{out}(t)$ Enthalpy, $h_{in}(t)$

† Only $h_{in}(t)$ is needed if $\dot{m}_{in}(t)$ and $\dot{m}_{out}(t)$ are both positive, i.e., inflow at inlet and outflow at exit. If $\dot{m}_{out}$ is negative, i.e., inflow is occurring from the outlet plenum, then $h_{out}(t)$ must also be specified.

pendent manner by local values of pressure specified at the boundaries. They depend only on the pressure difference applied across the channel. Hence, boundary set (1) becomes (1) *prescribed pressure difference between plena.* Similarly, for sets (2) and (3) the prescribed pressures do not influence the density since it is established at the reference pressure and enthalpy. In practice then, the inlet or the outlet pressure is considered equal to the reference pressure. Further, the mass flow rate is spatially uniform so that only a mass flow rate value which applies to the entire channel can be prescribed. Hence, sets (2), (3), and (4) reduce to (2) *prescribed mass flow rate.* These boundary condition sets, together with the reference pressure and enthalpy, define the unheated incompressible flow problem to be analyzed.

B Heated Channel

Now for heated, compressible flow, an enthalpy (or temperature) boundary condition must be added. In practical cases only channel cross-sectional averages are considered so that the problem is formulated as a lumped parameter, one-dimensional situation. For compressible flow, each of the sets of boundary conditions of Figure 1-4 can be specified. Table 1-3 summarizes the problem statement for a single, heated channel in subsonic, compressible flow. For this one-dimensional, lumped parameter case, any surface heat addition per unit length is represented by an equivalent volumetric energy generation rate. If the channel being considered is a coolant passage heated by fuel rods, the relevant constraints are the design limits of Section 2.4, Volume I. The solutions of this problem for the decoupled and coupled conservation equation cases are addressed in Chapter 13, Volume I and Chapter 2, respectively.

For heated, incompressible flow, the density can change locally with enthalpy, i.e., a thermally expandable flow. The mass flow rates at the inlet and exit can differ, but they are interdependent. The two boundary sets identified for unheated, incompressible flow are applicable here with the note that for set (2), the boundary at which the mass flow rate needs to be identified must be specified. Additionally, as stated above, an enthalpy boundary condition must be specified.

VI FLOW IN MULTIPLE, HEATED CHANNELS CONNECTED ONLY AT PLENA

Figure 1-5 illustrates this case which represents BWR and LMR fuel assemblies between inlet and outlet plena. This arrangement is an array of N one-dimensional channels so that the hydrodynamic boundary conditions can be obtained by adaptation of those already presented in Figure 1-4.

The pressure-pressure boundary condition set is directly applicable. The specification of either the inlet or outlet pressure for this set (or any other set)

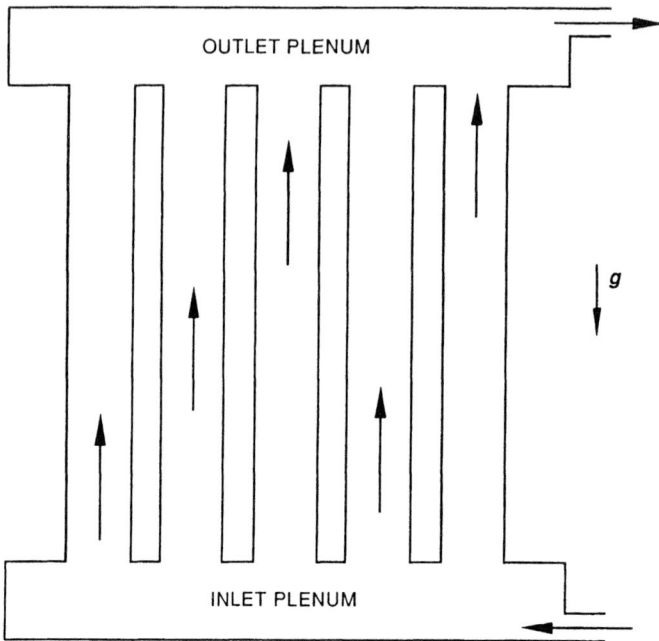

Figure 1-5 Array of channels connected only at plena.

requires the prior knowledge or assumption of the pressure distribution within both plena. The other boundary condition sets, (2) through (4), must now be expressed in terms of the total mass flow rate through a plenum and not the partition of this total mass flow rate among the channels. Thus, for an N channel array, $N - 1$ additional boundary conditions must be supplied at the inlet to supplement set (2), at the outlet for set (3), and at both boundaries for set (4). Considering set (2) as an example, the choice for this additional information which may initially come to mind is the inlet pressure level distribution. However, this provides N instead of $N - 1$ boundary conditions. Additionally, taken in combination with the outlet pressure level distribution, this approach would yield mass flow rates of all channels which, when summed up, may not necessarily equal the prescribed total mass flow rate. Therefore, designation of the inlet pressure level distribution, i.e., $p_{in}(r)$ would overprescribe the problem. The correct additional boundary conditions to be supplied are provided by specifying the inlet radial pressure distribution with respect to an arbitrary reference pressure, p^* (this reference p^* can be different from that specified to define density). Figure 1-6 illustrates the full set of boundary conditions for the new set (2). Hence, for any arbitrary value of p^*, the relative pressure drop among all channels is given by $N - 1$ equations, i.e.,

$$p_{1,in} - p_{2,in}; \; p_{1,in} - p_{3,in}; \; \cdots \; p_{1,in} - p_{N,in}$$

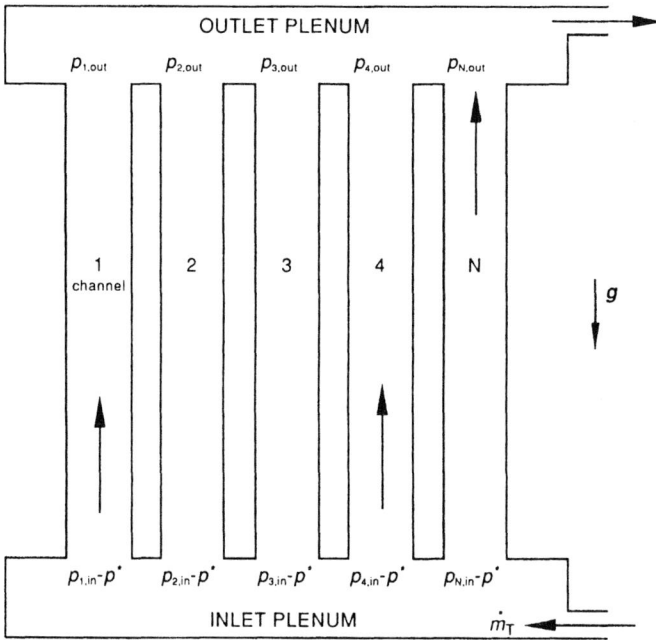

Figure 1-6 Boundary conditions for set (2).

This hydrodynamic set of boundary conditions consisting of the total inlet mass flow rate, inlet radial pressure gradient distribution, and outlet pressure level is designated set (2). Conversely interchanging the specification of radial pressure gradient distribution and pressure level between the inlet and outlet yields a new set labeled (3). These cases thus involve the total inlet mass flow rate and the radial pressure gradient between channels at the one plenum and the pressure level for each channel at the other plenum. For set (4) the radial pressure gradient distribution and the total mass flow rate are supplied at each plenum. Figure 1-7 illustrates these sets.

For incompressible flow in these heated channels, a reference pressure is prescribed, i.e., a thermally expandable flow. The reduction of boundary condition sets is analogous to the changes discussed for a thermally expandable flow in single channels. Set (1) reduces to a prescribed pressure difference. Sets (2) and (3) reduce respectively to a prescribed inlet or outlet radial pressure gradient and a total mass flow rate which can be prescribed at either the inlet or the outlet for each case. Sct (4) becomes identical to set (2) or to set (3).

For all boundary condition sets radial pressure gradients between channels in the same plenum can be accommodated. Again, in most practical cases these pressure gradients are taken equal to zero. Hence, in Figure 1-6 generally:

$$p_{in}(r) = \text{constant} \qquad (1\text{-}1)$$

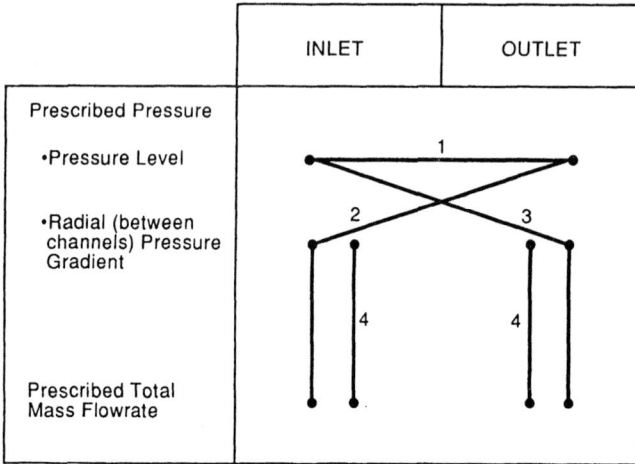

Figure 1-7 Boundary conditions for multiple channels connected only at plena in subsonic, compressible flow.

which is equivalent to:

$$\frac{\partial p(r,t)}{\partial r}\bigg|_{in} = 0 \tag{1-2}$$

and/or:

$$p_{out}(r) = \text{constant} \tag{1-3}$$

which is equivalent to:

$$\frac{\partial p(r,t)}{\partial r}\bigg|_{out} = 0 \tag{1-4}$$

leading to the condition of equal channel pressure drop, i.e.,

$$\Delta p_1 = \Delta p_2 = \ldots \Delta p_N \tag{1-5}$$

Table 1-4 summarizes the problem statement for multiple, heated channels connected only at plena except for boundary set (4) since it is not a very practical case.

The solution of the multiple, heated channel problem can be simplified if coolant property variations with enthalpy and pressure can be neglected. This is appropriate for many single-phase, incompressible coolant applications. In these cases continuity requires that the sum of the channel flow rates equal the total flow rate, and the momentum equation can be decoupled from the energy equation since the coupling arises only from property variations.

Table 1-4 Problem definition for multiple, heated channels connected only at plena in subsonic, compressible flow

	Specified total flow condition		Specified pressure condition
Goals	Enthalpy, $h_{coolant}(z,r,t)$† Void fraction, $\alpha(z,r,t)$ Channel flow rate, $\dot{m}(z,r,t)$ Pressure drop, $p_{in}(r,t) - p_{out}(r,t)$		Enthalpy, $h_{coolant}(z,r,t)$ Void fraction, $\alpha(z,r,t)$ Channel flow rate, $\dot{m}(z,r,t)$ Total mass flow rate, $\dot{m}_T(z,t)$
Constraints	Design limits (Section 2.4, Volume I) Geometry Materials		Design limits (Section 2.4, Volume I) Geometry Materials
Input conditions	Energy generation rate, $q'''_{coolant}(r,z,t)$ and/or surface heat flux, $q''(r,z,t)$ Friction factor, $f(h,p,G,q'')$		Energy generation rate, $q'''_{coolant}(z,r,t)$ and/or surface heat flux, $q''(r,z,t)$ Friction factor, $f(h,p,G,q'')$
	(3)	(2)	(1)
Boundary conditions‡	Pressure, $p_{in}(r,t)$ Pressure gradient, $\left.\dfrac{\partial p(r,t)}{\partial r}\right|_{out}$ Total outlet mass flow rate, $\dot{m}_T(t)_{out}$ Enthalpy, $h_{in}(r,t)$	Pressure, $p_{out}(r,t)$ Pressure gradient, $\left.\dfrac{\partial p(r,t)}{\partial r}\right|_{in}$ Total inlet mass flow rate, $\dot{m}_T(t)_{in}$ Enthalpy, $h_{in}(r,t)$	Pressure, $p_{in}(r,t)$ and $p_{out}(r,t)$ Enthalpy, $h_{in}(r,t)$

† r coordinate is used only to identify the position in the core; it does not represent cross-sectional position within a channel.
‡ Set (4) omitted since it is not a usual reactor case of interest. Only $h_{in}(t)$ is needed if $\dot{m}_{in}(t)$ and $\dot{m}_{out}(t)$ are both positive, i.e., inflow at inlet and outflow at exit. If $\dot{m}_{out}$ is negative, i.e., inflow is occurring from the outlet plenum, then $h_{out}(t)$ must also be specified.

Table 1-5 Solution method for problem of channels connected only at plena in subsonic, compressible flow

Relation between conservation equations	Coupled	Partial coupling	Uncoupled
Additional simplifying assumptions	None	Physical property dependence on temperature but not on pressure	No physical property variations with pressure and temperature
Method of solution	• Simultaneous solution of one-dimensional (axial) mass, momentum, and energy conservation equations for the coolant region • Solution of the one-dimensional (radial) energy equations for the fuel and clad at every axial position	• Simultaneous solution of one-dimensional (axial) mass and energy equations for the coolant if the mass flow rate at any location is known • Solution of the momentum conservation equation • Solution of the one-dimensional (radial) energy equations for the fuel and clad at every axial position	Sequential solution of the following equations: • The one-dimensional (axial) mass and momentum conservation equations for the coolant • The one-dimensional (axial) energy equation for the coolant • The one-dimensional (radial) energy equation for the clad and fuel

Analogously for two-phase flow problems in which density variations cannot be ignored, the mass and energy equations can still be partially decoupled from the momentum equation. For example, this can be done if it is assumed that the two phases are saturated, and that the pressure does not vary much along the channel; all properties are then calculated at a reference system pressure. This adoption of a heated, incompressible system approach is satisfactory if acoustic pressure propagation effects can be ignored.

However, the velocity at some location in the channel is still needed to proceed with the solution of the energy and mass equations since the density depends on the enthalpy, i.e., a thermally expandable flow. If additionally this density dependence on enthalpy can be neglected, then the conservation equations can be completely decoupled. This problem statement is the most basic one and is easily solved analytically for energy generation rates expressed by analytic functions and homogeneous (equal vapor and liquid velocities) flow. The methods of solution for all cases, which introduce much useful information on heat and momentum transfer, are presented in Chapter 13, Volume I and Chapter 2 for single-phase and two-phase coolants. Table 1-5 summarizes the method of solution for the conditions in which the conservation equations are coupled and uncoupled.

Example 1-1 Reactor boundary condition determination

PROBLEM A heated reactor core model with radially orificed, ducted sub-assemblies is instrumented as shown in Figure 1-8. A prediction of the steady-state enthalpy distribution test results is desired utilizing measured boundary conditions. Identify the boundary conditions that could be utilized.

SOLUTION Two sets of boundary conditions are possible. Set (1) inlet and outlet pressure distributions are specified from the pressure gauges in the inlet and outlet plena. (The distribution is usually assumed to be flat in these cores since the fluid moves short distances at relatively low velocities within the plena.) Inlet enthalpy is also specified from inlet pressure and temperature measurements. Set (2) outlet pressure and inlet radial pressure gradient, total mass flow rate, and inlet enthalpy are specified.

Note that the existence of different combinations of instrument readings allows us to check the unused reading against the analytic answer. It is possible, for example, to predict the flow through the reactor based on the two pressure measurements and then check this against the reading from the flow meter. Such an operation, called signal validation by use of analytic redundancy, is increasingly used to ensure that the information sent to reactor operators is accurate.

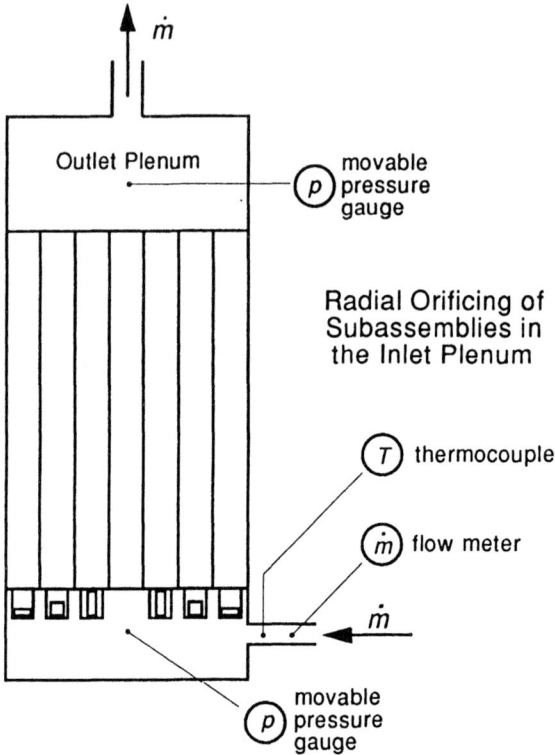

Figure 1-8 Instrumented reactor test model.

VII FLOW IN INTERCONNECTED, MULTIPLE HEATED CHANNELS

In this case communication between the channels is allowed. This necessitates the inclusion of a transverse momentum equation to describe the situation analytically. This flow arrangement is intrinsically two-dimensional and hence fundamentally different from the preceding one-dimensional cases. For this two-dimensional arrangement two conditions are needed at each inlet flow surface, and one is needed at each outflow surface.

Therefore, the inlet plenum boundary set for the one-dimensional case must be supplemented by the prescription of the lateral velocity. For the side surface the additional boundary conditions necessary are dependent upon whether inflow or outflow occurs. In either case let us only consider the case in which velocities are to be supplied as boundary conditions. For the side surface inflow case both the lateral and axial velocity components have to be prescribed as illustrated in Figure 1-9. For the outflow case only one velocity component has to be prescribed, and we propose it to be the lateral velocity.

The full set of boundary conditions is composed of the axial set for the one-dimensional array of channels (Figure 1-9) plus the appropriate set for the side surfaces. This combined set for the two-dimensional arrangement is illustrated

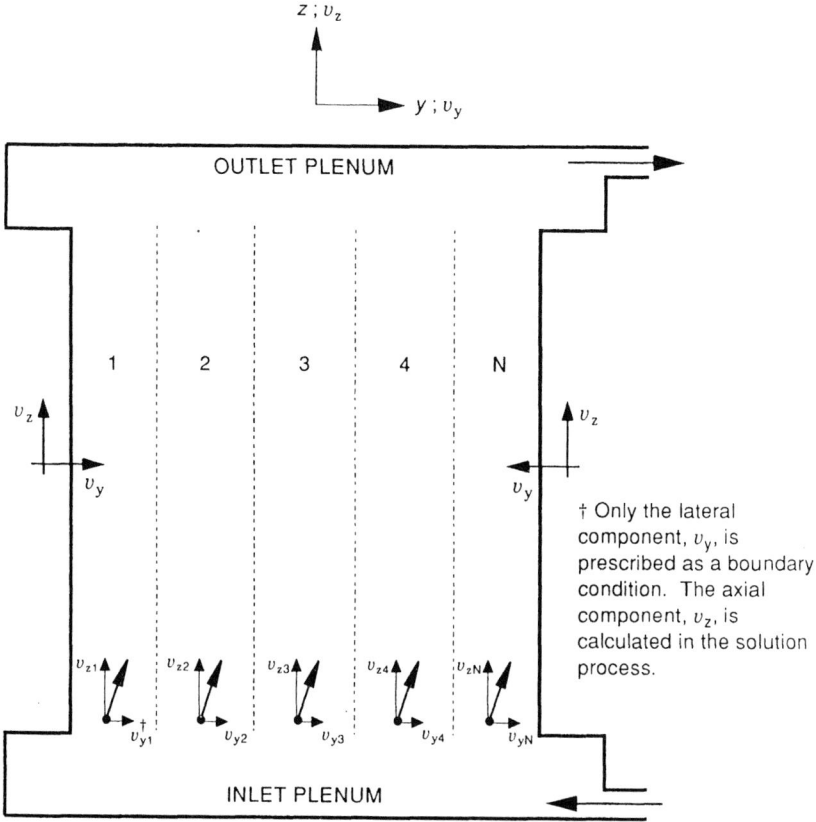

Figure 1-9 Lateral velocity boundary conditions for an array of interconnected heated channels with side boundary inlet flow.

in Figure 1-10. In the normal reactor case, these lateral and axial side surface velocities are zero.

In practice this problem is sometimes solved by specifying the inlet radial flow rate distribution from the plenum. This set of N boundary conditions replaces the $N - 1$ conditions from the inlet radial pressure gradient distribution and the total inlet mass flow rate (set (2)). However, it is found that the solution is not sensitive to the inlet boundary condition whether it be the radial flow rate distribution or the radial pressure gradient distribution because cross-flow between the interconnected channels largely obliterates the inlet boundary condition influence at a distance equal to 10 to 20 equivalent subchannel hydraulic diameters downstream of the inlet. Therefore the distinction between specifying individual channel inlet flows versus the total inlet flow which exists for the previous case of channels-connected-only-at-plena disappears here. Table 1-6 summarizes the problem statement for the case of interconnected chan-

Table 1-6 Problem definition for multiple, heated interconnected channels in subsonic, compressible flow

	Specified flow (total inlet) condition	Specified pressure condition
Goals	Enthalpy, $h_{coolant}(z,r,t)$† Void fraction, $\alpha(z,r,t)$ Channel flow rate, $\dot{m}(z,r,t)$ Pressure drop, $p_{in}(r,t) - p_{out}(r,t)$	Enthalpy, $h_{coolant}(z,r,t)$ Void fraction, $\alpha(z,r,t)$ Channel flow rate, $\dot{m}(z,r,t)$ Total mass flow rate, $\dot{m}_T(z,r,t)$
Constraints	Design limits (Section 2.4, Volume I) Geometry Materials	Design limits (Section 2.4, Volume I) Geometry Materials
Input conditions	Energy generation rate, $q'''_{coolant}(z,r,t)$ and/or surface heat flux, $q''(r,z,t)$ Friction factor, $f(h,p,G,q'')$ (3) Pressure, $p_{in}(r,t)$ Pressure gradient, $\left.\dfrac{\partial p}{\partial r}\right|_{out}(r,t)$ (2) Pressure, $p_{out}(r,t)$ Pressure gradient, $\left.\dfrac{\partial p}{\partial r}\right|_{in}(r,t)$	Energy generation rate, $q'''_{coolant}(z,r,t)$ and/or surface heat flux, $q''(r,z,t)$ Friction factor, $f(h,p,G,q'')$ (1) Pressure, $p_{in}(r,t)$ and $p_{out}(r,t)$
Boundary conditions‡	Total outlet mass flow rate, $\dot{m}_T(t)_{out}$ Total inlet mass flow rate, $\dot{m}_T(t)_{in}$ Inlet cross-flow velocity, $v_{yin}(r,t)$ Enthalpy, $h_{in}(r,t)$	Inlet cross-flow velocity, $v_{yin}(r,t)$ Enthalpy, $h_{in}(r,t)$

† r coordinate is used only to identify the position in the core; it does not represent cross-sectional position within a channel.
‡ Set 4 omitted and side domain velocities taken as zero since they are not the usual reactor case of interest. Only $h_{in}(t)$ is needed if $\dot{m}_{in}(t)$ and $\dot{m}_{out}(t)$ are both positive, i.e., inflow at inlet and outflow at exit. If $\dot{m}_{out}$ is negative, that is, inflow is occurring from the outlet plenum, then $h_{out}(t)$ must also be specified.

	INLET PLENUM	OUTLET PLENUM
Prescribed Pressure		
•Pressure Level	1	
•Radial (between channels) Pressure Gradient	2 4	3 4
Prescribed Total Mass Flowrate		
Lateral Velocity	•1,2,3,4	

Prescribed Side Velocities	Inlet Flow Across Side Boundaries	Outlet Flow Across Side Boundaries
•Axial Velocity	•1,2,3,4	
•Lateral Velocity	•1,2,3,4	•1,2,3,4

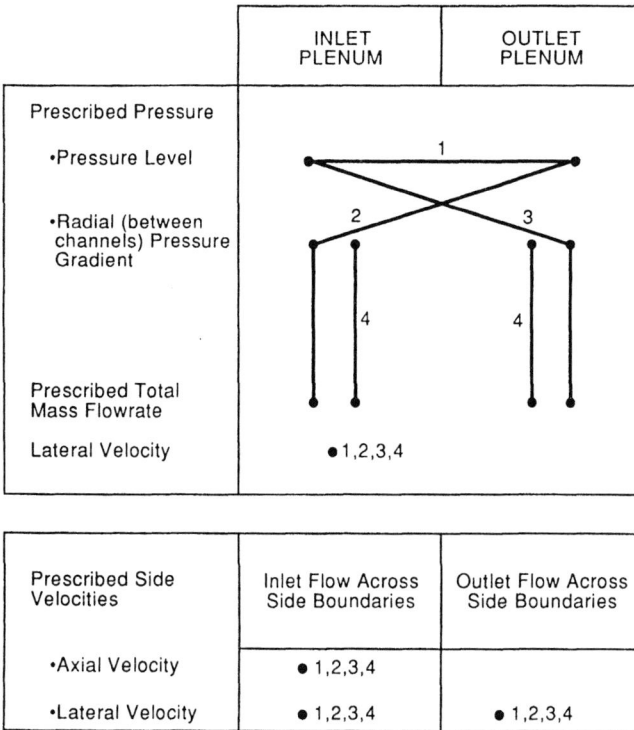

Figure 1-10 Boundary conditions for multiple interconnected channels in subsonic, compressible flow.

nels. The solution of this interconnected channel problem is addressed in Chapters 5, 6, and 7 for the cases of decoupled and coupled solutions of the conservation equations.

VIII APPROACHES FOR REACTOR ANALYSIS

The characteristics of each reactor type (and, to some extent the nature of the problem to be solved) determine which situation among those presented in Sections V through VII is the applicable modeling approach. The key characteristic is whether or not fuel rod arrays are bounded periodically by shrouds or ducts. If so, then a set of isolated channels in parallel flow between plena is created, and the isolated-channels-between-plena case is applicable if the cross-section of the flow channel is considered only on a homogenized basis. However, a subsequent or parallel solution of local conditions within any of these isolated channels requires the consideration of the interconnected array of fuel pins and coolant channels. Hence the interconnected case is applicable in such a subsequent or parallel solution.

Table 1-7 Approaches for power reactor core thermal hydraulic analyses

	Reactor	Fuel assembly
BWR/LMR	Isolated (array of fuel assemblies)	Interconnected (array of coolant subchannels)
PWR	Interconnected (array of fuel assemblies)	Interconnected (array of coolant subchannels)

Table 1-7 summarizes the approaches that have been adopted for power reactor core thermal-hydraulic analysis. These are discussed in the following subsections.

A BWR and LMR Core Analysis

In these cores, arrays of fuel elements are both mechanically and hydraulically grouped into assemblies. For overall core analysis these assemblies are homogenized and represented as single channels. Since the assemblies do not interact along their length, the core is represented as an array of single channels connected only at the inlet and outlet plena, as shown in *upper left region* of Figure 1-11. In this case the outlet and inlet conditions are strongly coupled. A typical boundary condition set adopted is a prescribed core inlet radial pressure gradient and outlet pressure level, both generally assumed to be uniform, and a total inlet mass flow rate (*set* (2) of Figure 1-7).

When analysis of an individual BWR or LMR assembly is desired, the assembly is modeled as an array of parallel, continuously interacting channels as shown in *lower right region* of Figure 1-11. This situation is equivalent to that of the PWR core discussed in the next section. Since the LMR utilizes a hexagonal rod array, it can be modeled as a set of concentric annuli in two-dimensional geometry if radial power gradients across the array can be ignored.

B PWR Core Analysis

Arrays of fuel elements are mechanically grouped for handling purposes into units called assemblies. However, PWR cores do not have shrouds or ducts that hydraulically isolate these arrays of fuel elements or assemblies between plena. All regions of coolant therefore interact continuously with their neighbors over the entire core length between the plena.

PWR analysis has traditionally been done in a two-staged manner. In both stages the situation is that of a group of interacting channels as Table 1-7 emphasizes. In the first stage entire assemblies are homogenized and represented as single channels. The homogenization process consists of representing a set of flow channels (e.g., entire fuel assembly or several assemblies) by a single channel having average properties such that it behaves macroscopically

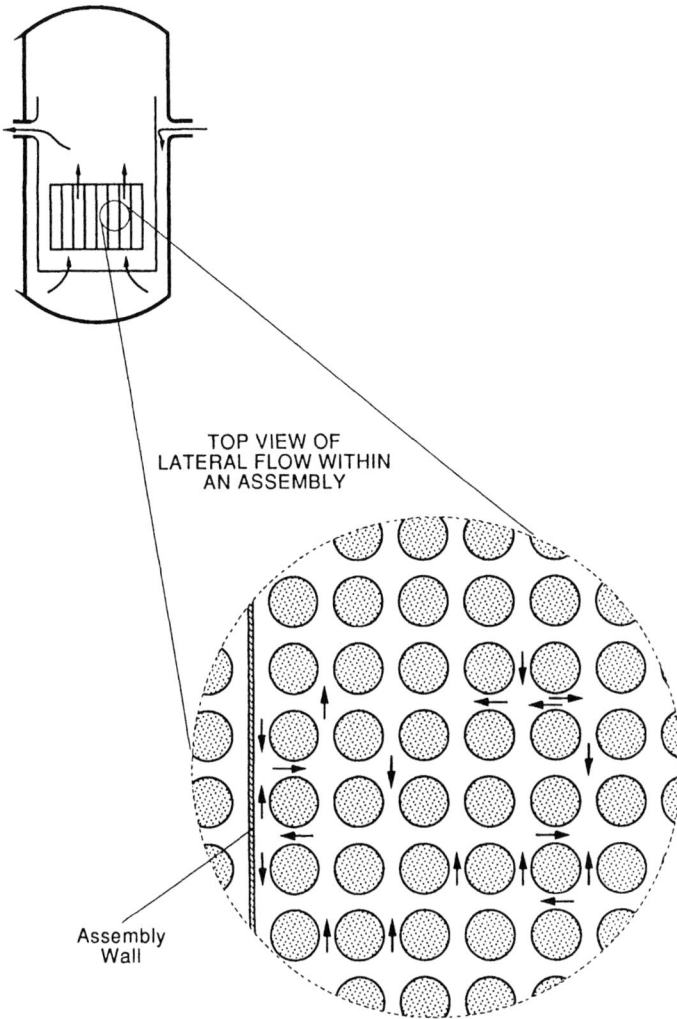

Figure 1-11 Fuel assembly within a BWR core (similar for a hexagonal array in an LMR core).

as the original set of flow channels. The microscopic details of the flow description within the assemblies are obviously lost, but a considerable reduction in computational effort results. In fact, it is practically impossible to consider, separately and simultaneously, all the flow channels in a reactor core because of their large number. The first stage of the analysis proceeds then by applying either set of boundary conditions to identify the macroscopic behavior of the core and the "hot" assembly.

The lateral exchange of mass, momentum, and energy through the faces of the hot assembly over its length is also obtained. The second, subsequent stage is an analysis of the hot assembly only, which utilizes the lateral boundary conditions obtained in the first stage. In this stage the detailed description of the flow over a more restricted core region is sought. It is important to keep in mind that this two-staged approach of identifying and then analyzing only the hot assembly makes the problem tractable at the cost of some approximation. Additionally, single-stage or parallel methods of analysis have been developed also which represent the region around the hot channel in detail but represent the remainder of the core by increasingly larger homogenized regions.

IX LUMPED AND DISTRIBUTED PARAMETER SOLUTION APPROACHES

Table 1-1 summarizes all the problem statements presented in this chapter and identifies the section in which the method of solution is presented. The multiple interconnected channel array is the configuration that presents the major analysis challenge. The direct solution of this interconnected channel problem, that is, a three-dimensional, transient, multi- but discrete-region problem, is the theoretically ideal approach. However, from an engineering point of view, such an approach is both unnecessary and impractical. A spectrum of simplifying assumptions can indeed be made, and it is possible to transform the problem into a tractable one. These simplified problems range in difficulty from statements that are handled by state-of-the-art methods on large-scale computers, e.g., subchannel analysis methods, to those which can be solved by hand computations, e.g., isolated unit-cell methods. In each case the strategy of identifying unit cells and homogenizing certain material zones within each unit cell is employed to varying degrees. These cases are classified and presented in this book in Chapters 5, 6, and 7.

The initial step in the solution process is the selection of either a thermodynamic system or control volume approach. Since the fuel assembly consists of stationary fuel rods and flowing coolant, it is inconvenient to utilize the system approach. Therefore the control volume approach is adopted in all the methods to be presented. The definition of the shape and size of the control volume varies and will be discussed for each method of analysis. These methods are most clearly categorized according to whether

- the fundamental conservation equations are applied to the continuum in a lumped or distributed manner, and whether
- the region of the fuel assembly selected for analysis is considered to be isolated from or interacting with its neighboring regions.

Since each point above identifies two paths for analysis, taken in combination, we find four methods for the analysis of rod bundles which we will present in the following sequence:

1. Lumped parameter-isolated region (Chapter 13, Volume I; Chapters 2, 3, 4)
2. Lumped parameter-interacting regions (Chapters 5, 6)
3. Distributed parameter-isolated region (Chapter 7)
4. Distributed parameter-interacting regions (Chapter 7).

The lumped method implies that the fuel assembly is divided into regions that are homogenized, and each homogenized region is characterized by one value of each property. For this case then, spatial property gradients within each homogenized region or control volume are not defined. Conversely the distributed manner means that the assembly properties have spatially dependent values that are defined for all spatial locations. These methods will be referred to as the lumped parameter (LP) and distributed parameter (DP) methods, respectively.

For the LP methods, the coolant region of the fuel assembly is subdivided into control volumes. The subchannel approach is a special case of the LP approach. In this case the fuel assembly is broken into symmetric channels. The fuel rod portions associated with each subchannel are included by either applying a heat flux boundary condition to the proper portion of the subchannel perimeter or creating fuel and clad control volumes that are in contact with each other and with their associated subchannels. However, it is the interaction of the adjacent homogenized coolant regions which introduces complexity in the analysis.

PROBLEMS

Problem 1-1 Boundary conditions for flow transients in test loops (Sections V and VII)
Identify the hydraulic and thermal boundary condition sets applicable for analysis of the three test sections described below in a loss of flow transient with constant power. Consider the cases of each test section located in the test loops sketched in Figures 1-12 and 1-13.

Test section 1. Single channel with a flow blockage that reduces the cross-sectional area for flow by 50% at the axial midposition.
Test section 2. Two interconnected channels, one of which has a flow blockage as described for test section 1.
Test section 3. Twenty interconnected channels, one of which has a flow blockage as described for test section 1.

Note that this problem requires six answers as illustrated below.

	Test section		
	Case 1	Case 2	Case 3
Loop A			
Loop B			

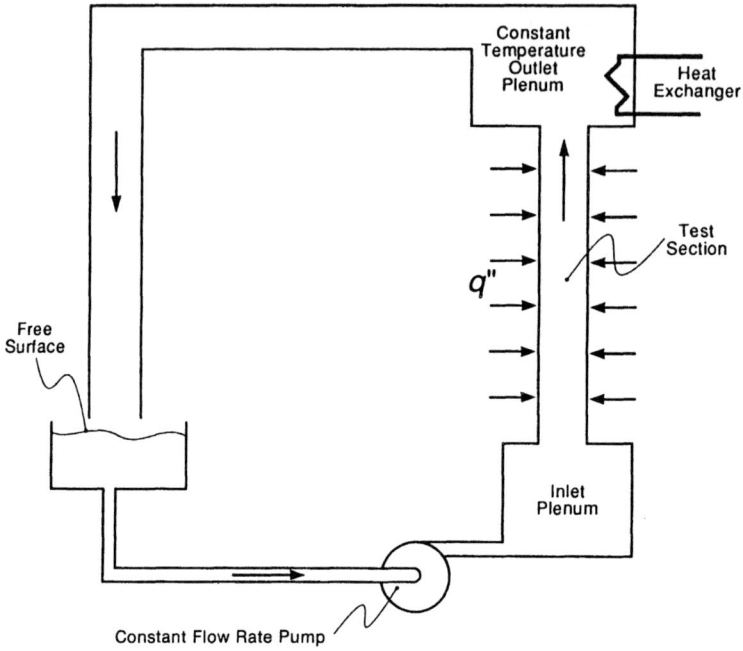

Figure 1-12 Loop A.

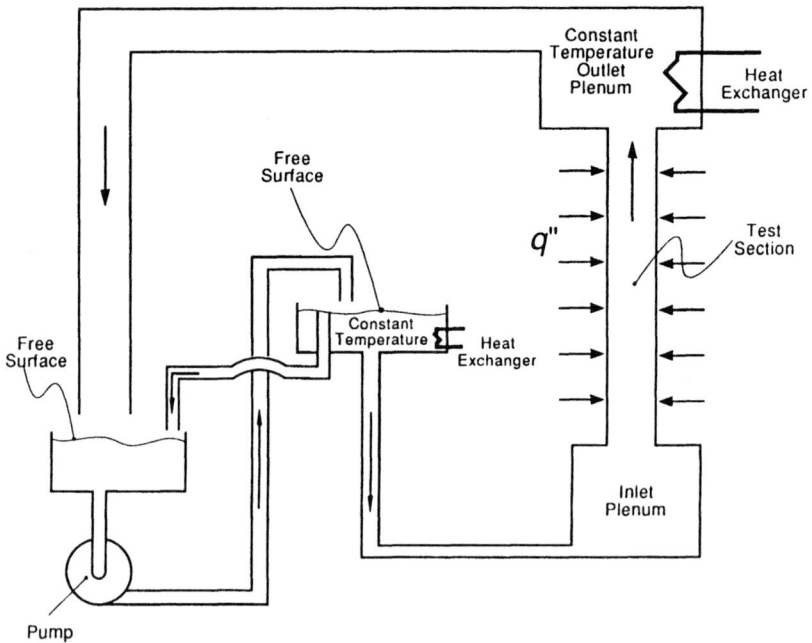

Figure 1-13 Loop B.

Problem 1-2 Boundary conditions for a multiple channel representation of a reactor system (Section VI)

An engineer proposes that a primary system flow analysis of a multiloop reactor plant should be done using only boundary condition set (2) or (3) instead of set (1) (see Figure 1-7) because the system loops are reducible to channels between plena and hence can be represented as such. With loops represented as channels, the continuity relation on total mass flow becomes $\dot{m}_{T\,in} = 0$ since the loops return the flow from the outlet to the inlet plena so that the algebraic sum of flows is zero at the inlet (and exit) plenum.

Questions: Is the proposal valid? If so, why is this representation in only limited use? Be sure to start your answer with a sketch defining the geometric situation to be analyzed.

SINGLE, HEATED CHANNEL
TRANSIENT ANALYSIS

I SIMPLIFICATION OF TRANSIENT ANALYSIS

The general transient equations are difficult to solve because of the coupling between the momentum and energy equations and the nonlinear nature of these general equations. However, it is possible to simplify these equations in two ways:

1. By decoupling the momentum and energy equations by approximating the velocity distribution in the channel. This approach encompasses the momentum integral method and other related methods described in Section II.
2. By transforming the partial differential equations into ordinary differential equations that can be solved separately in the time domain and the space domain. This approach is the foundation of the method of characteristics, which is described in Section III.

II SOLUTION OF TRANSIENTS WITH APPROXIMATIONS TO THE MOMENTUM EQUATION

Several approximations can be used to decouple the momentum and energy equations to facilitate solution of the transient problem. Additionally, the numerical solution of a transient problem would be particularly simplified if the compressibility of the fluid could be ignored. The nature and impact of several approximations will be discussed here following the work of Meyer [10] and specific applications from Lee and Kazimi [8].

The following initial approximations are made:

1. In two-phase flow the vapor and liquid move at the same velocity (no slip).
2. The lateral variations of properties in the flow channels can be neglected.

For the above conditions, Eqs. I,13-1, I,13-2a, and I,13-9b* are directly applicable:

$$\frac{\partial \rho_m}{\partial t} + \frac{\partial G_m}{\partial z} = 0 \tag{I,13-1}$$

$$\frac{\partial G_m}{\partial t} + \frac{\partial}{\partial z}\left(\frac{G_m^2}{\rho_m}\right) = -\frac{\partial p}{\partial z} - \frac{f G_m |G_m|}{2 D_e \rho_m} - \rho_m g \tag{I,13-2a}$$

$$\rho_m \frac{\partial h_m}{\partial t} + G_m \frac{\partial h_m}{\partial z} = \frac{q'' P_h}{A_z} + \frac{\partial p}{\partial t} + \frac{G_m}{\rho_m}\left[\frac{\partial p}{\partial z} + \frac{f G_m |G_m|}{2 D_e \rho_m}\right] \tag{I,13-9b}$$

Eqs. I,13-1, I,13-2a, and I,13-9b of Volume I, and appropriate constitutive relations, provide the solutions for $G_m(z,t)$, $p(z,t)$, and $h_m(z,t)$ for a given set of initial and boundary conditions. The initial distributions of these variables are assumed known from steady-state solutions.

The heat flux q'' can be specified as an input, whether constant or time dependent. However, in reality the heat flux in a reactor is dependent on the coolant and fuel thermal conditions. Hence, the specification of q'' assumes that the effects of the neutronic response and the transient heat conduction in the fuel can be specified.

The boundary conditions for solving the momentum equation are to be specified as $G_m(0,t)$ or $p(0,t)$ at the inlet and $G_m(L,t)$ or $p(L,t)$ at the outlet. As discussed in Chapter 1, a common approach is to specify $p(L,t)$ and either of $p(0,t)$ or $G_m(0,t)$. In the discussion that follows we shall assume the inlet and outlet pressures are specified. This corresponds to boundary condition set (1) in Chapter 1. This condition is suitable for transients in channels that are connected to large plena since the pressure in the plena would not be significantly affected by the transients in the channels themselves.

To solve the energy equation, the specific enthalpy of the fluid entering the channel should be specified, whether from the bottom or the top of the channel. In our case the flow is assumed initially upward everywhere in the channel, hence, the enthalpy $h_m(0,t)$ at the inlet will be specified. Also $q''(z,t)$ needs to be specified.

Furthermore, constitutive equations for ρ_m and f are required to complete definition of the problem. The equation of state for the density, assumed differ-

---

* The reader is referred to Volume I for review of equations designated.

entiable with respect to h_m and p, is specified as:

$$\rho_m = \rho_m(h_m, p) \tag{2-1}$$

The friction factor can be specified as:

$$f = f(h_m, p, G_m, q'') \tag{2-2}$$

The dependence of f on q'' comes about because the fluid properties, in particular the viscosity, are evaluated at the mid-film temperature, which depends on q''.

Now, we consider several approaches to solution of the specified set of equations.

A Sectionalized, Compressible Fluid (SC) Model

The most general approach involves numerical solution of a set of difference equations representing the differential transport equations, arranged to consider h_m, G_m, and p, and state variables at multiple points along the channel. The term *sectionalized* reflects the need to divide a channel into segments to execute the numerical solution. Using Eq. 2-1 we get:

$$\frac{\partial \rho_m}{\partial t} = \frac{\partial \rho_m}{\partial h_m}\bigg|_p \frac{\partial h_m}{\partial t} + \frac{\partial \rho_m}{\partial p}\bigg|_{h_m} \frac{\partial p}{\partial t} = R_h \frac{\partial h_m}{\partial t} + R_p \frac{\partial p}{\partial t} \tag{2-3}$$

where:

$$R_h \equiv \frac{\partial \rho_m}{\partial h_m}\bigg|_{p=\text{const}} \quad \text{and} \quad R_p \equiv \frac{\partial \rho_m}{\partial p}\bigg|_{h_m=\text{const}} \tag{2-4}$$

From Eq. I,13-1 and 2-3 we get:

$$R_h \frac{\partial h_m}{\partial t} + R_p \frac{\partial p}{\partial t} + \frac{\partial G_m}{\partial z} = 0 \tag{2-5}$$

Eqs. I,13-9b and 2-5 may be combined into two equations by eliminating $(\partial p/\partial t)$ and $(\partial h_m/\partial t)$:

$$\frac{\rho_m}{c^2} \frac{\partial p}{\partial t} + \rho_m \frac{\partial G_m}{\partial z} + \frac{R_h G_m}{\rho_m} \frac{\partial p}{\partial z} - R_h G_m \frac{\partial h_m}{\partial z} = -R_h \left[\frac{q'' P_h}{A_z} + \frac{f G_m^2 |G_m|}{2 D_e \rho_m^2} \right] \tag{2-6}$$

and

$$\frac{\rho_m}{c^2} \frac{\partial h_m}{\partial t} + \frac{\partial G_m}{\partial z} - \frac{R_p G_m}{\rho_m} \frac{\partial p}{\partial z} + R_p G_m \frac{\partial h_m}{\partial z} = R_p \left[\frac{q'' P_h}{A_z} + \frac{f G_m^2 |G_m|}{2 D_e \rho_m^2} \right] \tag{2-7}$$

where we have defined c^2 as:

$$c^2 \equiv \frac{\rho_m}{\rho_m R_p + R_h} \tag{2-8}$$

Note that c is the isentropic speed of sound in the fluid, generally given by:

$$c^2 \equiv \left.\frac{\partial p}{\partial \rho}\right|_s \qquad (2\text{-}9)$$

The equivalence of Eqs. 2-8 and 2-9 can be shown as follows:

$$\left.\frac{\partial \rho}{\partial p}\right|_s = \left.\frac{\partial \rho}{\partial p}\right|_h + \left.\frac{\partial \rho}{\partial h}\right|_p \left.\frac{\partial h}{\partial p}\right|_s = R_p + R_h \left.\frac{\partial h}{\partial p}\right|_s \qquad (2\text{-}10)$$

But since $dh = T\,ds + v\,dp$ then:

$$\left.\frac{\partial h}{\partial p}\right|_s = v = 1/\rho_m \qquad (2\text{-}11)$$

hence:

$$\left.\frac{\partial \rho}{\partial p}\right|_s = R_p + \frac{R_h}{\rho_m} \qquad (2\text{-}12)$$

Now Eqs. I,13-2a, 2-6, and 2-7 are partial differential equations in p, G_m, and h_m (i.e., the density does not appear as a differentiated variable). These equations can be written in pointwise difference form and solved in p, G_m, and h_m. Numerical considerations (i.e., the stability and/or accuracy) of the difference solution require that the time step of integration be less than the time interval for sonic wave propagation across the spatial grid points, i.e.,

$$\Delta t \le \Delta z/(c + |V_m|) \qquad (2\text{-}13)$$

where $V_m = G_m/\rho_m$ is the mean transport velocity. Compared with the transport velocity, the fluid sonic velocity is large, thus limiting the time step in most numerical schemes to very small values. This leads to a computationally expensive solution of this problem. The reader can consult Richtmyer and Morton [11] for a discussion of stability considerations of finite difference equations.

B Momentum Integral Model (MI) (Incompressible but Thermally Expandable Fluid)

To eliminate the computational limitations of the sonic effects, it is desirable to assume the fluid to be incompressible (i.e., $\partial \rho/\partial p = 0$). For this case Eq. 2-1 is replaced by:

$$\rho_m = \rho_m(h_m, p^*) \qquad (2\text{-}14)$$

where p^* is a system pressure assumed constant during the transient. This assumption is physically acceptable for classes of reactor transients such as operational transients that are not associated with loss of significant amounts of coolant. Because of the above assumption, the density becomes independent of the pressure p. However, the density is dependent on the enthalpy, which means the fluid is thermally expandable.

Furthermore, it can be assumed that the terms due to pressure changes and wall friction forces can be neglected in the energy equation. Thus Eq. I,13-9b is reduced to:

$$\rho_m \frac{\partial h_m}{\partial t} + G_m \frac{\partial h_m}{\partial z} = \frac{q'' P_h}{A_z}$$ (2-15)

Because of the assumption of fluid incompressibility, the local pressure gradient will not influence the mass flux of the fluid along the channel. In fact, for an isothermal incompressible fluid, the mass flux along the channel will equal the inlet mass flux so that the mass flux is determined by the inlet and outlet pressures only. For the present case of a heated channel, the momentum equation is only useful in determining the axially averaged mass velocity, $\hat{G}_m$, which can be obtained by solving the integral of the momentum Eq. I,13-2a:

$$\int_0^L \frac{\partial G_m}{\partial t} dz + \left(\frac{G_m^2}{\rho_m}\right)_{z=L} - \left(\frac{G_m^2}{\rho_m}\right)_{z=0}$$

$$= p_{z=0} - p_{z=L} - \int_0^L \frac{f|G_m|G_m}{2D_e\rho_m} dz - \int_0^L \rho_m g dz$$ (2-16)

If we define:

$$\hat{G}_m \equiv \frac{1}{L} \int_0^L G_m dz$$ (2-17)

Eq. 2-16 can be written as:

$$\frac{d\hat{G}_m}{dt} = \frac{1}{L}(\Delta p - F)$$ (2-18)

where:

$$\Delta p = p_{z=0} - p_{z=L}$$

$$F = \left(\frac{G_m^2}{\rho_m}\right)_{z=L} - \left(\frac{G_m^2}{\rho_m}\right)_{z=0} + \int_0^L \frac{f|G_m|G_m}{2D_e\rho_m} dz + \int_0^L \rho_m g dz$$ (2-19)

It is possible to include form loss terms in the definition of F to account for entrance, exit, and spacers. Here, these terms are ignored for simplicity. Thus, the momentum Eq. 2-18 provides the means to estimate $\hat{G}_m$. For the variation of G_m with z, we turn to the continuity and energy equations. The local mass velocity is given by the continuity equation:

$$\frac{\partial G_m}{\partial z} = -\frac{\partial \rho_m}{\partial t} = -\left(\frac{\partial \rho_m}{\partial h_m}\right)_{p^*}\left(\frac{\partial h_m}{\partial t}\right) = -R_h\left(\frac{\partial h_m}{\partial t}\right)$$ (2-20)

By combining Eqs. 2-15 and 2-20, the local mass velocity (due to local expansion) is given by:

$$\frac{\partial G_m}{\partial z} = -\frac{1}{\rho_m} R_h \left[\frac{q'' P_h}{A_z} - G_m \frac{\partial h_m}{\partial z}\right]$$ (2-21)

A difference approximation for the above equation provides the variation of the local mass velocity, G_m, about the average mass velocity, $\hat{G}_m$. Thus Eqs. 2-14, 2-15, 2-18, and 2-21 provide the needed equations for determining $\hat{G}_m(t)$, $G_m(z,t)$, $\rho_m(h_m, p^*)$, and $h_m(z,t)$ for given initial and boundary conditions.

As expected, the main advantage of the momentum integral method is that the numerical limitation of Eq. 2-13 is now replaced by the less stringent requirement that the time step does not exceed:

$$\Delta t \leq \Delta z / |V_m| \tag{2-22}$$

However, the approximations imposed on the density and therefore the momentum equation lead to loss of information within the time it takes for sonic waves to propagate through the channel.

C Single Mass Velocity (SV) Model

Further computational simplicity can be obtained if it is assumed that the mass velocity is constant. The implication of this assumption can be realized by considering the continuity equation of the MI model, Eq. 2-20.

Since
$$\frac{\partial G_m}{\partial z} = -\frac{\partial \rho_m}{\partial t} = -\frac{\partial \rho_m}{\partial h_m}\frac{\partial h_m}{\partial t}$$

then
$$\frac{\partial G_m}{\partial z} \simeq 0 \quad \text{if either}$$

$$\frac{\partial \rho_m}{\partial h_m} \simeq 0 \quad \begin{array}{l}\text{(i.e., neglect thermal expansion which is adequate} \\ \text{for single-phase flow within a moderate} \\ \text{temperature range)}\end{array} \tag{2-23}$$

or
$$\frac{\partial h_m}{\partial t} \simeq 0 \quad \text{(i.e., slow transient)}$$

With this assumption the mass flux at any location is then equal to the average flux, $\hat{G}_m$. Thus Eqs. 2-15 and 2-18 are to be solved in $\hat{G}_m(t)$ and $h_m(z,t)$.

D The Channel Integral (CI) Model

As an alternate simplification to the single mass velocity approach, it is possible to integrate the mass and the energy equations over the channel length. Thus, the conservation of mass equation would be given by:

$$\int_0^L \frac{\partial \rho_m}{\partial t}\,dz + \int_0^L \frac{\partial G_m}{\partial z}\,dz = \frac{\partial}{\partial t}\int_0^L \rho_m\,dz + \int_0^L \partial G_m = 0$$

or
$$\frac{dM}{dt} = G_{in} - G_{out} \tag{2-24}$$

where M is the total mass in the channel per unit area:

$$M = \int_0^L \rho_m dz; \; G_{in} = G_m(z = 0); \; G_{out} = G_m(z = L) \qquad (2\text{-}25)$$

The differential conservation of energy equation is given by:

$$\frac{\partial}{\partial t}(\rho_m h_m) + \frac{\partial}{\partial z}(G_m h_m) = \frac{q''P_h}{A_z} \qquad (2\text{-}26)$$

which can be integrated to give:

$$\frac{\partial}{\partial t}\int_0^L \rho_m h_m dz + \int_0^L \partial(G_m h_m) = \int_0^L \frac{q''P_h}{A_z} dz \qquad (2\text{-}27)$$

Note that in view of Eq. 2-24 we can write:

$$G_{out}h_{out} - G_{in}h_{in} = G_{out}h_{out} - \left(G_{out} + \frac{dM}{dt}\right)h_{in}$$

$$= G_{out}(h_{out} - h_{in}) - \frac{d}{dt}(Mh_{in}) \qquad (2\text{-}28)$$

where $h_{in} = h_m(z = 0)$, $h_{out} = h_m(z = L)$, and h_{in} is taken to be constant. Therefore, using Eq. 2-28 in Eq. 2-27 we get:

$$\frac{dE}{dt} = \bar{q} - G_{out}(h_{out} - h_{in}) \qquad (2\text{-}29)$$

where:

$$E = \int_0^L \rho_m(h_m - h_{in})dz \qquad (2\text{-}30)$$

and

$$\bar{q} = \int_0^L \frac{q''P_h}{A_z} dz \qquad (2\text{-}31)$$

To perform the integration of Eq. 2-30, an axial profile of the enthalpy is required. A shape factor $\beta(z)$ can be defined such as

$$h_m(z,t) = h_m(0,t) + \beta(z)[\hat{h}_m(t) - h_m(0,t)] \qquad (2\text{-}32)$$

where $h_m(0,t)$ is a constant equal to $h_m(0,0)$, $\hat{h}_m$ is the average enthalpy in the entire channel:

$$\hat{h}_m(t) = \frac{1}{L}\int_0^L h_m(z,t)dz \qquad (2\text{-}33)$$

and the shape factor $\beta(z)$ satisfies the relation:

$$\frac{1}{L}\int_0^L \beta(z)dz = 1 \qquad (2\text{-}34)$$

In practice $\beta(z)$ is chosen to represent the steady-state solution.

Also a profile for the mass flux can be assumed such that it can be related to the fluid thermal expansion as:

$$G_m(z,t) = G_m(0,t) + \gamma(z,\hat{h}_m)[\hat{G}_m(t) - G_m(0,t)] \qquad (2\text{-}35)$$

where:

$$\gamma(z,\hat{h}_m) = \frac{1}{\xi} \int_0^z - \beta \frac{d\rho_m}{dh_m} \, dz' \qquad (2\text{-}36)$$

and

$$\xi = \frac{1}{L} \int_0^L \left[\int_0^z \left(-\beta \frac{d\rho_m}{dh_m} \right) dz' \right] dz \qquad (2\text{-}37)$$

Thus, in the channel integral model the mass Eq. 2-24, the momentum Eq. 2-18 and the energy Eq. 2-29 are solved to provide the values of M, $\hat{G}_m$, and E, while Eqs. 2-32 and 2-35 provide the local values of $h_m(z,t)$ and $G_m(z,t)$, respectively.

This approach is a more acceptable approximation to the momentum integral model than the single mass velocity model, especially in cases of rapid boiling. However, since the enthalpy profile is preselected, this approximation should be avoided if there is a special interest in the enthalpy transport and distribution in the channel. An overview of the implications of the approximations discussed above is given in Table 2-1.

Example 2-1 PWR inlet pressure transient

PROBLEM Consider the case of a sudden pressure reduction at the inlet of a PWR channel whose outlet is maintained at constant pressure such that:

$$p_{in}(t) - p_{out} = 0.5[p_{in}(0) - p_{out}](1 + e^{-400t}) \qquad (2\text{-}38)$$

where t is time in seconds.

The geometry and operating conditions are given in Table 2-2.

SOLUTION The solutions by each of the above four methods will be discussed and the results will be compared. In these calculations the channel is divided into 10 axial segments. Details of the finite difference schemes used are found in Lee and Kazimi [8].

1. *Short-term channel flow response*
   a. *Sectionalized, compressible (SC) model: Figure 2-1.* At the inlet, due to the pressure decrease, the mass flux begins to decrease and this perturbation propagates into the channel as time elapses. The sonic velocity can be computed as about 900 m/s, and the channel transit time of this sonic wave is about 4.1 ms. For $t < 4.1$ ms, the rapid decrease of the mass flux has not yet reached the end of the channel; that is, the downstream region is not yet affected by the

Table 2-1 Implications of approximate solutions to momentum equations

	Sectionalized compressibility	Momentum integral	Single mass velocity	Channel integral
Sonic effects	Yes	No	No	No
Thermal expansion	Yes	Yes	No	Yes
Maximum allowable time step	$\dfrac{\Delta z}{c+|V|}$	$\dfrac{\Delta z}{|V|}$	$\dfrac{\Delta z}{|V|}$	$\dfrac{L}{\beta(L)|\bar{V}|}$
Limitations	Computer time too large	No information about condition before sonic propagation occurs	Problem for transients that result in large time rate of change in channel vapor content. Underpredicts frictional pressure drop due to boiling	$\dfrac{1}{L}\int_0^L \beta(z)dz = 1$ Problem where detailed enthalpy profile is important
Appropriate application	Very fast transients	Fast transients	Intermediate and slow transients	Fast transients

Table 2-2 Geometry and operating conditions for cases analyzed in examples of Chapter 2

Operating condition	PWR	BWR
Channel length (m)	3.66	3.05
Rod diameter (mm)	9.70	12.70
Pitch (mm)	12.80	15.95
Flow area for rod (mm²)	90.00	128.00
Equivalent diameter (mm)	12.00	13.00
q' Linear heat (kW/m) (axially constant)	17.50	16.40
Mass flux (kg/m²s)	4,125	2,302
Inlet pressure (MPa)	15.50	6.96
Outlet pressure (MPa)	15.42	6.90
Inlet enthalpy (kJ/kg)	1,337.2	1,225.5

Transient conditions:

1. Pressure drop decrease transient

$$p_{in}(t) - p_{out} = 0.5[p_{in}(0) - p_{out}](1 + e^{-400t}) \quad (2\text{-}38)$$

where t is time in seconds.

2. Heat flux increase transient

$$q'(t) = 1.1q'(0)$$

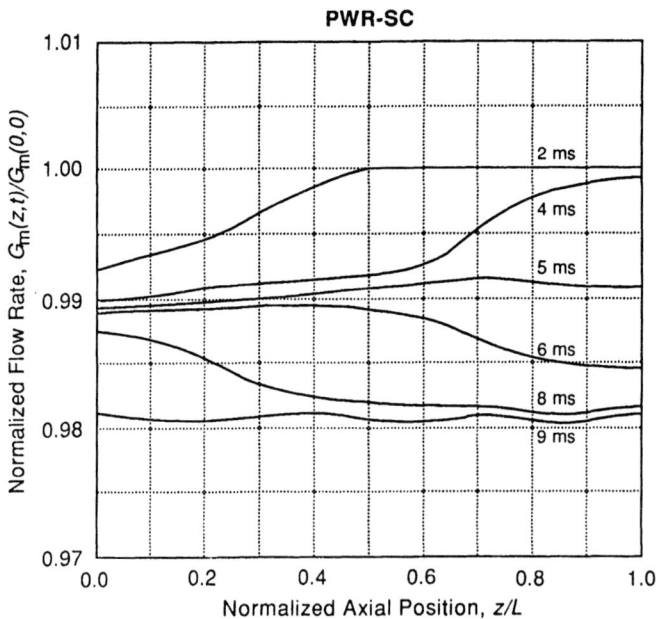

PWR-SC

Figure 2-1 Short-term response of the PWR inlet pressure transient using the SC model.

pressure wave. Figure 2-1 shows a decrease in $G(L,t)$ for $t < 4.1$ ms because of the finite difference segmentation.

For $t > 4.1$ ms, reflected pressure waves affect the mass flux profile for short times. Because of the assumption of constant outlet pressure, the incoming rarefaction wave will be reflected as a compression wave at the exit boundary. Thus a wave travels in the opposite direction with the same amplitude but opposite sign. The profiles at $t = 5$ ms and after show the effect of the reflected wave progressing toward the channel inlet from the exit. The net mass flux is the superposition of the forward wave and the reflected wave.

During this short period, the average mass flux ratio, $\hat{G}_m/G_m(0,0)$, decreases only from 1.00 to 0.98.

  b. *Momentum integral (MI) model: Figure 2-2.* The assumption of infinite sonic propagation results in the complete loss of local mass flux variation due to pressure perturbation. We only see the decreasing trend of the average mass flux.

Two reasons contribute to changing both the average mass flux and the slope from the steady-state profile:

  i. The decrease of inlet pressure which reduces the driving force, thus lessening the channel average mass flux.
  ii. The fluid thermal expansion due to heating which causes a small slope change, i.e., G_{in} becomes less than G_{out}.

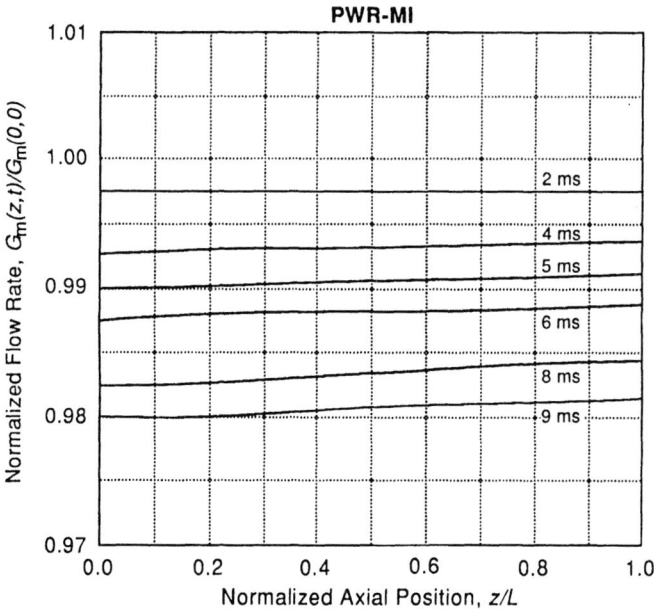

Figure 2-2 Short-term response of the PWR inlet pressure transient using the MI model.

PWR-CI

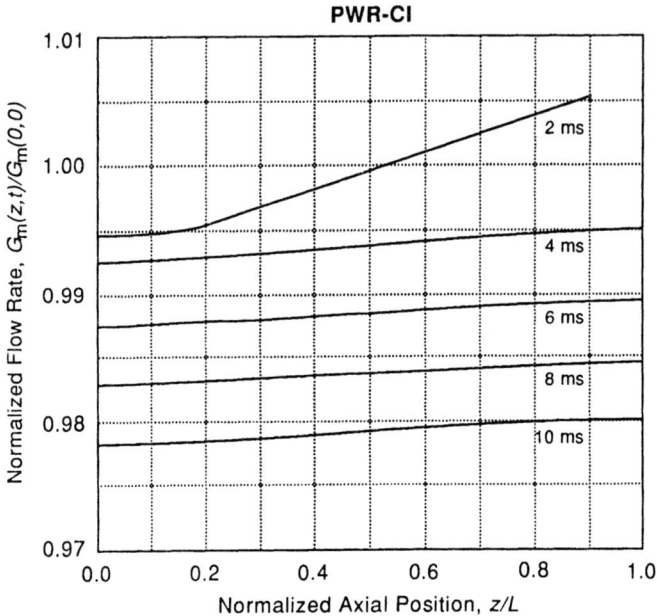

Figure 2-3 Short-term response of the PWR inlet pressure transient using the CI model.

c. *Channel integral (CI) model: Figure 2-3.* An abrupt change in the mass flux profile appears right after the transient begins. The reason is that the CI model applies global balances of mass, momentum, and energy based on the steady-state enthalpy profile. Since G_m near the inlet is decreased, h_m increases, which forces an expansion at the end of the channel.

The G_m profiles predicted by the CI model agree reasonably well with those of the MI model for $t \geq 3$ ms, which confirms that the transient enthalpy profile does not deviate much from the steady-state enthalpy profile. Thus we can conclude that the CI approach is a good approximation to the MI approach for a single-phase (liquid) flow transient.

d. *Single mass velocity (SV) model: Figure 2-4.* Here a single velocity prevails within the whole channel because of the rigid body type approximation.

The mass flux predicted by the SV model tends to be almost the same as the average mass fluxes predicted by MI and CI models, which is an expected result for liquid phase flow transients.

2. *Long-term channel flow response: Figures 2-5 to 2-8.* The predictions of the MI, CI, and SV methods for the inlet and exit mass flux histories are shown in Figures 2-5 and 2-6, respectively. Even after boiling begins ($t \approx 1.1$ s), the exit mass fluxes predicted by the three methods remain close.

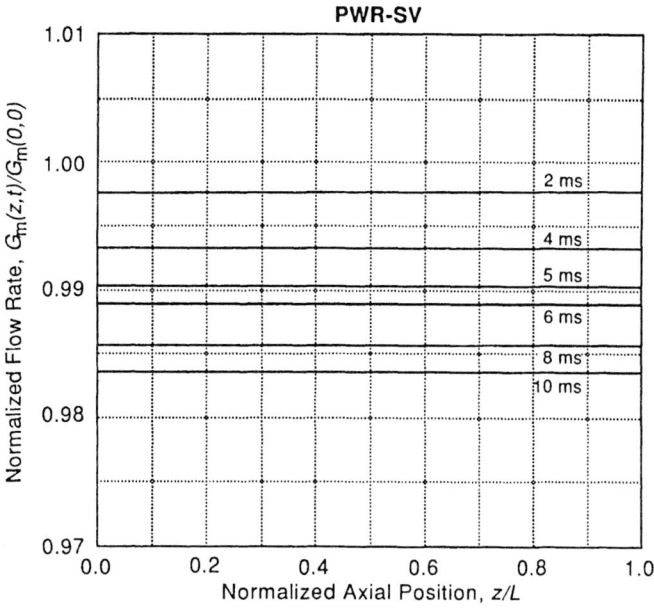

Figure 2-4 Short-term response of the PWR inlet pressure transient using the SV model.

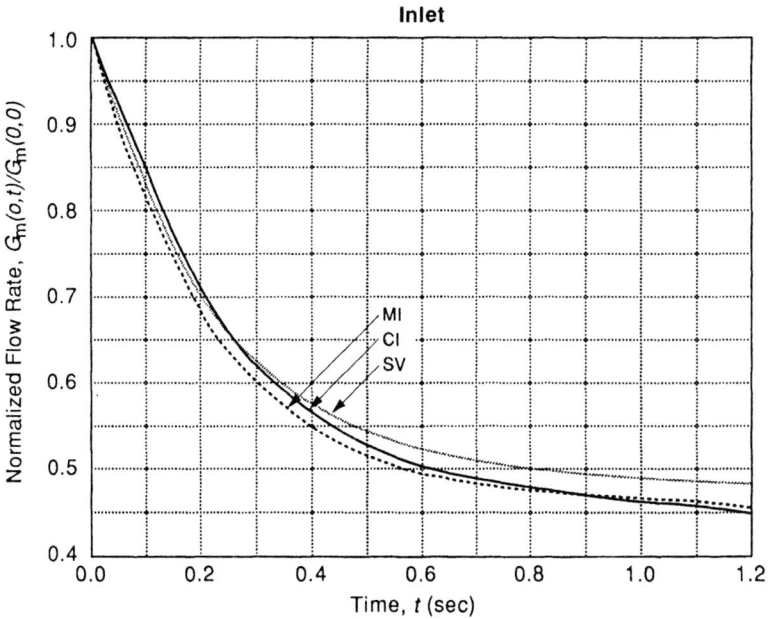

Figure 2-5 Inlet flow rate history of the PWR inlet pressure transient.

Outlet

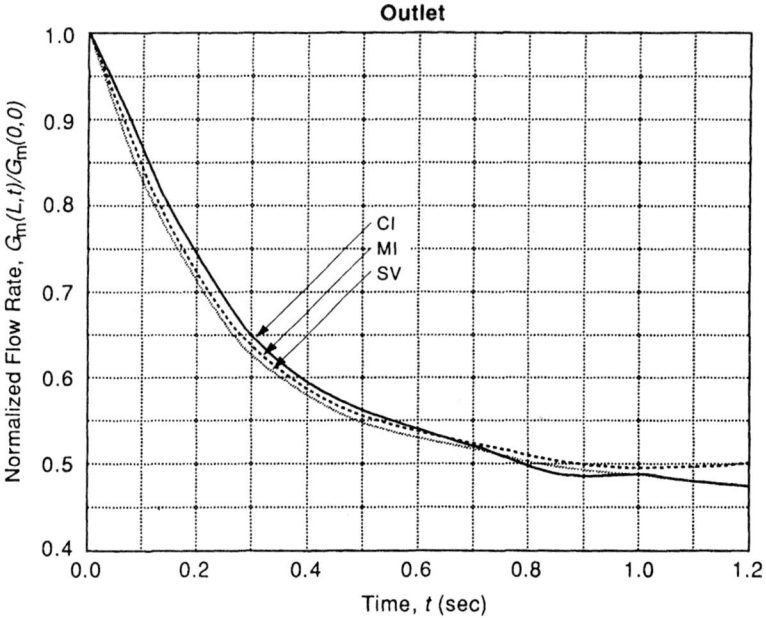

Figure 2-6 Outlet flow rate history of the PWR inlet pressure transient.

PWR-SC

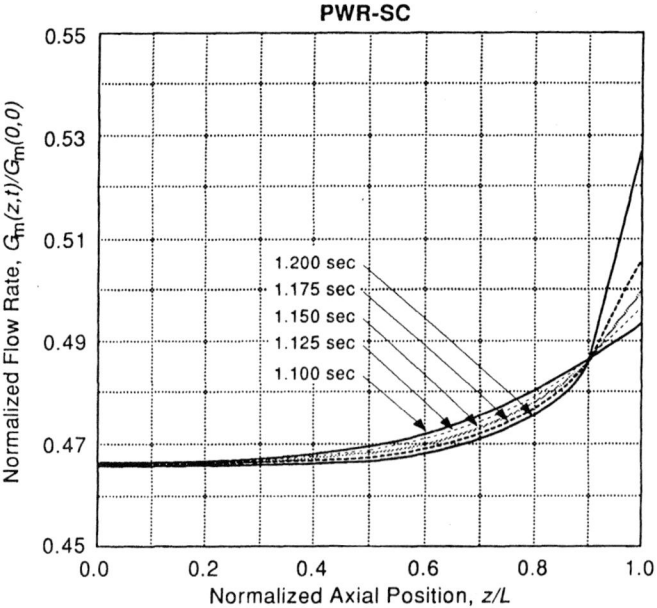

Figure 2-7 Flow rate profile of the PWR inlet pressure transient using the SC model.

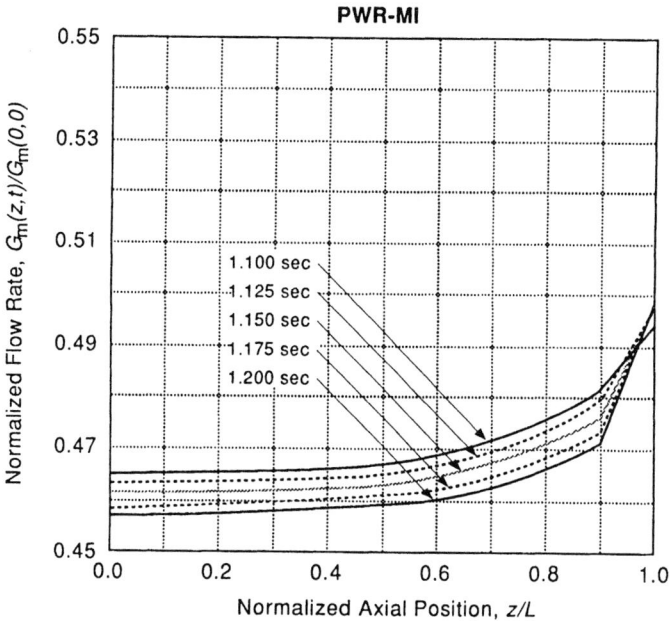

Figure 2-8 Flow rate profile of the PWR inlet pressure transient using the MI model.

The long-term response of the MI model is essentially a continuation of the short-term response. The average mass flux remains bounded as $G_{in} < \hat{G}_m < G_{exit}$ due to fluid thermal expansion, and the magnitude of $\hat{G}_m$ drops steadily but at a slower and slower rate as a new steady state is approached.

The result of the CI model shows surprisingly good agreement with the MI model throughout the transient. This demonstrates that the CI model has the potential for good prediction of PWR transients unless significant boiling occurs. The SV model also shows reasonably good agreement with the previous two models even though it significantly underpredicts the amount of boiling near the exit.

The G_m profiles by the SC and MI methods are shown with an expanded mass flow rate scale in Figures 2-7 and 2-8. It is clear that the effect of boiling in the last segment is to enhance the exit flow rate due to thermal expansion in the MI method but to a lesser extent than calculated by the SC method.

Example 2-2 PWR channel step heat flux transient

PROBLEM A second transient problem of interest is that of a rapid increase in the heat flux without change in the applied pressure drop. The PWR initial and geometry conditions are identified in Table 2-2. A 10% flux increase is applied in a step fashion.

SOLUTION The solution by the SC approach for a 10% heat flux step increase in the PWR channel is shown in Figure 2-9. It is seen that because the pressure begins to rise at internal channel points, a reduction in inlet flow rate and an accelerated exit flow rate occur. The magnitudes of inlet flow reduction and exit flow increase show an oscillatory behavior within the 20 ms shown. This is because of the sonic speed with which the pressure changes are propagated in the channel, allowing momentary cancellation of velocity variation due to interference between the various reflected waves.

The solution via the MI approach is shown in Figure 2-10. It is seen that the deviation of the local flow rate from the average flow rate remains within 0.5% from the very first instant until 20 ms.

The long-time behavior (not shown here) indicates that at boiling, the local mass velocity profile is changed (but not the average mass velocity) due to the more pronounced variation in the local expansion of the fluid. The increased friction at the exit limits the increase in the exit mass flow rate. A new steady state is reached at a lower average mass flux than the initial condition.

PWR-SC

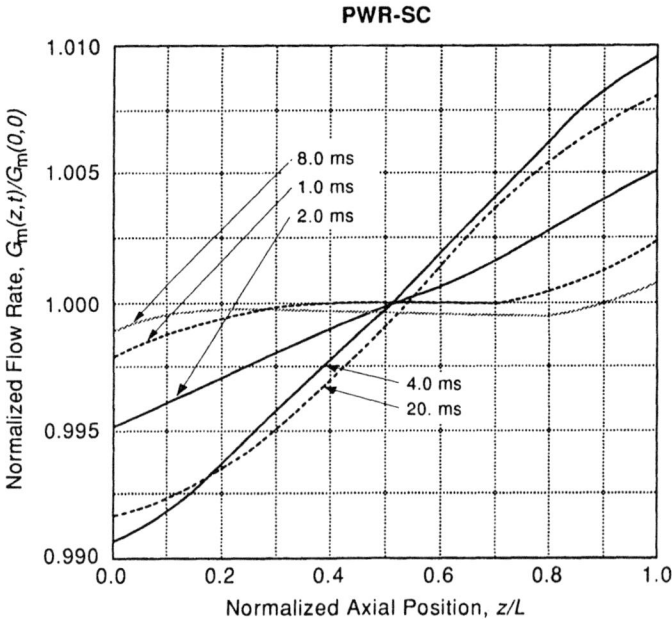

Figure 2-9 Flow rate profile of the PWR heat flux transient using the SC model.

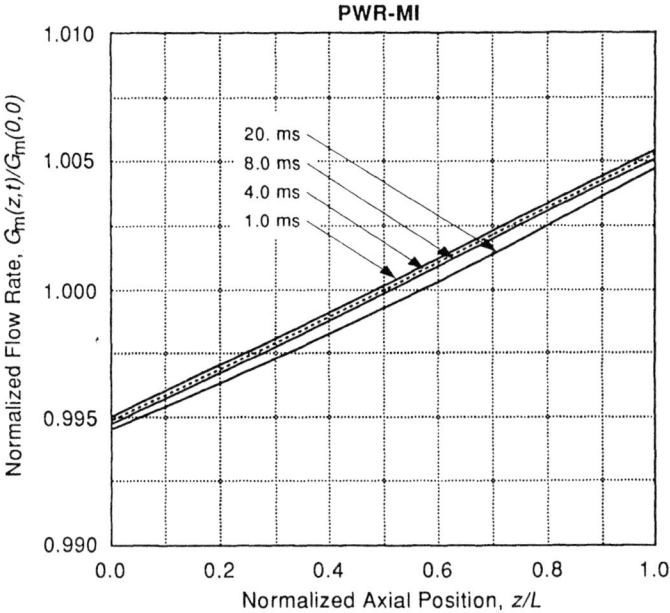

Figure 2-10 Flow rate profile of the PWR heat flux transient using the MI model.

Example 2-3 BWR inlet pressure decrease transient

PROBLEM Consider the same pressure decrease function of Example 2-1, Eq. 2-38, applied to the BWR channel whose conditions are defined in Table 2-2. Let us compare the results from the various models.

SOLUTION

1. *Short-term channel flow response*
   a. *SC model: Figure 2-11*. The inlet mass flux drops immediately, like in the PWR case, but more than half of the downstream channel maintains the initial pressure and mass flux. This situation, which is quite different from the PWR case, occurs because the sonic wave velocity is significantly lower in the two-phase flow region. Hence, the inlet pressure perturbation is not quickly transmitted into the two-phase region.

   Since the sonic velocity is about ~100 m/s in the two-phase region, the channel transit time of the sonic perturbation becomes ~20 ms. Consequently we cannot see any effect of reflected waves in Figure 2-11.

   The mass flux decrease in the front region is about 10%, which is much larger than in the PWR case. This is due to the fact that the

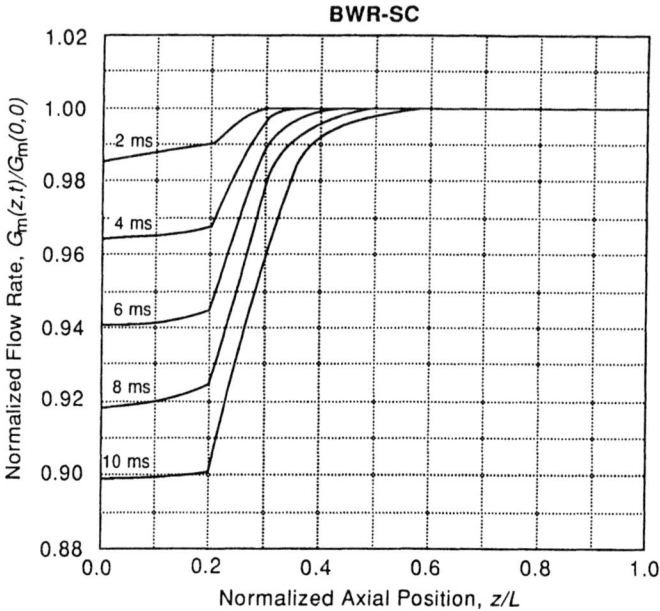

Figure 2-11 Short-term response of the BWR inlet transient using the SC model.

pressure gradient is confined to the front nodes during this short time.

b. *MI model: Figure 2-12.* We again see the averaging trend in this model, i.e., the severe local deviation of mass flux shown in the SC result tends to spread out across the whole channel. Thus, the mass flux in the single-phase region is overpredicted while that in the two-phase region is underpredicted. This deviation is the result of the infinite sonic propagation assumption and the application of the momentum balance in an integral sense. Based on this result, the MI model is not a good approximation for the few initial milliseconds in BWR channel flow transient where the amount of boiling extends over a significant fraction of the channel length.

c. *CI and SV models: Figures 2-13 and 2-14.* The predictions of the CI and SV models are much worse than the MI model in this case. The result of the CI model shows a large deviation from those of the MI and SC models for $t \leq 4$ ms, then improves and approaches that of the MI model for $t \geq 16$ ms. In this respect, the CI model is a better approximation than the SV model. The agreement of the CI and MI models for $t \geq 16$ ms implies that the CI model has a potential for good approximation to the MI model if an appropriate enthalpy profile is chosen for the given transient.

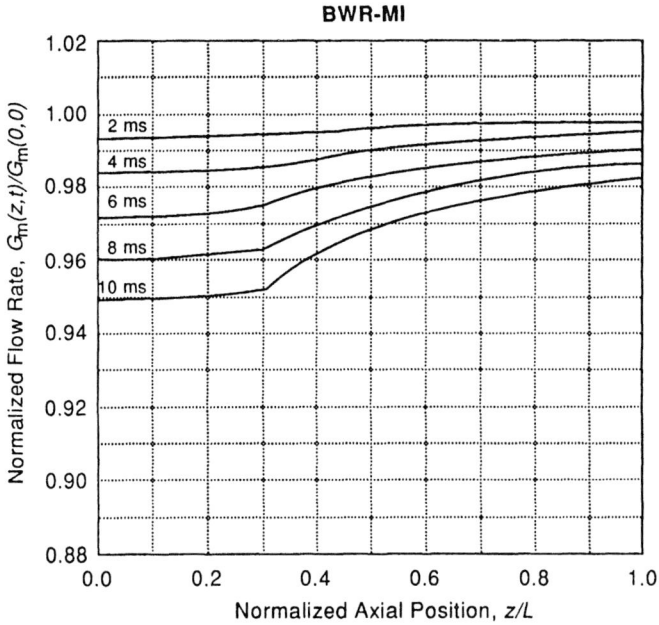

Figure 2-12 Short-term response of the BWR inlet pressure transient using the MI model.

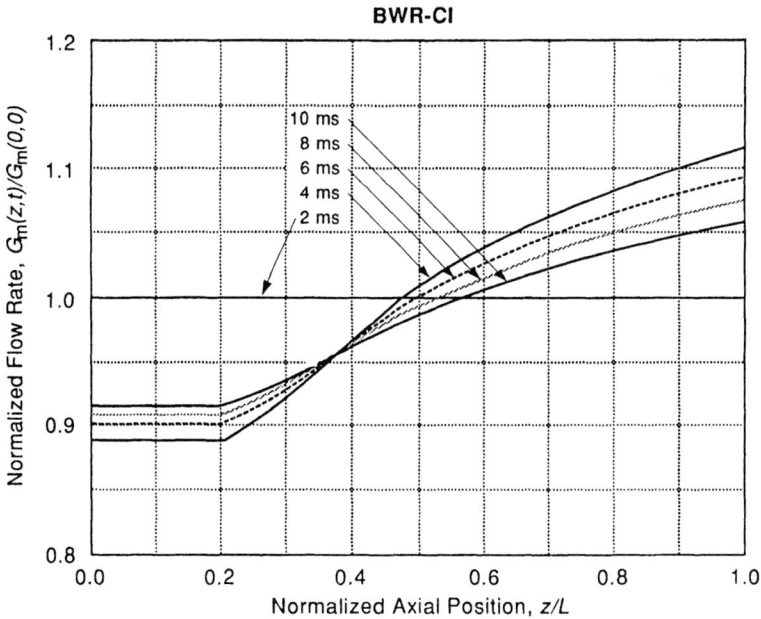

Figure 2-13 Short-term response of the BWR inlet pressure transient using the CI model.

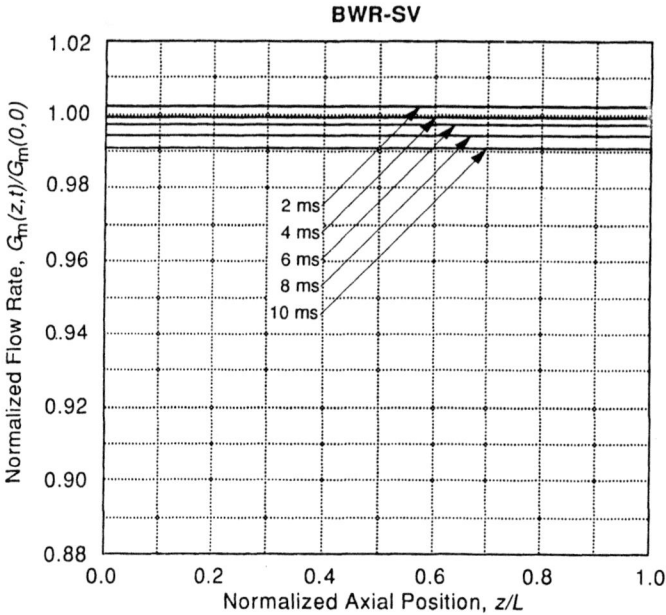

Figure 2-14 Short-term response of the BWR inlet pressure transient using the SV model.

2. *Long-term channel flow response: Figures 2-15 and 2-16.* The decrease in the flow rate at the inlet is very dramatic, reaching zero flow in 0.365 s. This is due to the tremendous increase in the pressure within the channel due to the increased boiling. The MI model predicts flow stagnation, or onset of flow reversal, in the first node at about $t = 0.3$ s. The only drawback of the MI model compared with the SC model is the neglect of information on the scale of the sonic channel transit time.

Calculations beyond flow stagnation were not possible by the numerical schemes used. The following is a plausible explanation for the occurrence of this local flow reversal:

If liquid is vaporized, the specific volume increases about 21 times ($v_g/v_f \approx 21$ at 6.9 MPa). This will cause a large flow resistance and, hence, a decrease of upstream mass flux. The decrease of G_{in} will further promote the vapor formation within the channel, driving the boiling boundary downward and increasing the channel flow resistance. Continuation of this chain process will eventually result in a flow reversal at the inlet node.

The result of the CI model is very close to the SC and MI results for the first 0.4 s. However, flow reversal does not occur as the results tend to overestimate the channel mass inlet flux. Beyond $t = 0.8$ s, the result of the CI model shows that the channel flow approaches a new

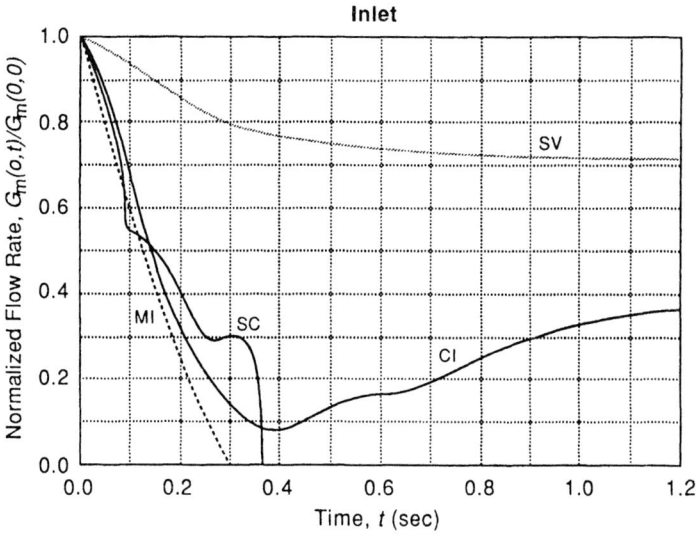

Figure 2-15 Inlet flow rate history of the BWR inlet pressure transient.

steady-state value after undergoing severe fluctuation in the axial mass flux profile and the quality profile.

The results of the SV model shown in Figure 2-15 consistently overpredict the mass flux and show large deviation from that of the CI model. This deviation seems to be caused by the inability to account for the large degree of boiling during a relatively short time.

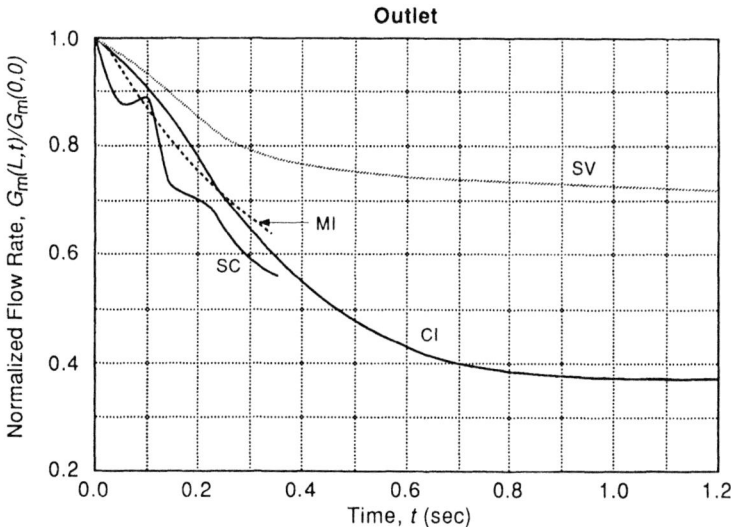

Figure 2-16 Outlet flow rate history of the BWR inlet pressure transient.

In summary, the inlet and exit mass flux histories predicted by the four methods, shown in Figures 2-15 and 2-16, respectively, illustrate clearly the failure of the SV method for a BWR channel transient. However, the other three methods predict close results during the period of inlet flow reduction.

III SOLUTION OF TRANSIENTS BY THE METHOD OF CHARACTERISTICS (MOC)

A Basics of the Method

The basic premise of the method of characteristics is the transformation of linear partial differential equations into ordinary differential equations. This can lead to analytical solution of the simple problems and more rapid numerical solution of the complicated problems. The versatility of the method of characteristics has been recognized in single-phase flow problems for a long time [12]. A review of its applications in the nuclear engineering field can be found in Weisman and Tentner [15].

Consider the linear first-order partial differential equation:

$$A \frac{\partial \psi(z,t)}{\partial t} + B \frac{\partial \psi(z,t)}{\partial z} = R \tag{2-39}$$

where A, B, and R are functions of t and z. It is well known that for Eq. 2-39 to be linear, A and B have to be independent of ψ, but R can be a linear function of ψ [5]. Abbott [1] has shown that for Eq. 2-39 to be linearly dependent on $d\psi$, i.e.,

$$d\psi = \frac{\partial \psi}{\partial t} dt + \frac{\partial \psi}{\partial z} dz \tag{2-40}$$

the variables A, B, and R should satisfy:

$$\frac{dt}{A} = \frac{dz}{B} = \frac{d\psi}{R} \tag{2-41}$$

Others have shown that Eq. 2-41 can be valid for quasi-linear partial differential equations in which A and B are also dependent on ψ [14].

The solution of any two equalities of Eq. 2-41 is equivalent to solving Eq. 2-39. Solving $dt/A = dz/B$ yields a relation between z and t starting from a specific value of z, say z_0 at $t = 0$. The set of curves representing the z–t relationship are called the *characteristic curves*. The *characteristic velocity* is given by the slope of the z–t curve:

$$\frac{dz}{dt} = \frac{B}{A} = C \tag{2-42}$$

Similarly, the relationship $dt/A = d\psi/R$ provides the value of ψ at time t starting from a specific value of t, say t_0 at $z = 0$. The relation $dz/B = d\psi/R$ provides the value of ψ as a function of z, starting from a known elevation z, say z_0 at $t = 0$.

B Applications to Single-Phase Transients

In a fluid system, a characteristic defines a path in the z–t space such as the path followed by a small packet of fluid or the path followed by a fluid disturbance propagating in the fluid. If we consider one-dimensional pipe flow, the characteristic is defined in a two-dimensional space having the axial location of the disturbance as one dimension and time as another. Note that the characteristics are related to the capacity for propagation in the particular physical system rather than to a particular propagation process. They are defined even when there are no physical disturbances present.

The MOC technique may be easily illustrated using a problem described by Tong and Weisman [14]. Consider the behavior during a flow transient with a single, heated channel. We assume that the inlet mass velocity follows the equation:

$$G_{in} = G_0/(1 + t) \tag{2-43}$$

for $t > 0$ and where G_0 is a constant. In addition, we assume (a) steady-state conditions at $t = 0$ and (b) the heat flux $q''(z)$ and inlet enthalpy, h_{in}, remain constant throughout the transient at their steady-state values. If the system pressure is taken constant, i.e., independent of z and t, the change in coolant enthalpy, h_m, is given by (same as Eq. 2-15 for a thermally expandable fluid):

$$\rho_m \frac{\partial h_m}{\partial t} + G_m \frac{\partial h_m}{\partial z} = \frac{q'' P_h}{A_z} \tag{2-44}$$

where P_h is the wall surface per unit length. When ρ_m is also taken constant, this becomes essentially the SV model of section II since $G_m(z,t)$ is now equal to $G_{in}(t)$. Eq. 2-44 has the same form as Eq. 2-39, and hence the characteristic equations are:

$$dt/\rho_m = dz/G_m = dh_m/(q'' P_h/A_z) \tag{2-45}$$

Note that even if ρ_m is dependent on h_m, Eq. 2-45 is still valid [14].

There are two solutions to Eq. 2-45. The first belongs to the packet of fluid within the reactor at a given position z_0 when the transient began. The subsequent positions of such a packet are shown by an appropriate line in region I of Figure 2-17. A second packet of fluid is one that had not yet entered the reactor when the transient began. This packet can be described in terms of t_0, the time interval between the beginning of the transient and the entrance of the packet into the reactor. The position-time history of such a packet is shown by a line within region II of Figure 2-17. Regions I and II are separated by the limiting characteristic, which corresponds to a fluid packet that was just at the reactor inlet when the transient began, i.e., for which $z_0 = 0$ and $t_0 = 0$.

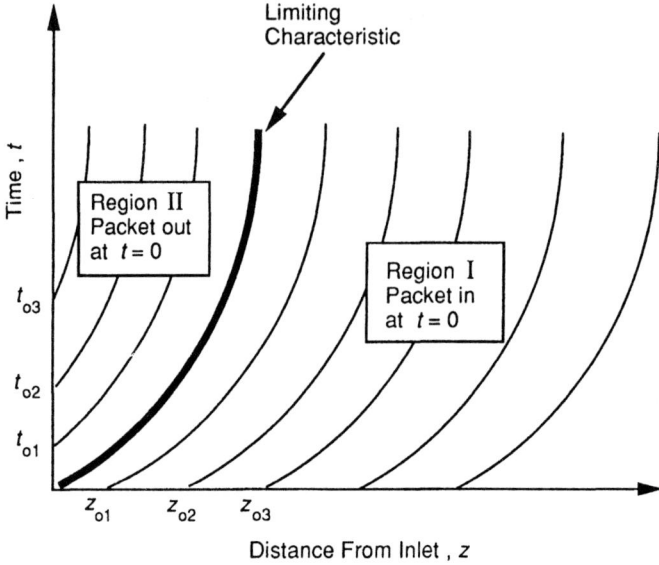

Figure 2-17 Time–distance relationships during a hypothetical flow transient.

The solution for region I is determined by first integrating $dz/G_m = dt/\rho_m$ to obtain:

$$z - z_0 = \int_0^t \frac{G_0}{\rho_m(1 + t)} \, dt = \frac{G_0}{\rho_m} \ell n(1 + t) \tag{2-46}$$

This is the equation describing the characteristic curves of region I and represents the path of a fluid packet. From the integration of $dz/G_m = dh_m/(q''P_h/A_z)$ we get:

$$h_m(z) - h_0(z_0) = \int_{z_0}^{z} \frac{1 + t}{G_0} \frac{q''P_h}{A_z} \, dz \tag{2-47}$$

where $h_0(z_0)$ is the coolant enthalpy at zero time and axial position z_0, and G_0 is the mass velocity at $t = t_0$. From steady-state conditions we evaluate $h_0(z_0)$ and substitute this in Eq. 2-47 to obtain:

$$h_m(z) = h_{in} + \frac{P_h}{A_z G_0} \int_0^{z_0} q'' dz + \frac{P_h}{A_z G_0} \int_{z_0}^{z} (1 + t) q'' dz \tag{2-48}$$

where h_{in} is the inlet enthalpy. After using Eq. 2-46 to solve for $(1 + t)$, we substitute into Eq. 2-48 and finally obtain:

$$h_m(z) = h_{in} + \frac{P_h}{A_z G_0} \int_0^{z_0} q'' dz$$

$$+ \frac{P_h}{A_z G_0} \int_{z_0}^{z} q'' \exp[\rho_m(z - z_0)/G_0] dz \tag{2-49}$$

Equation 2-49 may be integrated for a given flux shape to obtain the enthalpy at z for any packet of fluid starting at z_0. The corresponding time is obtained from Eq. 2-46.

To obtain the solution for region II, we integrate $dt/\rho_m = dz/G_m$ but now conduct the integration from t_0, the time when the fluid packet first entered the reactor (at that time $z = 0$), to t and obtain:

$$z = (G_0/\rho_m) \int_{t_0}^{t} \frac{dt}{1 + t} = (G_0/\rho_m)\ell n[(1 + t)/(1 + t_0)] \qquad (2\text{-}50)$$

This equation describes the characteristic curves in region II of Fig. 2-17. From integration of $dz/G_m = dh_m/(q''P_h/A_z)$ from 0 to z and substitution for $(1 + t)$, we have:

$$h_m(z) = h_{in} + \frac{P_h}{A_z} \int_0^z \frac{q''dz}{G_m} = h_{in} + \frac{P_h}{A_zG_0} \int_0^z q''(1 + t)dz$$

$$(2\text{-}51)$$

$$h_m(z) = h_{in} + \frac{P_h}{G_0A_z}(1 + t_0) \int_0^z q''\exp(\rho_m z/G_0)dz$$

The limiting characteristic curve is obtained from the solution of region I, Eq. 2-46 for $z_0 = 0$, or from the solution of region II, Eq. 2-50 at $t_0 = 0$. We then have:

$$\ell n(1 + t) = \frac{\rho_m z}{G_0}; \text{ limiting characteristic} \qquad (2\text{-}52)$$

The solution of the foregoing example is complete since the fluid enthalpy and flow velocity are known at each point in space and time. In region I, $h_m(z,t)$ is given by Eqs. 2-49 and 2-46, while in region II the analogous equations are 2-51 and 2-50. The flow velocity, by virtue of the assumption of constant density, is equal everywhere in the channel at any given time, i.e., $V(z,t) = V_{in}(t)$. This model is reasonable for a liquid whose flow velocity is much smaller than the acoustic propagation velocity. However, in many situations, it is important to evaluate the pressure disturbance that accompanies an abrupt change in local fluid velocity. To deal with such problems both the continuity and momentum conservation equations should be included in the analysis.

C Applications to Two-Phase Transients

1 General approach The method of characteristics has been applied to a variety of two-phase flow models. Tong and Weisman [14] consider the homogeneous flow model with vapor and liquid having the same velocity and in thermal equilibrium. Shiralkar et al. [13] allow for the existence of a drift flux but keep the assumption of thermal equilibrium. Ferch [2] derived characteristic equations that included thermal nonequilibrium but retained the assumption of equal phase velocities. Kroeger [6] included thermal nonequilibrium in his drift flux formulation of the conservation equations. He did assume vapor to be at saturation conditions and the liquid to have a constant density. Lyczkowski [9] applied the method of characteristics to the two-fluid model with the assumption that the pressure drop occurs in the vapor phase only. Gidaspow and Shin [3] introduced a two-fluid model that contains a specific constitutive equation for the relative velocity of the two phases. Their model leads to a "well-posed" set of equations with all real characteristics. Depending on the constitutive relations, a two-fluid model can be "ill-posed" in which case some of the characteristics become imaginary, leading to unstable numerical solutions.

To illustrate the application in two-phase flow problems we shall consider a boiling channel under the assumptions of constant system pressure and homogeneous flow with thermal equilibrium, drawing from the treatments of Zuber [16], Gonzales-Santalo and Lahey [4], and Lahey and Moody [7]. The fluid packets in the single-phase region of the channel can be described in the manner presented in Section IIIB.

Let us first define the boundary between the single-phase and two-phase regions. This boundary will be characterized by two parameters:

$\lambda(t)$ = The distance along the channel which it takes the fluid to reach the saturation condition. This distance is illustrated in Figure 2-18.

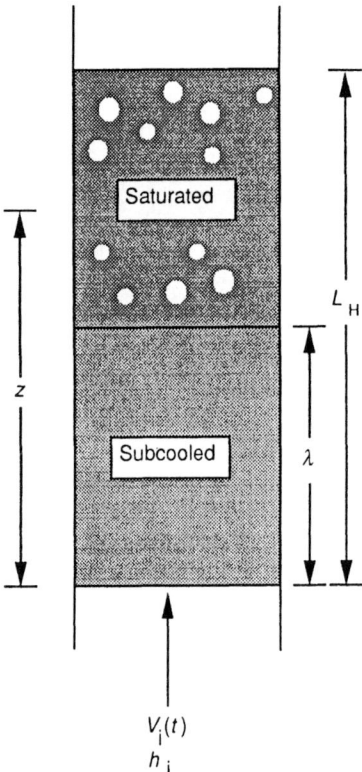

Figure 2-18 Basic regions in a heated channel.

and ν = The time it takes a fluid packet to lose its subcooling once it enters the channel.

For our assumptions, the mass flux in the single-phase region is constant, $G_m(z,t) = G_{in}(t)$. For the constant density liquid we have:

$$V_m(z,t) = V_\ell(z,t) = V_{in}(t) \text{ at the channel inlet} \tag{2-53}$$

Hence the parameter $\lambda(t)$ can be obtained from integration of $dz/G_m = dt/\rho_m$ as:

$$\lambda(t) = \int_0^\lambda dz = \int_{t-\nu}^t V_{in}(t)dt = \int_{t-\nu}^t \frac{G_{in}(t)}{\rho_\ell} dt \tag{2-54}$$

For a constant system pressure, just like the MI method of Section II, the energy equation (I,13-9b) can be written as:

$$\rho_m \frac{\partial h_m}{\partial t} + \rho_m V_m \frac{\partial h_m}{\partial z} = \frac{q''P_h}{A_z} \tag{2-55a}$$

or

$$\rho_m \frac{Dh_m}{Dt} = \frac{q''P_h}{A_z} \tag{2-55b}$$

In accordance with Eq. 2-41, Eq. 2-55a leads to the characteristic equation:

$$\frac{dh_m}{dt} = \frac{q'}{\rho_m A_z} \tag{2-56}$$

where: $q' = q''P_h$,

If the heat input along the channel is axially and temporally constant, Eq. 2-56 can be easily integrated between the inlet enthalpy and the saturation enthalpy:

$$\int_{h_{in}}^{h_f} dh_\ell = \frac{q'}{\rho_\ell A_z} \int_{t-\nu}^t dt \qquad t > \nu \tag{2-57}$$

or

$$\nu = (h_f - h_{in}) \left(\frac{\rho_\ell A_z}{q'} \right) \tag{2-58}$$

Thus, ν is the time it takes the fluid packet to lose its subcooling once it enters the channel. Note that since the volumetric heating (q'/A_z) is constant, if the inlet enthalpy is also constant, ν is constant. However, the distance traveled, $\lambda(t)$, varies according to the velocity as illustrated by Eq. 2-54. For a decreasing inlet flow transient, the position of $\lambda(t)$ is as shown in Figure 2-19. Note that ν is constant since the inlet enthalpy and the axial heat addition are held constant.

Let us now turn our attention to the two-phase domain (i.e., $z > \lambda$). In this domain, the continuity Eq. I,13-1 can be written as:

$$\frac{\partial \rho_m}{\partial t} + V_m \frac{\partial \rho_m}{\partial z} + \rho_m \frac{\partial V_m}{\partial z} = 0 \tag{2-59a}$$

or

$$\frac{D\rho_m}{Dt} = -\rho_m \frac{\partial V_m}{\partial z} \tag{2-59b}$$

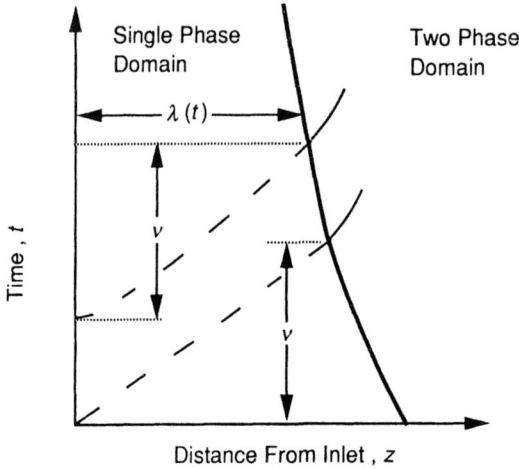

Figure 2-19 Major characteristics in a channel with decreasing inlet flow; the single-phase region.

The density and enthalpy are given by:

$$\rho_m = \frac{1}{v_f + x v_{fg}},$$

$$h_m = h_f + x h_{fg},$$

where x is the vapor flowing quality.

We can explicitly relate the change in the velocity to the vapor quality and phase-specific volumes as:

$$
\begin{aligned}
\frac{\partial V_m}{\partial z} &= -\frac{1}{\rho_m}\frac{D\rho_m}{Dt} = -\frac{1}{\rho_m}\frac{D}{Dt}\left(\frac{1}{v_f + x v_{fg}}\right) \\
&= \frac{1}{\rho_m(v_f + x v_{fg})^2}\left[\frac{D}{Dt}v_f + x\frac{D}{Dt}v_{fg} + v_{fg}\frac{Dx}{Dt}\right] \qquad (2\text{-}60) \\
&= \frac{1}{v_f + x v_{fg}}\left[\frac{D}{Dt}v_f + x\frac{D}{Dt}v_{fg} + v_{fg}\frac{Dx}{Dt}\right]
\end{aligned}
$$

Under the conditions applied here, namely incompressible fluids ($Dv_f/Dt = 0$) and a near constant pressure system ($Dv_{fg}/Dt = 0$) we get:

$$\frac{\partial V_m}{\partial z} = \frac{v_{fg}}{(v_f + x v_{fg})}\frac{Dx}{Dt} \qquad (2\text{-}61)$$

which can be integrated to yield:

$$V_m(z,t) = V_{in}(t) + \int_{\lambda(t)}^{z}\frac{v_{fg}}{v_f + x v_{fg}}\frac{Dx}{Dt}\,dz \qquad (2\text{-}62)$$

The energy equation can be manipulated to determine the flow quality. From Eq. 2-55b, for constant system pressure (h_{fg} and h_f are constant) we get:

$$h_{fg} \frac{Dx}{Dt} = \frac{q'}{\rho_m A_z} = \frac{q'(v_f + x v_{fg})}{A_z} \tag{2-63}$$

Define the parameter Ω, the characteristic boiling frequency [16] as:

$$\Omega \equiv \frac{q' v_{fg}}{A_z h_{fg}} \tag{2-64}$$

Eq. 2-63 reduces to:

$$\frac{Dx}{Dt} - \Omega x = \Omega \frac{v_f}{v_{fg}} \tag{2-65}$$

The rate of change of quality can be eliminated between Eq. 2-62 and Eq. 2-65 so that we obtain:

$$V_m(z,t) = \frac{dz}{dt} = \frac{G_{in}(t)}{\rho_\ell} + \int_{\lambda(t)}^{z} \frac{v_{fg}}{v_f + x v_{fg}} \Omega \left(\frac{v_f}{v_{fg}} + x \right) dz \tag{2-66}$$

or

$$\frac{dz}{dt} = \frac{G_{in}(t)}{\rho_\ell} + \int_{\lambda(t)}^{z} \Omega dz \tag{2-67}$$

Eq. 2-65 is the desired general relation for the quality change with time, while Eq. 2-67 is the general equation for the position change with time. It should be noted that the value of ρ_ℓ at the inlet is nearly equal to the saturation value ρ_f, and in practice, therefore, ρ_ℓ is taken as ρ_f.

There are two types of fluid packets to be considered: namely those that were outside the two-phase region initially and those that were in the two-phase region initially. The path of the liquid initially at the saturation position defines the main path characteristic line, as illustrated in Figure 2-20.

Considering the packets initially present in the two phase region (region II in Figure 2-20), the initial conditions are defined by:

at $t = 0$

$x_0 = f(z)$ established by a steady-state balance and $\tag{2-68}$

$z_0 > \lambda(0)$

The packets initially outside the two-phase region, including those outside the channel (region I in Figure 2-20), reach the two-phase region at a time t_B. The conditions for x and z at the time, t_B, that a packet of fluid starts boiling are:

at $t = t_B > 0$

$x = 0$ and $\tag{2-69}$

$z = \lambda(t)$

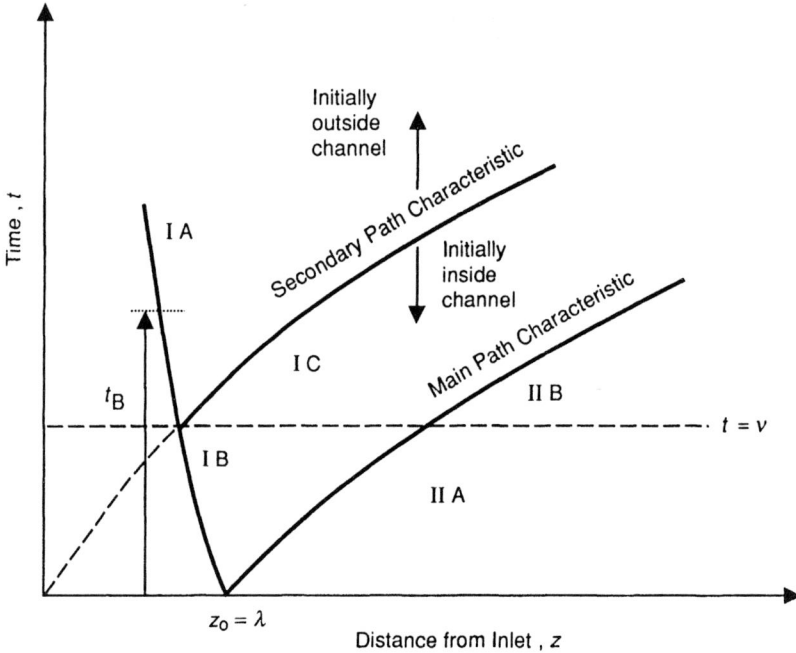

Figure 2-20 Major characteristics in a channel with a two-phase region and a decreasing fluid flow or increasing heat input.

Under our assumptions Ω is a constant, and so for all $G_{in}(t)$ of practical significance, both Eqs. 2-65 and 2-67 are integrable. Hence, for any constant pressure flow transient, we can obtain an exact solution for the fluid velocity and quality as a function of space and time along the heated length. This will be seen through an application to an example situation, namely an exponential flow decay in the channel.

2 The case of an exponential flow decay, constant heat flux Let us assume that:

$$G_{in} = G_0 e^{-Kt} \qquad \text{for } t > 0 \qquad (2\text{-}70)$$

along with the previously stated assumptions of constant inlet enthalpy and axial heat addition rate q'. First we define the suitable $\lambda(t)$ for the distinct time domains, $t < \nu$ and $t > \nu$, in Figure 2-20.

In the single-phase region and for packets of fluid not initially in the channel (which will reach the two-phase region at $t > \nu$) from Eq. 2-54, when $\rho_\ell = \rho_f$ per our assumptions:

$$\lambda(t) = \frac{G_0}{\rho_f K} e^{-Kt}(e^{K\nu} - 1); \ t > \nu \qquad (2\text{-}71)$$

For packets initially in the single-phase region but reaching the two-phase region at $t < \nu$, then:

$$\lambda(t) = \int_{t-\nu}^{t} \frac{G_{in}}{\rho_f} \, dt = \int_{t-\nu}^{0} \frac{G_0}{\rho_f} \, dt + \int_{0}^{t} \frac{G_0 e^{-Kt}}{\rho_f} \, dt$$

so that:

$$\lambda(t) = \frac{G_0}{\rho_f} (\nu - t) + \frac{G_0}{\rho_f K} (1 - e^{-Kt}); \ t < \nu \qquad (2\text{-}72)$$

The value of $\lambda(t)$ applicable in each of the two time domains is summarized in Table 2-3, along with the significance of the time–space regions (IA, IB, IC, IIA, and IIB).

Now we define the time–space relation. For the two-phase region, we have to consider regions I and II separately. Region I is the region of fluid packets entering the two-phase regions at $t > 0$.

For region I To obtain the quality we integrate Eq. 2-65 under the condition $x = 0$ at $t = t_B$. This leads to:

$$x = \frac{v_f}{v_{fg}} [e^{\Omega(t-t_B)} - 1] \qquad (2\text{-}73)$$

Hence the parameter t_B is required to complete the specification of the quality. It is obtained from integrating Eq. 2-67, which for constant Ω and the assumed exponential flow decay yields:

$$\frac{dz}{dt} - \Omega(z - \lambda(t)) = \frac{G_0 e^{-Kt}}{\rho_f} \qquad (2\text{-}74)$$

Table 2-3 Tabulation of the boiling boundary for the various two-phase regions in the channel

Time	Zone	Expression for boiling boundary	Significance of region
$t > \nu$	IA	$\lambda(t) = \dfrac{G_0}{\rho_f K} \exp(-Kt)[\exp(K\nu) - 1]$	Fluid packets that had not yet entered the heater at $t = 0$.
	IC	$\lambda(t)$ Same as IA	The packets that at $t = 0$ were in the single-phase region of the heater.
	IIB	$\lambda(t)$ Same as IA	The packets that at $t = 0$ were in the two-phase portion of the heater.
$t < \nu$	IB	$\lambda(t) = \dfrac{G_0}{\rho_f} (\nu - t) + \dfrac{G_0}{\rho_f K} [1 - \exp(-Kt)]$	The packets that at $t = 0$ were in the single-phase portion of the heater.
	IIA	$\lambda(t)$ Same as IB	The packets in the two-phase portion of the heater for $0 \leq t \leq \nu$.

To obtain the parameter t_B, Eq. 2-74 is integrated using the appropriate value of $\lambda(t)$ from Table 2-3 and the appropriate initial condition. The results for the three subregions are:

For subregion IA: $t > t_B > v$; using $\lambda(t_B > v)$ and the initial condition at $t = t_B$, $z = \lambda(t_B)$ results in:

$$t_B = \frac{\Omega}{\Omega + K} t - \frac{1}{\Omega + K} \ell n \left[\frac{1}{K_9} \left(z + \frac{G_0 K_3}{\Omega + K} e^{-Kt} \right) \right] \qquad (2\text{-}75a)$$

For subregion IB: $v > t > t_B$; using $\lambda(t_B < t)$ and the initial condition at $t = t_B$, $z = \lambda(t_B)$ results in:

$$t_B = t - \frac{1}{\Omega} \ell n \left\{ \Omega \left[\frac{\rho_f}{G_0} z - \frac{K_4}{\Omega} + \frac{1}{\Omega} + t + \frac{e^{-Kt}}{K} \right] \right\} \qquad (2\text{-}75b)$$

For subregion IC: $t > v > t_B$; using $\lambda(t_B > v)$ and the initial condition at $t = v$, $z = z_{IB}$ (since a fluid packet crosses from IB to IC at $t = v$):

$$t_B = t - \frac{1}{\Omega} \ell n \left[\frac{\rho_f}{G_0} \Omega \left\{ z - K_7 e^{\Omega t} + \frac{G_0 K_3}{\Omega + K} e^{-Kt} \right\} \right] \qquad (2\text{-}75c)$$

Note that for any combination of z and t, Eqs. 2-75a through 2-75c can be used to specify t_B. The K_i constants are defined as:

$$K_3 \equiv \frac{1}{\rho_f} \left[1 - \frac{\Omega}{K} (e^{Kv} - 1) \right]$$

$$K_4 \equiv \frac{\Omega}{K} + \Omega v$$

$$K_7 \equiv \frac{G_0}{\rho_f} e^{-\Omega v} \left[\frac{1}{K} - \frac{1}{\Omega} - e^{-Kv} \left(\frac{1}{K} - \frac{\rho_f K_3}{\Omega + K} \right) \right]$$

$$K_9 \equiv \frac{G_0 e^{Kv}}{\rho_f (K + \Omega)}$$

For region II This is the region where the fluid packets were in the two-phase region at $t \leq 0$. In this case $x = x_0$ and $z = z_0$ at $t = 0$, so that Eq. 2-65 can be integrated to give:

$$x(t, z_0) = x_0(z_0) e^{\Omega t} + \frac{V_f}{V_{fg}} [e^{\Omega t} - 1] \qquad (2\text{-}76)$$

The value of x_0 is determined from the steady-state energy balance as:

$$x_0(z_0) = \frac{1}{h_{fg}} \left[\frac{q' z_0}{G_0 A_z} - (h_f - h_{in}) \right] \qquad (2\text{-}77)$$

Note that Eq. 2-74 can now be integrated using the appropriate values of λ from Table 2-3 to yield two subregions. For region IIA ($t < v$), at $t = 0$, $z = z_0$

leading to:

$$z(t,z_0) = \frac{G_0}{\rho_f}\left[\frac{K_4}{\Omega} - \frac{1}{\Omega} - t - \frac{e^{-Kt}}{K} - K_5 e^{\Omega t}\right] + z_0 e^{\Omega t} \qquad (2\text{-}78a)$$

and for region IIB $(t > v)$ at $t = v$, $z = z_{IIA}$ leading to:

$$z(t,z_0) = z_0 e^{\Omega t} + K_6 e^{\Omega t} - \frac{G_0 K_3}{\Omega + K} e^{-Kt} \qquad (2\text{-}78b)$$

where: $K_5 \equiv \dfrac{K_4}{\Omega} - \dfrac{1}{\Omega} - \dfrac{1}{K} = v - \dfrac{1}{\Omega}$

$$K_6 \equiv \frac{G_0 e^{-\Omega v}}{\rho_f}\left[\frac{1}{K} - \frac{1}{\Omega} - e^{-Kv}\left(\frac{1}{K} - \frac{\rho_f K_3}{\Omega + K}\right) - K_5 e^{\Omega v}\right]$$

The solutions to this problem and the significance of each region are summarized in Tables 2-3 and 2-4.

Example 2-4 Time to boiling inception due to flow decay problem

PROBLEM Assume liquid flows axially through a heated tube. At $t = 0$, the inlet flow starts to decrease exponentially such that:

$$G_{in}(t) = G_{in}(0)e^{-t}$$

where t is in seconds. Let the initial flow velocity at the inlet be: $V_{in}(0) = 3.05$ m/s.

Determine the time boiling starts in the channel. Ignore wall friction and liquid compressibility. Assume that inlet temperature and the heating rate remain constant.

Table 2-4 Expressions for the quality and time-to-boil in the two-phase portion of the channel

	Region	Expression for quality	Expression for t_B and/or z
$t > v$	IA	$x = \dfrac{v_f}{v_{fg}}\left[\exp[\Omega(t - t_B)] - 1\right]$	$t_B = \dfrac{\Omega}{\Omega + K}\left(t - \dfrac{1}{\Omega}\ell n\left[\dfrac{1}{K_9}\left(z + \dfrac{G_0 K_3}{\Omega + K}e^{-Kt}\right)\right]\right)$
	IC	Same as IA	$t_B = t - \dfrac{1}{\Omega}\ell n\left(\dfrac{\rho_f \Omega}{G_0}\left[z - K_7 e^{\Omega t} + \dfrac{G_0 K_3}{\Omega + K}e^{-Kt}\right]\right)$
	IIB	$x = x_0 \exp(\Omega t) + \dfrac{v_f}{v_{fg}}(e^{\Omega t} - 1)$	$z = z_0 e^{\Omega t} + K_6 e^{\Omega t} - \dfrac{G_0 K_3}{\Omega + K}e^{-Kt}$
$t < v$	IB	Same as IA	$t_B = t - \dfrac{1}{\Omega}\ell n\left(\Omega\left[\dfrac{\rho_f z}{G_0} - \dfrac{K_4 + 1}{\Omega} + t + \dfrac{e^{-Kt}}{K}\right]\right)$
	IIA	Same as IIB	$z = \dfrac{G_0}{\rho_f}\left[\dfrac{K_4}{\Omega} - \dfrac{1}{\Omega} - t - \dfrac{e^{-Kt}}{K} - K_5 e^{\Omega t}\right] + z_0 e^{\Omega t}$

Input information:
  Tube diameter = 12.7 mm
  Tube heated length, L = 3.05 m
  Heating rate, q' = 16.4 kW/m
  Water inlet temperature = 204°C
  For water at 6.9 MPa: ρ_f = 740 kg/m³
  $\qquad\qquad\qquad c_p$ = 4.2 kJ/kg K
  $\qquad\qquad\qquad T_{sat}$ = 285°C
  Take $\rho_\ell \simeq \rho_f$ = 740 kg/m³

SOLUTION We are looking for the time t that it takes the coolant enthalpy at the channel exit to reach boiling.

Using the method of characteristics, the value of t is first assumed less than ν since particles are initially in the channel and heating.

Integrating Eq. 2-54 for $\lambda(t < \nu)$ one obtains Eq. 2-72 where $V_{in}(0)$ is substituted for G_0/ρ_f and K is set equal to unity:

$$\lambda(t < \nu) = \int_{t-\nu}^{t} V_{in}(t')dt' = \int_{t-\nu}^{0} V_{in}(0)dt' + \int_{0}^{t} V_{in}(0)e^{-t'}dt'$$

$$\lambda(t < \nu) = V_{in}(0)[(\nu - t) + (1 - e^{-t})] \qquad (2\text{-}79)$$

Using Eq. 2-58:

$$\nu = \frac{h_f - h_{in}}{q'/(A_z\rho_\ell)} = \frac{c_p(T_{sat} - T_{in})A_z\rho_\ell}{q'} = 1.95 \text{ s}$$

Setting $\lambda(t < \nu) = L = 3.05$ m since boiling initially occurs at the exit, one may obtain t from Eq. 2-79. Note that if the time t is found to be negative, the assumption of $t < \nu$ is incorrect, and λ must be obtained for $t > \nu$.

Solving Eq. 2-79 for t gives:

$$3.05 = 3.05[2.95 - t - e^{-t}]$$

$$t = 1.78 \text{ s}$$

Example 2-5 Determining the exit enthalpy after flow decay

PROBLEM For the heated tube described in Example 2-4, use the method of characteristics to determine the exit enthalpy of the fluid packets that enter the tube at saturated liquid conditions. Assume the system pressure remains constant and inlet flow rate, $G_{in}(t)$ is described by:

$$G_{in}(t) = G_0 \qquad\qquad \text{for } t < 0$$

$$G_{in}(t) = G_0 \exp(-t) \qquad \text{for } t \geq 0$$

where:

$$G_0 = 2260 \text{ kg/m}^2 \text{ s}$$

SOLUTION There are two cases that must be considered in solving this problem:

Case 1. Fluid packets in the channel before the flow decay starts (at $t = 0$, these packets will be located at $z(0) = z_0$).

Case 2. Fluid packets entering the channel at $t = t_0$, $t_0 > 0$ (at $t = t_0$, these packets will be located at $z(t_0) = 0$).

An expression for Dx/Dt is given in Eq. 2-65, which is repeated below:

$$\frac{Dx}{Dt} - \Omega x = \Omega \frac{v_f}{v_{fg}} \tag{2-65}$$

where Ω is determined from Eq. 2-64. Using properties of water at 6.9 MPa, we get:

$$\Omega = \frac{q' v_{fg}}{A_z h_{fg}} = \frac{(16.4 \text{ kW/m})(26.5 \times 10^{-3} \text{ m}^3/\text{kg})}{(1.27 \times 10^{-4} \text{ m}^2)(1511.6 \text{ kJ/kg})} = 2.26 \text{ s}^{-1}$$

The position of a fluid, $z(t)$, may be obtained from Eq. 2-67 with $\lambda(t) = 0$:

$$\frac{dz}{dt} - \Omega z = \frac{G_{in}(t)}{\rho_f} \tag{2-80}$$

For case 1, which describes fluid packets already in the channel at $t = 0$, Eq. 2-65 is integrated to obtain Eq. 2-76:

$$x(t, z_0) = \left[x_0(z_0) + \frac{v_f}{v_{fg}} \right] \exp(\Omega t) - \frac{v_f}{v_{fg}} \tag{2-76}$$

Substituting values for Ω and water properties at 6.9 MPa gives:

$$x(t, z_0) = (x_0(z_0) + 0.051) \exp(2.26t) - 0.051 \tag{2-81}$$

The quality at $t = 0$, $x_0(z_0)$ may be determined from a heat balance that results in Eq. 2-77. In our case $h_f - h_{in} = 0$:

$$x_0(z_0) = \frac{q' z_0}{G_0 A_z h_{fg}} = 3.78 \times 10^{-2} z_0 \tag{2-82}$$

Eq. 2-80 may then be integrated to find the time required for the packet to reach the channel exit, t_{exit}, using the boundary conditions at $t = 0$, $z(0) = z_0$ and at $t = t_{exit}$, $z(t_{exit}) = L$ represents the channel length:

$$\frac{dz}{dt} - \Omega z = \frac{G_0}{\rho_f} \exp(-t)$$

$$\int_{z_0}^{L \exp(-\Omega t_{exit})} d[z(t) \exp(-\Omega t)] = \frac{G_0}{\rho_f} \int_0^{t_{exit}} \exp[-t - \Omega t] dt \tag{2-83}$$

$$L = z_0 \exp(\Omega t_{exit}) + \frac{G_0}{\rho_f(\Omega + 1)} \{\exp(\Omega t_{exit}) - \exp(-t_{exit})\}$$

Substituting values from Example 2-4 and water at 6.9 MPa, this simplifies to:

$$3.05 = z_0 \exp(2.26t_{exit}) + 0.937\{\exp(2.26t_{exit}) - \exp(-t_{exit})\}$$

(2-84)

For a given initial fluid packet position, z_0, one may then obtain $x(z_0, 0)$ and t_{exit} from Eqs. 2-82 and 2-84. These values may then be substituted into Eq. 2-81 to obtain the exit quality from which the exit enthalpy may be obtained utilizing Eq. 2-85 below:

$$h_{exit} = h(L, t_{exit}) = h_f + x(L, t_{exit})h_{fg}$$
$$= 1262 + x(L, t_{exit})(1511.6) \text{ kJ/kg}$$

(2-85)

Values of the exit enthalpy for different initial fluid packet positions are shown in Figure 2-21.

To obtain a solution for case 2 fluid packets that enter the channel at a time t_0 after the transient begins, Eq. 2-73 must be used. Given that in this case $t_0 = t_B$ and we are interested in the particular time $t = t_{exit}$ at which the fluid reaches the channel exit, we get:

$$x(t_{exit}, t_B) = \frac{v_f}{v_{fg}} \{\exp \Omega(t_{exit} - t_B) - 1\}$$

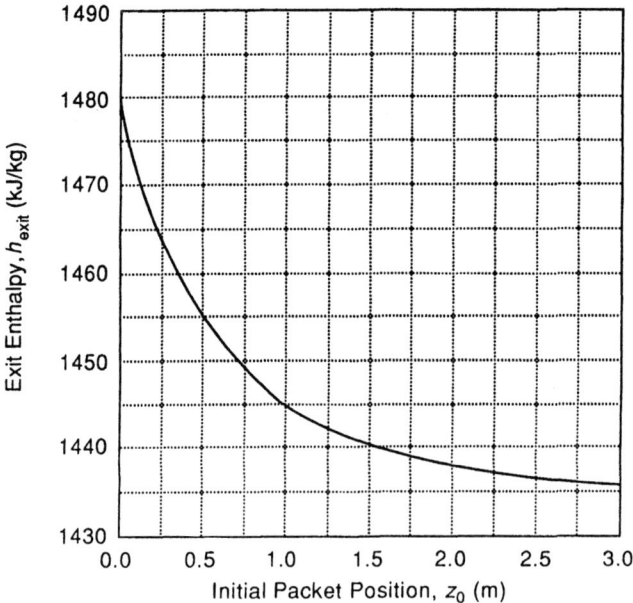

Figure 2-21 Exit enthalpies for various initial fluid packet positions (Case 1 packets).

Substituting values for Ω and water at 6.9 MPa and t_0 for t_B gives:

$$x(t_{exit}, t_0) = 0.051\{\exp[2.26(t_{exit} - t_0)] - 1\} \qquad (2\text{-}86)$$

To determine the time required for the packets to reach the channel exit, Eq. 2-80 must be integrated with the boundary conditions at $t = t_0$, $z(t_0) = 0$, and at $t = t_{exit}$, $z(t_{exit}) = L$:

$$\frac{dz}{dt} - \Omega z = \frac{G_0 \exp(-t)}{\rho_f}$$

$$\int_0^{L \exp(-\Omega t_{exit})} d[z(t) \exp(-\Omega t)] = \int_{t_0}^{t_{exit}} \frac{G_0}{\rho_f} \exp[-t(\Omega + 1)]dt$$

$$\frac{L\rho_f(\Omega + 1)}{G_0} = \exp[\Omega(t_{exit} - t_0) - t_0] - \exp(-t_{exit})$$

Substituting values for the channel described in Example 2-4 gives:

$$3.25 = \exp[\Omega(t_{exit} - t_0) - t_0] - \exp(-t_{exit}) \qquad (2\text{-}87)$$

For a given time at which the fluid packet entered the channel, t_0, the channel exit time, t_{exit}, may be obtained from Eq. 2-87. These times may be substituted in Eq. 2-86 to obtain the exit quality, which may then be utilized in Eq. 2-85 to obtain the exit enthalpy. Values of the exit enthalpy for different inlet times are shown in Figure 2-22.

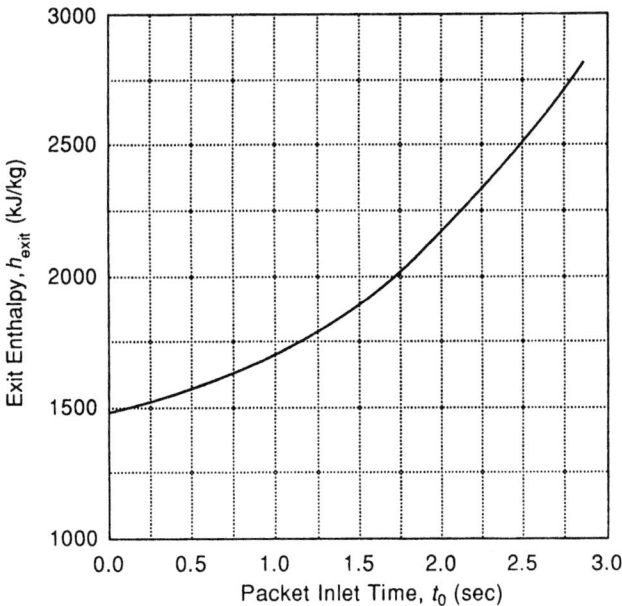

Figure 2-22 Exit enthalpies for packets entering at various times, t_0, after the start of the transient (Case 2 packets).

REFERENCES

1. Abbott, M. B. *An Introduction to the Method of Characteristics*. London: Thames and Hudson Ltd., 1966.
2. Ferch, R. L. Methods of characteristics solutions or non-equilibrium transient flow boiling. *J. Multiphase Flow* 5:265–279 1979.
3. Gidaspow, D., and Shin Y. W. *Blowdown-Six Equation Model by the Method of Characteristics,* ANL-80-3. Argonne National Laboratory, 1980.
4. Gonzalez-Santalo, J. M., and Lahey, R. T., Jr. *An Exact Solution of Flow Decay Transients in Two-Phase Systems by the Method of Characteristics,* New York: ASME Paper 72 WA/HT-47. 1972.
5. Hildebrand, F. B. *Advanced Calculus for Applications*. Englewood Cliffs, NJ: Prentice-Hall, 1962.
6. Kroeger, P. G. *Application of Non-Equilibrium Drift Flux Model to Two-Phase Blowdown Experiments,* BNL-NUREG-21506-R. Upton, MA: Brookhaven National Laboratory, 1977.
7. Lahey, R. T., and Moody, F. J. *The Thermal-Hydraulics of a Boiling Water Reactor*. Hinsdale, IL: American Nuclear Society, 1977.
8. Lee, M., and Kazimi M. S. *Transient Response of a Single Heated Channel,* MITNE-271, July 25, 1985.
9. Lyczkowski, R. M. Transient propagation behavior of two-phase flow equations. In *Heat Transfer Research and Applications,* AIChE Symposium 74, 1978.
10. Meyer, J. E. Hydrodynamic models for the treatment of reactor thermal transients. *Nucl. Sci. Eng.* 10:269–277, 1961.
11. Richtmyer, R. D., and Morton, K. W. *Difference Methods for Initial Value Problem*. New York: Wiley Interscience Publishers, 1967.
12. Shapiro, A. H. *The Dynamics and Thermodynamics of Compressible Fluid Flow,* Vol. I. New York: Ronald Press Co, 1953.
13. Shiralkar, B. S., Schnelbly, L. E., and Lahey, R. T., Jr. Variation of the vapor volumetric fraction during flow and power transients. *Nucl. Eng. Des.* 25:350–368, 1973.
14. Tong, L. S., and Weisman, J. *Thermal Analysis of Pressurized Water Reactors,* 2nd Ed. Hindsdale, IL: American Nuclear Society, 1979.
15. Weisman, J., and Tentner, A. Application of the method of characteristics to solution of nuclear engineering problems. *Nucl. Sci. Eng.* 78:1–29, 1981.
16. Zuber, N., *Flow Excursions and Oscillations in Boiling, Two-Phase Flow Systems with Heat Addition,* Einhoven Symposium on Two-Phase Flow Dynamics, EUR-4288. Einhoven, Holland: Euratom, 1967.

PROBLEMS

Problem 2-1 Comparison of approximate channel transient analysis methods (Section II)

The momentum integral (MI) model or the single velocity (SV) model under certain conditions may be an acceptable alternative to the sectionalized, compressible flow (SC) model. Consider a transient for a BWR channel where the inlet pressure decays as:

$$p_{in}(t) = 0.5[p_{in}(0) + p_{out}(0)]$$

$$+ 0.5[p_{in}(0) - p_{out}(0)] \exp(-t/\tau)$$

where:

$$p_{in}(0) = 6.96 \text{ MPa}$$

$$p_{out}(t) = 6.90 \text{ MPa for all } t.$$

$$\tau = 5 \text{ ms}$$

a. Why would you consider using either the SV or MI models instead of the SC model? What are the important trade-offs?

SINGLE VELOCITY MODEL

b1. Calculate the channel-averaged mass flux at some time $t > 0$ using the SV model. Assume that acceleration, frictional and gravitational effects are negligible. Comment on the physical reality of the result.

b2. What is the axial mass flux distribution predicted by the SV model?

b3. Describe briefly how you would calculate the axial enthalpy distribution.

MOMENTUM INTEGRAL MODEL

c1. Calculate the channel-averaged mass flux at some time $t > 0$ using the MI model. Again, assume that acceleration, frictional and gravitational effects are negligible.

c2. Identify the relevant equations you would use to calculate the axial mass flux and enthalpy distributions. For these equations identify the unknowns and the inputs for the solution.

Problem 2-2 Verification of the equations for the method of characteristics (Section III)

Derive Eq. 2-75a, b, c and 2-78a and b from the characteristic equation:

$$\frac{dz}{dt} - \Omega(z - \lambda(t)) = \frac{G_{in}(t)}{\rho_f} \tag{2-74}$$

for the case of an exponential flow decay given by:

$$G_{in} = G_0 e^{-Kt} \qquad \text{for } t > 0 \tag{2-70}$$

Simplify your expressions using the definitions for K_3, K_4, K_5, K_6, K_7, K_9 given in Section III.

Problem 2-3 Application of MOC to an inlet enthalpy change (Section III)

Water flows upward in a uniformly heated channel that is being tested under transient conditions. In one transient the inlet enthalpy is being decreased at a rate given by:

$$h_i = h_{io} e^{-t/\tau} = h_{io} e^{-0.02t}$$

while the inlet flow rate and the heating rate are kept constant. For the initial conditions given below, evaluate the time it takes to eliminate all evaporation in the channel. Use the method of characteristics, and ignore the pressure drop along the channel.

BWR CHANNEL CONDITIONS

Heated length	$L = 3.0$ m
Flow area	$A_f = 3.5 \times 10^{-3}$ m^2
Inlet mass flux	$G_{in} = 1,900$ kg/m^{2}s
Linear heat generation	$q' = 1,000$ kW/m (axially uniform)
Inlet enthalpy at	$h_{io} = 1225$ kJ/kg
Water properties:	liquid density $= 760$ kg/m^3
	saturated liquid enthalpy $= 1254$ kJ/kg

Answer: $t = 22.3$ s

Problem 2-4 Application of MOC to an increase in the inlet mass flux (Section III)

The flow in a uniformly heated channel is being increased as a function of time from a steady-state value of G_0 as

$$G_{in} = G_0 e^{bt}$$

where b is constant. The inlet enthalpy is kept constant. Derive an equation to determine the time it will take for the enthalpy rise in the channel to equal half the original value.

THREE

FLOW LOOPS

I INTRODUCTION

The operation of a nuclear reactor plant requires an understanding of the plant response to both expected and unlikely transients. To ensure that the thermal and pressure design limits are not exceeded, the plant protection system is to be actuated at certain set points when the flow conditions approach alarming values but while still below the design limits. Thus it is desirable to be able to analyze the transients of the entire coolant primary and secondary loops in order to prejudge the sequence of events and provide for any intervening actions at sufficiently early times so as to avert reaching the limiting conditions.

The methods of coolant loop analysis can be of various levels of complication. For application in reactor operation and control, rather rapid calculational schemes and simplified models are needed. For safety analysis, detailed models, particularly of the core, may be needed. Very often the loop or system analysis can be done in a simplified way to obtain boundary conditions of pressure and flow for the core or steam generator. Then detailed analysis of these components are performed utilizing the boundary conditions obtained from the system analysis. System analysis methods are thus used to describe the macrobehavior of the plant under such events as loss of flow, steam line break accidents, and turbine trip events. Then component codes can be used to analyze the local behavior of the hottest fuel assembly.

Another important application of the loop analysis is the optimization of the geometry of the plant so as to ensure adequate shutdown heat removal capacity by maximizing the natural circulation potential. An interesting objective for plant designs is to remove the postshutdown heat by passive means without the use of externally provided power to the pump. Often, however, active cooling of the core is pursued for some time (few days) after shutdown. In boiling water reactors, because of the buoyancy head associated with the extensive change of phase in the core, the coolant is even capable of passively removing a high level of operating power amounting to hundreds of megawatts.

A number of sophisticated thermal-hydraulic codes have been developed for power reactor flow loops (including the systems codes such as RETRAN, RELAP, and TRAC for light water reactors (LWRs) and SSC for the LMRs). However, the basic trends of coolant and fuel temperatures during transient events can be observed from simplified approaches to loop analysis as will be presented here. This is demonstrated in a simplified one-dimensional model.

In this chapter we shall consider a loop with a single, heated channel representing the core and one or more cooled channels representing the heat exchangers. One loop in a PWR system with a U-tube steam generator is schematically shown in Figure 3-1. Although this loop represents the geometry

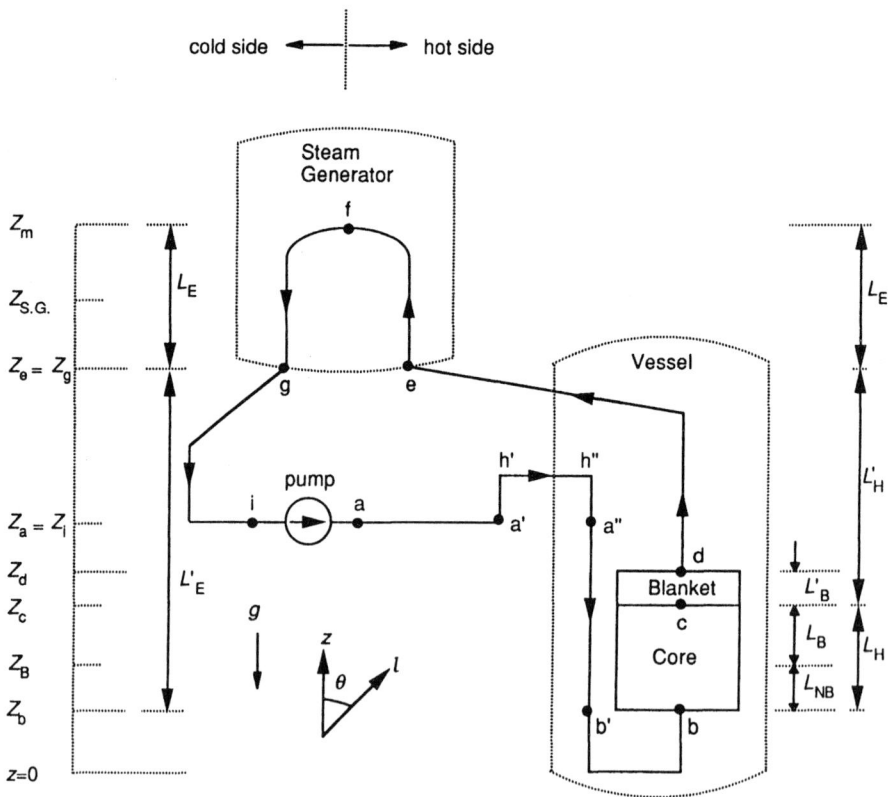

Figure 3-1 Schematic of a simplified PWR loop with a U-tube steam generator.

of a PWR, its treatment can be generalized to all reactors in which the primary coolant is not also the working fluid in the turbine.

II LOOP FLOW EQUATIONS

Let us develop general, one-dimensional equations for the loop as shown in Figure 3-1. The loop has the essential features of interest for application to single-phase as well as two-phase core conditions, namely:

1. Heat is added in the "hot leg" over a segment length L_H representing the core and starting from height $z = Z_b$ from an arbitrary datum.
2. Boiling occurs within L_H such that the boiling length L_B starts at a height Z_B from the datum of our system.
3. A portion of the hot leg of length L_B' above the core does not provide heat input, thus representing a reflector or gas plenum in the fuel rods.
4. Cooling occurs over the length $2L_E$ of the heat exchanger, which represents a simplified U-tube steam generator in this loop.

The one-dimensional equations for a channel of axially constant cross-sectional area for flow at an angle θ with the positive z axis (see Figure 3-1) can be stated as follows:

For momentum:

$$\frac{\partial G_m}{\partial t} + \frac{\partial}{\partial l}\left(\frac{G_m^2}{\rho_m}\right) = -\frac{\partial p}{\partial l} - \frac{fG_m|G_m|}{2D_e\rho_m} - \rho_m g \cos\theta \tag{3-1}$$

and for energy:

$$\rho_m \frac{\partial h_m}{\partial t} + G_m \frac{\partial h_m}{\partial l} - \frac{\partial p}{\partial t} = \frac{q''P_h}{A} + \frac{G_m}{\rho_m}\left[\frac{\partial p}{\partial l} + \frac{fG_m|G_m|}{2D_e\rho_m}\right] \tag{3-2}$$

In the above equations l is the dimension in the flow direction along the channel length, which may not always be in the vertical direction, and f is the appropriate friction factor for the existing single or two-phase flow regime.

Let us integrate each term of Eq. 3-1 around the loop in Figure 3-1 from the pump outlet position, a, to pump inlet position, i, starting with the terms of the right side of Eq. 3-1.

The static pressure is integrated to give:

$$\int_a^i -\frac{\partial p}{\partial l}\, dl = p_a - p_i = \Delta p_{\text{pump}} \tag{3-3}$$

where Δp_{pump} is positive under operating conditions and is negative for a stationary pump due to losses.

The friction term is integrated over the core and blanket, over the steam generator and the rest of the loop to give:

$$\int_a^i -\frac{fG_m|G_m|}{2D_e\rho_m}\,dl = \int_b^d -\frac{fG_m|G_m|}{2D_e\rho_m}\,dl + \int_e^g -\frac{fG_m|G_m|}{2D_e\rho_m}\,dl + \int_a^b -\frac{fG_m|G_m|}{2D_e\rho_m}\,dl$$

$$+ \int_d^e \ldots dl + \int_g^i \ldots dl = -\Delta p_{core} - \Delta p_{S.G.} - \Delta p_{ex} \tag{3-4}$$

where Δp_{ex} is the pressure drop external to the core and steam generator. Note that Δp_{core}, $\Delta p_{S.G.}$, and Δp_{ex} are positive quantities when representing the magnitude of the frictional pressure losses in the flow direction around the loop. When the phasic density is assumed to vary only in the gravitational term, f and G_m become position independent within the constant-area heated channel, then:

$$\int_b^d -\frac{fG_m|G_m|}{2D_e\rho_m}\,dl = -\left(\frac{f_{\ell o}G_m|G_m|}{2D_e\rho_\ell}\right)_{core}(L_{NB} + \overline{\phi_{\ell o}^2}(L_B + L_B')) \tag{3-5}$$

Note that if form losses exist within the heated channel then Eq. 3-5 may be modified to:

$$\int_b^d -\frac{fG_m|G_m|}{2D_e\rho_m}\,dl$$

$$= -\frac{f_{\ell o}G_m|G_m|(L_{NB} + \overline{\phi_{\ell o}^2}(L_B + L_B')) + \sum_j K_j(G_m|G_m|)_j D_e}{2D_e\rho_\ell} \tag{3-6}$$

where K_j takes on the appropriate value of the form loss coefficient at position j within the channel. Position j can involve either single phase or two-phase conditions.

The gravity term is integrated as follows:

$$\int_a^i -\rho_m g \cos\theta\,dl = -\int_a^i \rho_m g\,dz$$

$$= -\int_a^{a'} \rho_m g\,dz - \int_{a'}^{a''} \rho_m g\,dz - \int_{a''}^{b'} \rho_m g\,dz \tag{3-7}$$

$$- \int_{b'}^b \rho_m g\,dz - \int_b^f \rho_m g\,dz - \int_f^i \rho_m g\,dz$$

where $dz = \cos\theta\,dl$.

It can be easily seen that when the effect of pressure drop on the density of each phase is neglected, the value of ρ_m changes only due to heating or cooling in the core and steam generator. Hence in the adiabatic sections $a' - a''$ and $b' - b$ the upward and downward flow sections have equal gravity heads. Thus:

$$\int_{a'}^{a''} \rho_m g\,dz = 0 \tag{3-8}$$

and

$$\int_{b'}^{b} \rho_m g \, dz = 0 \qquad (3\text{-}9)$$

Also, there is no change in z between the positions a and a'. Therefore, Eq. 3-7 reduces to:

$$\int_{a}^{i} - \rho_m g \, dz = - \int_{b}^{f} \rho_m g \, dz - \int_{f}^{i} \rho_m g \, dz - \int_{a''}^{b'} \rho_m g \, dz$$

$$= - \int_{b}^{f} \rho_m g \, dz - \int_{f}^{b'} \rho_m g \, dz \qquad (3\text{-}10)$$

so that the integration around the whole loop can be written as:

$$\int_{a}^{i} - \rho_m g \cos\theta \, dl = -(\bar{\rho}_m)_{b-f} g [Z_m - Z_b]$$

$$- (\bar{\rho}_m)_{f-b'} g [Z_b - Z_m]$$

or

$$\int_{a}^{i} - \rho_m g \cos\theta \, dl = [(\bar{\rho}_m)_{f-b'} - (\bar{\rho}_m)_{b-f}] g [Z_m - Z_b] \qquad (3\text{-}11)$$

where:

$$(\bar{\rho}_m)_{b-f} = \frac{\int_{b}^{f} \rho_m \, dz}{Z_m - Z_b}; \quad (\bar{\rho}_m)_{f-b'} = \frac{\int_{f}^{b'} \rho_m \, dz}{Z_b - Z_m} \qquad (3\text{-}12)$$

Thus, the net buoyancy term is due to the difference between the average densities of the cold side and hot side multiplied by the height over which the density variation occurs.

Now, considering the left hand side of Eq. 3-1, we integrate each term so that we get the mass temporal acceleration for the coolant in the loop:

$$\int_{a}^{i} \left(\frac{\partial G_m}{\partial t} \right) dl = \sum_{k} L_k \frac{\partial (G_m)_k}{\partial t} = \sum_{k} \frac{L_k}{A_k} \frac{\partial \dot{m}_k}{\partial t} \qquad (3\text{-}13)$$

where L_k is the length of the kth flow section of constant area A_k.

As for the spatial acceleration:

$$\int_{a}^{i} \frac{\partial}{\partial l} \left(\frac{G_m^2}{\rho_m} \right) dl = \left(\frac{G_m^2}{\rho_m} \right)_i - \left(\frac{G_m^2}{\rho_m} \right)_a \qquad (3\text{-}14)$$

If the pipes on both sides of the pump are of equal cross-sectional area and the fluid density does not change through the pump, then

$$\left(\frac{G_m^2}{\rho_m} \right)_i - \left(\frac{G_m^2}{\rho_m} \right)_a = 0 \qquad (3\text{-}15)$$

The loop momentum equation can be written neglecting form losses as:

$$\sum_k L_k \frac{\partial (G_m)_k}{\partial t} = \Delta p_{pump} - \Delta p_f + \Delta p_B \tag{3-16}$$

where:

Δp_{pump} = pressure head provided by the pump

Δp_f = frictional pressure drop around the entire loop which can be divided into core, steam generator, and external loop pressure drops so that:

$$\Delta p_f \equiv \left(\frac{fG_m|G_m|}{2D_e\rho_m} L \right)_{core} + \left(\frac{fG_m|G_m|}{2D_e\rho_m} L \right)_{S.G.} + \Delta p_{f,ex} \tag{3-17}$$

and

Δp_B = buoyancy pressure head given by;

$$\Delta p_B \equiv [(\bar{\rho}_m)_{cold\ leg} - (\bar{\rho}_m)_{hot\ leg}] g(Z_m - Z_b) \tag{3-18}$$

Let us now examine the energy equation, Eq. 3-2. The last two terms involving the pressure changes and frictional dissipation along the loop are usually negligible relative to the heat addition and enthalpy change terms.† Therefore, the loop energy equation can be written as the integral of Eq. 3-2 around the complete loop (dropping the last two terms on the right side):

$$\int A\rho_m \frac{\partial h_m}{\partial t} dl + \int AG_m \frac{\partial h_m}{\partial l} dl - \int A \frac{\partial p}{\partial t} dl = \int q''P_h dl \tag{3-19}$$

For small changes in pressure, the term involving $\partial p/\partial t$ can be neglected. Because enthalpy changes outside the core and the steam generator are negligible, Eq. 3-19 may be written as:

$$\int_b^c A\rho_m \frac{\partial h_m}{\partial t} dl + \int_e^g A\rho_m \frac{\partial h_m}{\partial t} dl + \int_b^c AG_m \frac{\partial h_m}{\partial l} dl + \int_e^g AG_m \frac{\partial h_m}{\partial l} dl \tag{3-20}$$

$$= \int_b^c q''P_h dl + \int_e^g q''P_h dl; \text{ negligible pressure variation}$$

Although the above equations are written in terms that can be applied to two-phase as well as single-phase coolant conditions, it is important to note that in many reactor systems the primary loop consists of a single phase coolant (PWR, LMR, and HTGRs). Only under severe transients would the primary side of a steam generator of a PWR or a heat exchanger in the LMR involve a two-phase primary coolant. Therefore, analysis of single-phase conditions is first presented followed by an analogous analysis of two-phase conditions.

---

† Bau and Torrence [3] have theoretically examined single-phase laminar free convection loops. They found the frictional dissipation and pressure work terms to be of comparable magnitude but opposing effects. The frictional dissipation enhances the flow, while the pressure work retards it.

III STEADY-STATE, SINGLE-PHASE, NATURAL CIRCULATION

Under natural circulation conditions, a loop operates without a pump, and the flow is driven entirely by the buoyancy-generated pressure head. Properly designed reactor plants generally benefit from the natural circulation potential in obtaining a decay power removal capability. The exact power that may be removed from the core varies according to plant design.

A Dependence on Elevations of Thermal Centers

Under steady-state conditions all the time-dependent terms disappear from Eqs. 3-16 and 3-20. For single-phase flow conditions, $L_{NB} = L_H$ and the boiling length L_B and the averaging subscript m (for G and h) can be dropped.

Thus the momentum equation, Eq. 3-16, can be reduced to:

$$\Delta p_f = \Delta p_B \tag{3-21}$$

From the equations defining Δp_f and Δp_B, Eq. 3-17 and 3-18, Eq. 3-21 can also be written as:

$$\left(\frac{fG|G|L}{2D_e\rho_\ell}\right)_{core} + \left(\frac{fG|G|L}{2D_e\rho_\ell}\right)_{S.G.} + \Delta p_{f,ex}$$

$$= (\bar{\rho}_{\text{cold leg}} - \bar{\rho}_{\text{hot leg}})g(Z_m - Z_b) \tag{3-22}$$

The density dependence on temperature can be assumed linear so that in general:

$$\rho = \rho_0[1 - \beta(T - T_0)] \tag{3-23}$$

where ρ_0 is a reference density at a reference temperature T_0 and β is the expansion coefficient, which equals $(-\partial\rho/\partial T)/\rho$.

Consider, for simplicity, uniform axial heat addition and extraction in the core and the steam generator, respectively. The energy balance across the core is given by:

$$\dot{m}(h_c - h_b) = q''_H(P_h)_H L_H = \dot{Q}_H \tag{3-24}$$

Thus, the coolant temperature rise across the core (heated channel) is given by:

$$\Delta T_H = T_c - T_b = \frac{\dot{Q}_H}{\bar{c}_p \dot{m}} \tag{3-25}$$

where $\bar{c}_p$ is the coolant average specific heat at constant pressure for the core temperature range (i.e., $dh = \bar{c}_p dT$). For uniform axial heat addition, the axial temperature rise in the hot leg $(b - e)$ is given by:

$$T - T_b = \Delta T_H \frac{z - Z_b}{Z_c - Z_b} \qquad Z_b \leq z \leq Z_c$$

$$= \Delta T_H \qquad Z_c \leq z \leq Z_e \tag{3-26}$$

In terms of the distance from the beginning of the heated channel, l, we can recast Eq. 3-26 in the form:

$$T - T_b = \Delta T_H \frac{l}{L_H}; \qquad 0 < l < L_H \tag{3-27}$$
$$= \Delta T_H \qquad ; \qquad L_H < l < L_H + L'_H$$

At steady state, the temperature drop across the steam generator, ΔT_E, is equal to ΔT_H.

It is possible to rearrange the buoyancy term into a form that depends on the maximum density difference and the vertical height between the center of heating and the center of cooling in the loop. Since the buoyancy head is given by the integral of the gravity term around the loop, it is possible to write it as:

$$\Delta p_B = - \int_b^c \rho g dz - \int_c^e \rho g dz - \int_e^g \rho g dz - \int_g^b \rho g dz \tag{3-28}$$

$$\text{heating} \qquad \text{adiabatic} \qquad \text{cooling} \qquad \text{adiabatic}$$

But dT/dz is constant within the first integral in the above equation, and $d\rho/dT$ can be taken approximately constant in single-phase flows, then the density varies linearly with z, and the first integral can be written as:

$$- \int_b^c \rho g dz = - \frac{\rho_b + \rho_c}{2} g L_H \tag{3-29a}$$

For the second integral, ρ is constant. Therefore:

$$- \int_c^e \rho g dz = -\rho_c g L'_H \tag{3-29b}$$

For the third integral, we can assume a linear increase of ρ with z between the points e and f, and similarly between the points f and g. Therefore:

$$- \int_e^g \rho g dz = - \int_e^f \rho g dz + \int_g^f \rho g dz$$

$$= - \frac{\rho_e + \rho_f}{2} g L_E + \frac{\rho_f + \rho_g}{2} g L_E = \frac{\rho_g - \rho_e}{2} g L_E$$

But since $\rho_e = \rho_c$ and $\rho_g = \rho_b$, we get:

$$- \int_e^g \rho g dz = \frac{\rho_b - \rho_c}{2} g L_E \tag{3-29c}$$

The last term in Eq. 3-28 can be given by:

$$- \int_g^b \rho g dz = \rho_b g L'_E \tag{3-29d}$$

Substituting Eqs. 3-29a through d into Eq. 3-28 yields:

$$\Delta p_B = \rho_b \left(\frac{L_E}{2} + L'_E - \frac{L_H}{2} \right) g - \rho_c \left(\frac{L_E}{2} + L'_H + \frac{L_H}{2} \right) g \qquad (3\text{-}30)$$

But since:

$$L'_E - \frac{L_H}{2} = L'_H + \frac{L_H}{2} \qquad (3\text{-}31)$$

Eq. 3-30 becomes:

$$\Delta p_B = (\rho_b - \rho_c) g \Delta L \qquad (3\text{-}32)$$

where:

$$\Delta L = \frac{L_E}{2} + L'_H + \frac{L_H}{2} \qquad (3\text{-}33)$$

That is, ΔL is the vertical distance between the center of the core and the center of the steam generator

$$\Delta p_B = (\rho_{in} - \rho_{out})_{core} \, g \Delta L \qquad (3\text{-}34a)$$

or

$$= (\rho_{out} - \rho_{in})_{S.G.} \, g \Delta L \qquad (3\text{-}34b)$$

and using Eq. 3-23:

$$\Delta p_B = \beta \rho_0 \Delta T_H g \Delta L \qquad (3\text{-}34c)$$

The physical interpretation of Eqs. 3-34a and b is that the buoyancy head is equal to the difference between the maximum coolant density and minimum coolant density along the loop times the difference in elevation between the thermal "center" of heat extraction and the thermal "center" of heat addition. This general statement holds for other axial profiles of heat addition and extraction.

The frictional losses may also be written in the form:

$$\Delta p_f = \left(\frac{fG|G|}{2D_e \rho_\ell} L \right)_{core} + \left(\frac{fG|G|}{2D_e \rho_\ell} L \right)_{S.G.} + \Delta p_{f,ex} \equiv C_R \frac{\dot{m}^2}{2\rho_\ell} \qquad (3\text{-}35a)$$

where $\dot{m}$ is the mass flow rate in the loop

$C_R \equiv R(\dot{m})^{-n}$ is the hydraulic resistance coefficient, where R is the proportionality constant which can also include form losses. For highly turbulent flow, $n = 0.2$. For laminar flow, $n = 1$.

Therefore:

$$\Delta p_f = \frac{1}{2} R \frac{(\dot{m})^{2-n}}{\rho_\ell} \qquad (3\text{-}35b)$$

Substituting from Eqs. 3-34c and 3-35b into Eq. 3-21 and assuming that the temperature variation around the loop is such that the density variation is relatively small so that $\rho_0 \simeq \rho_\ell$, we get:

$$\frac{1}{2} R \frac{(\dot{m})^{2-n}}{\rho_0} = \beta \rho_0 \Delta T_H g \Delta L \qquad (3\text{-}36)$$

where ρ_0 denotes the reference coolant density. Therefore for a given temperature difference the mass flow rate may be obtained from

$$\dot{m} = \left(\frac{2\beta \Delta T_H g \Delta L}{R} \rho_0^2 \right)^{1/(2-n)} \qquad (3\text{-}37)$$

For a given heating power $\dot{Q}_H$, the flow rate is determined by substituting for ΔT_H from Eq. 3-25 into Eq. 3-36 to obtain

$$\dot{m} = \left(\frac{2\beta \dot{Q}_H g \Delta L}{\bar{c}_p R} \rho_0^2 \right)^{1/(3-n)} \qquad (3\text{-}38)$$

It should be clear that in actual reactor systems the temperature increase in the core may not be linear, and the temperature decrease in the heat exchanger may be close to exponential. Thus, the solution of the exact equations in reactor systems may have to be achieved numerically. Zvirin et al. [16] have shown that the difference between the exact solutions and the linear temperature models is small, on the order of 5% in $\dot{m}$ and ΔT_H.

For a given allowable ΔT_H, the maximum power that can be removed by natural convection for a given system configuration is given by substituting Eq. 3-38 into Eq. 3-37 and rearranging the result:

$$\dot{Q}_H = \bar{c}_p \left(\frac{2\beta g \Delta L \rho_0^2}{R} \right)^{1/(2-n)} (\Delta T_H)^{(3-n)/(2-n)} \qquad (3\text{-}39)$$

It is seen that the heat removal capability is most affected by ΔT_H, $\bar{c}_p$, and ρ_0, and to a lesser extent by ΔL and R.

B Friction Factors in Natural Convection

It is important to note that the friction factor associated with natural convection in small tube experiments has been observed to be higher than that of forced convection at the same Reynolds number. Figure 3-2 illustrates this observation. It is seen that transition between laminar and turbulent flow was observed to occur around Re = 1,500, similar to forced convection flow conditions. It is also seen that the friction factor is dependent on the channel geometry.

When a fluid is forced to flow at a rather low velocity in a heated channel, the selection of the proper value of the friction factor depends on the dominant flow regime in the loop. One of three regimes may exist: forced convection, mixed convection or natural convection. Several dimensionless groups have been proposed to characterize the flow regime boundaries. Reference 10 proposes Re and GrPr. The higher the Re number, the higher is the value of GrPr

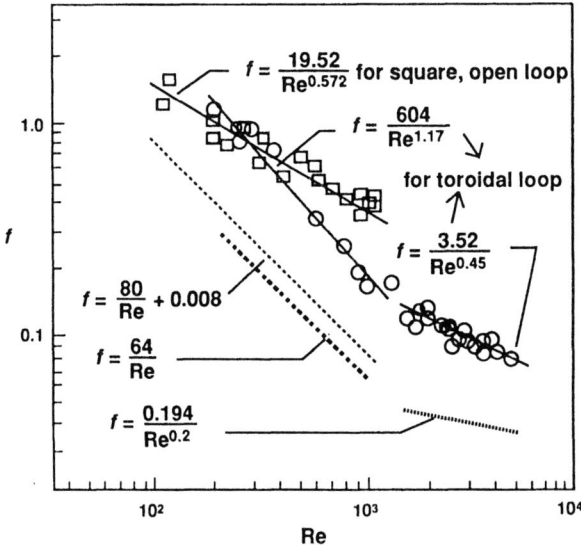

Figure 3-2 Friction coefficient as a function of the Reynolds number for water loops. *Solid lines* are for natural convection. *Dotted lines* are for forced convection. *(After Zvirin et al. [17] based on the work of H. F. Creveling et al. [4] and Bau and Torrence [2].)*

needed to reach a purely natural convection flow regime. The geometry of the channel will also influence the boundaries between the regimes.

Example 3-1 Natural circulation in a PWR

PROBLEM Consider the PWR system, shown schematically in Figure 3-3, when operating at natural circulation conditions given in Table 3-1. Compute:

1. The steady-state natural circulation flow rate as a percentage of the full power flow rate
2. The power level.

Table 3-1 PWR system parameters

Geometry	Full power operating conditions	Natural circulation operating conditions
See Fig. 3-3	$\dot{Q}_H = 2772$ MW thermal	$\dot{Q}_{H,n} =$ unspecified
$D_{cold\ leg} = 0.661$ m	$T_{in} = 291.7\ °C$	$T_{in} = 291.7\ °C$
	$T_{out} = 321.1\ °C$	$T_{out} = 326.7\ °C$
	$p = 15.17$ MPa	$p = 15.17$ MPa
	$\Delta p_{pump} = 0.62$ MPa	$\Delta p_{pump} = 0.0$

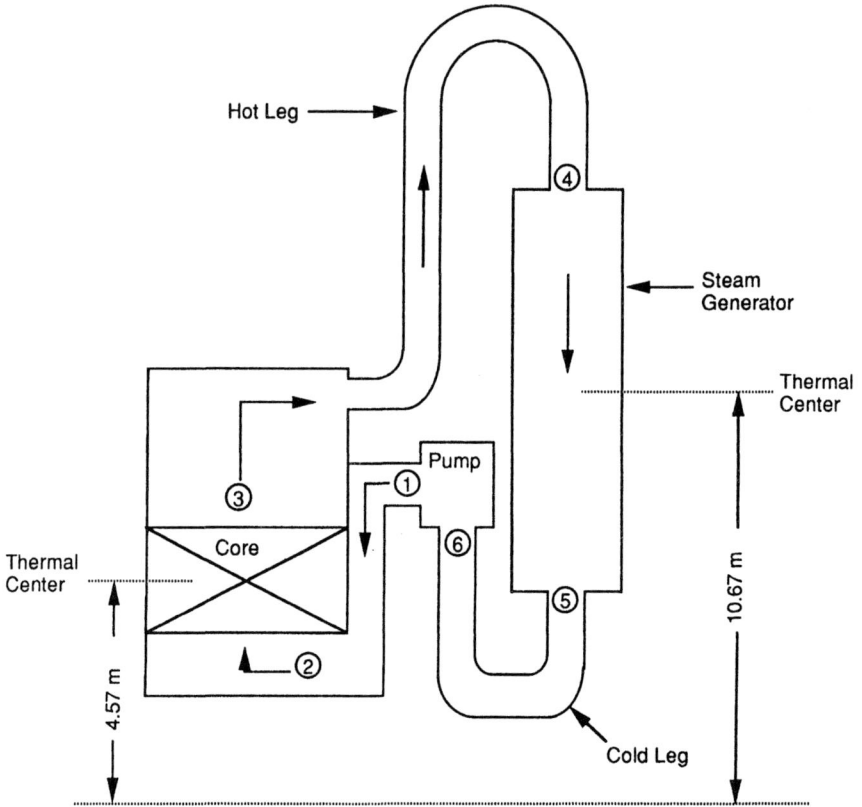

Figure 3-3 Schematic diagram of primary system model (not to scale).

Water properties are given in Table 3-2. Assume that the friction factor for both steady state and full power can be approximated by:

$$f = 0.184 \, Re^{-0.2}$$

SOLUTION Let us first evaluate the hydraulic resistance proportionality constant R by considering the normal operating conditions.

From Eq. 3-16 applied to steady-state conditions:

$$\Delta p_f = \Delta p_{pump} + \Delta p_B$$

Under normal conditions the buoyancy pressure head is estimated by

$$\Delta p_B = (\rho_{in} - \rho_{out})_{core} \, g\Delta L \qquad (3\text{-}34a)$$

ρ_{in} is ρ at 291.7 °C = 740.0 kg/m³, ρ_{out} = ρ at 321.1 °C = 674.2 kg/m³

$$\Delta L = \frac{z_4 + z_5}{2} - \frac{z_3 + z_2}{2} = 10.67 - 4.57 = 6.10 \text{ m}$$

$$\therefore \Delta p_B = (740.0 - 674.2)(9.8)(6.1) = 3.93 \times 10^3 \text{ Pa}$$

Table 3-2 Water properties at 15.17 MPa

T (°C)	ρ(kg/m³)	c_p(kJ/kg K)	μ(kg/m·s)
287.8	747.9	5.23	9.714×10^{-5}
293.3	736.7	5.32	9.549×10^{-5}
298.9	725.5	5.44	9.342×10^{-5}
304.4	712.7	5.61	9.177×10^{-5}
310.0	699.8	5.78	8.970×10^{-5}
315.6	687.0	6.03	8.764×10^{-5}
321.1	674.2	6.28	8.598×10^{-5}
326.7	659.8	6.53	8.350×10^{-5}
332.2	645.4	7.08	8.102×10^{-5}

Therefore:

$$\Delta p_f = \Delta p_{pump} + \Delta p_B = 6.2 \times 10^5 + 3.93 \times 10^3 = 6.24 \times 10^5 \text{ Pa}$$

Now, the loop hydraulic resistance constant R is obtained by rearranging Eq. 3-35b:

$$R = \frac{\Delta p_f}{1/2 \rho_\ell^{-1} \dot{m}^{2-n}}$$

but

$$\dot{m} \left(\frac{kg}{s}\right) = \frac{\dot{Q}_H}{\bar{c}_p \Delta T} = \frac{\dot{Q}_H(kW)}{(5.785 kJ/kg \ K)(29.4K)} = 5.88 \times 10^{-3} \dot{Q}_H(kW)$$

$$= 5.88 \times 10^{-3}(2.77 \times 10^6) = 1.629 \times 10^4 \text{ kg/s}$$

$$\therefore R = \frac{6.24 \times 10^5}{(1/2)(740.0)^{-1}(1.629 \times 10^4)^{1.8}} = 24.2$$

Next consider natural circulation condition for $\Delta T = 326.7 - 291.7 = 35 \ ^\circ C$:

$$\dot{m} = \left(\frac{2\beta \Delta T_H g \Delta L}{R} \rho_0^2\right)^{1/(2-n)} \tag{3-37}$$

Now, β between 291.7 and 326.7 °C can be obtained from Table 3-2:

$$\beta = -\frac{1}{\rho} \left(\frac{d\rho}{dT}\right) = \frac{|\Delta\rho|}{\rho_{ave}\Delta T}$$

$$= \frac{|659.8 - 740.0|}{\left(\dfrac{659.8 + 740.0}{2}\right)(326.7 - 291.7)}$$

$$= \frac{80.2}{(699.9)(35.0)} = 3.27 \times 10^{-3} \ ^\circ C^{-1}$$

$$\therefore \dot{m} = \left(\frac{2(0.00327)(35.0)(9.8)(6.1)(740)^2}{24.2}\right)^{1/1.8} = 1123 \text{ kg/s}$$

$$= 6.89\% \text{ of full-power rate}$$

The energy equation for both decay and full power cases can be used to obtain the decay power level:

$$\dot{Q}_{H,n} = \dot{m}_n \bar{c}_{p,n}(T_{out,n} - T_{in})$$

$$\dot{Q}_H = \dot{m}\bar{c}_p(T_{out} - T_{in})$$

$$\dot{Q}_{H,n} = \dot{Q}_H \frac{\dot{m}_n}{\dot{m}} \frac{\bar{c}_{p,n}}{\bar{c}_p} \frac{T_{out,n} - T_{in}}{T_{out} - T_{in}} \qquad (3\text{-}40)$$

where:

$$\bar{c}_{p,n} = \frac{c_{p_{in,n}} + c_{p_{out,n}}}{2}$$

$$\bar{c}_p = \frac{c_{p_{in}} + c_{p_{out}}}{2}$$

Numerical calculation:

$$\bar{c}_{p,n} = \frac{5.29 + 6.53}{2} = 5.91 \text{ kJ/kg K}$$

$$\bar{c}_p = \frac{5.29 + 6.28}{2} = 5.785 \text{ kJ/kg K}$$

Substituting the above values in Eq. 3-40 we get:

$$\dot{Q}_{H,n} = \dot{Q}_H \left(\frac{1123}{16290}\right)\left(\frac{5.91}{5.785}\right)\left(\frac{35.0}{29.4}\right)$$

$$= 0.0838 \, \dot{Q}_H$$

$$= 232.4 \text{ MWth}$$

Example 3-2 Thermosyphon analysis

PROBLEM In this example the flow loop principles are applied to the thermosyphon phenomena. As shown in Figure 3-4 (*top*), a loop of tubing full of fluid is heated by a uniform heat flux, q'', throughout its lower half, whereas the upper half rejects heat at the same rate. A slight disturbance would cause fluid to circulate forming a "thermosyphon." Show that the natural circulation in laminar flow is established such that:

$$\text{Re} \simeq \frac{1}{64} (\text{Gr})_{\Delta T_f} \qquad (3\text{-}41)$$

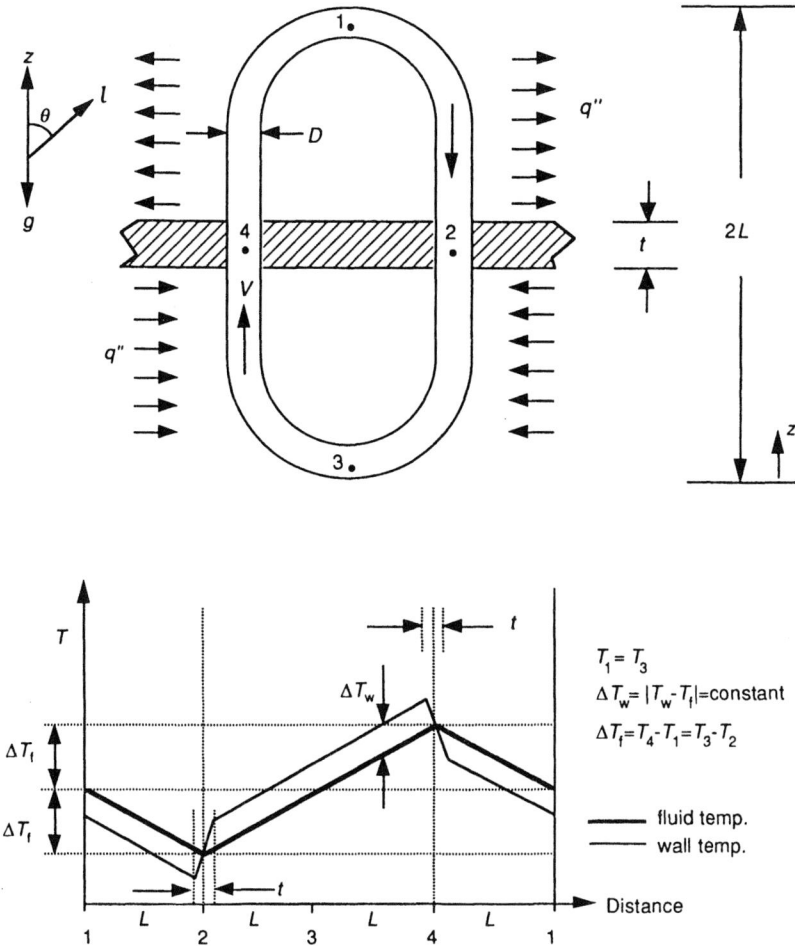

Figure 3-4 Flow loop schematic and temperature variation along the loop.

where:

$$\text{Re} = \frac{\rho VD}{\mu} \tag{3-42}$$

and

$$(\text{Gr})_{\Delta T_f} = \frac{\rho^2 g \beta \Delta T_f D^3}{\mu^2} \tag{3-43}$$

The fluid and wall temperature variations around the loop are ΔT_f and ΔT_w. Show that the ratio of the two characteristic temperature differences in Figure 3-4 (*bottom*) is given by:

$$\frac{\Delta T_f}{\Delta T_w} = \frac{4L}{D} \text{St} \tag{3-44}$$

where:

$$St = \frac{h}{\rho V c_p} \qquad (3\text{-}45)$$

where h is the wall heat transfer coefficient.

To simplify the analysis, the following assumptions may be introduced:

• The tube diameter is small such that $L \gg D$.
• The tube vertical length is much larger than the horizontal length.
• The effect of gradual rather than sudden wall temperature variation near points 2 and 4 on Figure 3-4 (*top*) may be neglected.
• Laminar fully developed flow exists throughout the loop.
• Fluid properties are temperature independent except for the density which has a temperature dependence given by Eq. 3-23.
• Pressure losses at the end bends 1 and 3 on Figure 3-4 (*top*) are negligible.
• The effect of the transverse temperature distribution on the velocity profile may be neglected.

With these assumptions, the velocity V and the internal heat transfer coefficient are essentially constant around the loop so that the temperature distribution around the loop can be simplified as shown in Figure 3-4 (*bottom*).

SOLUTION The flow rate can be related to the temperature difference along the loop by Eq. 3-37:

$$\dot{m} = \left(\frac{2\beta \Delta T_H g \Delta L}{R} \rho_0^2\right)^{1/(2-n)} \qquad (3\text{-}37)$$

For this example $\Delta T_H = 2\Delta T_f$, the difference between the maximum and minimum temperatures in the loop, and $\Delta L = L$, the elevation of the center of heat addition above the center of heat extraction. For laminar flow $n = 1$. Taking $\rho_0 \simeq \rho_\ell = \rho$, therefore:

$$\dot{m} = \frac{4\beta \Delta T_f g L}{R} \rho^2$$

In this system, if wall friction is assumed to dominate over all form losses, Eq. 3-35b can be manipulated to get:

$$R = \frac{2\rho}{\dot{m}^{2-n}} \Delta p_f = \frac{2\rho}{\dot{m}} \Delta p_f$$

Thus:

$$R = \frac{2\rho}{\dot{m}} f \frac{4L}{D} \frac{1}{2\rho} \frac{\dot{m}^2}{(\pi D^2/4)^2} = f \frac{4L}{D} \frac{\dot{m}}{(\pi D^2/4)^2}$$

For laminar flow in a circular pipe:

$$f = \frac{64}{\text{Re}} = \frac{64\mu}{\dot{m}D} (\pi D^2/4)$$

$$\therefore R = \frac{64\mu(\pi D^2/4)}{\dot{m}D} \frac{4L}{D} \frac{\dot{m}}{(\pi D^2/4)^2} = \frac{64\mu(4L)}{D^2(\pi D^2/4)}$$

and

$$\dot{m} = \frac{4\beta\Delta T_f g L}{64\mu(4L)} \rho^2 D^2(\pi D^2/4)$$

But

$$\text{Re} = \frac{\dot{m}D}{(\pi D^2/4)\mu} = \frac{\beta\Delta T_f g \rho^2 D^3}{64\mu^2}$$

(3-41)

$$\therefore \text{Re} = \frac{1}{64} (\text{Gr})_{\Delta T_f}$$

To derive the second relation, the definition of Stanton number may be employed:

$$\text{St} = \frac{h}{\rho V c_p} = \frac{q''/\Delta T_w}{\rho V c_p}$$

(3-46)

In Eq. 3-46 q'' can be replaced by performing a heat balance on the loop from point 4 to point 3:

$$q''(\pi DL) = \dot{m}c_p(T_3 - T_4)$$

(3-47)

or alternatively:

$$q''(\pi DL) = \rho V \left(\frac{\pi D^2}{4}\right) c_p \Delta T_f$$

which can be rearranged as:

$$\frac{q''}{\rho V c_p} = \frac{1}{4} \frac{D}{L} \Delta T_f$$

(3-48)

Substitution in Eq. 3-46 yields Eq. 3-44:

$$\text{St} = \frac{1}{4} \frac{\Delta T_f}{\Delta T_w} \left(\frac{D}{L}\right)$$

IV STEADY-STATE, TWO-PHASE, NATURAL CIRCULATION

From Eqs. 3-16, 3-17 and 3-18, the momentum equation of a natural circulation loop can be written for steady state with boiling in the core as:

$$\left[\frac{f_{\ell o}G_m^2}{2D_e\rho_\ell} (L_{NB} + \overline{\phi_{\ell o}^2}(L_B + L_B'))\right]_{core} + \Delta p_{f,S.G.} + \Delta p_{f,ex}$$

$$= [(\bar{\rho}_m)_{cold\ leg} - (\bar{\rho}_m)_{hot\ leg}]g(Z_m - Z_b)$$

(3-49)

The energy equation, Eq. 3-2 when integrated over the core for steady-state conditions (and neglecting the work terms) yields:

$$\dot{Q}_H = \int_{Z_b}^{Z_c} q''P_h dz = \int_{Z_b}^{Z_c} AG_m \frac{dh_m}{dz} dz \qquad (3\text{-}50)$$

where the elevations are defined in Figure 3-1. The solution of the above two equations will be illustrated now for idealized conditions. The following assumptions are made:

1. The heat sink is capable of instantaneous condensation of the arriving steam such that the cold leg contains liquid only. Consequently:

$$(\bar{\rho}_m)_{\text{cold leg}} \simeq \rho_f \qquad (3\text{-}51)$$

The condensation effectively occurs at the mid-elevation level, at $z = Z_{\text{S.G.}}$, where $Z_{\text{S.G.}} - Z_e = L_E/2$.

2. The friction and form pressure drops are negligible outside the core:

$$\Delta p_{f,\text{ex}} = 0 \quad \text{and} \quad \Delta p_{f,\text{S.G.}} = 0 \qquad (3\text{-}52)$$

3. The axial heat flux is uniform so that:

$$q''(z) = \text{constant} = \frac{\dot{Q}_H}{L_H P_h}; \quad Z_c > z > Z_b \qquad (3\text{-}53)$$

Under these conditions the position for boiling inception Z_B is defined from the energy balance over the single phase length L_{NB}:

$$q''P_h(Z_B - Z_b) \equiv q''P_h L_{\text{NB}} = \dot{m}(h_f - h_{\text{in}}); \quad h_{\text{in}} < h_f \qquad (3\text{-}54)$$

It is easier to describe the average enthalpy in terms of equilibrium vapor quality x so that at any position the enthalpy is given by:

$$h = h_f + xh_{fg} \qquad (3\text{-}55)$$

If the inlet quality is x_{in} and the pressure along the core is approximately constant (i.e., $h_{fg} = $ constant), Eq. 3-54 can be written as:

$$q''P_h(z - Z_b) = \dot{m}(x - x_{\text{in}})h_{fg} \qquad (3\text{-}56)$$

The coolant heating starts for $z > Z_b$, and for a uniform heat addition the axial change in enthalpy is linear so that:

$$x - x_{\text{in}} = 0 \qquad\qquad 0 < z < Z_b$$

$$x - x_{\text{in}} = \frac{\dot{Q}_H}{\dot{m}h_{fg}} \frac{z - Z_b}{L_H} \qquad Z_b < z < Z_c \qquad (3\text{-}57)$$

$$x - x_{\text{in}} = \frac{\dot{Q}_H}{\dot{m}h_{fg}} \qquad\qquad Z_c < z < Z_e + \frac{L_E}{2}$$

The mixture density ρ_m for saturated conditions is given by Eq. 5-50b of Volume I:

$$\rho_m = \alpha\rho_g + (1 - \alpha)\rho_f \qquad \text{(I,5-50b)}$$

$$= \rho_f - \alpha(\rho_f - \rho_g) \qquad \text{(3-58)}$$

where α is the cross-sectional average void fraction, which may be estimated using the homogeneous equilibrium model (HEM) by:

$$\alpha = \cfrac{1}{1 + \cfrac{1 - x}{x}\cfrac{\rho_g}{\rho_f}} \qquad \text{(I,11-30)}$$

From Eqs. 3-58 and I,11-30, ρ_m can also be written in terms of quality as:

$$\rho_m = \rho_f - \cfrac{\rho_f - \rho_g}{1 + \cfrac{1 - x}{x}\left(\cfrac{\rho_g}{\rho_f}\right)} = \cfrac{\rho_f}{1 + x\left(\cfrac{\rho_f}{\rho_g} - 1\right)} \qquad \text{(3-59)}$$

so that for $\rho_f \gg \rho_g$:

$$\rho_m \simeq \cfrac{\rho_f}{1 + x\cfrac{\rho_f}{\rho_g}} \qquad \text{(3-60)}$$

Thus, the average density in the heated channel can be estimated from Eq. 3-60 as:

$$(\bar{\rho}_m)_{\text{hot leg}} = \frac{1}{Z_{\text{S.G.}} - Z_b} \int_{Z_b}^{Z_{\text{S.G.}}} \cfrac{\rho_f}{1 + x\cfrac{\rho_f}{\rho_g}} \, dz \qquad \text{(3-61)}$$

$$= \frac{1}{Z_{\text{S.G.}} - Z_b} \left[\rho_f(Z_B - Z_b) + \int_0^{x_0} \cfrac{\rho_f}{1 + x\cfrac{\rho_f}{\rho_g}}\left(\frac{dz}{dx}\right) dx \right.$$

$$\left. + \cfrac{\rho_f}{1 + x_0\cfrac{\rho_f}{\rho_g}} (Z_{\text{S.G.}} - Z_c)\right] \qquad \text{(3-62)}$$

where x_0 is the core outlet quality, and x_{in} is assumed less than zero.

Since

$$\frac{dx}{dz} = \frac{x_0}{L_B} = \frac{x_0 - x_{\text{in}}}{L_H} = \text{constant} \qquad \text{(3-63)}$$

Equation 3-62 may be written as:

$$(\bar{\rho}_m)_{\text{hot leg}} = \frac{\rho_f}{Z_{\text{S.G.}} - Z_b}\left[L_{NB} + \frac{\ell n(1 + \gamma)}{\gamma} L_B + \frac{1}{1 + \gamma}\left(L'_H + \frac{L_E}{2}\right)\right] \qquad \text{(3-64)}$$

where

$$\gamma = x_0 \frac{\rho_f}{\rho_g} \tag{3-65}$$

Now we can write the buoyancy pressure term using Eqs. 3-49 and 3-64 along with assumption 1 as:

$$(\bar{\rho}_{m,cold} - \bar{\rho}_{m,hot})(Z_m - Z_b) = (\bar{\rho}_{m,cold} - \bar{\rho}_{m,hot})(Z_{S.G.} - Z_b)$$

$$= \rho_f(Z_{S.G.} - Z_b) - \rho_f \left[L_{NB} + \frac{\ell n(1 + \gamma)}{\gamma} L_B + \frac{1}{1 + \gamma} \left(L'_H + \frac{L_E}{2} \right) \right] \tag{3-66}$$

The friction multiplier $\phi_{\ell o}^2$ according to the HEM model is given by:

$$\phi_{\ell o}^2 \approx \frac{\rho_f}{\rho_m} \tag{I,11-81}$$

Therefore, the average multiplier over the boiling length is:

$$\overline{\phi_{\ell o}^2} = \frac{1}{L_B + L'_B} \int_{Z_B}^{Z_d} \frac{\rho_f}{\rho_m} dz \tag{3-67}$$

Using Eq. 3-60, we substitute for ρ_m in Eq. 3-67 and get after integration:

$$\overline{\phi_{\ell o}^2} = \frac{L_B}{L_B + L'_B} \left[1 + \frac{\gamma}{2} \right] + \frac{L'_B}{L_B + L'_B} [1 + \gamma] \tag{3-68}$$

Equations 3-52, 3-66, and 3-68 can be used to rewrite Eq. 3-49 in the form:

$$\left(\frac{f_{\ell o} G_m^2}{2 D_e \rho_f} \right)_{core} \left\{ L_{NB} + \left(\frac{L_B}{L_B + L'_B} \left[1 + \frac{\gamma}{2} \right] + \frac{L'_B}{L_B + L'_B} [1 + \gamma] \right) (L_B + L'_B) \right\}$$

$$= \rho_f \left(\frac{L_E}{2} + L'_E \right) g - \rho_f \left[L_{NB} + \frac{\ell n(1 + \gamma)}{\gamma} L_B + \frac{1}{1 + \gamma} \left(L'_H + \frac{L_E}{2} \right) \right] g \tag{3-69}$$

where it was noted that

$$Z_{S.G.} - Z_b = \frac{L_E}{2} + L'_E$$

Equation 3-69 may be rearranged to get:

$$\rho_f \left(\frac{L_E}{2} + L'_E \right) g = \rho_f \left\{ L_{NB} + \frac{\ell n(1 + \gamma)}{\gamma} L_B + \frac{1}{1 + \gamma} \left(L'_H + \frac{L_E}{2} \right) \right\} g$$

$$+ \left(\frac{f_{\ell o} G_m^2}{2 D_e \rho_f} \right)_{core} \left\{ L_{NB} + L_B \left[1 + \frac{\gamma}{2} \right] + L'_B[1 + \gamma] \right\} \tag{3-70}$$

The left side of the Eq. 3-70 represents the static pressure in the cold leg, which can be viewed as an external pressure head driving the flow in the hot leg. The first term on the right side represents the hot leg static pressure, and the second term represents the pressure needed to overcome frictional and form losses in the hot channel. Thus for a given cold leg, the left side is a constant, independent of G_m. On the other hand, the right side is a set of curves depending on the exit quality x_0 (or γ). The steady-state G_m is determined from the positions of intersection of the curves expressing the right side terms and the horizontal line expressing the left side term on a Δp versus G_m figure. The potential for multiple intersections of these curves reflects the potential for oscillatory behavior in natural circulation loops [7].

Eq. 3-70 may be expanded to include the friction in any part other than those explicitly considered here. For example a friction length L_1^+ in the single-phase part of the loop and an equivalent length L_1 due to form losses can be added so that Eq. 3-70 can be written as:

$$\rho_f \left(\frac{L_E}{2} + L_E' \right) g = \rho_f \left\{ L_{NB} + \frac{\ell n(1 + \gamma)}{\gamma} L_B + \frac{1}{1 + \gamma} \left(L_H' + \frac{L_E}{2} \right) \right\} g$$

$$+ \left(\frac{f_{\ell o} G_m^2}{2 D_e \rho_f} \right)_{core} \left\{ L_{NB} + L_1 + L_1^+ + L_B \left[1 + \frac{\gamma}{2} \right] + L_B'[1 + \gamma] \right\} \tag{3-71}$$

where:

$$\gamma = x_0 \frac{\rho_f}{\rho_g}$$

and where the boiling length and nonboiling length can be determined for given x_{in} and x_0 from the relations:

$$L_B = L_H - L_{NB}$$

$$\frac{L_{NB}}{L_H} = \frac{0 - x_{in}}{x_0 - x_{in}}$$

$$x_0 - x_{in} = \frac{\dot{Q}_H}{\dot{m} h_{fg}} = \frac{q''(P_h L_H)}{\dot{m} h_{fg}}$$

Example 3-3 Stability of natural circulation in a boiling loop: A homogeneous flow approach

PROBLEM A boiling water test loop at low pressure has been operated to simulate LMR natural circulation when boiling occurs in the core. Examine the flow/pressure drop characteristics for the test loop operating on natural circulation. The numerical values of the fixed parameters involving hot channel geometry and fluid properties are as follows:

$$\rho_f = 957.9 \text{ kg/m}^3 \ (T = 100 \text{ °C}, p = 101.3 \text{ kPa})$$

$$x_{in} = -0.146$$

$$\frac{\rho_f}{\rho_g} = 1603$$

$$f = .0055$$

Table 3-3 Comparison of test section parameters with typical LMR fuel assembly designs

Parameter: Heated surface material and geometry	Test apparatus: Electrically heated tube	Fast test reactor: 316 SS clad fuel rod bundle	LMR: 316 SS clad fuel rod bundle (without spacers)
L_1 (mm)	177.8	177.8	355.6
$L_1 + L_1^+$ (mm)	266.6	322.7	355.6
L_H (mm)	914.4	914.4	914.4
L_B' (mm)	1,219.2	1,219.2	2,082.8
$L_B' + L_B'^+$ (mm)	1,397.0	1,447.8	2,290.0
D_e hydraulic diameter (mm)	3.86	3.25 for average subchannel	4.06
L channel length (mm)	2,311.4	2,311.4	2,844.8
A flow area (mm²)	11.68	9.94 for average subchannel	9.33

$$D = 3.86 \text{ mm}$$

$$A = 11.68 \text{ mm}^2$$

$$L_1 = 177.8 \text{ mm, single-phase length before entering the heated length}$$

$$L_1 + L_1^+ = 266.6 \text{ mm, effective length to account for } L_1 \text{ and form losses}$$

$$L_H = 914.4 \text{ mm, heated length}$$

$$L_H' = L_B' = 1.219 \text{ m, effective length after the heated length}$$

$$L_B' + L_B'^+ = 1.397 \text{ m (due to form losses)}$$

$$L_E = 0$$

These loop parameters are compared with the LMR reactor parameters in Table 3-3. Table 3-4 compares fluid conditions for the water test loop and the LMR.

Table 3-4 Comparison of fluid conditions for water tests at atmospheric pressure with those for LMR at inception of boiling during loss of piping integrity (LOPI) accident

Condition	Tests	Fast test reactor	LMR
Fluid	Water	Na	Na
p (kPa)	101.3	172.4	172.4
T_{sat} (°C)	100	948.9	948.9
ρ_f (kg/m³)	957.9	727.3	727.3
ρ_g (kg/m³)	0.5975	0.4406	0.4406
(ρ_f/ρ_g)	1603	1651	1651
h_{fg} (kJ/kg)	2.257×10^3	3.814×10^3	3.814×10^3
c_p (kJ/kg K)	4.186	1.298	1.298
T_{in} (°C)	21.1	422.2	385.0
x_{in} or $\dfrac{c_p(T_{in} - T_{sat})}{h_{fg}}$	−0.146	−0.179	−0.192
T_2 (°C)	43.3	593.3	535
x_2 or $\dfrac{c_p(T_{sat} - T_2)}{h_{fg}}$	−0.105	−0.121	−0.141
μ_f (kg/m·s)	2.839×10^{-4}	1.500×10^{-4}	1.500×10^{-4}
σ_f (N/m)	0.058	Approximately 0.175	Approximately 0.175
k_f (W/m K)	0.682	54.5	54.5

SOLUTION Solutions for Eq. 3-71 for these conditions are shown in Figure 3-5 for various values of x_0.

The left side of Eq. 3-71 can be added to Figure 3-5 as:

$$\rho_f(L_H + L_H')g = 957.9(0.9144 + 1.219)9.8 = 2.00 \times 10^4 \text{ Pa}$$

Since the test loop is operated at a constant heat flux, it is useful to convert Figure 3-5 from a set of curves representing fixed values of x_0 to a set of curves representing fixed q''. This can be done by using Eq. 3-56 to determine for a given q'' the values of $\dot{m}$ corresponding to the various values of x_0 presented in Figure 3-5. For example at $q'' = 175$ kW/m², we can find the value of $\dot{m}$ as a function of x_0 using:

$$\dot{m} = (1.75 \times 10^5)\pi(0.00386)(0.9144)/[(x_0 - x_{in})(2.257 \times 10^6)]$$

$$= 8.54 \times 10^{-4}/(x_0 - x_{in})$$

Therefore, for $x_0 = 0.04$, $\dot{m} = 4.59 \times 10^{-3}$ kg/s and for $x_0 = 0.4$, $\dot{m} = 1.56 \times 10^{-3}$ kg/s. Thus the graph for $q'' = 175$ kW/m² can be generated by plotting the calculated values of Δp for each combination of $\dot{m}$ and x_0.

The general results are shown in Figure 3-6 for values of q'' from 30 to 300 kW/m² (approximately 1 to 10×10^4 BTU/h ft²).

For a fixed heat flux, $\Delta p - \dot{m}$ curves exhibit S shapes at low pressures. Note, however, that in this case the loop Δp does not include an acceleration component. Hence, the S shape is solely due to the trend of increasing friction factor, ϕ_{lo}^2, as the mass flux, G_m, decreases while Δp is a product of ϕ_{lo}^2 and G_m^2.

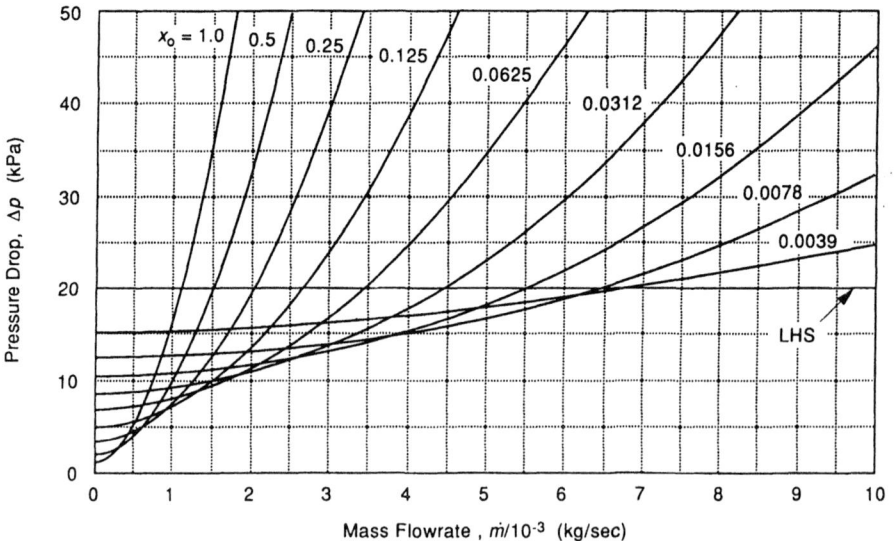

Figure 3-5 Flow/pressure drop characteristics of test loop with exit quality as a parameter.

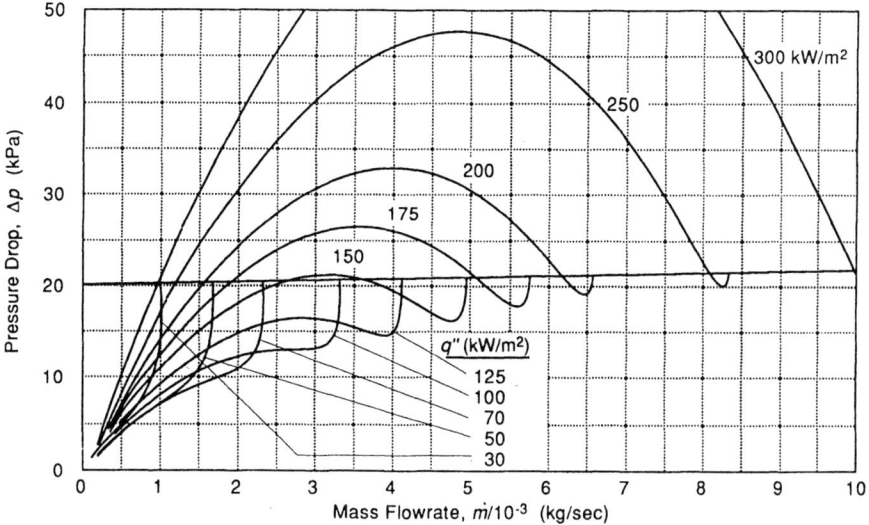

Figure 3-6 Flow/pressure drop characteristics of test loop with heat flux as a parameter: The analytical solution.

The only operating values for the loop are those corresponding to $\Delta p = 20$ kPa. Thus, from Figure 3-6 at the heat flux of 175 kW/m² this may occur at three different flow rates: for a mass flow of 5.22×10^{-3} kg/s, (and from Figure 3-5 the exit quality is about 0.022), a mass flow of 1.9×10^{-3} kg/s (for $x_0 = 0.307$), or a mass flow of 5.75×10^{-3} kg/s (for $x_0 = 0.0037$). The first of these intersections is for an unstable situation, as will be explained below.

From Figure 3-6 one can observe the following:

1. For heat fluxes below approximately 144 kW/m² there is only one steady-state flow rate satisfying Eq. 3-71. This flow rate corresponds to a value of $x_0 < 0$ (subcooled) and is stable. (That is, a momentary decrease in flow rate at constant q'' causes a decrease in Δp.)
2. At heat fluxes between 144 and 250 kW/m² there are three steady-state flow rates satisfying Eq. 3-71. These three flow rates correspond to values of x_0 between zero and 1.0. However, only the flow rates corresponding to the lowest and highest x_0 are stable. At a flow rate of 5.22×10^{-3} kg/s a momentary decrease in flow rate at constant $q'' = 175$ kW/m² causes an increase in Δp which causes flow to decrease further until the lowest of the three stable flow rates is reached. An increase in flow rate is associated with a lower Δp which would lead to furthering the increase in the flow rate.

Example 3-4 Stability of natural circulation in a boiling loop: A nonhomogeneous flow approach

PROBLEM For the conditions of the test loop given in Example 3-3 examine the natural circulation flow rate/pressure drop characteristics using the

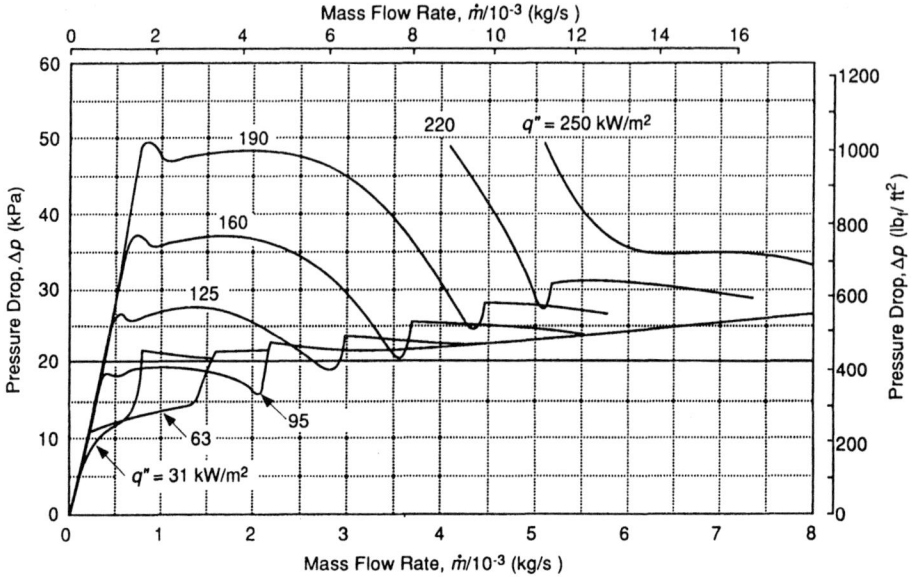

Figure 3-7 Flow/pressure drop characteristics of test tube: The computer solution.

nonhomogeneous flow model of Example 13.5 of Volume I (the PEDROP routine).

Since in this case the water properties are no longer held constant, use an inlet pressure of 117.2 kPa to lead to channel average pressure of about 101.3 kPa (which was the basis for calculating the water properties in Example 3-3).

SOLUTION The results of the PEDROP calculation are shown in Figure 3-7. It is seen that in this case:

1. For heat fluxes below about 96 kW/m² only single flow rate is stable at a subcooled exit condition.
2. At the heat fluxes between 96 and 160 kW/m² three possible flow rates exist.
3. Above the heat flux 160 kW/m² only superheated exit conditions will exist.

V LOOP TRANSIENTS

In describing the loop transients we shall use a simplified approach, which renders the problem analytically solvable. The purpose of our treatment will be to identify the major time constants in thermal-hydraulic analysis and the physical parameters controlling these time constants. Essentially three time constants are of interest:

τ_1: the time constant for flow decay
τ_2: the time constant for primary system temperature response
τ_3: the time constant for fuel temperature transients.

If any of these time constants is small relative to the time period involving significant changes of system parameters during a transient, the process associated with that time constant may be treated using a quasi-steady-state process. This often helps simplify the treatment of the system. On the other hand if the time constant is much longer than the time involved in the transient, the treatment can also be simplified by considering some of the system parameters as constants. For example, if τ_3 is on the order of a few seconds when a power transient of only a few milliseconds is being analyzed, the fuel temperature can be evaluated with constant coolant temperature.

A Single-Phase Loop Transients

1 Hydraulic considerations Under normal PWR and LMR plant operation flow conditions, the buoyancy effects can be neglected. As the pumping power changes, the buoyancy effects on the transient can be neglected until the flow is reduced to only a few percent of the normal level. Thus, even in the event of pump coastdown, the flow behavior is dominated initially by the inertia term and the hydraulic resistance term (see Figure 3-8).

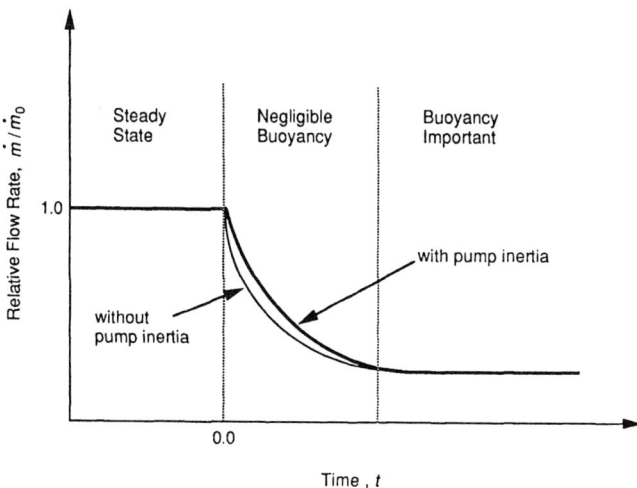

Figure 3-8 Flow coastdown stages.

In describing the loop transients a simplified pump relation is used here to keep the problem analytically solvable. Later in Section VC, a rigorous treatment of the centrifugal pumps commonly used in nuclear power plants will be described.

From Eq. 3-16, by defining the friction pressure drop using the approach of Eq. 3-35b, the momentum equation for pump dominated transients can be written as:

$$\left(\sum_k \frac{L_k}{A_k} \right) \frac{d\dot{m}}{dt} = \Delta p_{\text{pump}} - \frac{1}{2} \frac{R}{\rho_\ell} (\dot{m})^{2-n} \tag{3-72}$$

Thus the rate of change of the fluid momentum (the left side) is controlled by the resistive forces of the loop and the pressure change at the pump.

The pump head will vary according to the product of the square of the rotor speed, ω, and a function depending on the ratio $\dot{m}/\omega$ so that:

$$\Delta p_{\text{pump}} = \omega^2 f \left(\frac{\dot{m}}{\omega} \right) \tag{3-73}$$

For single-phase flow, the flow rate may be assumed directly proportional to the pump rotor angular speed:

$$\dot{m} = k\omega \tag{3-74}$$

For a first approximation, $f(\dot{m}/\omega)$ in Eq. 3-73 may be taken constant so that:

$$\Delta p_{\text{pump}} \simeq \Delta p_R \left(\frac{\dot{m}}{\dot{m}_R} \right)^2 \tag{3-75}$$

where the subscript R denotes rated conditions. Thus, the pump head-flow relation has been simplified to a parabolic behavior. This is adequate under quasi-steady-state conditions, but considerable deviation from this relation may occur under transient conditions, as will be discussed in Section VC.

Substituting from Eq. 3-75 into Eq. 3-72 we get an explicit equation for $\dot{m}$:

$$\left(\sum_k \frac{L_k}{A_k} \right) \frac{d\dot{m}}{dt} = \frac{\Delta p_R}{\dot{m}_R^2} \dot{m}^2 - \frac{1}{2} \frac{R}{\rho_\ell} (\dot{m})^{2-n} \tag{3-76}$$

which can be solved to obtain the transient flow rate, given appropriate values for the constants.

An important time constant in determining the plant response to possible pump failures is the time rate of flow decay if the pumping power was lost. If the pump failure is assumed to result in pump-free coastdown (i.e., $\Delta p_{\text{pump}} = 0$), flow is governed by the equation:

$$\left(\sum_k \frac{L_k}{A_k} \right) \frac{d\dot{m}}{dt} = - \frac{1}{2} \frac{R}{\rho_\ell} (\dot{m})^{2-n} \tag{3-77}$$

An analytical solution for the initial flow behavior (prior to significant buoyancy effects) for pump-free coastdown can be obtained as long as R can be assumed independent of the flow. The constant R assumption is adequate if transitions from turbulent flow to laminar flow do not occur during the transient. A constant ρ assumption is appropriate for single-phase liquids and high-pressure gases, i.e., $\rho_\ell = \rho = $ constant. In this case the flow transient is obtained as:

$$\frac{\dot{m}}{\dot{m}_R} = \frac{1}{\left[1 + \dfrac{(1-n)R(\dot{m}_R)^{(1-n)}}{2\rho \sum\limits_k (L_k/A_k)}\, t\right]^{1/(1-n)}}$$

$$= \frac{1}{[1 + t/(\tau_1)_{\text{tur}}]^{1/(1-n)}} \tag{3-78a}$$

Note that for laminar flow, $n = 1$, Eq. 3-78a is inapplicable, and the flow coastdown is described by:

$$\frac{\dot{m}}{\dot{m}_R} = \exp\left[-\frac{Rt}{2\rho \sum\limits_k \dfrac{L_k}{A_k}}\right] = \exp[-t/(\tau_1)_{\text{lam}}] \tag{3-78b}$$

Therefore, flow decay time constants for pump free coastdown are given by:

$$(\tau_1)_{\text{tur}} = \frac{2\rho}{(1-n)R(\dot{m}_R)^{(1-n)}} \sum_k \frac{L_k}{A_k} \tag{3-78c}$$

$$(\tau_1)_{\text{lam}} = \frac{2\rho}{R} \sum_k \frac{L_k}{A_k} \tag{3-78d}$$

From the above equations it is obvious that the flow decay time constant is inversely proportional to the hydraulic resistance R but is directly proportional to the density ρ.

2 Primary coolant temperature The primary coolant energy equation Eq. 3-19 may be written as:

$$\int \rho c_p A \frac{\partial T}{\partial t}\, dl + \int \dot{m} c_p \frac{\partial T}{\partial l}\, dl - \int A \frac{\partial p}{\partial t}\, dl = \dot{Q}_H - \dot{Q}_E \tag{3-79}$$

where dh_m is approximated by $c_p dT$, $\dot{Q}_H$ is the heat addition rate to the coolant in the core, and $\dot{Q}_E$ is the heat extraction rate in the heat exchanger. The second term on the left side can be dropped for a closed loop, and the third term on the left side can be neglected when the temporal pressure changes are negligible (such as when the system is equipped with a pressurizer). This reduces the primary coolant energy equation to:

$$\rho c_p V \frac{\partial \bar{T}}{\partial t} = \dot{Q}_H - \dot{Q}_E \tag{3-80}$$

where:

$$\bar{T} = \frac{\int TA\,dl}{\int A\,dl} = \text{average primary coolant temperature in the loop} \quad (3\text{-}81)$$

and

$$V = \sum_k A_k L_k = \text{the entire coolant volume} \quad (3\text{-}82)$$

Let us now obtain the desired time-dependent behavior of $\bar{T}$ by solving Eq. 3-80 under simplified conditions. The coolant heat addition and heat extraction rates, $\dot{Q}_H$ and $\dot{Q}_E$, will be expressed without considering the thermal inertia of the core fuel and structure and the heat exchanger structure, respectively. The core fuel and structural thermal inertia will be considered when their temperatures are to be evaluated in Section VC3. Further, $\dot{Q}_H$ will vary with time by virtue of its dependence on $\bar{T}$. The dependence will be derived next for substitution into Eq. 3-80.

For a steady secondary flow in the heat exchanger equal to $\dot{m}_s$, and neglecting structural thermal inertia:

$$\dot{Q}_E = \dot{m}_s (h_{se} - h_{si}) \quad (3\text{-}83)$$

If the secondary flow is only single phase, the value of $\dot{Q}_E$ may be given by:

$$\dot{Q}_E = \dot{m}_s c_{p,s} (T_{se} - T_{si}) \quad (3\text{-}84)$$

where T_{si} is the secondary coolant inlet temperature and T_{se} is the exit temperature of the secondary flow. In PWRs, the inlet subcooling of the secondary flow is small, so that Eq. 3-83 is more appropriate. However, for simplicity we shall use Eq. 3-84 to describe the energy removal process.

Then the energy equation Eq. 3-80 may be written as:

$$\rho c_p V \frac{\partial \bar{T}}{\partial t} = \dot{Q}_H - \dot{m}_s c_{p,s} (T_{se} - T_{si}) \quad (3\text{-}85)$$

Ignoring energy storage in the heat exchanger structures and assuming constant mass flow rate $\dot{m}_s$, the heat balance between the primary and secondary coolants may be given by:

$$hS(\bar{T} - \bar{T}_s) = \dot{m}_s c_{p,s} (T_{se} - T_{si}) \quad (3\text{-}86)$$

where:

$$\bar{T}_s \equiv \frac{T_{se} + T_{si}}{2} \quad (3\text{-}87)$$

and where h and S are the heat transfer coefficient and surface area within the heat exchanger, respectively.

From the last two equations, T_{se} may be defined by:

$$T_{se} = \left[hS\bar{T} + \left(\dot{m}_s c_{p,s} - \frac{hS}{2} \right) T_{si} \right] \Big/ \left[\dot{m}_s c_{p,s} + \frac{hS}{2} \right] \quad (3\text{-}88)$$

Substituting from Eq. 3-88 into Eq. 3-85 we get:

$$\rho c_p V \frac{\partial(\bar{T} - T_{si})}{\partial t} + \frac{\dot{m}_s c_{p,s} hS}{\dot{m}_s c_{p,s} + \frac{hS}{2}} (\bar{T} - T_{si}) = \dot{Q}_H \qquad (3\text{-}89)$$

The solution to Eq. 3-89 when T_{si} as well as $\dot{Q}_H$ is assumed constant provides an insight into the parameters controlling the coolant temperature response:

$$(\bar{T} - T_{si}) = (\bar{T}_0 - T_{si})e^{-t/\tau_2} + \frac{\dot{Q}_H \tau_2}{\rho c_p V} [1 - e^{-t/\tau_2}] \qquad (3\text{-}90)$$

where $\bar{T}_0$ is the value of $\bar{T}$ at $t = 0$ and:

$$\tau_2 = \frac{\rho c_p V(\dot{m}_s c_{p,s} + hS/2)}{\dot{m}_s c_{p,s} hS} \qquad (3\text{-}91)$$

The time constant τ_2 illustrates the significance of both the secondary flow rate and the heat transfer coefficient in the steam generator for the rate at which the primary coolant temperature will change. For a small heat transfer coefficient, the time constant is given by:

$$\tau_2 \simeq \rho c_p V/hS; \qquad hS \ll \dot{m}_s c_{p,s} \qquad (3\text{-}92a)$$

For a small secondary flow rate, the time constant becomes:

$$\tau_2 \simeq \frac{\rho c_p V}{2\dot{m}_s c_{p,s}}; \qquad hS \gg \dot{m}_s c_{p,s} \qquad (3\text{-}92b)$$

The value of the product hS, the inverse of the thermal resistance in the heat exchanger, can be approximated as follows. The value of h, the thermal conductance per unit surface area, is obtained from the thermal resistances of the primary coolant, the tube wall, the secondary coolant and fouling in series:

$$\frac{1}{h} = \frac{1}{h_p} + \frac{\delta}{k} + \frac{1}{h_s} + \sum \frac{1}{h_i} \qquad (3\text{-}93)$$

For simplicity, we may ignore the term, $\sum 1/h_i$, which is due to fouling in the heat exchanger, and the conduction term across the tubes, δ/k. Hence, the total thermal resistance in the heat exchanger is obtained from:

$$\frac{1}{hS} = \frac{1}{h_p S} + \frac{1}{h_s S} \qquad (3\text{-}94)$$

Let us treat the effects of power transients in two cases: (a) constant primary coolant flow rate and (b) change of primary flow rate. Recall that the secondary flow rate $\dot{m}_s$ was taken constant.

To a first approximation, the ratio h_p/h_s may be assumed proportional to the mass flow rates:

$$\frac{h_p}{h_s} = \frac{\dot{m}}{\dot{m}_s} \qquad (3\text{-}95)$$

Substituting from Eq. 3-95 into Eq. 3-94, we can express the thermal resistance in the heat exchanger as a function of the secondary to primary mass flow ratio:

$$\frac{1}{hS} = \frac{1}{h_sS \left(\dfrac{\dot{m}}{\dot{m}_s}\right)} + \frac{1}{h_sS} = \frac{1}{h_sS} \left(\frac{\dot{m}_s}{\dot{m}} + 1\right) \tag{3-96a}$$

which can be used for a constant primary flow rate.

For a change in the primary flow rate, express h_p as:

$$\frac{h_p}{h_{p0}} = \left(\frac{\dot{m}}{\dot{m}_0}\right)^{0.8} \text{(from Dittus-Boelter Eq.)} \tag{3-97}$$

Substituting Eq. 3-97 into 3-94 leads to the result for the second case:

$$\frac{1}{hS} = \frac{1}{h_{p0}S \left(\dfrac{\dot{m}}{\dot{m}_0}\right)^{0.8}} + \frac{1}{h_sS} \tag{3-96b}$$

In a U-tube PWR steam generator, it is possible that the transient involves a two-phase level drop on the secondary side which divides the region into one with a good heat transfer coefficient and the other with a smaller heat transfer coefficient. The analysis can then be carried out for a two-zone steam generator. The simple analysis presented here for the primary average temperature was found satisfactory for both one- and two-loop natural convection experiments as long as the flow is not unstable [15 and 17].

Example 3-5 Calculation of primary system temperature time constants

PROBLEM For a Babcock and Wilcox PWR, calculate the time constant τ_2 and the mean primary system temperature at $t = 60$ s for the case of:

1. A step decrease in the power to the coolant, $\dot{Q}_H$, to 10% of full power.
2. A step decrease of power to 10% full power and instantaneous flow decrease to 10% nominal flow. (Ignore time variant mass flow due to temperature/buoyancy effects.)

Refer to Table 3-5 for specifications of the Babcock and Wilcox PWR.

SOLUTION The time constant is given by Eq. 3-91:

$$\tau_2 = \frac{\rho V \left(\dot{m}_s c_{p,s} + \dfrac{hS}{2}\right)}{\dot{m}_s hS}$$

where it is assumed that the specific heats of the primary and secondary coolants are equal, $c_p = c_{p,s}$.

Table 3-5 Typical characteristics for a Babcock and Wilcox PWR

Gross thermal power	3,818 MW(th)
Thermal hydraulics	
Primary coolant	
Pressure, p	15.5 MPa
Core inlet temperature, T_{ci}	301 °C
Average core exit temperature, T_{ce}	332 °C
Core flow rate, $\dot{m}$	21.0 Mg/s
Volume, V	402 m³
Secondary coolant	
Pressure, p_s	7.83 MPa
Steam generator inlet temperature, T_{si}	244 °C
Steam generator exit temperature, T_{se}	313 °C

The mean temperature is given by Eq. 3-90:

$$\bar{T} - T_{si} = (\bar{T}_0 - T_{si}) \exp(-t/\tau_2) + \frac{\dot{Q}_H \tau_2}{\rho c_p V} (1 - \exp(-t/\tau_2))$$

Evaluate fluid properties at the mean temperature of the extremes of the system:

$$\bar{T}_{sys} = \frac{T_{ce} + T_{si}}{2}$$

where T_{ce} is the core exit temperature and T_{si} is the secondary coolant inlet temperature, and assume both are constant over the transient.

$$\bar{T}_{sys} \simeq \frac{332 + 244}{2} = 288 \text{ °C}$$

$$\bar{\rho} = 748 \text{ kg/m}^3 \quad \text{(Table 3-2)}$$

$$\bar{c}_p = 5.23 \text{ kJ/kg K}$$

1. To find the values of hS and $\dot{m}_s$ the assumption is made that h is temperature invariant. Eq. 3-86 can be evaluated at steady full-power conditions.

$$hS(\bar{T}_0 - \bar{T}_s) = \dot{m}_s c_{p,s}(T_{se} - T_{si}) = 3818 \text{ MW}$$

where:

$$T_{se} = 313 \text{ °C}, \ T_{si} = 244 \text{ °C}$$

$$T_{ci} = 301 \text{ °C}, \ T_{ce} = 332 \text{ °C}$$

$$\bar{T}_0 = \frac{332 + 301}{2} \qquad \bar{T}_s = \frac{313 + 244}{2}$$

$$\simeq 317 \text{ °C} \qquad\qquad \simeq 279 \text{ °C}$$

Hence:

$$\dot{m}_s = 10.5 \text{ Mg/s} = 10.5 \times 10^3 \text{ kg/s}$$

$$hS = 10^5 \text{ kW/°C}$$

This yields:

$$\tau_2 = \frac{750(402)\left((10.5 \times 10^3)(5.23) + \dfrac{10^5}{2}\right)}{10.5 \times 10^3(10^5)} = 30 \text{ s}$$

$$\bar{T} = 244 + 73e^{-t/30} + \frac{\dot{Q}_H(30)}{750(5.23)402}(1 - e^{-t/30})$$

so that $\bar{T} = 260 \text{ °C}$ for $t = 60$ s at $\dot{Q}_H = 381,800$ kW

2. Solving Eqs. 3-96a and 3-95 for h_sS and h_pS at the initial conditions yields:

$$(h_sS)_0 = (hS)_0\left(1 + \frac{\dot{m}_s}{\dot{m}}\right)$$

$$= 10^5\left(1 + \frac{10.5}{21.0}\right) = 1.5 \times 10^5 \text{ kW/°C}$$

$$(h_pS)_0 = (h_sS)_0\frac{\dot{m}}{\dot{m}_s} = 1.5 \times 10^5\left(\frac{21.0}{10.5}\right) = 3 \times 10^5 \text{ kW/°C}$$

The desired value of $1/hS$ is obtained from Eq. 3-96b where h_sS, which is constant, is evaluated at the initial conditions:

$$\frac{1}{hS} = \frac{1}{h_{p0}S\left(\dfrac{\dot{m}}{\dot{m}_0}\right)^{0.8}} + \frac{1}{(h_sS)_0}$$

Now, substituting the numerical values for the decreased primary flow:

$$\frac{1}{hS} = \frac{1}{3 \times 10^5(0.1)^{0.8}} + \frac{1}{1.5 \times 10^5}$$

$$hS = 3.61 \times 10^4$$

and

$$\tau_2 = \frac{750(402)\left((10.5 \times 10^3)(5.23) + \dfrac{3.61 \times 10^4}{2}\right)}{10.5 \times 10^3(3.61 \times 10^4)} = 58 \text{ s}$$

Also:

$$\frac{\dot{Q}_H \left(\dot{m}_s c_{p,s} + \frac{hS}{2} \right)}{\dot{m}_s c_{p,s} hS} = \frac{\dot{Q}_H \tau_2}{\rho c_p V} = 3.67 \times 10^{-5} \dot{Q}_H$$

$$= 14 \text{ °C when } \dot{Q}_H = 381{,}800 \text{ kW}$$

$$\bar{T} = 244 + 73 e^{-t/58} + 14(1 - e^{-t/58})$$

at $\quad t = 60 \text{ s} \quad \bar{T} = 279 \text{ °C.}$

The results of both parts 1 and 2 of this example are plotted in Figure 3-9. It is clear that for these cases of power reduction the decrease in the primary flow leads to a more gradual reduction in the primary temperature than does the case with a constant primary flow rate.

3 Thermal time constants of the core Thermal response of the reactor fuel and coolant transients is sufficiently intricate to require detailed numerical analysis of the three-dimensional distribution of the heat flow. However, the simple lumped parameter approach will again be used here to illustrate the basic parameters governing the rate of fuel temperature change.

It is possible to write the rate of heat addition to the coolant from the core, $\dot{Q}_H$, as dependent on an overall heat transfer coefficient h_c and on the overall heat transfer area in the core S_c:

$$\dot{Q}_H = h_c S_c (\bar{T}_f - \bar{T}) \qquad (3\text{-}98)$$

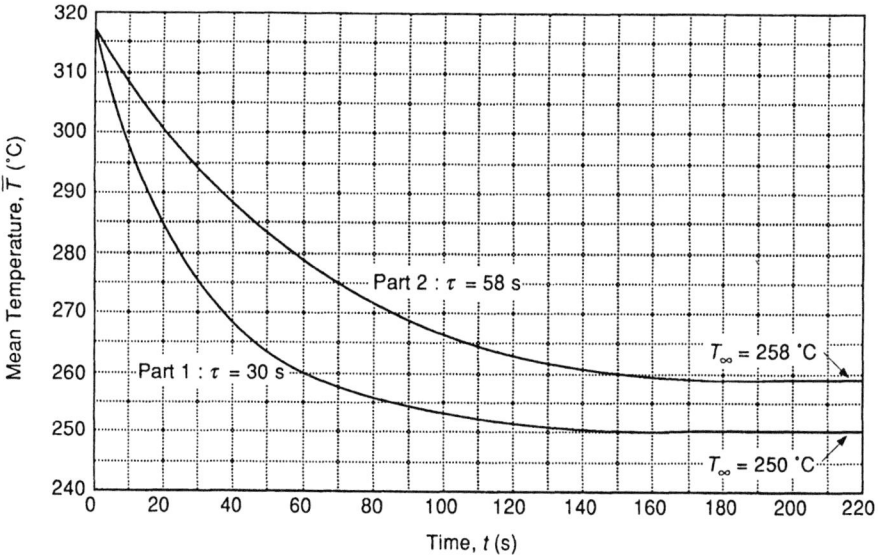

Figure 3-9 Mean temperature decay curves.

where $\bar{T}_f$ is the core average fuel element temperature and $\bar{T}$ is the core average coolant temperature.

At steady state, $\dot{Q}_H$ is equal to the heat generated in the core. For transient conditions, however, the core fuel and structure heat capacity have to be considered so that in general:

$$\dot{Q}_H = \dot{Q}_{core} - M_f c_f \frac{d\bar{T}_f}{dt} \tag{3-99}$$

where M_f is the mass of solid materials in the core, and c_f is their effective heat capacity. Then substituting Eq. 3-98 into Eq. 3-99 we get:

$$M_f c_f \frac{d\bar{T}_f}{dt} = \dot{Q}_{core} - h_c S_c (\bar{T}_f - \bar{T}) \tag{3-100}$$

The energy convected by the coolant outside the core is given by:

$$\dot{Q}_R = \dot{m} c_p (T_{ce} - T_{ci}) \tag{3-101}$$

where T_{ce} and T_{ci} are the core exit and inlet temperatures of the primary coolant. Now, the core average coolant temperature may be approximated by:

$$\bar{T} = \frac{T_{ce} + T_{ci}}{2} \tag{3-102}$$

So that the coolant average temperature in the core is given by:

$$M_c c_p \frac{d\bar{T}}{dt} = \dot{Q}_H - \dot{Q}_R$$

or

$$M_c c_p \frac{d\bar{T}}{dt} = h_c S_c (\bar{T}_f - \bar{T}) - 2\dot{m} c_p [\bar{T} - T_{ci}] \tag{3-103}$$

where M_c is the coolant mass in the core. Eqs. 3-100 and 3-103 provide a simple representation of the core average response of the fuel temperature and the coolant temperature, respectively. From Eq. 3-100 it is possible to define a fuel time constant τ_3 such that:

$$\tau_3 = M_f c_f / h_c S_c \tag{3-104}$$

It is a measure of the time required to transport heat from fuel to coolant. For UO_2-Zr-clad fuel elements τ_3 is typically few seconds. For HTGRs with much larger core heat capacity the value of τ_3 is much larger.

The significance of τ_3 can be illustrated by the heat removal from the core by a constant temperature coolant when $\dot{Q}_{core} = 0$ (hypothetically). In this case, Eq. 3-100 can be solved to yield:

$$\bar{T}_f(t) = \bar{T} + [\bar{T}_{f0} - \bar{T}] e^{-t/\tau_3} \tag{3-105}$$

Therefore, half the sensible heat in the fuel is removed in 0.693 τ_3.

It is desirable to decouple the two energy equations 3-100 and 3-103 for core average fuel element and coolant temperatures, respectively. This can be done by observing that in many reactors the following conditions can be reasonably met:

1. The heat capacity of the solid fuel is much larger than that of the coolant in the core:

$$M_f c_f \gg M_c c_p \tag{3-106}$$

2. The average fuel temperature drop is much larger than the average temperature rise of the coolant:

$$\bar{T}_f - \bar{T} \gg \bar{T} - T_{ci} \tag{3-107}$$

Because of the second assumption we can write Eq. 3-100 as:

$$\frac{d\bar{T}_f}{dt} = \frac{\dot{Q}_{core}}{M_f c_f} - \frac{1}{\tau_3} (\bar{T}_f - T_{ci}) \tag{3-108}$$

Then, the first assumption can be used to ignore thermal inertia of the coolant, thereby obtaining a quasi-steady relation from Eq. 3-103 of the form:

$$h_c S_c (\bar{T}_f - \bar{T}) = 2\dot{m} c_p (\bar{T} - T_{ci}) \tag{3-109}$$

Therefore, differentiating Eq. 3-109:

$$\frac{d\bar{T}_f}{dt} = \left(\frac{2\dot{m} c_p}{h_c S_c} + 1\right) \frac{d\bar{T}}{dt} + \frac{2c_p}{h_c S_c} (\bar{T} - T_{ci}) \frac{d\dot{m}}{dt} \tag{3-110}$$

From the fact that $\bar{T}_f - \bar{T} \gg \bar{T} - T_{ci}$, Eq. 3-109 yields:

$$\frac{2\dot{m} c_p}{h_c S_c} \gg 1$$

Hence:

$$\frac{d\bar{T}_f}{dt} \simeq \frac{2\dot{m} c_p}{h_c S_c} \frac{d\bar{T}}{dt} + \frac{2c_p (\bar{T} - T_{ci})}{h_c S_c} \frac{d\dot{m}}{dt} \tag{3-111}$$

Rearranging Eq. 3-111 and utilizing Eq. 3-108 yields:

$$\frac{d\bar{T}}{dt} = \frac{h_c S_c}{2\dot{m} c_p} \left\{ \frac{\dot{Q}_{core}}{M_f c_f} - \frac{1}{\tau_3} (\bar{T}_f - T_{ci}) \right\} - \frac{2c_p}{2\dot{m} c_p} (\bar{T} - T_{ci}) \frac{d\dot{m}}{dt}$$

Finally, again utilizing Eq. 3-109 and the second assumption, the above relation reduces to:

$$\frac{d\bar{T}}{dt} = \frac{\dot{Q}_{core}}{2\dot{m} c_p \tau_3} - \left(\frac{1}{\tau_3} + \frac{1}{\dot{m}} \frac{d\dot{m}}{dt}\right) (\bar{T} - T_{ci}) \tag{3-112}$$

Equations 3-108 and 3-112 provide the desired two uncoupled equations in $\bar{T}_f$ and $\bar{T}$, for fixed T_{ci}. It is clear that the core time constant τ_3 is dominant unless a coolant transient were so rapid that:

$$\frac{1}{\tau_3} < \frac{1}{\dot{m}} \frac{d\dot{m}}{dt} \tag{3-113}$$

Thus for reactor power transients, with $\dot{m}$ essentially unchanged, the core response is only dependent on τ_3. If the power transient were very rapid with respect to τ_3, the core may be treated as adiabatic. On the other hand, if the power transient were slow, then steady-state temperature distributions in the fuel may be assumed.

B Two-Phase Loop Transients

The major difference in this case from the single-phase case is the inclusion of the buoyancy effects that are much more significant here as well as modification of the friction term to reflect the changing boiling distance and the friction multiplier $\bar{\phi}_{\ell o}^2$ with quality. Again, we impose the assumptions:

1. Axially uniform heat generation rate in core
2. The HEM model is applicable
3. The friction losses are negligible outside the hot leg
4. Form losses are small compared with friction pressure losses.

Using Eqs. 3-16 and 3-70, the momentum equation may be written as:

$$\left(\sum_k \frac{L_k}{A_k} \right) \frac{d\dot{m}}{dt} = \Delta p_{\text{pump}} - \frac{f_{\ell o} G_m^2}{2D_e \rho_\ell} \left\{ L_{\text{NB}} + \left[1 + \frac{\gamma}{2} \right] L_B + [1 + \gamma] L_B' \right\}$$

$$+ \rho_f g \left\{ \frac{L_E}{2} + L_E' - \left[L_{\text{NB}} + \frac{\ell n (1 + \gamma)}{\gamma} L_B + \frac{1}{1 + \gamma} \left(L_H' + \frac{L_E}{2} \right) \right] \right\} \tag{3-114}$$

where only γ and L_B are dependent on G_m. Note that as G_m decreases, both γ and L_B increase. Therefore, the friction term and the buoyancy term increase, and the net effect could be an oscillatory behavior, as encountered in the steady state discussed in Example 3-3.

C Detailed Pump Representation

Flow loop transients are substantially influenced by pump transients and often initiated by them. The start-up and stoppage of pumps are among the routine operations of reactor plants. Pump failures, due to loss of power or inadvertent operator action, are events of relatively common occurrence in all power plants.

Accurate prediction of pump performance includes specification of its head (H), torque (π), discharge or volumetric flow rate (Q), and rotor speed (ω). The pump motor, by exerting a torque on the rotating shaft, provides energy to the impeller which creates the flow associated with a head increase from the suc-

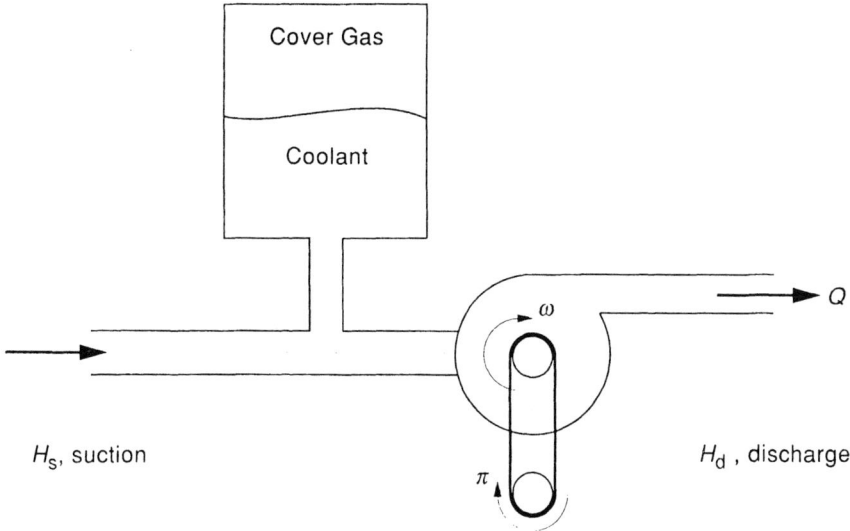

Figure 3-10 Schematic diagram of a pump with a reserve tank.

tion side to the discharge side (see Figure 3-10). The pump head H is used here to define the available head at the discharge side, $H_d - H_s$.

During any transient, the motor provides torque for the following:

1. Overcoming frictional resistance and form losses in the loop (piping, valves, core, etc.)
2. Overcoming frictional losses in the pump rotating parts
3. Acceleration of fluid in loop
4. Acceleration of pump rotating parts.

At steady state, no acceleration is involved, and only the first two items need be considered. Detailed transient analysis can be found in Wylie and Streeter [13] and Grover and Koranne [5].

Also, under certain transient conditions, a pump impeller may be forced to spin in a direction opposite to normal. This may be caused by the gravity head and the head provided by other pumps in a loop which may overcome the frictional resistances. Thus positive and negative values of ω and Q may have to be modeled.

Finally, it should be recognized that in principle the hydraulic representation of a turbine resembles a reverse pump operation since the flow will lose a "head" but generate a torque at the rotor.

For a pump, only two operation parameters can be considered independent among H, Q, ω, and π. The other two are determined from the pump characteristics. It is commonly assumed that the pump steady-state characteristics also hold for transient conditions. For a pump operating at any two conditions I and

2, the characteristics are described by specific relationships (often referred to as homologous relations) among the four operating parameters:

$$\frac{H_1}{\omega_1^2} = \frac{H_2}{\omega_2^2} \tag{3-115}$$

and

$$\frac{Q_1}{\omega_1} = \frac{Q_2}{\omega_2} \tag{3-116}$$

Equations 3-115 and 3-116 can be manipulated to yield the relationship between head and volumetric flow:

$$\frac{H_1}{Q_1^2} = \frac{H_2}{Q_2^2} \tag{3-117}$$

A more realistic relationship, usually available from tests, relates the head H to both the volumetric flow Q and the angular velocity ω:

$$\frac{H}{\omega^2} = a_1 + a_2 \left(\frac{Q}{\omega}\right) + a_3 \left(\frac{Q}{\omega}\right)^2 \tag{3-118}$$

where a_1, a_2, and a_3 are empirical constants.

The common approach uses nondimensional parameters with respect to the rated conditions:

$$h \equiv \frac{H}{H_R}, \; \nu \equiv \frac{Q}{Q_R}; \; \alpha \equiv \frac{\omega}{\omega_R}; \; \beta \equiv \frac{\pi}{\pi_R} \tag{3-119}$$

where the subscript R refers to the rated quantities, which represent the point of best performance.

Introduction of the nondimensional parameters in Eq. 3-118 results in the general relation:

$$\frac{h}{\alpha^2} = b_1 + b_2 \frac{\nu}{\alpha} + b_3 \left(\frac{\nu}{\alpha}\right)^2 \text{ for } 0 < \left|\frac{\nu}{\alpha}\right| < 1 \tag{3-120a}$$

where:

$$b_1 = \frac{a_1 \omega_R^2}{H_R}, \; b = \frac{a_2 Q_R \omega_R}{H_R}, \; b_3 = \frac{a_3 Q_R^2}{H_R} \tag{3-120b}$$

Similarly, because of the linear relation between the angular velocity and the volumetric flow rate, the head may be related to the square of the flow rate in the form:

$$\frac{h}{\nu^2} = b_3 + b_2 \left(\frac{\alpha}{\nu}\right) + b_1 \left(\frac{\alpha}{\nu}\right)^2 \tag{3-121}$$

The above equations are often used in nuclear reactor system codes.

If the pump efficiency is assumed independent of operating conditions, then the ratio of the rotational energy $\pi\omega$ to the flow work QH is constant so that:

$$\frac{\pi_1\omega_1}{Q_1H_1} = \frac{\pi_2\omega_2}{Q_2H_2} \tag{3-122}$$

which, using Eq. 3-116, leads to:

$$\frac{\pi_1}{H_1} = \frac{\pi_2}{H_2} \tag{3-123}$$

Similar equations to the head equations, Eqs. 3-120a and 3-121 can therefore be established for the dependence of β/α^2 on (ν/α) and β/ν^2 on (α/ν). The polynomial of Eq. 3-120a is often expanded to the third degree. For example, the Main Yankee pump curves for positive rotation as provided in the final safety analysis report [9] are:

$$\frac{h}{\alpha^2} = 1.8 - 0.3 \left(\frac{\nu}{\alpha}\right) + 0.35 \left(\frac{\nu}{\alpha^2}\right) - 0.85 \left(\frac{\nu}{\alpha^3}\right) \tag{3-124}$$

$$\frac{\beta}{\alpha^2} = 1.37 - 1.28 \left(\frac{\nu}{\alpha}\right) + 1.61 \left(\frac{\nu}{\alpha^2}\right) - 0.70 \left(\frac{\nu}{\alpha^3}\right) \tag{3-125}$$

For zero rotor speed, the pump head can be described as

$$h = -0.00418|\nu|\nu \tag{3-126}$$

For negative rotor speed, Rust [12] provides representative fits for a centrifugal pump:

$$\frac{h}{\alpha^2} = 0.5 + 0.51 \left(\frac{\nu}{\alpha}\right) - 0.26 \left(\frac{\nu}{\alpha^2}\right) + 0.25 \left(\frac{\nu}{\alpha^3}\right) \tag{3-127}$$

$$\frac{\beta}{\alpha^2} = -0.65 + 1.90 \left(\frac{\nu}{\alpha}\right) - 1.28 \left(\frac{\nu}{\alpha^2}\right) + 0.54 \left(\frac{\nu}{\alpha^3}\right) \tag{3-128}$$

Equations 3-127 and 3-128 can be used to represent negative rotation. It is necessary in a numerical analysis to allow transitions from positive rotor speed to negative speed to be smoothed out to avoid numerical difficulties. Linear interpolations between 3-124 and 3-127 can be used between $-0.1 \le \alpha \le 0.1$ [8]. It should also be noticed that the required work should always be greater than the delivered work to force the coolant flow. Hence:

$$\omega\pi > Q(\Delta p)_{\text{pump}} \tag{3-129a}$$

$$\omega\pi > Q(\rho g H) \tag{3-129b}$$

So that if during a calculation, Eq. 3-129a or b is not satisfied, the calculated torque can be arbitrarily fixed, assuming the pump has fixed efficiency:

$$\omega\pi = \frac{Q(\rho g H)}{\eta_{\text{pump}}} \tag{3-130}$$

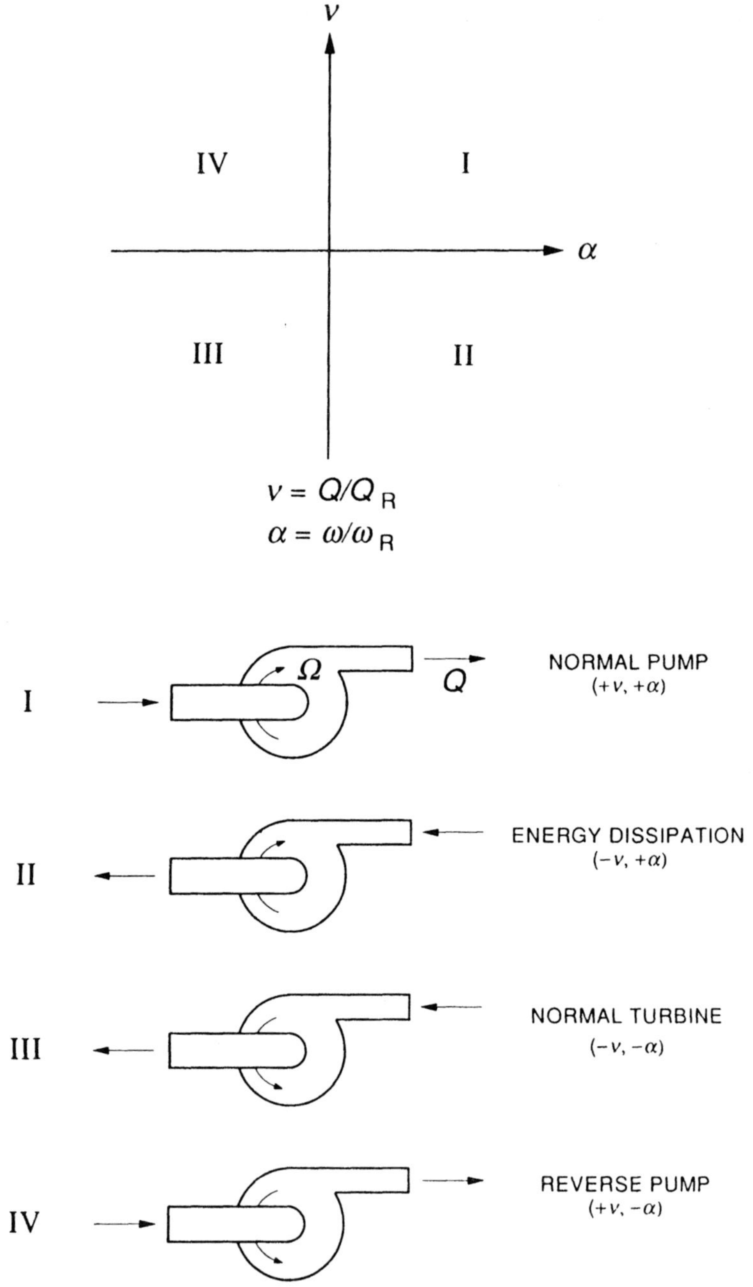

$$v = Q/Q_R$$
$$\alpha = \omega/\omega_R$$

Figure 3-11 Pump configurations under different regimes of operation.

Additionally, the pump head calculated above should be greater than the pump suction head specified for a given pump by the manufacturer. For Maine Yankee [9], the suction head is given by:

$$\frac{h}{\alpha^2} = 1.15 \left(\frac{\nu}{\alpha^2}\right)^2 - 0.149 \tag{3-131}$$

If the head of Eq. 3-124 for positive flow or Eq. 3-127 for negative flow falls below that of Eq. 3-131, the pump is tripped automatically.

Figure 3-11 illustrates the various regimes of pump operations, some of which are usually allowed in system codes [1,11]. During a coastdown transient the pump may pass from the normal pumping region, quadrant I, through a reverse flow but positive rotation region, quadrant II, to reverse flow and rotation, region III, unless the rotor shaft is equipped with an antireverse ratchet to avoid regions III as usually is the case for most PWRs. Figures 3-12, 3-13, and 3-14 provide the general shape of the homologous relations as described in the final analysis safety report of Maine Yankee [9] and presented in [1,8].

Several experiments have been performed to investigate the generality of the homologous representation for two-phase as well as single-phase characteristics [14]. Figure 3-15 illustrates the general behavior for single-phase tests. Figure 3-16 illustrates the head degradation in case of two-phase flow with increasing void fraction up to a void fraction $\alpha \simeq 0.8$, with head recovery for higher voids. The model of Zarechnak et al. [14] does not represent the smooth head recovery observed in the two-phase experiments.

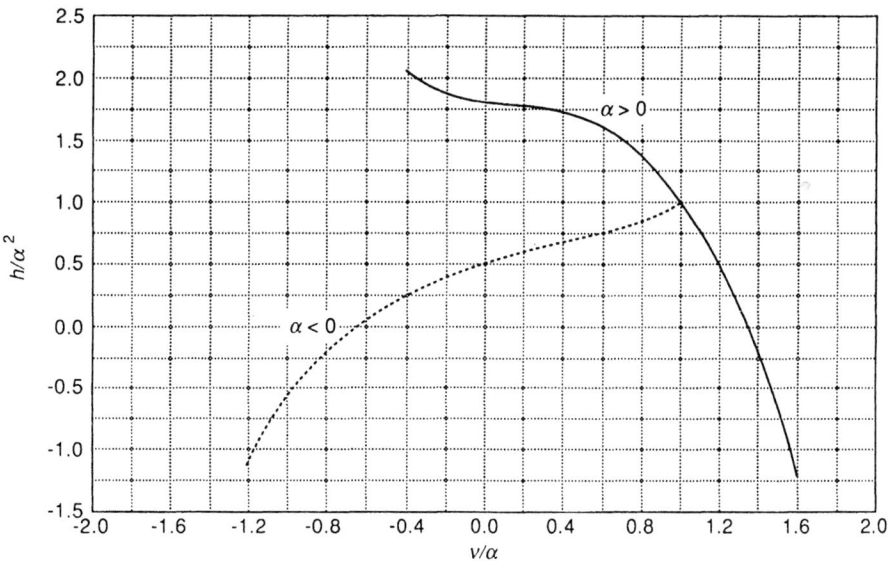

Figure 3-12 Homologous head curves for a centrifugal pump. *(After Agrawal et al. [1].)*

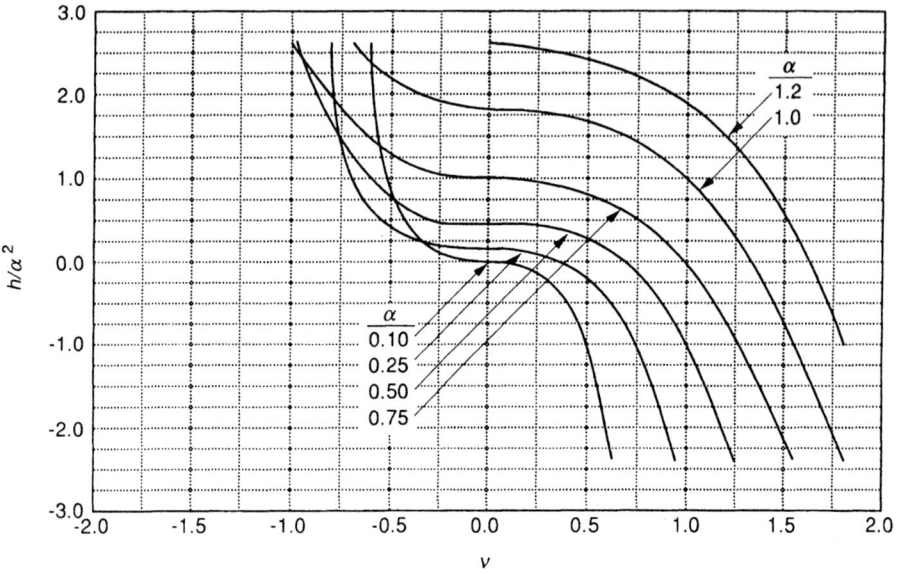

Figure 3-13 Maine Yankee pump head capacity curves for $\alpha > 0$. *(After Kao [8].)*

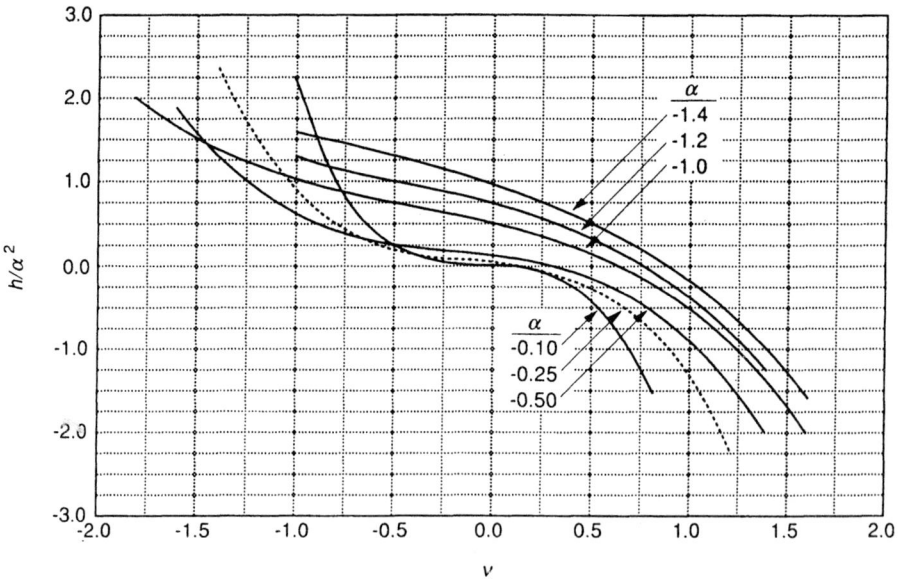

Figure 3-14 Pump head capacity curves for $\alpha < 0$. *(After Kao [8].)*

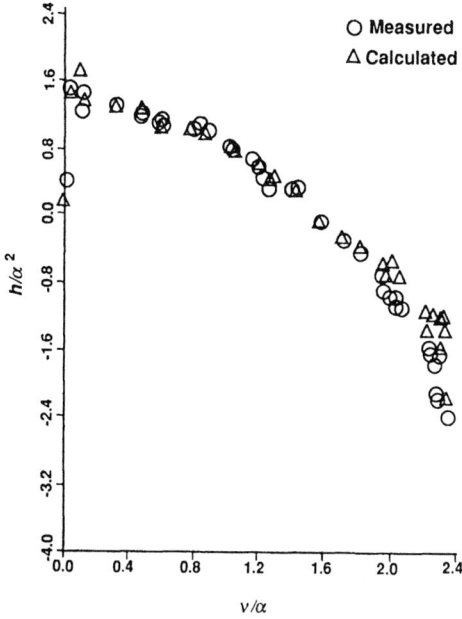

Figure 3-15 Calculated and measured homologous heads (h/α^2) versus flow/speed (v/α) Plots for single-phase (water) EPRI/CE data. *(From Zarechnak et al. [14].)*

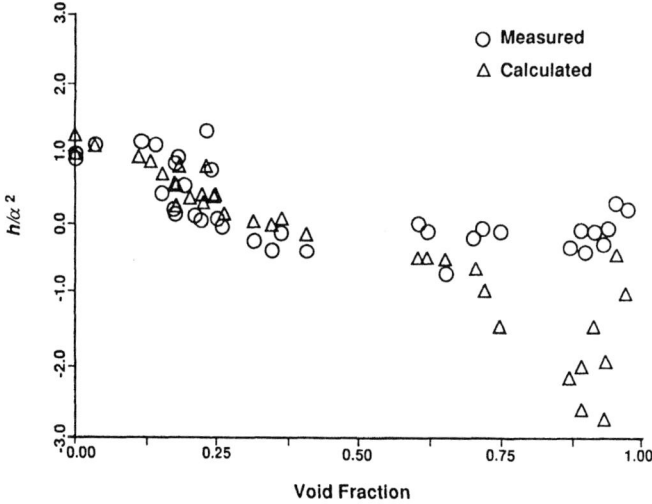

Figure 3-16 Calculated and measured homologous heads (h/α^2) versus void fraction plots for semiscale data at both 2,700 and 1,600 rpm for flow/speed (v/α) between 0.75 and 1.25. *(From Zarechnak et al. [14].)*

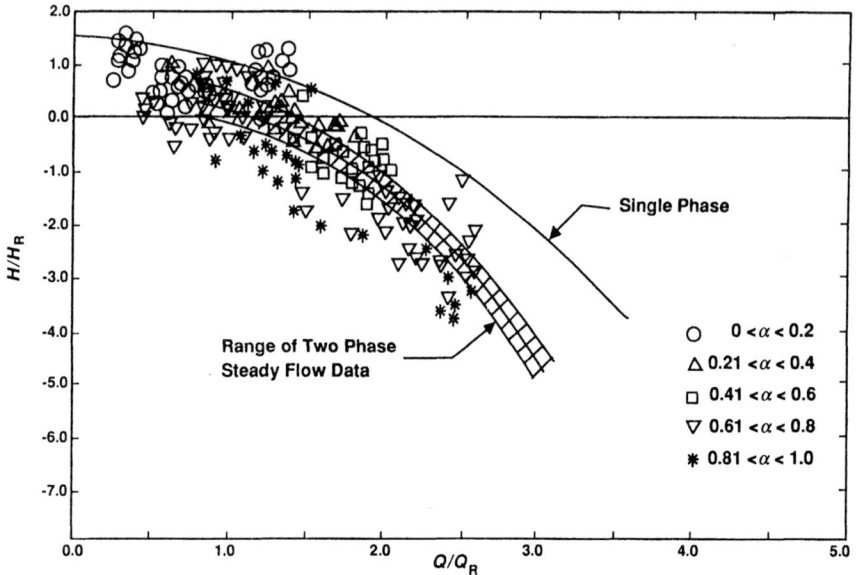

Figure 3-17 Pump performance data in transient two-phase flows, forward flow direction. $N = 1180$ rpm; $p_0 = 5.5$ MPa. *(From Heidrick et al. [6].)*

Finally, the pump characteristics under severe transient flow conditions have also been tested and found to yield results well represented by the steady-state conditions, as seen in Figure 3-17 where the substantial scatter is probably due to measurement imprecision [6]. It must be noted that normal operation of a liquid pump (as in a PWR main pump) would avoid the conditions leading to creation of bubbles due to cavitation to prevent pump damage. Therefore, the two-phase flow tests described in this figure are motivated by the conditions that may arise in unlikely events such as a loss of coolant and not the anticipated transients.

REFERENCES

1. Agrawal, A. K., et al. *An Advanced Thermohydraulic Simulation Code for Transients in LMFBRs SSC-L Code,* BNL-NUREG-50773, 1978.
2. Bau, H. H., and Torrence, K. E. *Transient and Steady Behavior of an Open, Symmetrically Heated Free Convection Loop,* E80-03. Ithaca, NY: Cornell University, 1980.
3. Bau, H. H., and Torrence, K. E. On the effects of viscous dissipation and pressure work in free convection loops. *Int. J. Heat Mass Transfer* 26:727–734, 1983.
4. Creveling, H. F., et al. Stability characteristics of a single-phase free convection loop. *J. Fluid Mechanics* 67:65–84, 1975.
5. Grover, R. B., and Koranne, S. M. Analysis of pump transients. *Nucl. Eng. Design* 67:137–141, 1981.
6. Heidrick, T. R. Rajon, V. S. V., and Nguyen, D. M. The behavior of centrifugal pumps in

steady and transient steam-water flows. *International Topics of Meeting on Nuclear Reactor Thermal Hydraulics,* NUREG/CP-0014, Nuclear Regulatory Commission. Washington, D.C. Vol. 1, 1980.

7. Jain, K. C., Petrick, M., Miller, D., and Bankoff, S. G. Self-sustained hydrodynamic oscillations in a natural circulation boiling water loop. *Nucl. Eng. Des.* 4:223–252, 1966.
8. Kao, Shih-Ping. *A Multiple-Loop Primary System Model for Pressurized Water Reactor Plant Sensor Validation.* CSDL-T-857. The Charles Stark Draper Laboratory, Cambridge, Mass. 1984.
9. Maine Yankee Atomic Power Station Final Safety Analysis Report, Docket 30309. Framingham, MA: Yankee Atomic Co., 1971.
10. Metais, B., and Eckert, E. R. G. Forced, mixed and free convection regimes. *J. Heat Transfer* 86:295–292, 1964.
11. Ransom, V. H., et al. *RELAP5/MOD1 Code Manual Vol. 1: System Models and Numerical Methods,* NUREG/CR-1826. Washington, DC: Nuclear Regulatory Commission, 1980.
12. Rust, J. H. *Nuclear Power Plant Engineering.* Buchanan, GA: Haralson Publishing Co., 1979.
13. Wylie, E. G., and Streeter, V. L. *Fluid Transients.* New York: McGraw-Hill, 1978.
14. Zarechnak, A., Damerell, P., Chapin, D., and Wu, D. Thermodynamic model of centrifugal pump performance in two-phase flow. *International Topics of Meeting on Nuclear Reactor Thermal Hydraulics,* NUREG/CP-0014, Vol. 1. Washington, DC: Nuclear Regulatory Commission, 1980.
15. Zvirin, Y. A review of natural circulation loops in pressurized water reactors and other systems. *Nucl. Eng. Des.* 67:203–225, 1981.
16. Zvirin, Y., Shitzer, A., and Grossman, G. The natural circulation solar heat models with linear and non-linear temperature distributions. *Int. J. Heat Mass Transfer* 20:997–999, 1977.
17. Zvirin, Y., Jeuck, P. R., Sullivan, C. W., and Duffey, R. B. Experimental and analytical investigation of a natural circulation system with parallel loops. *J. Heat Transfer* 103:645–652, 1981.

PROBLEMS

Problem 3-1 Natural circulation in a boiling channel with steam separation (Section IV)

Consider a small BWR being designed to rely completely on natural circulation in the vessel to achieve heat removal from the core, as depicted in Figure 3-18. Assume that the separator is 100%

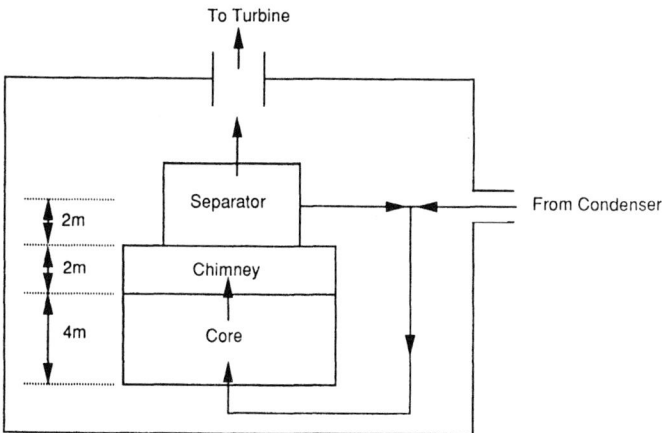

Figure 3-18

efficient (i.e., all vapor is allowed to pass, and all liquid is returned to the downcomer) and that the pressure losses in the separator are negligible. The chimney region can be assumed to have the same geometry as the core region. The steam emerging from the vessel expands through a turbine, is condensed, pumped up, and returned at the top of the downcomer.

Estimate the maximum allowable core thermal power for the system and fuel bundle design conditions given in the Table such that the core exit average steam quality is 15%.

You may use the HEM model. Assume the single-phase friction factor in the core and the chimney to be $f = 0.02$ and the water thermodynamic properties to be pressure independent. Ignore friction pressure loss outside the core and the chimney and acceleration pressure drop.

BWR DESIGN CONDITIONS

System pressure = 7.1 MPa
Fuel assembly flow area = 9.55×10^{-3} m^2
Fuel assembly equivalent hydraulic diameter = 0.013 m
Number of fuel assemblies = 400
Condensate vessel inlet subcooling = 25 °C
Length of core = 4 m
Length of chimney = 2 m
Separator effective length = 2 m
Radial and axial heat generation profiles = uniform

Answer: Q = 1311 MWth

Problem 3-2 Regions of instability in a boiling channel (Section IV)
Consider the boiling water loop test apparatus described in Example 3-3. Evaluate the possibility of unstable heat fluxes in the loop if the system pressure was raised to $p = 100$ p.s.i. such that:

$$\frac{\rho_f}{\rho_g} = 250$$

MULTIPLE HEATED CHANNELS CONNECTED ONLY AT PLENA

I INTRODUCTION

In this chapter the solution of the continuity, momentum, and energy transport equations of the coolant in a vertical, parallel array of heated channels connected only at plena will be discussed. These solutions are applicable to a fuel assembly modeled as an array of laterally isolated subchannels. We will start with the one-dimensional transient transport equations of the coolant. The solution procedure will be presented for conditions according to the problem statement for these channel arrays as outlined in Table 1-4. The physical basis for channel behavior will be stressed since the problem formulation allows multiple solutions for the flow in individual channels. Instability mechanisms will be pointed out. Sample results for steady-state conditions available in the literature will also be presented.

Sections I through V present the governing equations, boundary conditions, and general solution procedure. The low flow rate situation is treated in Sections VI and VII and requires inclusion of gravity effects that are neglected at high flow rates. This regime is applicable to decay heat conditions with impaired circulation through the primary system in which substantial internal circulation occurs within the reactor vessel. These internal circulation flows involve both upflow and downflow zones and are instrumental in minimizing peak temperatures in the core. Section VIII presents the solution procedure for the case of decoupled conservation equations which is applicable to high flow rate conditions. These results provide a useful indication of coolant flow and temperature distribution for assessment of assembly thermal performance. The isolated channel geometry of this chapter is relaxed in Chapters 5 and 6 to allow subchannels to communicate laterally along their length.

II GOVERNING ONE-DIMENSIONAL, STEADY-STATE FLOW EQUATIONS

Figure 4-1 illustrates the geometry of multiple, heated channels that are connected only at plena. It will be assumed that the one-dimensional mass, momentum, and energy transport equations derived in Volume I, Chapter 5 in the form stated in Chapter 13 are applicable. For transients in which time-dependent boundary conditions vary slowly compared with the fluid transit time in the channel, a quasi-steady-state approach is adequate. For simplicity we shall only treat the steady-state equations, although most numerical solution schemes can treat both the steady-state and transient forms of the conservation equations. The relevant conservation equations will be rewritten to specifically allow for downflow as well as upflow within the array of channels. The sign convention adopted is positive for upflow and negative for downflow. We assume, in order to simplify the notation, that for our system of N channels, any one of which is n, 1 through M are in upflow, and $(M + 1)$ through N are in downflow.

A Continuity Equation

The relevant continuity equation is:

$$\frac{\partial \rho_m}{\partial t} + \frac{\partial}{\partial z} (G_m) = 0 \qquad (I,13\text{-}1)$$

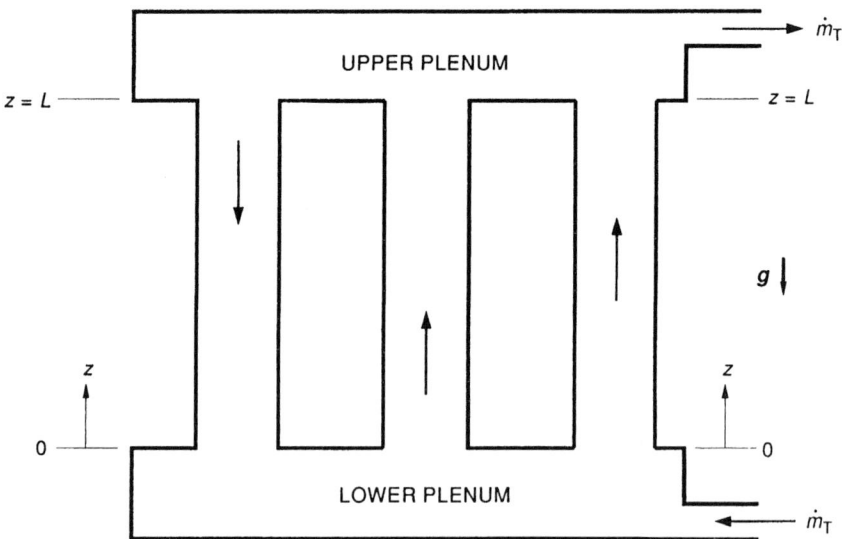

Figure 4-1 Flow in vertical, parallel channels connected only at plena.

Integrating along a channel under steady-state conditions and multiplying by the constant cross-sectional area yield:

$$G_{mn}A_n = \text{constant} \tag{4-1}$$

where $G_{mn}A_n \equiv \dot{m}_n$. Additionally, the total net flow through the system is the algebraic sum of all the channel flows:

$$\sum_{n=1}^{N} G_{mn}A_n = \dot{m}_T \tag{4-2}$$

If the net flow is from the lower plenum to the upper plenum, $\dot{m}_T$ is positive, whereas if the net flow is from the upper plenum into the lower plenum, $\dot{m}_T$ is negative.

It is possible to take the system return path between plena as simply another channel. In that case this channel flow rate would be $-\dot{m}_T$, and the right side of Eq. 4-2 would be zero. That approach is not taken here, i.e., the system return path is not considered as one of the N total channels; rather it is represented as a separate path for each plenum.

B Momentum Equation

The relevant momentum equation is:

$$\frac{\partial G_m}{\partial t} + \frac{\partial}{\partial z}\left(\frac{G_m^2}{\rho_m^+}\right) = -\frac{\partial p}{\partial z} - \frac{fG_m|G_m|}{2D_e\rho_m} - \rho_m g \cos\theta \tag{I,13-2a}$$

This momentum equation can be rewritten in terms of the channel pressure loss, Δp_n. If all channels are vertically oriented and are of length L, we can define Δp_n for each channel as:

$$\Delta p_n \equiv p_n(\text{lower}) - p_n(\text{upper}) \tag{4-3}$$

For steady-state conditions and local losses the momentum equation becomes:

$$\Delta p_n = \int_0^L \rho_{mn}g\,dz + \int_0^L \frac{f_n G_{mn}|G_{mn}|}{2D_{en}\rho_{mn}}\,dz + \sum_i \frac{K_{in}G_{mn}|G_{mn}|}{2\rho_{mn}}$$
$$+ G_{mn}^2\left[\frac{1}{\rho_{mn}^+(L)} - \frac{1}{\rho_{mn}^+(0)}\right] \tag{4-4}$$

If in this equation ρ_{mn}^+ is replaced by ρ_{mn} and f_n is taken for single phase, it becomes strictly valid for single-phase fluids as well as homogeneous multiphase flow. The third term on the right side of Eq. 4-4 represents the local pressure losses in the channel. Included are entrance and exit losses as well as form losses along the channel. In some applications (a heavily orificed channel, for example) this term is the largest contributor to the overall channel pressure drop and therefore should not be neglected.

Let us examine the application of this result for a system of channels, in both upflow and downflow, connected to common plena. Figure 4-2 illustrates the direction of the friction and gravity forces acting on the control volume for both cases.

For upflow, the first three terms on the right side of Eq. 4-4 are positive. The fourth term, the acceleration pressure drop term, is also positive for the heated case with fluids whose density decreases with temperature since this implies:

$$\rho_{mn}(0) > \rho_{mn}(L) \qquad \text{(heated upflow)}$$

For downflow, however, the channel has as its inlet the upper plenum and discharges to the lower plenum.

Examining each term of Eq. 4-4, we observe that:

- The gravity term is positive, independent of flow direction.
- The friction and local loss terms are negative as indicated by our sign convention on G_{mn} because these terms cause the pressure to decrease in the direction of flow.

The acceleration term takes its sign from the difference of its two components. Two cases are of interest: heated and cooled downflow:

$$\rho_{mn}(L) > \rho_{mn}(0) \qquad \text{(heated downflow)}$$

$$\rho_{mn}(L) < \rho_{mn}(0) \qquad \text{(cooled downflow)}$$

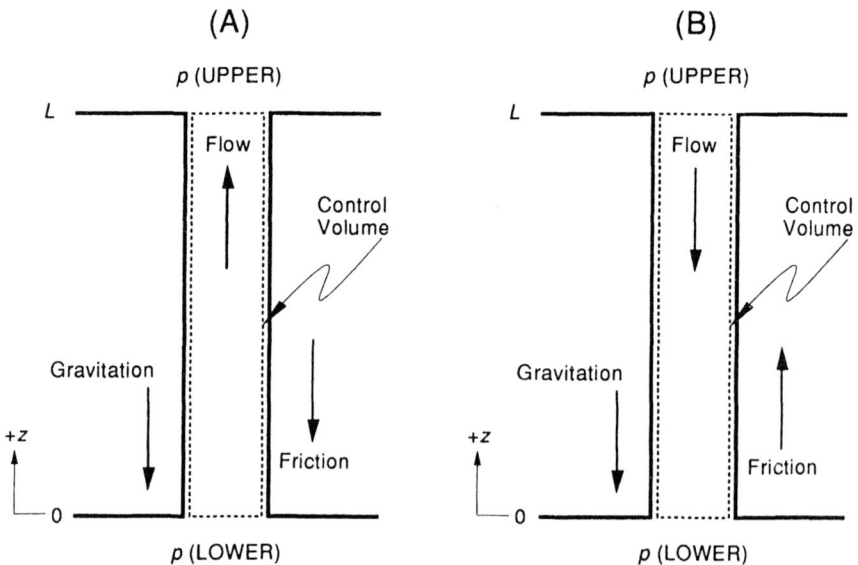

Figure 4-2 External forces exerting on control volume for upflow (*a*) and downflow (*b*) cases.

Thus the acceleration term for heated downflow is negative, whereas for cooled downflow it is positive. For downflow, then, the pressure drop Δp_n, defined by Eq. 4-3, is composed of a positive gravity component and negative friction and local loss components, while the acceleration component is negative for heated downflow and positive for cooled downflow.

C Energy Equation

The relevant equation is:

$$\dot{m}\frac{d}{dz} h_m = q'(z) \tag{I,13-13c}$$

Integrating the equation along the constant area channel n to position z yields, for steady-state upflow:

$$h_{mn}(z) = h_{mn}(0) + \frac{1}{\dot{m}_n} \int_0^z q'_n dz \qquad 1 \le n \le M \tag{4-5}$$

For heated upflow conditions, the second term is positive. For downflow we integrate Eq. I,13-13c from position L to position z, yielding:

$$h_{mn}(z) = h_{mn}(L) + \frac{1}{\dot{m}_n} \int_L^z q'_n dz \qquad M + 1 \le n \le N$$

$$= h_{mn}(L) - \frac{1}{\dot{m}_n} \int_z^L q'_n dz \qquad M + 1 \le n \le N \tag{4-6}$$

where the second term is positive for heated downflow and negative for cooled downflow by the sign conventions on $\dot{m}_n$ and q'_n. Considering that the total heat input rate over the channel length is $\dot{q}_n$, both of these equations yield:

$$h_{mn}(L) = h_{mn}(0) + \frac{\dot{q}_n}{\dot{m}_n} \tag{4-7}$$

where the second term on the right side will be positive for cooled downflow and heated upflow, and negative for heated downflow and cooled upflow.

III STATE EQUATION

To complete the equation set, we write the equation of state, i.e.:

$$\rho = \rho(h,p) \tag{4-8}$$

which for the mixture is written as:

$$\rho_m = \rho_m(h_m,p) \tag{4-9}$$

all variables being functions of position z. Now we can explicitly relate the parameters in Eqs. 4-4, 4-5, and 4-6 by the following equations:

$$\rho_m = \rho_\ell(p) + \alpha\rho_{\ell v}(p) \tag{4-10}$$

$$h_m = h_\ell(p) + xh_{\ell v}(p) \tag{4-11}$$

where the relation between the void fraction, α, and the flow quality, x, for homogeneous flow has been presented earlier, i.e.:

$$\alpha = \frac{1}{1 + \left(\dfrac{1-x}{x}\right)\left(\dfrac{\rho_v}{\rho_\ell}\right)} \tag{I,11-30}$$

and

$$\rho_{\ell v} \equiv \rho_v - \rho_\ell$$

$$h_{\ell v} \equiv h_v - h_\ell$$

The state equations, Eqs. 4-10, 4-11, and I,11-30 can be rearranged to yield:

$$\rho_m = \rho_\ell(p) + \rho_{\ell v}(p)\left[1 + \frac{\rho_v(p)}{\rho_\ell(p)}\left(\frac{h_{\ell v}(p) - h_m + h_\ell(p)}{h_m - h_\ell(p)}\right)\right]^{-1} \tag{4-12}$$

IV APPLICABLE BOUNDARY CONDITIONS

Boundary conditions are specified for the channels and the plena in this situation.

A Channel Boundary Conditions

Table 1-4 summarizes the sets of boundary conditions applicable to the channels of this problem. The hydrodynamic components of these sets are as follows:

Set (1). Prescribed pressure levels at the inlet and outlet plena. In this case the net flow between plena is the algebraic sum of the individual channel flows. This net flow is a consequence of the prescribed pressure boundary conditions.

Sets (2) and (3). Prescribed total inlet flow rate, radial pressure gradient at one plenum, and the pressure level at the other plenum. Depending on plena specification, this yields two sets of pressure boundary conditions:

    Set (2). Lower plenum pressure gradient/upper plenum pressure level.

    Set (3). Upper plenum pressure gradient/lower plenum pressure level.

In these cases also, the net flow rate between plena is the algebraic sum of the individual channel flows; however, here this net flow rate is specified. Although these boundary sets do not specify numerical values for the axial pressure drop

Δp_n they do establish the relationships between radial pressure drops in the plena, i.e.:

$$p_n(\text{upper}) = f(p_1(\text{upper})) \text{ for } n = 2, 3, 4, \ldots N \tag{4-13}$$

or

$$p_n(\text{lower}) = f(p_1(\text{lower})) \text{ for } n = 2, 3, 4, \ldots N \tag{4-14}$$

Special cases of particular interest are:

1. *Equal pressure drop for each channel.* This is usually the case if the plena flow Reynolds numbers are low and the plena equivalent diameters are large. For all boundary sets it follows that:

$$p_n(\text{upper}) = p_1(\text{upper}) \text{ for } n = 2, 3, 4, \ldots N \tag{4-15}$$

and

$$p_n(\text{lower}) = p_1(\text{lower}) \text{ for } n = 2, 3, 4, \ldots N \tag{4-16}$$

2. *Zero net inlet mass flow rate.* This situation, which physically represents an isolated reactor vessel, is of safety-related interest since it describes a situation in which the primary loop heat exchangers (i.e., steam generators for an LWR) are not available. For this case Eq. 4-2 becomes:

$$\sum_{n=1}^{N} G_{mn} A_n = 0 \tag{4-17}$$

B Plena Heat Transfer Boundary Conditions

For this geometry of plena connected by channels, an additional consideration is heat addition or extraction from the plena. A fundamental common approximation is that of fully mixed plena which obviates the need for an analysis of the velocity and temperature fields within the plena. For steady-state, fully mixed plena conditions, the general case is that of plena at unequal but constant temperatures. For this case, heated or cooled channels can be in upflow or downflow; however, plenum temperatures are maintained constant by heat extraction or addition from external reservoirs.

The relevant fully mixed plenum energy balances are written next for each plenum. The plena flow rates and heat addition rates $\dot{Q}$ are illustrated in Figure 4-3. Our sign convention is that heat addition is positive and upflow is positive. The heat addition rate $\dot{Q}$ in each plenum at steady state is equal to:

$$\dot{Q}(\text{upper}) = -\sum_{n=1}^{M} \dot{m}_n h_n(L) - \sum_{n=M+1}^{N} \dot{m}_n h_n(\text{upper}) + \dot{m}_T h_T(\text{upper}) \tag{4-18}$$

and

$$\dot{Q}(\text{lower}) = \sum_{n=1}^{M} \dot{m}_n h_n(\text{lower}) + \sum_{n=M+1}^{N} \dot{m}_n h_n(0) - \dot{m}_T h_T(\text{lower}) \tag{4-19}$$

Since:

$$\dot{m}_n(\text{downflow}) < 0 \quad \text{and} \quad \dot{m}_n(\text{upflow}) > 0$$

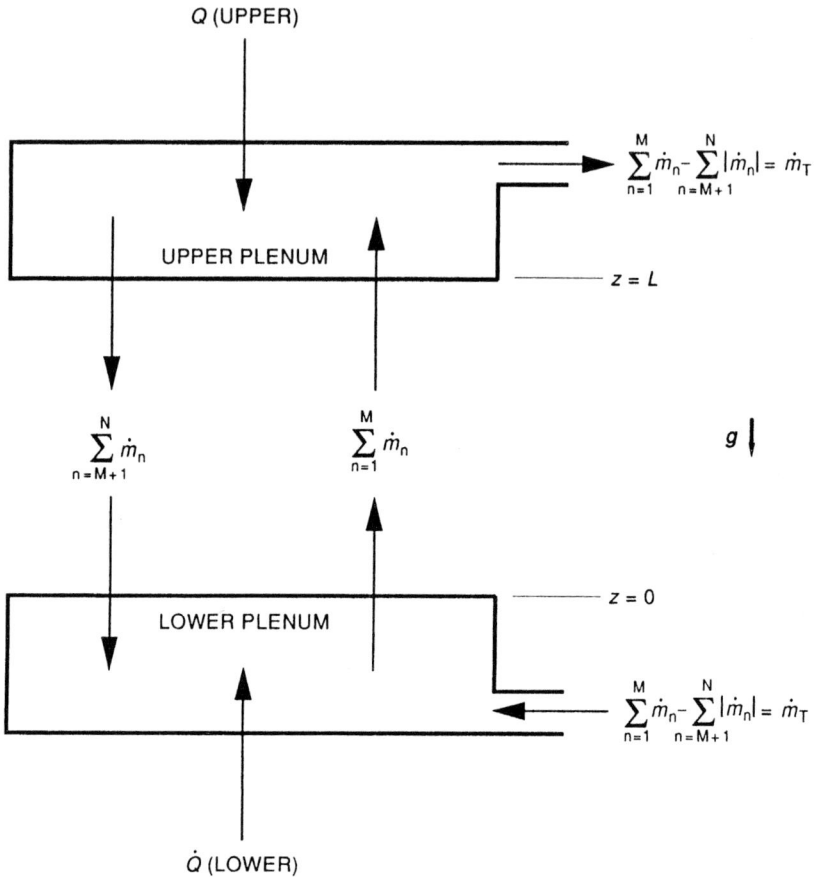

Figure 4-3 Plena mass and energy balances.

For positive $\dot{m}_T$, h_T(upper) equals the upper plenum enthalpy, h(upper), and h_T(lower) must be a prescribed inlet enthalpy; whereas for negative $\dot{m}_T$, h_T(lower) $= h$(lower) and h_T(upper) must be a prescribed inlet enthalpy.

For illustration, consider the special case of zero net flow between two plena connected by three otherwise noncommunicating channels (i.e., $\dot{m}_T = 0$). Let two channels be heated but with channel 2 at a lower rate than 1 and the third be cooled or adiabatic. In this case, illustrated in Figure 4-4, channel 1 will be in upflow, channel 3 in downflow, whereas channel 2 can flow in either direction. For steady-state conditions, heat must be extracted from or added to the plena to maintain the prescribed plena conditions of T(upper) and T(lower). Consider two cases for single-phase coolants: channel 2 in upflow and channel 2 in downflow.

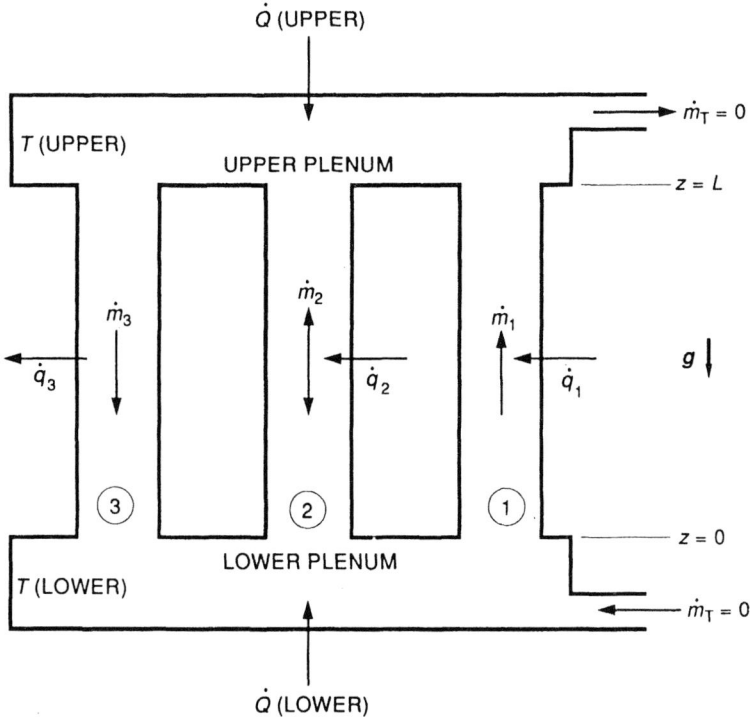

Figure 4-4 Flow in a three-channel array with zero net flow between the plena.

1 For channel 2 in upflow:

$$h_1(0) = h_2(0) = h(\text{lower}) \tag{4-20}$$

and

$$h_3(L) = h(\text{upper}) \tag{4-21}$$

Eq. 4-18 becomes:

$$\dot{Q}(\text{upper}) = -\dot{m}_1 h_1(L) - \dot{m}_2 h_2(L) - \dot{m}_3 h(\text{upper})$$

By continuity;

$$\dot{m}_1 + \dot{m}_2 + \dot{m}_3 = 0 \tag{4-22}$$

Hence:

$$\dot{Q}(\text{upper}) = \dot{m}_1[h(\text{upper}) - h_1(L)] + \dot{m}_2[h(\text{upper}) - h_2(L)] \tag{4-23}$$

Similarly, Eq. 4-19 becomes:

$$\dot{Q}(\text{lower}) = \dot{m}_1 h(\text{lower}) + \dot{m}_2 h(\text{lower}) + \dot{m}_3 h_3(0)$$
$$= -\dot{m}_3[h(\text{lower}) - h_3(0)] \tag{4-24}$$

Under the special case that T(upper) equals T(lower):

$$h(\text{upper}) = h(\text{lower}) \tag{4-25}$$

Consequently:

$$h_1(0) = h_2(0) = h(\text{upper})$$

and

$$h_3(L) = h(\text{lower})$$

Hence $\dot{Q}$(upper) from Eq. 4-23 becomes:

$$\dot{Q}(\text{upper}) = \dot{m}_1[h_1(0) - h_1(L)] + \dot{m}_2[h_2(0) - h_2(L)]$$
$$\dot{Q}(\text{upper}) = -\dot{q}_1 - \dot{q}_2 \tag{4-26}$$

and $\dot{Q}$(lower) from Eq. 4-24 becomes

$$\dot{Q}(\text{lower}) = -\dot{m}_3[h_3(L) - h_3(0)] = -\dot{q}_3 \tag{4-27}$$

2 For channel 2 in downflow

Analogously:

$$h_1(0) = h(\text{lower}) \tag{4-28}$$
$$h_2(L) = h_3(L) = h(\text{upper}) \tag{4-29}$$

Eq. 4-18 becomes:

$$\dot{Q}(\text{upper}) = -\dot{m}_1 h_1(L) - \dot{m}_2 h(\text{upper}) - \dot{m}_3 h(\text{upper})$$

By continuity again:

$$\dot{m}_1 + \dot{m}_2 + \dot{m}_3 = 0 \tag{4-22}$$

Hence:

$$\dot{Q}(\text{upper}) = \dot{m}_1[h(\text{upper}) - h_1(L)] \tag{4-30}$$

Similarly, Eq. 4-19 becomes:

$$\dot{Q}(\text{lower}) = \dot{m}_1 h(\text{lower}) + \dot{m}_2 h_2(0) + \dot{m}_3 h_3(0)$$
$$\dot{Q}(\text{lower}) = -\dot{m}_2[h(\text{lower}) - h_2(0)] - \dot{m}_3[h(\text{lower}) - h_3(0)] \tag{4-31}$$

Again for the special case that T(upper) equals T(lower), for single phase:

$$h_1(0) = h(\text{upper})$$

and

$$h_2(L) = h_3(L) + h(\text{lower})$$

Hence $\dot{Q}$(upper) from Eq. 4-30 becomes:

$$\dot{Q}(\text{upper}) = \dot{m}_1[h_1(0) - h_1(L)] = -\dot{q}_1 \tag{4-32}$$

and $\dot{Q}$(lower) from Eq. 4-31 becomes:

$$\dot{Q}(\text{lower}) = -\dot{m}_2[h_2(L) - h_2(0)] - \dot{m}_3[h_3(L) - h_3(0)]$$
$$\dot{Q}(\text{lower}) = -\dot{q}_2 - \dot{q}_3$$
(4-33)

Therefore, for this illustration case of zero net flow and equal plenum temperatures, external heat addition or extraction from the plena may be required as dictated by Eqs. 4-26 and 4-27 for channel 2 in upflow and by Eqs. 4-32 and 4-33 for channel 2 in downflow.

Now if there is no external plena heat addition or extraction, $\dot{Q}$(upper) $\equiv$ $\dot{Q}$(lower) $\equiv$ 0. Then for the case of zero net flow under steady-state conditions, the heat added in channels 1 and 2 must be extracted in channel 3. Hence channel 3 cannot be adiabatic. Additionally the plena temperatures cannot be maintained equal. This temperature inequality condition follows from the observation that for adiabatic plena the plenum temperatures are equal to the mass averaged temperatures of the flow streams entering the plena. Specifically for the case of channel 2 in upflow and for $\dot{Q}$(upper) and $\dot{Q}$(lower) equal to zero, Eqs. 4-23 and 4-24 become:

$$h(\text{upper}) = \frac{\dot{m}_1 h_1(L) + \dot{m}_2 h_2(L)}{\dot{m}_1 + \dot{m}_2}$$
(4-34)

$$h(\text{lower}) = h_3(0)$$
(4-35)

Now an energy balance in channel 3 by Eq. 4-6 yields:

$$h_3(0) = h_3(L) - \frac{1}{\dot{m}_3} \int_0^L q_3' dz$$
(4-6a)

By definition for downflow in channel 3, $h_3(L) \equiv h(\text{upper})$ so that combining the above results into Eq. 4-6a we obtain:

$$h(\text{lower}) = h(\text{upper}) - \frac{1}{\dot{m}_3} \int_0^L q_3' dz$$

Since q_3' is nonzero, the enthalpies and hence the temperatures of the plena are unequal in steady state.

V THE GENERAL SOLUTION PROCEDURE

The continuity equation, Eq. 4-1, indicates that G_{mn} and $\dot{m}_n$ are constant for each channel. In general, the density and the friction factor will vary along the channel. To account for these variations, we will use the axially averaged density and friction factor defined by:

$$\bar{\rho}_{mn} \equiv \frac{1}{L} \int_0^L \rho_{mn}(z) dz$$
(4-36)

$$\bar{f}_n \equiv \frac{1}{L} \int_0^L f_n(z)dz \qquad (4\text{-}37)$$

For a single-phase fluid, the axial variation of the friction factor is due to the changes of fluid viscosity along the channel, whereas for two-phase fluids, both the viscosity and two-phase multiplier, $\phi_{\ell o}^2$, will change along the length. Since the energy and momentum equations are integrated over the channel length, we can only state the fluid conditions entering and leaving the channel. We shall

Table 4-1 Coupled momentum–energy solution procedures for parallel channel arrays connected only at plena

	Boundary set (1)	Boundary set (2)	Boundary set (3)
Unknown variables			
G_{mn}, p_n(lower), p_n(upper)	$3N$	$3N$	
$\rho_{mn}(0)$, $\rho_{mn}(L)$, $\bar{\rho}_{mn}$	$3N$	$3N$	
$f_n(0)$, $f_n(L)$, $\bar{f}_n$	$3N$	$3N$	
$h_{mn}(0)$, $h_{mn}(L)$	$2N$	$2N$	
Q(lower), Q(upper)	2	2	
$\dot{m}_T$	$-$?	1	
	$11N + 2$	$11N + 3$	
Governing equations			
Conservation equations			
• Momentum, Eq. 4-4	N	N	
• Energy, Eq. 4-7	N	N	
Constraint equation			
• Total flow, Eq. 4-2	$-$?	1	
State eqs.: Density as a function of pressure and enthalpy evaluated at:			
• $\rho_{mn}(0)$, $\rho_{mn}(L)$, Eq. 4-12	$2N$	$2N$	
• $\bar{\rho}_{mn}$, Eq. 4-38	N	N	
Constitutive equations			
• $f_n(0)$, $f_n(L)$, Eq. I,9-13 or Eq. I,9-20	$2N$	$2N$	
• $\bar{f}_n$, Eq. 4-39	N	N	
	$8N$	$8N + 1$	
Boundary conditions			
p_n(upper)	N	N	$-$
p_n(lower)	N	$-$	N
$h_{mn}(0)$ or $h_{mn}(L)$	N	N	N
$\dot{m}_T$	$-$	1	1
p_n(upper) $= f(p_1$(upper)), Eq. 4-13	$-$	$-$	$N - 1$
p_n(lower) $= f(p_1$(lower)), Eq. 4-14	$-$	$N - 1$	$-$
Q(upper), Eq. 4-18	1	1	1
Q(upper), Eq. 4-19	1	1	1
	$3N + 2$	$3N + 2$	$3N + 2$

$-$ = not applicable.

thus approximate the average density and friction factor for single-phase fluid by:

$$\bar{\rho}_{mn} \simeq \frac{\rho_{mn}(0) + \rho_{mn}(L)}{2} \tag{4-38}$$

$$\bar{f}_{mn} \simeq \frac{f_n(0) + f_n(L)}{2} \tag{4-39}$$

The solution procedures for boundary sets (1), (2), and (3) differ. These three cases have been specified as:

Set (1). Prescribed inlet and exit pressure levels.
Set (2). Upper plenum pressure level/lower plenum pressure gradient and $\dot{m}_T$.
Set (3). Lower plenum pressure level/upper plenum pressure gradient and $\dot{m}_T$.

Table 4-1 illustrates the governing equations and boundary conditions necessary to balance the unknown variables for each case. In each case the total heat input rate for each channel, $\dot{q}_n$, is a given input condition. As Table 4-1 illustrates, the number of equations and boundary conditions necessary for closure is $11N + 2$ for boundary set (1) and $11N + 3$ for boundary sets (2) and (3).

These sets of equations are coupled through the dependence of the density ρ_m on both the energy equation through the enthalpy h and the momentum equation through the pressure p. The solution procedure can be simplified by use of a reference pressure p^* that eliminates the dependence of ρ_m on local variations of ρ throughout the system but does not affect the dependence of ρ_m on h, i.e.:

$$\rho_m(z) = \rho_m(h_m(z), p^*) \tag{2-14}$$

This approach is an approximation that is valid as long as the density variation with pressure is small and local pressures are close to the reference pressure p^*. Although the use of a reference pressure p^* does simplify the solution procedure by reducing the effort expended in evaluating $\rho_m(z)$, the equations are still coupled and in general require numerical solution techniques.

VI CHANNEL HYDRAULIC CHARACTERISTICS

The general solution procedure presented in Section V in principle is sufficient for problem solutions. However, the momentum equation in particular leads to complexities because the channel pressure drop/flow rate behavior when extended over the entire flow rate range is not a simple linearly increasing function. In fact this characteristic curve has a shape that can lead to multiple solutions and instabilities. Familiarity with these characteristics provides the physical basis essential to understanding and analyzing complex system behavior.

We will discuss the channel pressure drop/flow rate characteristic in two stages to emphasize the friction-dominated versus the gravity-dominated flow regimes. For this section, we will assume that the channel power, flow rate, and inlet subcooling are such that two-phase conditions are encountered in the friction-dominated regime as well as in the gravity-dominated regime. For generality, conditions of upflow and downflow will be included. In Section VII a single-phase example is considered in which single-phase conditions exist well within the gravity-dominated regime.

A The Friction-Dominated Regime

The S-shaped pressure drop/flow rate characteristic for the friction-dominated regime has been described in Section 13.VB of Volume I.

For multiple, identical channels operating at different heat fluxes, the S-curve takes different shapes. In general, in the friction-dominated regime, an increase in $\dot{q}$ for identical channel geometry and constant inlet subcooling causes the S-curve to change as shown in Figure 4-5. The onset of nucleation and the achievement of complete vaporization, *points B and C*, respectively, occur at higher $\dot{m}$ for increased $\dot{q}$ due to simple energy balance considerations. Additionally, for increased $\dot{q}$, the overall pressure drop for fixed $\dot{m}$ is higher in both the liquid and vapor regions due to property variations with the temperature. These property variations influence the pressure drop as given below:

$$\Delta p \text{ is proportional to } \left[\frac{\mu^n}{\rho}\right] \text{ for fixed } \dot{m} \qquad (4\text{-}40)$$

For liquid water, both μ and ρ decrease approximately equally with temperature increase, but the effect of ρ dominates because the exponent n is only

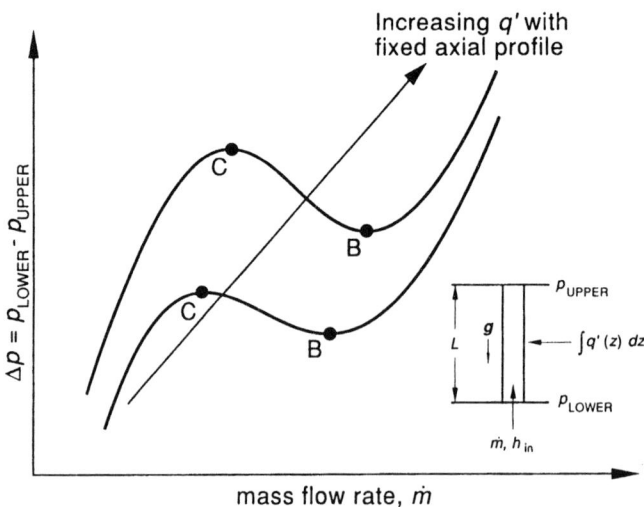

Figure 4-5 The effect of channel power on the channel hydraulic characteristic.

about 0.2. For the vapor region, μ increases while ρ decreases as temperature increases. Hence in both regions at a fixed $\dot{m}$, Δp increases as $\dot{q}$ increases as shown in Figure 4-5. From this equation one can appreciate that an array of channels of arbitrary geometry and heat flux can have a variety of possible specific S-curves. For a fixed boundary Δp_{ex}, a variety of flow rate distributions will exist among the channels in the array.

Extending consideration to the downflow case, a mirror image S-shaped curve results as shown in Figure 4-6 which illustrates the friction-dominated regime. A consequence of the curve of Figure 4-6 for the friction-dominated region is that for a fixed boundary condition, Δp_{ex}, only all positive or all negative flows are predicted in a multiple channel array. No situation exists for the friction-dominated region in which the flow is upward in some channels and downward in others. As will be shown, this is not true in the gravity-dominated regime.

B The Gravity-Dominated Regime

The pressure drop/flow rate characteristic for the gravity-dominated regime is governed by Eq. 4-41, which is the gravity term of the general momentum

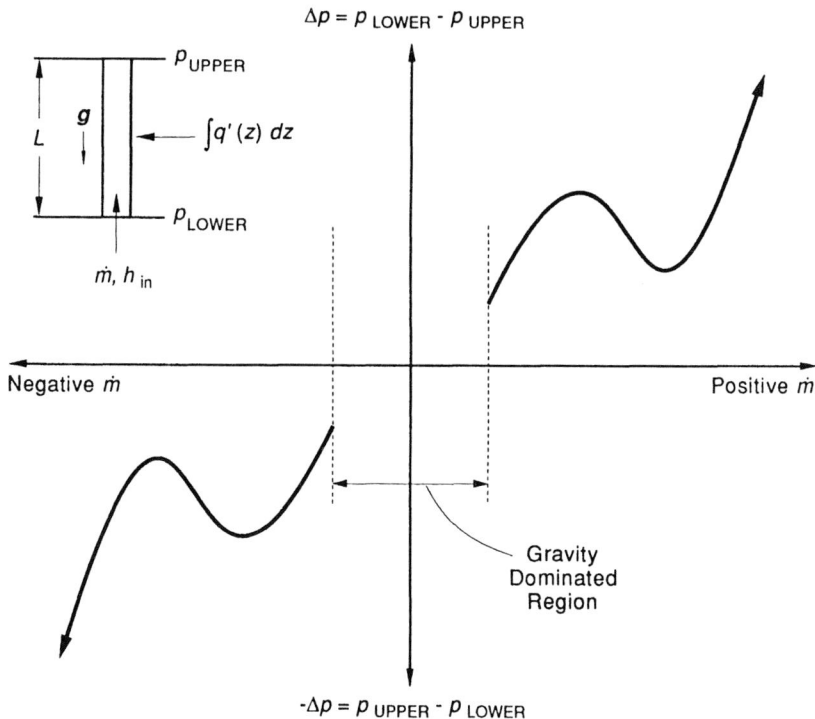

Figure 4-6 The friction-dominated region for upflow and downflow in a heated channel.

balance of Eq. 4-4 expressed in terms of a two-phase density:

$$\Delta p_{\mathrm{n}} = \int_0^L [\rho_\ell(1 - \alpha) + \rho_v\alpha)gdz \tag{4-41}$$

At low flow rates, the velocity levels become low enough so that friction and acceleration losses become negligible compared with gravity effects. Hence this is the gravity-dominated regime. The method of computing a characteristic curve has been demonstrated in Section 13.V of Volume I. Utilizing this procedure, which includes both friction and gravity effects for three heat flux conditions, yields the results presented in Figures 4-7 and 4-8. Figure 4-8 illustrates the crossover behavior in the gravity-dominated regime of very low mass flow rates in which pressure drop is inversely proportional to heat flux.

Let us focus now around the zero flow rate location. For positive (upward) mass flow rates, both the liquid and gas phases flow upward. The pressure drop is given mainly by the gravity head as represented by Eq. 4-41 where α is determined by both the mass flow rate and the channel power. The zero flow rate condition is a discontinuity in the curve since the heated channel with zero flow rate would not exhibit a fixed Δp. Any initial liquid present would vaporize, and the vapor would rise upon heating, thus creating an upflow. Also, as the zero flow rate condition is approached, the fluid condition depends on the channel geometric characteristics and the channel power. Figure 4-9 illustrates the region around zero flow rate just described.

For negative mass flow rate or liquid downflow, Figure 4-9 illustrates an initial decrease in pressure drop. In this region the vapor and liquid are in

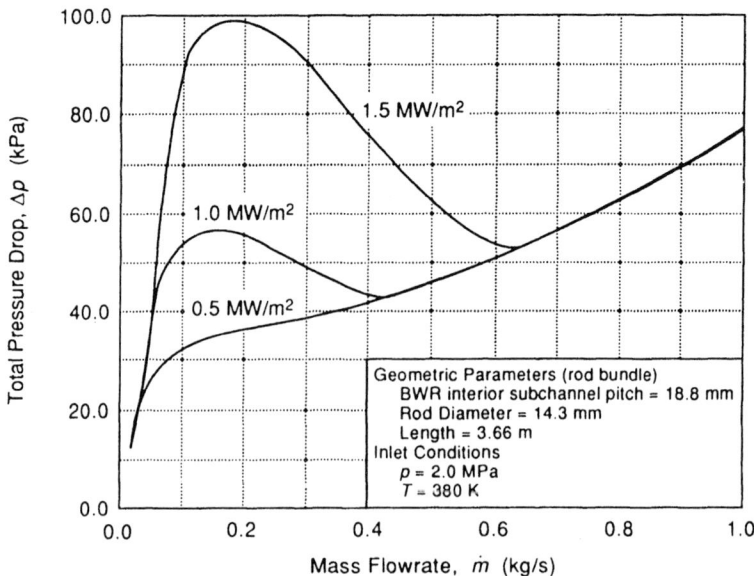

Figure 4-7 The effect of channel power on the characteristic curves: Full range of mass flow rate.

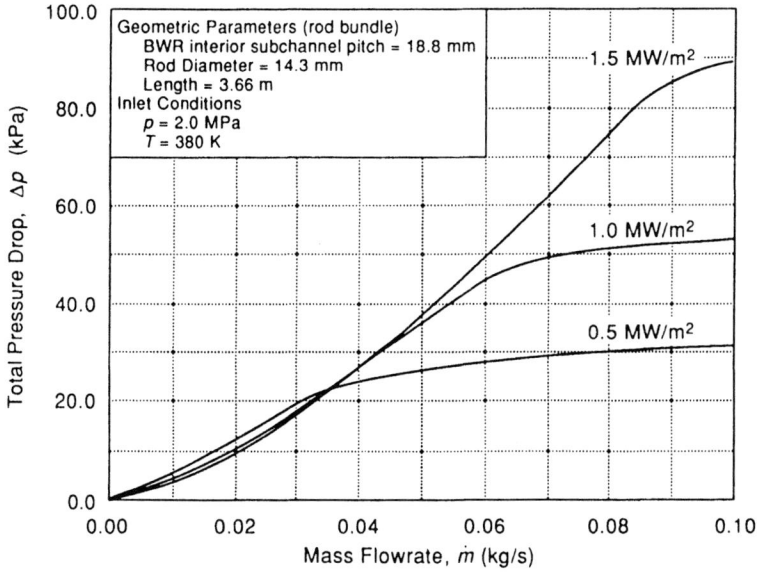

Figure 4-8 The effect of channel power on the characteristic curves: Low mass flow rate domain.

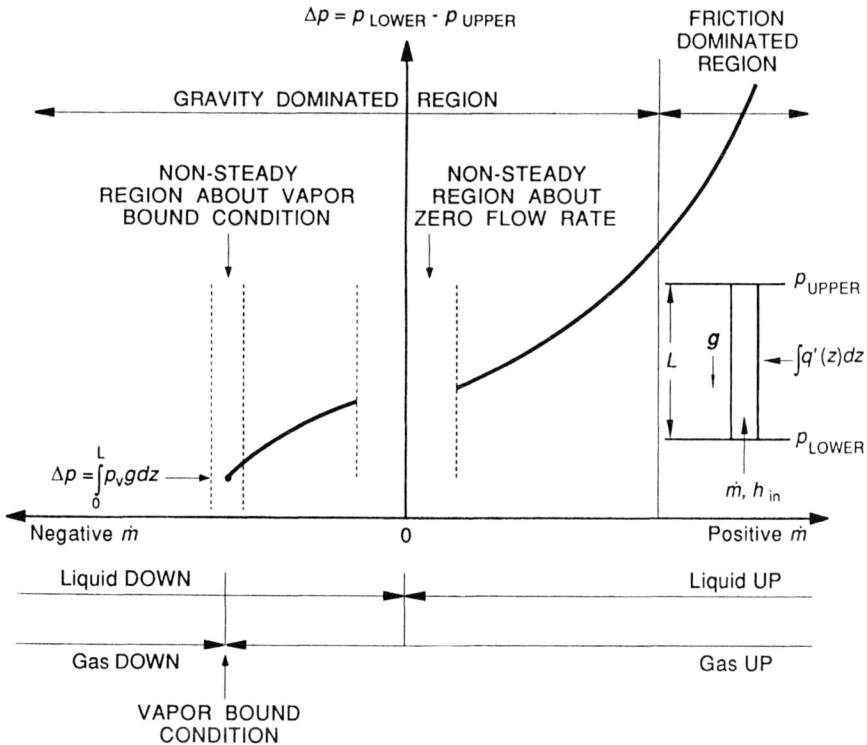

Figure 4-9 The gravity-dominated region for upflow and downflow in a heated channel.

counterflow. For the case of constant channel power, increased liquid down-flow will slightly reduce net vapor generation but will also retard the rising vapor produced. The retarding effect is dominant, yielding an increase in the void fraction. Hence the gravity-dominated channel pressure drop given by Eq. 4-41 initially decreases since α increases. With increasing liquid flow rate this trend first leads to a stagnation of the generated vapor and filling of the channel with vapor. This condition is not a definable steady state since the channel will first fill with vapor and then probably the vapor will be expelled. For an all-vapor condition the Δp is not zero and is equal to:

$$\Delta p_n = \int_0^L \rho_v g dz \qquad (4\text{-}42)$$

Figure 4-9 shows this all-vapor Δp value at the point labeled the *vapor-bound condition*.

For further increasing negative mass flow rates, the vapor flows concurrently downward with the liquid. The void fraction trend reverses to a decreasing α, and the pressure drop increases as dictated by Eq. 4-41. The maximum pressure drop in the gravity-controlled region is the all-liquid channel condition. For the heated channel this condition is generally not reached in the gravity-dominated regime because the channel power will create some vaporization. Further, the increase in pressure drop is bounded by the onset of a friction drop component which, for negative flow, is opposite to the gravity component. This balance between gravity and friction components leads to a maximum in the pressure drop/flow rate characteristic. For further negative flow rate increases, friction dominates, and the Δp decreases. This behavior is shown in Figure 4-10, which illustrates performance over the full range of mass flow rates. Characteristics of this type have been determined experimentally by Singh and Griffith [3] who found that slight variations from this idealized behavior occur about the flooding point due to bubble trapping at sharp entrance corners unless the entrance is streamlined. These authors worked in terms of a superficial liquid velocity, j_f, as Figure 4-10 reflects, in place of $\dot{m}$, but these parameters are almost proportional since:

$$\dot{m} \equiv (\rho V A)_f + (\rho V A)_g = (j_f \rho_f + j_g \rho_g) A_T \qquad (4\text{-}43)$$

$$= j_f \rho_f A_T \left(1 + \frac{\rho_g}{\rho_f} S \frac{\alpha}{1 - \alpha} \right) \qquad (4\text{-}44)$$

where:

$$j_f \equiv V_f (1 - \alpha) \qquad (4\text{-}45)$$

and S is the slip ratio, the gas-to-liquid velocity ratio. Note that typical parameters for low void, high pressure systems, where $\alpha \le 0.5$, $\rho_g/\rho_f \le 1/20$ and $S \le 2$:

$$\frac{\rho_g}{\rho_f} S \frac{\alpha}{1 - \alpha} \le 0.1$$

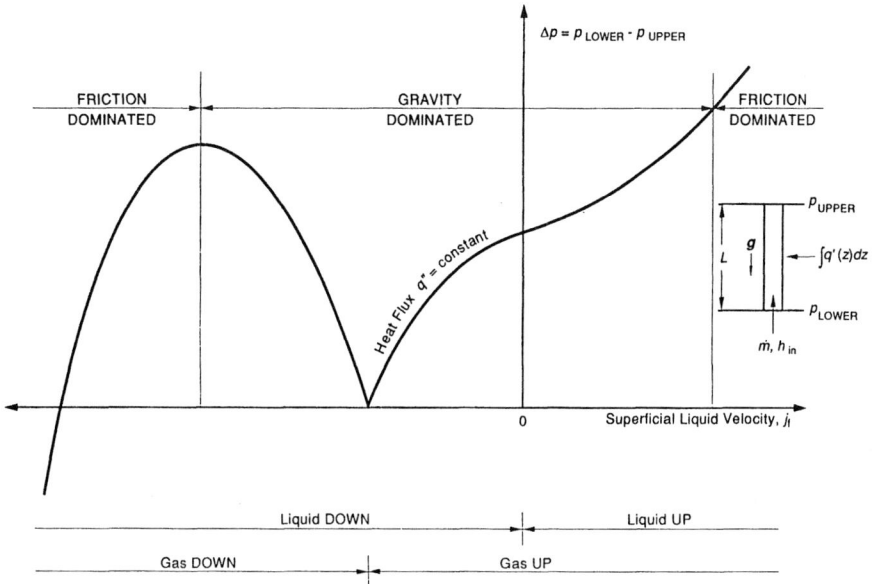

Figure 4-10 Pressure drop versus flow rate curve schematic. *(After Singh and Griffith [3].)*

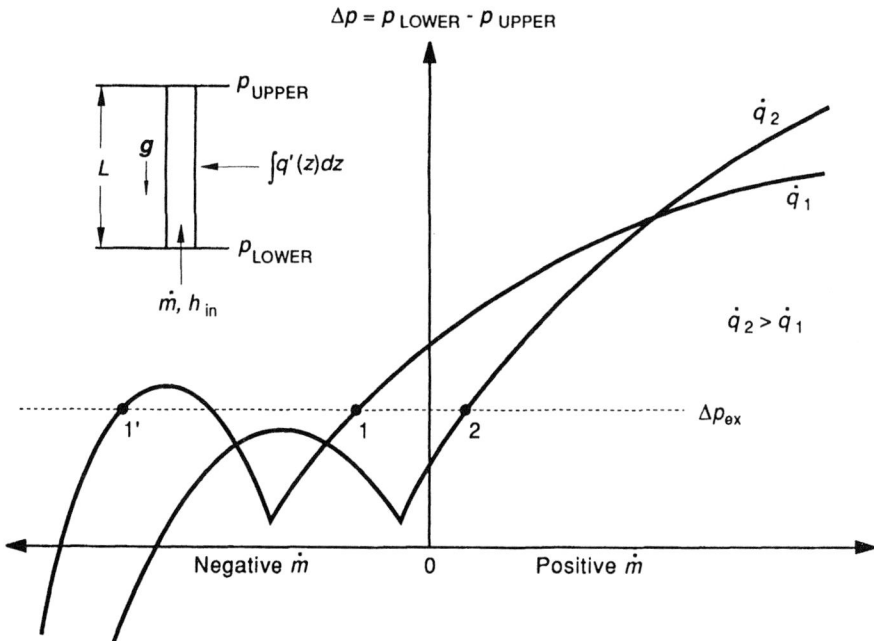

Figure 4-11 Flow in multiple channels in the gravity-dominated region.

and hence Eq. 4-44 becomes:

$$\dot{m} \approx j_f \rho_f A_T \qquad (4\text{-}46)$$

For multiple, identical channels operating at different channel powers, different individual curves of the shape of Figure 4-10 exist. The pressure drop across the array is almost independent of the behavior of any one channel, and thus each channel can be thought of as having a constant Δp_{ex} boundary condition as shown in Figure 4-11. It is possible to have upflow and downflow simultaneously in different channels. This is illustrated schematically in Figure 4-11 in which downflow occurs in channel 1 (*intersections 1* and *1'*) but upflow only in channel 2 (*intersection 2*).

Let us now turn to analytical solutions of the general equation set. We shall solve two single-phase examples. A coupled conservation equation solution covering the gravity-dominated regime is covered in Section VII. A decoupled solution of the momentum equation covering the friction-dominated regime is covered in Section VIII.

VII COUPLED CONSERVATION EQUATION: SINGLE-PHASE, NONDIMENSIONAL SOLUTION PROCEDURE

As stated in Section V the momentum and energy equations are in general coupled since the density ρ_m, which is a function of the enthalpy h_m and pressure p, appears in the momentum equation. If the dependence of ρ_m on h_m and p is not strong, then an average reference density evaluated at a reference enthalpy and pressure can be used in the momentum equation to solve for the channel flow rate. This flow rate can then be substituted into the energy equation. This is the procedure used in Section VIII.

There are situations in which the local dependence of ρ_m on h_m and p is important (for example, if the gravity pressure drop is significant or if two-phase conditions exist). It then becomes necessary to alter our approach. Instead, we will use the reference pressure approach described in Section V in which the fluid is allowed to expand thermally but is considered to be incompressible, i.e.:

$$\left(\frac{\partial \rho_m}{\partial p}\right)_{h_m} = 0$$

In this section, the general solution procedure for the coupled energy and momentum equations for the single-phase case will be discussed following [1,2]. Included will be a nondimensional analysis of the integrated momentum and energy equations and examples illustrating the implications of some of the channel hydraulic characteristics discussed in Section VI.

A Derivation of a Single, Coupled Momentum–Energy Equation

In order to simplify the general solution procedure, the following assumptions are made:

1. The plena are fully mixed. Thus, the analysis of temperature and velocity fields in the plena is not necessary.
2. No radial (transverse) pressure gradient exists in either plenum. The pressure drops of all the channels are therefore equal, i.e., $\Delta p_n = \Delta p$.
3. Fluid viscosity, specific heat, and thermal expansivity are independent of temperature and pressure.
4. Fluid density is independent of temperature and pressure except for the gravity term in the momentum equation, in which density is a linear function of temperature (the Boussinesq hypothesis).
5. Channels are all in either laminar or turbulent flow and of similar geometry so that the friction factor dependence on Reynolds number is the same for all channels, i.e., $f = C \mathrm{Re}^{-n}$.
6. Spatial acceleration is neglected. This is actually stated by assumption 4 since spatial acceleration is caused by density variations along the channel.

We start our derivation by applying assumptions 2, 3, 4, and 6 to momentum Eq. 4-4 yielding:

$$\Delta p = \int_0^L \rho_n g\, dz + f_n \frac{L}{D_{en}} \frac{G_n|G_n|}{2\rho^*} + \sum_i \frac{K_{in} G_n|G_n|}{2\rho^*} \qquad (4\text{-}47)$$

where the reference density ρ^* is used in the friction and local loss terms. The energy equations, Eqs. 4-5 and 4-6, when written in terms of temperature, become:

$$T_n(z) - T(\text{lower}) = \int_0^z \frac{q_n' dz}{G_n A_n c_p} \qquad (\text{upflow}) \qquad (4\text{-}48)$$

$$T_n(z) - T(\text{upper}) = -\int_z^L \frac{q_n' dz}{G_n A_n c_p} \qquad (\text{downflow}) \qquad (4\text{-}49)$$

Because a reference density is used in the friction and local loss terms in Eq. 4-47, the momentum and energy equations are now only coupled through the gravity (integral) term of Eq. 4-47. By assumption 4, we can write this term as:

$$\int_0^L \rho_n g\, dz = \int_0^L \rho^* g(1 - \beta(T_n - T^*))\, dz \qquad (4\text{-}50)$$

Let us first examine the upflow case. Using Eq. 4-48 to substitute for T_n in Eq. 4-50 yields:

$$\int_0^L \rho_n g\, dz = \int_0^L \rho^* g \left[1 - \beta \left(\int_0^z \frac{q_n' dz'}{G_n A_n c_p} + T(\text{lower}) - T^* \right) \right] dz \quad (\text{upflow}) \quad (4\text{-}51)$$

Expanding Eq. 4-51 we obtain:

$$\int_0^L \rho_n g dz = \rho^* g \int_0^L dz - \frac{\rho^* g \beta}{G_n A_n c_p} \int_0^L \int_0^z q_n' dz' dz$$

$$- \rho^* g \beta [T(\text{lower}) - T^*] \int_0^L dz \quad (\text{upflow}) \tag{4-52}$$

Combining the first and third terms on the right side gives:

$$\int_0^L \rho_n g dz = \rho^* g L (1 - \beta[T(\text{lower}) - T^*)]$$

$$- \frac{\rho^* g \beta}{G_n A_n c_p} \int_0^L \int_0^z q_n' dz' dz \quad (\text{upflow}) \tag{4-53}$$

We can simplify Eq. 4-53 by defining:

$$\dot{q}_n = \int_0^L q_n' dz \tag{4-54}$$

and

$$S_n \equiv \frac{\int_0^L \int_0^z q_n' dz' dz}{\dot{q}_n L} \tag{4-55}$$

so that Eq. 4-53 becomes:

$$\int_0^L \rho_n g dz = \rho^* g L \left[1 - \beta[T(\text{lower}) - T^*] - \frac{\dot{q}_n \beta S_n}{G_n A_n c_p} \right] \quad (\text{upflow}) \tag{4-56}$$

S_n is a dimensionless quantity that reflects the symmetry of the axial heat flux shape of the channel. For the case of uniform heat flux, Eq. 4-55 simplifies to:

$$S_n = \frac{\int_0^L q' z dz}{\dot{q}_n L} = 0.5 \tag{4-57}$$

This represents the case in which the average fluid density in the channel is the arithmetic average of the fluid entering and leaving the channel. For a given total channel power $\dot{q}_n$ as S_n decreases, the average fluid density in the channel increases as can be seen from Eq. 4-56.

Our next step is to write the single momentum–energy equation. Substituting the result for the gravity term into the momentum equation, Eq. 4-47, we get:

$$\Delta p = \rho^* g L \left[1 - \beta[T(\text{lower}) - T^*] - \frac{\dot{q}_n \beta S_n}{G_n A_n c_p} \right] + f_n \frac{L}{D_{en}} \frac{G_n |G_n|}{2\rho^*}$$

$$+ \sum_i \frac{K_{in} G_n |G_n|}{2\rho^*} \quad (\text{upflow}) \tag{4-58}$$

Similarly, for downflow:

$$\Delta p = \rho^* g L \left[1 - \beta[T(\text{upper}) - T^*] + \frac{\dot{q}_n \beta(1 - S_n)}{G_n A_n c_p} \right] + f_n \frac{L}{D_{en}} \frac{G_n|G_n|}{2\rho^*}$$

$$+ \sum_i \frac{K_{in} G_n |G_n|}{2\rho^*} \; (\text{downflow}) \tag{4-59}$$

Eqs. 4-58 and 4-59 are the desired combined momentum–energy equations that describe the system of N parallel channels connected only at plena. Note that separate equations are written for upflow and downflow since the gravity components have different temperatures, i.e., $T(\text{lower})$ versus $T(\text{upper})$ even though the mass velocity terms are written as $G_n|G_n|$ consistent with the convention that G_n is positive for upflow and negative for downflow.

Example 4-1 Calculation of the channel power shape factor

PROBLEM A channel of length L has a linear heat generation rate of:

$$q' = q_0' \sin \frac{\pi z}{L}$$

Calculate S_n for this case.

SOLUTION S_n is defined by Eq. 4-55. For this case:

$$S_n = \frac{\int_0^L \int_0^z q_0' \sin\left(\frac{\pi z'}{L}\right) dz' dz}{L \int_0^L q_0' \sin\left(\frac{\pi z}{L}\right) dz} = \frac{-q_0' \frac{L}{\pi} \int_0^L \left(\cos\left(\frac{\pi z}{L}\right) - \cos(0)\right) dz}{-q_0' \frac{L^2}{\pi} (\cos(\pi) - \cos(0))}$$

$$= \frac{q_0' \frac{L^2}{\pi}}{2 q_0' \frac{L^2}{\pi}} = 0.5$$

Thus, a channel with a sine-shaped axial power distribution has the same value of S_n as one with uniform heat flux. In general this is true for any channel power distribution that is symmetrical about the axial midpoint. For these cases, the average density in the channel can be calculated by the arithmetic average of the densities entering and leaving the channel.

B Nondimensional Equations

We start the dimensionless analysis by dividing the combined momentum–energy equations, Eqs. 4-58 and 4-59, by $\rho^* g L$ and subtracting unity from each to obtain:

$$\frac{\Delta p}{\rho^* g L} - 1 = -\beta[T(\text{lower}) - T^*] - \frac{\dot{q}_n \beta S_n}{G_n A_n c_p} + f_n \frac{L}{D_{en}} \frac{G_n|G_n|}{2\rho^{*2} g L}$$

$$+ \sum_i \frac{K_{in} G_n|G_n|}{2\rho^{*2} g L} \quad (\text{upflow}) \tag{4-60}$$

$$\frac{\Delta p}{\rho^* g L} - 1 = -\beta[T(\text{upper}) - T^*] + \frac{\dot{q}_n \beta(1 - S_n)}{G_n A_n c_p} + f_n \frac{L}{D_{en}} \frac{G_n|G_n|}{2\rho^{*2} g L}$$

$$+ \sum_i \frac{K_{in} G_n|G_n|}{2\rho^{*2} g L} \quad (\text{downflow}) \tag{4-61}$$

We now introduce the following dimensionless variables:

$$\text{Re}_n = \frac{G_n D_{en}}{\mu}$$

$$\Delta p^+ = \frac{\Delta p}{\rho^* g L} - 1$$

$$X_U = \beta[T(\text{upper}) - T^*]$$

$$X_L = \beta[T(\text{lower}) - T^*]$$

$$\Theta_n = \frac{\dot{q}_n \beta D_{en}}{A_n c_p \mu}$$

$$f_n = C|\text{Re}_n|^{-n}$$

$$\delta_n = \frac{C\mu^2}{2D_{en}^3 \rho^{*2} g}$$

$$\gamma_n = \frac{\sum_i K_{in} \mu^2}{2D_{en}^2 L \rho^{*2} g}$$

The new dimensionless momentum–energy equations can now be written† as:

$$\Delta p^+ = \left[-X_L - \frac{S_n \Theta_n}{\text{Re}_n} \right] + \delta_n \text{Re}_n^{2-n} + \gamma_n \text{Re}_n^2 \quad (\text{upflow}) \tag{4-62}$$

$$\Delta p^+ = \left[-X_U + \frac{(1 - S_n)\Theta_n}{\text{Re}_n} \right] + \delta_n \text{Re}_n |\text{Re}_n|^{1-n}$$

$$+ \gamma_n \text{Re}_n |\text{Re}_n| \quad (\text{downflow}) \tag{4-63}$$

† The consistency in nomenclature maintained throughout the book requires using the letter n as both an exponent (the Reynolds number exponent associated with the friction factor) and a subscript (channel number designation) throughout Sections VII and VIII.

The first term on the right side (in brackets) is the dimensionless gravity pressure drop. X_L and X_U account for differences in the reference density (which can be defined at any temperature) from the densities in the plena. The second term in the brackets accounts for the density change caused by heating (or cooling) the fluid. It is proportional to $\dot{q}_n$ (through Θ_n) and inversely proportional to G_n (through Re_n). The second and third terms on the right side represent the friction and local losses, respectively, where S_n and γ_n are functions of fluid properties and the channel geometry.

The next step is to express the mass constraint equation in terms of the channel Reynolds number as defined below and to non-dimensionalize this constraint equation.

$$\mathrm{Re}_n = \frac{G_n D_{en}}{\mu}$$

We start by defining the total flow area, A_T, and hydraulic diameter, D_{eT}, of our system of N channels as:

$$A_T = \sum_{n=1}^{N} A_n \tag{4-64}$$

$$D_{eT} = \frac{4 \sum_{n=1}^{N} A_n}{\sum_{n=1}^{N} P_{wn}} \neq 4 \sum_{n=1}^{N} \frac{A_n}{P_{wn}} \tag{4-65}$$

where P_{wn} is the wetted perimeter of channel n. Note that D_{eT} is not equal to the sum of the individual channel hydraulic diameters.

Per our notation, channels 1 through M are in upflow, and $(M + 1)$ through N are in downflow. Substituting the channel Reynolds number into the mass constraint Eq. 4-2 yields:

$$\sum_{n=1}^{M} \frac{\mathrm{Re}_n \mu}{D_{en}} A_n + \sum_{n=M+1}^{N} \frac{\mathrm{Re}_n \mu}{D_{en}} A_n - \dot{m}_T = 0 \tag{4-66}$$

Dividing through by $\mu A_T / D_{eT}$ results in:

$$\sum_{n=1}^{M} \mathrm{Re}_n \left(\frac{A_n D_{eT}}{A_T D_{en}} \right) + \sum_{n=M+1}^{N} \mathrm{Re}_n \left(\frac{A_n D_{eT}}{A_T D_{en}} \right) - \dot{m}_T \frac{D_{eT}}{\mu A_T} = 0 \tag{4-67}$$

Defining the dimensionless parameters:

$$r_n = \frac{A_n D_{eT}}{A_T D_{en}} \tag{4-68}$$

and

$$\mathrm{Re}_T = \frac{\dot{m}_T D_{eT}}{\mu A_T} \tag{4-69}$$

we can write Eq. 4-67 as:

$$\sum_{n=1}^{M} \text{Re}_n r_n + \sum_{n=M+1}^{N} \text{Re}_n r_n - \text{Re}_T = 0 \tag{4-70}$$

Eqs. 4-62, 4-63, and 4-70 comprise a set of $(N + 1)$ equations. The $2N + 1$ unknowns are Δp^+, Re_n, and Θ_n. For closure, either Re_n or Θ_n must be specified. For nuclear reactor system applications, typically Θ_n (heat flux) is specified. Additionally, r_n, Re_T, S_n, X_L, X_U, δ_n, and γ_n are specified.

Example 4-2 Flow characteristics of a three-channel system

PROBLEM A water test section consists of three identical channels connected only at the upper and lower plena. Channel 1 is heated, channel 2 is adiabatic, and channel 3 is cooled where $\dot{q}_1 = 500$ J/s, $\dot{q}_2 = 0$, $\dot{q}_3 = -200$ J/s. The relevant channel parameters are:

$D_e = 6.350 \times 10^{-3}$ m

$L = 2$ m

$A = 1.58 \times 10^{-4}$ m^2

$S_1 = S_3 = 0.5780$

$\sum_i K_i = 0$ (i.e., no form losses)

$C = 95$

$n = 1$ (i.e., laminar flow).

The water properties for the test are:

$T(\text{upper}) = 20\ °\text{C}$

$T(\text{lower}) = 20\ °\text{C}$

$T^* = 25\ °\text{C}$

$\rho^* = 993.1$ kg/m^3

$c_p = 4186$ J/kg °C

$p^* = 1.01325 \times 10^5$ Pa

$\mu = 8.62 \times 10^{-4}$ Pa-s

$\beta = 4.75 \times 10^{-4}/°\text{C}$

Calculate and plot Δp^+ versus Re_n for the three channels on the same graph for $-1{,}000 < \text{Re}_n < 1{,}000$, where $\text{Re}_n < 0$ indicates downflow. What happens as Re_n approaches 0?

SOLUTION By definition:

$$X_U = \beta(T(\text{upper}) - T^*) = (4.75 \times 10^{-4})(20 - 25) = -2.375 \times 10^{-3}$$

$$X_L = \beta(T(\text{lower}) - T^*) = (4.75 \times 10^{-4})(20 - 25) = -2.375 \times 10^{-3}$$

$$\delta = \frac{C\mu^2}{2D_e^3\rho^{*2}g} = \frac{(95)(8.62 \times 10^{-4})^2}{(2)(6.35 \times 10^{-3})^3(993.1)^2(9.8)} = 1.426 \times 10^{-5}$$

$$\gamma = \frac{\sum^i K_i\mu^2}{2D_e^2L\rho^{*2}g} = 0$$

$$\Theta_n = \frac{\dot{q}_n\beta D_e}{Ac_p\mu}$$

$$\Theta_1 = \frac{(500)(4.75 \times 10^{-4})(6.35 \times 10^{-3})}{(1.58 \times 10^{-4})(4,186)(8.62 \times 10^{-4})} = 2.645$$

$$\Theta_2 = 0$$

$$\Theta_3 = \frac{(-200)(4.75 \times 10^{-4})(6.35 \times 10^{-3})}{(1.58 \times 10^{-4})(4,186)(8.62 \times 10^{-4})} = -1.058$$

The Δp^+ versus Re_n curve for each channel is given by Eqs. 4-62 and 4-63:

$$\Delta p^+ = -X_L - \frac{S_n\Theta_n}{\text{Re}_n} + \delta\text{Re}_n^{2-n} + \gamma\text{Re}_n^2 \quad \text{(upflow)} \qquad (4\text{-}62)$$

$$\Delta p^+ = -X_U + \frac{(1 - S_n)\Theta_n}{\text{Re}_n} + \delta\text{Re}_n|\text{Re}_n|^{1-n}$$
$$+ \gamma\text{Re}_n|\text{Re}_n| \quad \text{(downflow)} \qquad (4\text{-}63)$$

Substituting the values calculated above, we obtain:

$$\Delta p^+ = 2.375 \times 10^{-3} - 0.578\frac{\Theta_n}{\text{Re}_n} + 1.426 \times 10^{-5}\text{Re}_n \quad \text{(upflow)}$$

$$\Delta p^+ = 2.375 \times 10^{-3} + 0.422\frac{\Theta_n}{\text{Re}_n} + 1.426 \times 10^{-5}\text{Re}_n \quad \text{(downflow)}$$

For Re_n ranging from 10^3 in downflow to 10^3 in upflow the resulting Δp^+ − Re plots for each channel are shown in Figure 4-12.

We see that for a prescribed Δp^+, there can be more than one combination of flow rates which satisfies the momentum–energy equation. Also, note that the curves for heated and cooled channels are not defined for a very low Reynolds number. This is because our linear relationship between density and temperature is not true in this region. Integrating Eqs. 4-48 and 4-49 we find $T_n(L)$ as a function of Re_n, and as can be seen in Figure 4-13 for the upflow condition, channel 1 will boil when Re = 70 corresponding to $\Delta p^+ = -18.0 \times 10^{-3}$, and channel 3 will freeze at Re = 111 corresponding

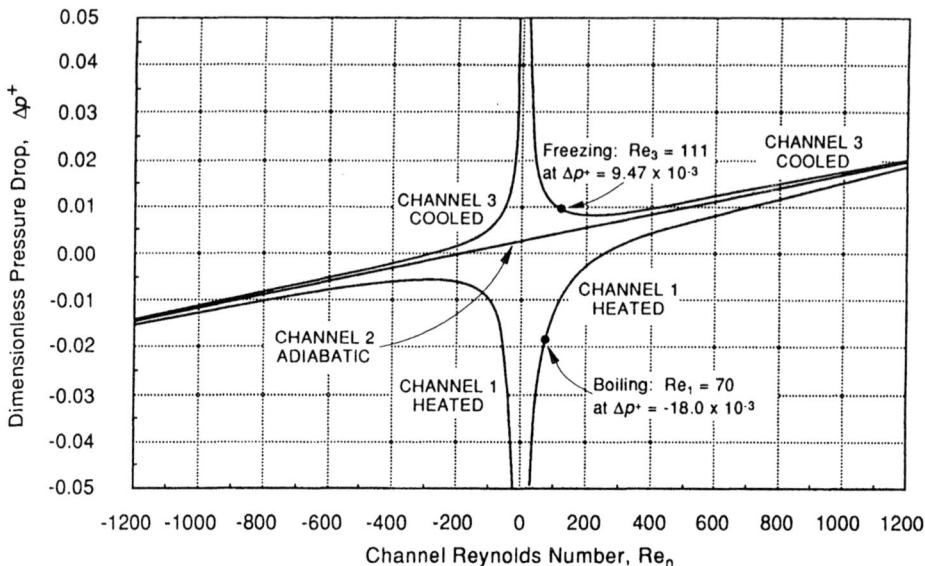

Figure 4-12 Flow characteristics of the three-channel system of Example 4-2. (Boiling and freezing conditions shown for upflow cases only.)

to $\Delta p^+ = +9.47 \times 10^{-3}$. These conditions are also plotted in Figure 4-12 at the corresponding Δp^+ values calculated from Eq. 4-62. Freezing and boiling also occur under downflow conditions that can be calculated from Eq. 4-63. Depending on the specific channel conditions these phase change conditions can be so located on a plot as Figure 4-12 to complicate the

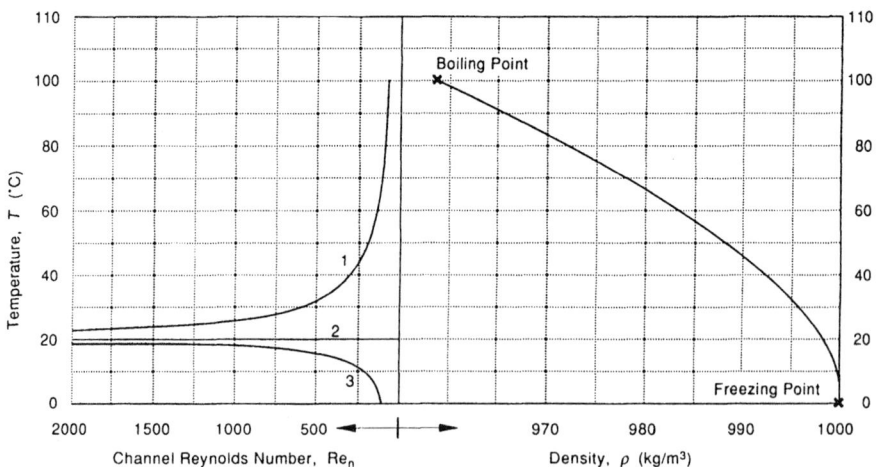

Figure 4-13 Temperature and density characteristics for upflow in the three-channel system of Example 4-2.

orderly transition from upflow to downflow and vice versa (see Problem 4-2).

Figure 4-14 illustrates the behavior of the system as the total plenum inlet Reynolds number is varied. This result is obtained from algebraically summing the channel flows of Figure 4-12 over the range of dimensionless pressure drop. The double-valued results at large positive Δp^+ reflect the multivalued behavior of channel 3. Only stable operating conditions of channel 3 are considered. These vary depending on whether channel 3 is operated initially in downflow or upflow. The horizontal jump (i.e., decrease in Re at fixed Δp^+) for a decreasing positive Reynolds number represents the transition of channel 3 from upflow to downflow. The analogous behavior is exhibited for negative Reynolds number (i.e., inlet flow to upper plenum) due to flow reversal in channel 1. Figure 4-14 illustrates that for zero external flow into the system, Δp^+ is single valued at approximately 1.4×10^{-3}.

C Onset of Mixed Convection (Upflow)

Typically, a reactor core operates in the friction-dominated (forced convection) regime. There are circumstances, however, under which the flow rate decreases such that the fluid buoyancy becomes significant. A reactor coolant pump coastdown is one such event. Since at high power-to-flow ratios, the momentum equation is coupled to the energy equation, it is desirable to calcu-

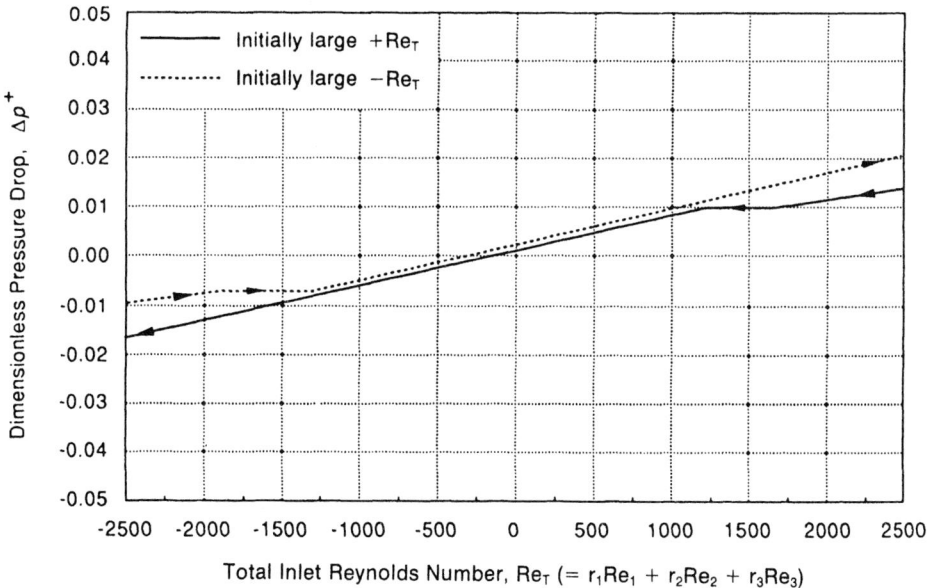

Figure 4-14 Two-pressure drop-flow trajectories for the three-channel system of Example 4-2.

late at what point the fluid buoyancy is significant. We will call this the onset (upper limit) of mixed convection.

The momentum equation for upflow is:

$$\Delta p^+ = \left[-X_L - \frac{S_n \Theta_n}{\mathrm{Re}_n} \right] + \delta_n \mathrm{Re}_n^{2-n} + \gamma_n \mathrm{Re}_n^2 \text{ (upflow)} \tag{4-62}$$

For mixed convection, the buoyancy term:

$$- \frac{S_n \Theta_n}{\mathrm{Re}_n}$$

(not the total gravity term), must be greater than some fraction of the friction and local pressure drop terms. If we let this fraction be F_{mc}, then the onset of mixed convection occurs when:

$$F_{mc}[\delta_n \mathrm{Re}_n^{2-n} + \gamma_n \mathrm{Re}_n^2] = \frac{S_n \Theta_n}{\mathrm{Re}_n} \tag{4-71}$$

For the special case of no local pressure loss ($\gamma_n = 0$) and laminar flow ($n = 1$), Eq. 4-71 simplifies to:

$$F_{mc}[\delta_n \mathrm{Re}_n] = \frac{S_n \Theta_n}{\mathrm{Re}_n} \tag{4-72}$$

Solving for Re_n:

$$\mathrm{Re}_n = \left[\frac{S_n \Theta_n}{F_{mc} \delta_n} \right]^{1/2} \tag{4-73}$$

where the positive root only should be selected since upflow is being considered.

This analysis shows that the onset of mixed convection in a channel with no form loss and experiencing laminar flow occurs when:

$$\frac{1}{\mathrm{Re}_n^2} \left[\frac{S_n \Theta_n}{\delta_n} \right] \geq F_{mc} \tag{4-74}$$

A typical value of F_{mc} which reflects a reasonable buoyancy influence is 0.10.

We can also solve for the nondimensional pressure drop at the onset of mixed convection. Substituting Eq. 4-73 into Eq. 4-62 and setting $n = 1$ (laminar flow) and $\gamma_i = 0$ (no local pressure loss), we obtain:

$$\Delta p^+ = -X_L - S_n \Theta_n \left[\frac{F_{mc} \delta_n}{\Theta_n S_n} \right]^{1/2} + \delta_n \left[\frac{S_n \Theta_n}{F_{mc} \delta_n} \right]^{1/2} \tag{4-75}$$

which can be simplified to:

$$\Delta p^+ = -X_L + [S_n \Theta_n \delta_n]^{1/2} [F_{mc}^{-1/2} - F_{mc}^{1/2}] \tag{4-76}$$

D Adiabatic Channel Flow Reversal

For a system of parallel channels operating initially in upflow of which some are adiabatic and some are heated, there will be a certain inlet flow rate below which the flow will reverse in the adiabatic channels. For laminar flow and no local pressure loss Eq. 4-62 becomes:

$$\Delta p^+ = -X_L - \frac{S_n \Theta_n}{Re_n} + \delta_n Re_n \quad \text{(upflow)} \tag{4-77}$$

Now, for the adiabatic channels ($\Theta_n = 0$) immediately before flow reversal, the flow will stagnate. At this point, the dimensionless pressure drop in the adiabatic channels (and all the other channels, by assumption) from Eq. 4-77 is equal to $-X_L$ since $Re_n = 0$. Now, for the heated channels Eq. 4-77 becomes:

$$\Delta p^+ = -X_L = -X_L - \frac{S_n \Theta_n}{Re_n} + \delta_n Re_n \tag{4-78}$$

at flow stagnation in an adiabatic channel.

This implies that for the heated channels:

$$\frac{S_n \Theta_n}{Re_n} = \delta_n Re_n \tag{4-79}$$

which can be solved for Re_n for the heated channels as:

$$Re_n = \left[\frac{S_n \Theta_n}{\delta_n} \right]^{1/2} \tag{4-80}$$

where the positive root only should be selected.

The inlet flow rate $\dot{m}_T$ at which the flows in the adiabatic channels reverse is derived from the mass constraint:

$$\dot{m}_T = \sum_{n=1}^{N} G_n A_n \tag{4-2}$$

We can solve the definition of Re_n for G_n to yield:

$$G_n = \frac{Re_n \mu}{D_{en}} \tag{4-81}$$

which can be substituted into Eq. 4-2, resulting in:

$$\dot{m}_T = \sum_{n=1}^{N} \frac{Re_n \mu A_n}{D_{en}} \tag{4-82}$$

Our final substitution is to replace Re_n with the expression derived for the Reynolds number in the heated channels, Eq. 4-80. This substitution yields:

$$\dot{m}_T = \sum_{n=1}^{N} \left[\frac{S_n \Theta_n}{\delta_n} \right]^{1/2} \frac{\mu A_n}{D_{en}} \tag{4-83}$$

Equation 4-83 gives the inlet flow rate to the system of parallel channels at which the flows in the adiabatic channels reverse.

E Stability of Cooled Upflow

In Volume I, Section 13V it was shown that for stable flow in a channel, the channel must operate in a region in which:

$$\left(\frac{\partial \Delta p}{\partial \dot{m}_n}\right) > \left(\frac{\partial \Delta p_{ex}}{\partial \dot{m}_n}\right)$$

which for a purely gravity-driven system reduces to:

$$\left(\frac{\partial \Delta p}{\partial \dot{m}_n}\right) > 0$$

Similarly, it can be shown that an equivalent requirement is that:

$$\left(\frac{\partial \Delta p^+}{\partial \mathrm{Re}_n}\right) > 0$$

which is the relation to be used for our analyses.

Figure 4-15 shows the relation between Δp^+ and Re for the cooled upflow conditions given by Eq. 4-62. Proceeding from a high flow rate to a lower flow rate (*point A*), we see an initial decrease in Δp^+ as expected in the friction-dominated regime. For further decrease in the flow rate, the average density in the channel increases (because the channel is cooled). This causes an increase

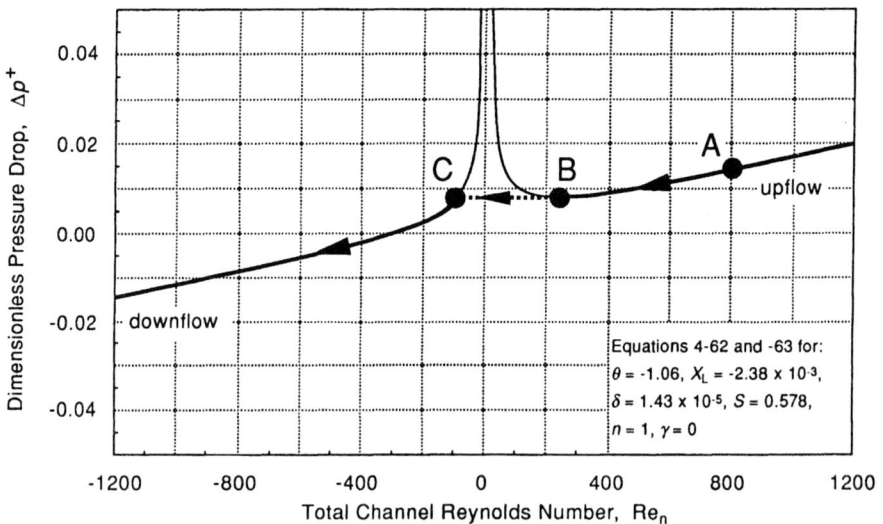

Figure 4-15 Behavior for initially cooled upflow.

in the gravity component of Δp^+. At some critical Reynolds number:

$$\left(\frac{\partial \Delta p^+}{\partial \mathrm{Re_n}}\right) = 0 \ (point \ B)$$

An infinitesimal further decrease in Δp^+ will cause the flow in the channel to reverse (*point C*). The *dotted line* of Figure 4-15 shows this behavior for a flow coastdown. Notice that stable downflow can exist for any Δp^+ so that once the channel reverses to downflow it may not reverse again to upflow.

We can derive an expression for the $\mathrm{Re_n}$ and Δp^+ at which the flow will reverse from upflow to downflow. For upflow in a channel with no local pressure loss and if laminar flow exists, Eq. 4-62 reduces to:

$$\Delta p^+ = -X_L - \frac{S_n \Theta_n}{\mathrm{Re_n}} + \delta_n \mathrm{Re_n}$$

Notice that Θ_n is negative for cooled channels. At the point of flow reversal:

$$\frac{\partial \Delta p^+}{\partial \mathrm{Re_n}} = \frac{S_n \Theta_n}{\mathrm{Re_n^2}} + \delta_n = 0 \tag{4-84}$$

Solving for $\mathrm{Re_n}$:

$$\mathrm{Re_n} = \left[\frac{-\Theta_n S_n}{\delta_n}\right]^{1/2} \tag{4-85}$$

where the positive root only should be selected. For the cooled channel of Example 4-2 this yields:

$$\mathrm{Re_3} = \left[\frac{-(-1.058)(0.578)}{(1.426 \times 10^{-5})}\right]^{1/2} = 207.1$$

as *point B* of Figure 4-15 illustrates.

Substituting back into Eq. 4-78 for Δp^+ at point B:

$$\Delta p^+ = -X_L - \Theta_n S_n \left[\frac{\delta_n}{-\Theta_n S_n}\right]^{1/2} + \delta_n \left[\frac{-\Theta_n S_n}{\delta_n}\right]^{1/2} \tag{4-86}$$

which can be simplified to:

$$\Delta p^+ = -X_L + 2[-\Theta_n \delta_n S_n]^{1/2} \tag{4-87}$$

The cooled channel of Example 4-2 will reverse from upflow to downflow when:

$$\Delta p^+ = 2.375 \times 10^{-3} + (2)[-(-1.058)(1.426 \times 10^{-5})(0.578)]^{1/2} = 8.28 \times 10^{-3}$$

as *point B* of Figure 4-15 illustrates.

This result indicates that for a quasi-steady flow coastdown, a cooled channel in upflow will reverse itself to downflow when the Δp^+ is less than that given by Eq. 4-87. The flow rate at which this occurs has been calculated using Eq. 4-85.

F Stability of Heated Downflow

Figure 4-16 shows the relation between Δp^+ and Re for Eq. 4-63 for heated downflow. As we proceed from a very negative (down) flow rate to a more positive one (*point A*), we see an initial increase in Δp^+. This is because we are in the friction-dominated regime. As the magnitude of the downflow decreases further, the average density in the channel is decreasing (because the channel is heated). This causes a decrease in the gravity component of Δp^+. At some critical Reynolds number, $(\partial \Delta p^+ / \partial \text{Re}_n) = 0$ (*point B*). An infinitesimal further increase in Δp^+ will cause the flow to reverse to upflow (*point C*).

An expression can be derived for the Re and Δp^+ of this reversal to upflow. The pressure drop equation for downflow Eq. 4-63 applies and can be simplified to:

$$\Delta p^+ = -X_U + \frac{(1 - S_n)\Theta_n}{\text{Re}_n} + \delta_n \text{Re}_n \tag{4-88}$$

for the case of no local pressure loss and with laminar flow present. The point of reversal is given by:

$$\frac{\partial \Delta p^+}{\partial \text{Re}_n} = -\frac{(1 - S_n)\Theta_n}{\text{Re}_n^2} + \delta_n = 0 \tag{4-89}$$

which can be solved for Re_n to yield:

$$\text{Re}_n = -\left[\frac{(1 - S_n)\Theta_n}{\delta_n}\right]^{1/2} \tag{4-90}$$

where the negative root only should be selected.

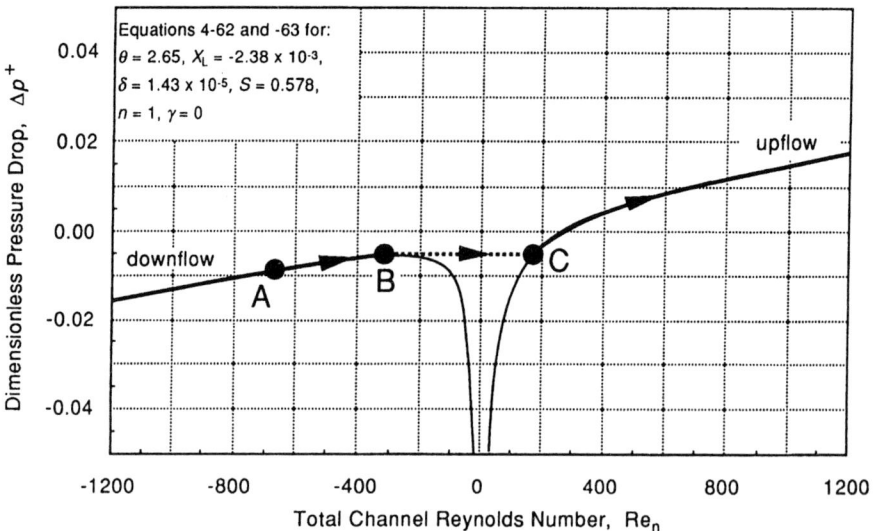

Figure 4-16 Behavior for initially heated downflow.

For the heated channel of Example 4-2, the flow will reverse from down-flow to upflow when the Reynolds number in the channel is greater than:

$$\text{Re}_1 = -\left[\frac{(0.422)(2.645)}{(1.426 \times 10^{-5})}\right]^{1/2} = -279.8$$

as illustrated in Figure 4-16 by *point B*.

Substituting back into our equation for Δp^+:

$$\Delta p^+ = -X_U - (1 - S_n)\Theta_n \left[\frac{\delta_n}{(1 - S_n)\Theta_n}\right]^{1/2} - \delta_n \left[\frac{(1 - S_n)\Theta_n}{\delta_n}\right]^{1/2} \quad (4\text{-}91)$$

which can be simplified to:

$$\Delta p^+ = -X_U - 2[(1 - S_n)\Theta_n\delta_n]^{1/2} \quad (4\text{-}92)$$

For Example 4-2, the reversal will occur at:

$$\Delta p^+ = 2.375 \times 10^{-3} - (2)[(0.422)(2.645)(1.426 \times 10^{-5})]^{1/2} = -5.60 \times 10^{-3}$$

as illustrated in Figure 4-16 by *point B*.

This result implies that a heated channel in downflow will reverse itself to upflow when the Δp^+ is greater than that given by Eq. 4-92. The flow rate in the channel at this condition has been calculated by Eq. 4-90.

G Preference for Upflow

Imagine a situation in which all the heated channels in a parallel channel system are in downflow. As the pressure drop is increased, the channels will reverse one by one to upflow. We can say that the channel that reverses first has the strongest "preference for upflow." This preference is related to Eq. 4-92 since this equation determines the Δp^+ at which the channel reverses. Each channel of our array has a critical Δp^+ from Eq. 4-92. The first to reverse has the most negative Δp^+ value. Therefore, a more negative Δp^+ given by Eq. 4-92 indicates a higher upflow preference. Hence, an upflow preference number must decrease with increasing Δp^+. Based on these observations and the form of Eq. 4-92, we define an upflow preference number for channel n by:

$$U_n = \frac{1}{2}[-X_U - \Delta p^+] \quad (4\text{-}93)$$

where Δp^+ is given by Eq. 4-92. Note that X_U will not affect U_n since all channels have the same value of X_U. Substituting Eq. 4-92 into Eq. 4-93 yields:

$$U_n = [\Theta_n\delta_n(1 - S_n)]^{1/2} \quad (4\text{-}94)$$

By definition:

$$\Theta_n \quad \text{is proportional to} \quad \frac{\dot{q}_n D_{en}}{A_n} \quad (4\text{-}95)$$

and

$$\delta_n \quad \text{is proportional to} \quad \frac{1}{D_{en}^3} \qquad (4\text{-}96)$$

so that:

$$U_n \quad \text{is proportional to} \quad \frac{\dot{q}_n{}^{1/2}(1 - S_n)^{1/2}}{A_n^{1/2}D_{en}} \qquad (4\text{-}97)$$

Yahalom and Bein [4] suggested that preference for downflow in a channel can be determined by an increasing value of the dimensionless number Y defined by:

$$Y_n \equiv \frac{\left(\dfrac{D_{en}}{D_{el}}\right)^{\alpha} \left(\dfrac{A_n}{A_1}\right)^{\beta}}{\left(\dfrac{\dot{q}_n}{\dot{q}_1}\right)^{\gamma}} \qquad (4\text{-}98)$$

when channel 1 is taken as that channel with the greatest value of:

$$\frac{\dot{q}_n^{\gamma}}{D_{en}^{\alpha}A_n^{\beta}}$$

The parameters α, β and γ are constants. Then, for any other channel the larger its value of Y, the greater is its preference for downflow. The form of the expression:

$$\frac{\dot{q}_n^{\gamma}}{D_{en}^{\alpha}A_n^{\beta}}$$

agrees with the expression for U_n, Eq. 4-97.

Moreover the measures of the preference for upflow, i.e., a large U_n, and the preference for downflow, i.e., a large Y_n, are also consistent. This follows by observing that the channel with largest value of:

$$\frac{\dot{q}_n^{\gamma}}{D_e^{\alpha}A^{\beta}}$$

has the largest value of U_n and the smallest value of Y_n ($= 1$). Therefore it has the largest preference for upflow and consistently the smallest preference for downflow.

H Limits of the Solution Procedure of Section VII

It must be stressed that the results of Section VI have been obtained from a one-dimensional laminar analysis of an array of subchannels communicating through plena idealized by the assumptions stated in Section VIA. These assumptions cause the plena to function only to impose the same axial pressure drop across all channels and the same inlet and exit temperatures to all channels. More accurate analysis of the behavior of this channel plena system requires more realistic modeling of plenum–channel interaction and the multi-dimensional nature of the flow in the channels particularly for conditions of flow reversal [2].

VIII DECOUPLED CONSERVATION EQUATION: ANALYTICAL SOLUTION PROCEDURE FOR HIGH FLOW RATE CASES

We shall next introduce the necessary assumptions to permit an analytical solution in the friction-dominated region but at low enough power-to-flow ratios so that multiple solutions and instabilities are not of concern. Analytical determination of the channel velocity and enthalpy fields for specified boundary conditions requires decoupling of the momentum and energy equations. This is done by neglecting variations of density with pressure and enthalpy and assigning the density a value at a selected, average reference pressure p^* and enthalpy h^*, i.e., the reference value ρ^*:

$$\rho_m(z) \equiv \rho^* = f(p^*, h^*) \tag{4-99}$$

Such an assumption restricts the analysis to channels having single-phase coolants that do not experience large changes in density through the channel. This applies to PWR and LMR assemblies in which the coolant temperature rise is typically 36 °C (64 F) and 160 °C (288 F), respectively, which corresponds to a density variation of 8.8 and 4.9%, respectively.

With this assumption, Eq. 4-4 becomes:

$$\Delta p_n = \bar{\rho}_{mn} g L + \bar{f}_n \frac{L}{D_{en}} \frac{G_{mn}|G_{mn}|}{2\rho^*} + \sum_i \frac{K_{in} G_{mn}|G_{mn}|}{2\rho^*} \tag{4-100}$$

For high flow rates the gravitation term is much smaller than the other terms and can be neglected. Our system of equations considering only mass and momentum conservation now consists of N equations for pressure drop:

$$\Delta p_n = \bar{f}_n \frac{L}{D_{en}} \frac{G_{mn}|G_{mn}|}{2\rho^*} + \sum_i \frac{K_{in} G_{mn}|G_{mn}|}{2\rho^*} \tag{4-101}$$

and one overall system continuity equation:

$$\dot{m}_T = \sum_{n=1}^{N} G_{mn} A_n \tag{4-2}$$

The boundary condition sets (1), (2), and (3) can be expressed more simply for our reference pressure condition now as:

Set (1). Prescribed pressure drop, Δp_n
Sets (2) and (3). Prescribed total flow rate, $\dot{m}_T$.

Table 4-2 Momentum solution procedures for parallel channel arrays connected only at plena

	Prescribed pressure drop (Δp_n)	Prescribed total flow rate ($\dot{m}_T$)
Unknown variables		
G_{mn}	N	N
Δp_n	N	N
$\bar{f}_n$	N	N
$\dot{m}_T$	–	1
	$3N$	$3N + 1$
Governing equations		
Conservation equation: Momentum, Eq. 4-101	N	N
Constraint equation: Total flow, Eq. 4-2	–	1
Constitutive equation: Friction factor $\dot{f}_n$, Eq. 4-39	N	N
	$2N$	$2N + 1$
Boundary conditions		
Δp_n	N	–
$\dot{m}_T$	–	1
$p_n(\text{upper}) = f(p_1(\text{upper}))$, Eq. 4-13		
or	–	$N - 1$
$p_n(\text{lower}) = f(p_1(\text{lower}))$, Eq. 4-14		
	N	N

– = not applicable.

Table 4-2 illustrates the governing equations and boundary conditions necessary to balance the unknowns for each case.

A Prescribed Channel Pressure Drop Condition: Solution Procedure

For the pressure drop condition, $\dot{m}_n$ is directly obtained from Eq. 4-101 after it is rewritten with the friction factor represented as:

$$f_n = \frac{C}{\left[\dfrac{\dot{m}D_e}{\mu A}\right]^n_n} = \frac{C}{\left[\dfrac{\rho^* V D_e}{\mu}\right]^n_n} \qquad \text{(4-102 and 4-103)}$$

where C and exponent n are constants that are a function of channel geometry and Reynolds number. Specifically, if we can neglect local pressure losses, Eq. 4-101 becomes:

$$\Delta p_n = C\left[\frac{\mu A}{\dot{m}D_e}\right]^n_n \left[\frac{L\dot{m}^2}{2D_e\rho^* A^2}\right]_n \qquad \text{(4-104)}$$

where subscript $n = 1, 2, \ldots N$.

B Prescribed Total Flow Condition: Solution Procedure

For the prescribed total flow condition, the solution procedure is more complex. This is the commonly encountered situation of determining the flow distribution among fuel assembly subchannels to the first order by neglecting crossflows. This discussion will be based on LWR and LMR assemblies which, as described in Volume I, Chapter 1, are comprised of three different subchannel types.

Eq. 4-17 is thus rewritten as:

$$\dot{m}_T = N_1 \dot{m}_1 + N_2 \dot{m}_2 + N_3 \dot{m}_3 \qquad (4\text{-}105)$$

where N_1, N_2, N_3 are the number of subchannels of types 1, 2 and 3 for the bundle under consideration.

Rearranging Eq. 4-105 we obtain:

$$\dot{m}_1 = \frac{\dot{m}_T}{N_1 + N_2 \dfrac{\dot{m}_2}{\dot{m}_1} + N_3 \dfrac{\dot{m}_3}{\dot{m}_1}} \qquad (4\text{-}106)$$

Similar equations for $\dot{m}_2$ and $\dot{m}_3$ can be written. The subchannel flow rate ratios:

$$\frac{\dot{m}_2}{\dot{m}_1} \text{ and } \frac{\dot{m}_3}{\dot{m}_1}$$

are found by equating the pressure drop in each subchannel type as given by equations of the type 4-104, where the friction factors have been represented as in Eq. 4-103. Even if geometry effects on the constants are neglected, differences in C and n can exist because each subchannel type can operate in a different Reynolds number range. We will use the constant set C,n for the conditions as yet unknown, of subchannel type 1, and C',n' for subchannel type 2.

Now expressing Eq. 4-104 first for subchannel type 1 and then for subchannel type 2 and taking the ratio of these equations and applying the equal pressure drop constraint of Eqs. 4-15 and 4-16, we obtain:

$$\frac{\left[C' \left(\dfrac{\mu A}{\dot{m} D_e} \right)^{n'} \dfrac{L \dot{m}^2}{D_e 2 \rho^* A^2} \right]_2}{\left[C \left(\dfrac{\mu A}{\dot{m} D_e} \right)^{n} \dfrac{L \dot{m}^2}{D_e 2 \rho^* A^2} \right]_1} = 1 \qquad (4\text{-}107)$$

After some algebraic simplification we obtain:

$$\frac{\dot{m}_2^{2-n'}}{\dot{m}_1^{2-n}} = \frac{C}{C'} \frac{A_2^{2-n'}}{A_1^{2-n}} \frac{D_{e_2}^{1+n'}}{D_{e_1}^{1+n}} \frac{\mu_1^{n}}{\mu_2^{n'}} \qquad (4\text{-}108)$$

where ρ_2^* has been taken to equal to ρ_1^* by virtue of Eq. 4-99. Variations in viscosity can also be neglected (approximately 11 and 34% for typical PWR and LMR cases) for simplicity. Now the ratio

$$\frac{\dot{m}_3}{\dot{m}_1}$$

can be obtained in a similar manner. After making provisions for subchannel type 3 to operate at a Reynolds number significantly different from subchannel types 1 and 2 by introducing the constants C'' and n'', the result is:

$$\frac{\dot{m}_3^{2-n''}}{\dot{m}_1^{2-n}} = \frac{C}{C''} \frac{A_3^{2-n''}}{A_1^{2-n}} \frac{D_{e_3}^{1+n''}}{D_{e_1}^{1+n}} \frac{\mu_1^n}{\mu_3^{n''}} \tag{4-109}$$

An explicit expression for $\dot{m}_1$ from Eq. 4-106 using Eqs. 4-108 and 4-109 requires either that the exponents n, n', and n'' be equal or that they be preselected for different flow regimes. This preselection requirement leads to an iterative solution procedure since the flow regimes are unknown. In the next sections we will examine both cases.

When $\dot{m}_1$ is available, the energy conservation equation, Eq. 4-5, can be directly solved for $h_n(z)$ since the linear heat generation rate q_n' and the inlet enthalpy h_n(lower) are available for each subchannel type.

1 Prescribed total flow condition: Fuel assembly flow split for all-turbulent or all-laminar conditions The simplest calculation for flow split between subchannels is possible when all subchannels are operating in the same flow regime—either all-laminar or all-turbulent. Under that case and assuming the differences in subchannel geometries cause negligible differences in the constants, we can take:

$$C = C' = C'' \tag{4-110}$$

and

$$n = n' = n'' \tag{4-111}$$

Eqs. 4-108 and 4-109, neglecting viscosity differences between subchannels, become:

$$\frac{\dot{m}_2}{\dot{m}_1} = \left(\frac{D_{e_2}}{D_{e_1}}\right)^{(1+n)/(2-n)} \left(\frac{A_2}{A_1}\right) \tag{4-112}$$

and

$$\frac{\dot{m}_3}{\dot{m}_1} = \left(\frac{D_{e_3}}{D_{e_1}}\right)^{(1+n)/(2-n)} \left(\frac{A_3}{A_1}\right) \tag{4-113}$$

and the explicit relation for $\dot{m}_1$ obtained by substituting Eqs. 4-112 and 4-113 into Eq. 4-106 is:

$$\dot{m}_1 = \frac{\dot{m}_T}{N_1 + N_2 \dfrac{A_2}{A_1} \left(\dfrac{D_{e_2}}{D_{e_1}}\right)^{(1+n)/(2-n)} + N_3 \dfrac{A_3}{A_1} \left(\dfrac{D_{e_3}}{D_{e_1}}\right)^{(1+n)/(2-n)}} \tag{4-114}$$

Note that since the density is taken constant, the above relations for the ratios of subchannel mass flow rates, Eqs. 4-112 and 4-113, can be transformed to velocity ratios yielding:

$$\frac{V_2}{V_1} = \left(\frac{D_{e_2}}{D_{e_1}}\right)^{(1+n)/(2-n)} \tag{4-115}$$

$$\frac{V_3}{V_1} = \left(\frac{D_{e_3}}{D_{e_1}}\right)^{(1+n)/(2-n)} \tag{4-116}$$

Now to obtain the associated subchannel enthalpy or temperature increases, we express Eq. 4-5 for each subchannel as:

$$h_n(z) - h_n(\text{in}) = \frac{P_{hn}}{\dot{m}_n} \int_0^z q_n''(z)dz \tag{4-117}$$

where P_h is the heated perimeter of the subchannel. Assuming that the axial heat flux distribution is equal for the rods that provide heat for channel types 1 and 2 and recalling that the density has been assumed constant, we can write Eq. 4-117 for both a subchannel of type 1 and 2. Dividing these equations we obtain:

$$\frac{[(h(z) - h(\text{in})]_1}{[h(z) - h(\text{in})]_2} = \frac{P_{h1}}{P_{h2}} \frac{(VA)_2}{(VA)_1} = \frac{P_{h1}Q_2}{P_{h2}Q_1} \tag{4-118}$$

where Q_2 and Q_1 are volumetric flow rates.

Using relations 4-115, 4-116, and 4-118, the ratio of the velocities and temperature rises in various type subchannels can be calculated.

Significant decrease in flow rate passing through edge subchannels versus interior subchannels results from a reduction in the outer rod to duct wall spacing. This decrease in flow rate, i.e., in Q_2/Q_1, results in an increase in relative temperature rise, i.e., $\Delta T_2/\Delta T_1$. In hexagonal LMR bundles this temperature rise increase due to the reduction of peripheral spacing would be beneficial in reducing the temperature gradient around the periphery of edge fuel pins and hence the induced stress in the clad. Of course a ratio $\Delta T_2/\Delta T_1$ equal to 1.00 is optimum.

These parameter ratios are strong functions of the Reynolds number, pitch to diameter ratio, and edge spacing. Figure 4-17 presents a full set of results for the half-spacing case. It should be remembered that these results are for a uniform radial power distribution. Imbalances in temperature rise ratios can be aggravated by a power skew across the bundle.

Example 4-3 Subchannel relative velocities and enthalpies for single-phase flow in a PWR

PROBLEM Compute the ratio of mass flow rates for an edge subchannel relative to an interior subchannel for a PWR core in a single-phase turbulent flow conditions. Neglect fuel pin spacers and fluid property changes with temperature. Assume that the fuel assemblies are arranged such that

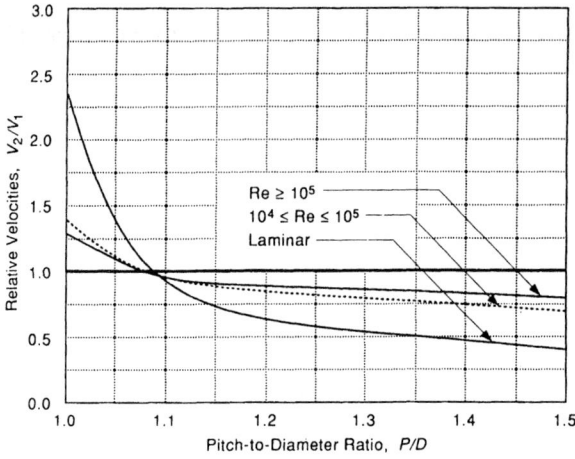

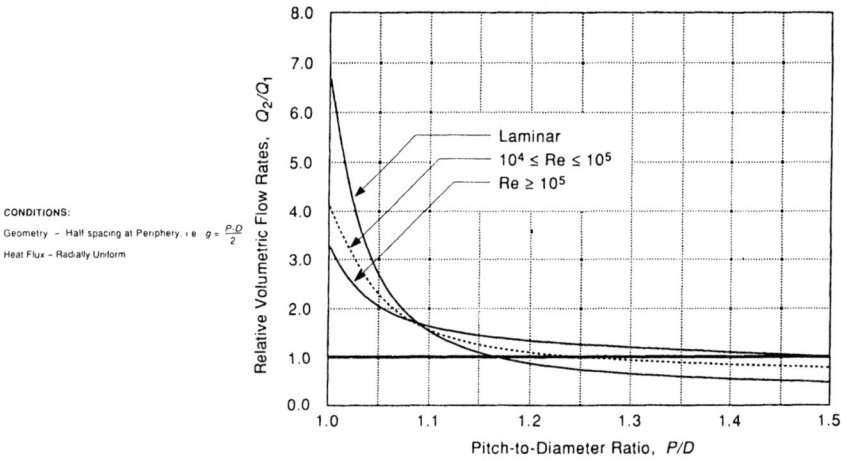

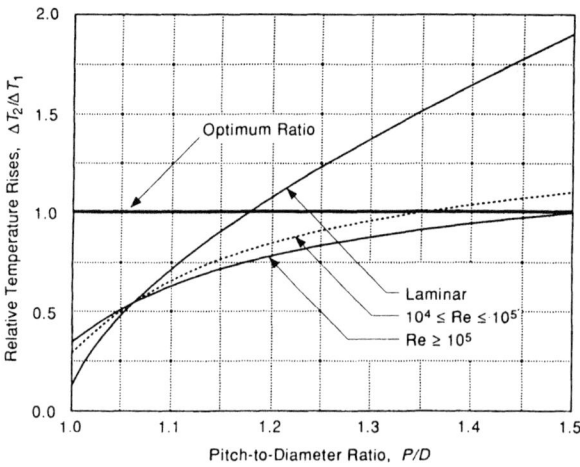

Figure 4-17 Relative performance of edge (type 2) and interior (type 1) subchannels of bare pins in a hexagonal array. *(After Zhukov et al. [5].)*

156

the spacing between the centerlines of the edge pins in adjacent assemblies is 20% greater than that of the pins within an assembly.

Note that the consequences to channel enthalpy rise would also have to consider the rod power levels of this assumed arrangement in addition to this spacing effect.

SOLUTION In the friction factor–Reynolds relation, take $n = 0.2$ since the flow is turbulent.

From Eq. 4-112:

$$\frac{\dot{m}_2}{\dot{m}_1} = \left[\frac{D_{e_2}}{D_{e_1}}\right]^{(1+n)/(2-n)} \left(\frac{A_2}{A_1}\right) = \left(\frac{D_{e_2}}{D_{e_1}}\right)^{0.67} \left(\frac{A_2}{A_1}\right)$$

Consistent with Volume I, Chapter 1, take subchannel type 2 as the edge or side channel and subchannel type 1 as the interior channel. From Table J-1 the area of the interior channel is:

$$A_1 = P^2 - \frac{\pi D^2}{4}$$

From the problem statement, the area of the edge subchannel is:

$$A_2 = (1.2P)^2 - \frac{\pi D^2}{4}$$

$$P_{w_1} = \pi D$$

also:

$$P_{w_2} = \pi D$$

Hence:

$$D_{e_1} = \frac{4A_1}{P_{w_2}} = D\left[\frac{4}{\pi}\left(\frac{P}{D}\right)^2 - 1\right]$$

$$D_{e_2} = D\left[\frac{5.76}{\pi}\left(\frac{P}{D}\right)^2 - 1\right]$$

From Table I-3 in Volume I the geometrical parameters are:

$$P = 12.6 \text{ mm}$$

$$\frac{P}{D} = \frac{12.6}{9.5} = 1.326$$

Hence:

$$A_1 = 87.9 \text{ mm}^2$$

$$A_2 = 157.7 \text{ mm}^2$$

$$\frac{D_{e_1}}{D} = 1.239$$

$$\frac{D_{e_2}}{D} = 2.224$$

Therefore:

$$\frac{\dot{m}_2}{\dot{m}_1} = \left(\frac{2.224}{1.239}\right)^{0.67} \left(\frac{157.7}{87.9}\right) = 2.655$$

2 Prescribed total flow condition: Flow split and temperature rise in the transition flow regime. The analysis of Section VIIIB1 assumed that all subchannels in the bundle operated in laminar flow or that all the subchannels operated in turbulent flow. In certain instances some of the subchannels are in laminar flow and others are in turbulent flow. In these cases, the functional dependence of the friction factor changes between subchannels in the bundle. In this analysis we will take the following values for characterizing the friction factor:

Re	C	n
Re < 2100	80	1.0
Re > 2100	0.316	0.25

Note that for Re < 2100 (laminar flow), C is usually given as 64, which is derived for flow in a round tube. For any given subchannel, C and n will vary with pitch-to-diameter ratio in a manner that causes the friction factor to start below the equivalent round tube value at $P/D = 1.0$ and increase above the round tube value at P/D approximately 1.1 and greater. Numerical results are available for some subchannel geometries under some flow conditions as discussed in Chapter 7. For illustrative purposes here, nominal values of $C = 80$ and $n = 1$ will be used for every subchannel in laminar flow, and nominal values of $C = 0.316$ and $n = 0.25$ will be used for every subchannel in turbulent flow.

Regardless of the flow regime the pressure drop along each of the subchannels is the same. Hence each subchannel type has the same pressure drop, i.e.:

$$\Delta p_1 = \Delta p_2 = \Delta p_3 \qquad \text{(4-15 and 4-16)}$$

and the total mass flow rate is the sum of the mass flows through the different subchannels. Relations for subchannel pressure drop and total mass flow rate have been written already as Eqs. 4-104 and 4-105, respectively, but we will now recast them for convenience in terms of velocities as:

$$V_T = N_1 \frac{A_1}{A_T} V_1 + N_2 \frac{A_2}{A_T} V_2 + N_3 \frac{A_3}{A_T} V_3 \qquad \text{(4-119)}$$

and

$$\Delta p = g_1 V_1^a = g_2 V_2^a = g_3 V_3^a \qquad \text{(4-120)}$$

where the functional relations g_1, g_2, and g_3 depend on flow regime in each subchannel and the subchannel type. In Eq. 4-119 the densities are assumed uniformly constant throughout all subchannels.

Example 4-4 Flow distribution in a hexagonal bundle

PROBLEM Compute the ratio of the subchannel velocity to the total bundle velocity for each subchannel, and subchannel Reynolds numbers for a 61-pin hexagonal bundle. Assume water at 27 °C (80 F) is flowing through the bundle to hydrodynamically simulate LMR conditions. The bundle geometric and operating conditions are given in Table 4-3.

SOLUTION The function g_1 for an interior subchannel in laminar flow is therefore:

$$\Delta p_{1L} = \frac{C_1}{Re_1^n} \frac{L}{D_{e_1}} \frac{\rho V_1^2}{2} = 80 \frac{\mu}{\rho V_1 D_{e_1}} \frac{L}{D_{e_1}} \frac{\rho V_1^2}{2}$$

$$= \frac{80}{D_{e_1}^2} \left(\frac{\mu}{\rho}\right) \frac{L\rho}{2} V_1 \qquad (4\text{-}121a)$$

$$= (6.1417 \text{ s}^{-1}) \left(\frac{L\rho}{2}\right) V_1$$

where s^{-1} reflects the units used in evaluating the first two terms of the preceding equation.

Hence $g_1 V_1^a$ for laminar flow is $6.1417 \text{ s}^{-1}(L\rho/2)V_1$. Table 4-4 gives g_1, g_2, and g_3 for both laminar and turbulent flow conditions. In turbulent flow,

Table 4-3 Hexagonal bundle geometric and operating conditions

Channel parameters derived using formula of Table I-J3

	Interior	Edge	Corner
N_j	96	24	6
D_e	3.3555 mm	4.1131 mm	3.1025 mm
A	10.4600 mm²	20.9838 mm²	7.4896 mm²
P_w	12.4690 mm	20.4070 mm	9.6562 mm

Geometry:
Number of pins, N_p = 61
Diameter of fuel rod, D = 6.35 mm (0.25 inches)
Diameter of wire, D_s = 1.588 mm (0.0625 inches) = g
Pitch = 7.9380 mm (0.3125 inches)
Lead length = 304.8 mm (12 inches)
Angle of wire inclination = 86°
Total flow area = 1552.7 mm²
Bundle hydraulic diameter = 3.56 mm
Water T = 27 °C (80 F)
ρ = 997 kg/m³ (62.2 lbm/ft³)
μ = 8.618 × 10⁻⁴ kg/m·s (2.083 lbm/hr ft)

Table 4-4 Pressure drop coefficients for hexagonal subchannel types (61-pin LMR assembly example)

	Interior subchannel	Edge subchannel	Corner subchannel
Laminar	$g_{1L} = (6.1417 \text{ s}^{-1}) \dfrac{L\rho}{2}$	$g_{2L} = (4.0876 \text{ s}^{-1}) \dfrac{L\rho}{2}$	$g_{3L} = (7.1842 \text{ s}^{-1}) \dfrac{L\rho}{2}$
Turbulent	$g_{1T} = (11.9308 \text{ m}^{-0.75} \text{ s}^{-0.25}) \dfrac{L\rho}{2}$	$g_{2T} = (9.2504 \text{ m}^{-0.75} \text{ s}^{-0.25}) \dfrac{L\rho}{2}$	$g_{3T} = (13.1591 \text{ m}^{-0.75} \text{ s}^{-0.25}) \dfrac{L\rho}{2}$

$\Delta p_n = g_{n,j} V_n^a$ (Eq. 4-120).
$n = 1, 2, 3; j = L, T$.
$a = 1$ for $j = L$.
$a = 1.75$ for $j = T$.

using values of $C = 0.316$ and $n = 0.25$, Eq. 4-121a becomes:

$$\Delta p_{1T} = \frac{0.316}{D_{e_1}^{1.25}} \left(\frac{\mu}{\rho}\right)^{0.25} \left(\frac{L\rho}{2}\right) V_1^{1.75} \tag{4-121b}$$

Now considering first a low flow situation with all the subchannels in laminar flow, we use the laminar function to express Eq. 4-120 as:

$$(6.1417 \text{ s}^{-1})V_1 = (4.0876 \text{ s}^{-1})V_2 = (7.1842 \text{ s}^{-1})V_3 \tag{4-122a}$$

where the lengths and fluid densities of the channels are taken equal. Applying these velocity relations and the geometry characteristics of our bundle from Table 4-3 to Eq. 4-119 yields:

$$\frac{V_2}{V_T} \equiv X_2 = 1.2966 \tag{4-123}$$

The other relative velocity or velocity fractions are readily determined from Eq. 4-122a as:

$$X_1 \equiv \frac{V_1}{V_T} = 0.8386 \tag{4-124a}$$

and

$$X_3 \equiv \frac{V_3}{V_T} = 0.7378 \tag{4-124b}$$

These calculated relative velocities hold until the first subchannel type enters turbulent flow. This condition is assumed to occur when the subchannel Reynolds number reaches 2,100. Therefore we next express the subchannel Reynolds numbers as a function of bundle velocity considering the geometry of each subchannel.

These relations are:

$$Re_1 = \rho \frac{V_1 D_{e_1}}{\mu} = \left(3,881.9 \frac{s}{m}\right) X_1 V_T \tag{4-125a}$$

$$Re_2 = \rho \frac{V_2 D_{e_2}}{\mu} = \left(4,758.3 \frac{s}{m}\right) X_2 V_T \tag{4-125b}$$

$$Re_3 = \rho \frac{V_3 D_{e_3}}{\mu} = \left(3,589.2 \frac{s}{m}\right) X_3 V_T \tag{4-125c}$$

$$Re_T = \rho \frac{V_T D_{e_T}}{\mu} = \left(4,118.3 \frac{s}{m}\right) V_T \tag{4-125d}$$

Now using the laminar velocity ratios of Eqs. 4-123 and 4-124a and b find that the edge subchannel becomes turbulent first ($Re_2 = 2,100$) at an average bundle velocity V_T of 0.340 m/s.

At this point, because of our assumptions of a sudden transition, the edge subchannel friction characteristic g_2 changes, and the velocity ratio changes discontinuously. Under this condition, Eq. 4-122a is rewritten at the new steady state as:

$$(6.1417 \text{ s}^{-1})V_1 = (9.2504 \text{ m}^{-0.75} \text{ s}^{-0.25})V_2^{1.75} = (7.1842 \text{ s}^{-1})V_3 \quad (4\text{-}122b)$$

and Eq. 4-119 becomes:

$$V_T = \frac{9.2504}{6.1417}(0.6467)V_2^{1.75} + 0.3243V_2$$

$$+ \left(\frac{9.2504}{7.1842}\right)(0.0289)V_2^{1.75} \quad (4\text{-}122c)$$

which, when rewritten to yield a convergent iterative form, becomes:

$$V_2 = \left(\frac{V_T - 0.3243V_2}{1.0113}\right)^{1/1.75} \quad (4\text{-}126a)$$

and from Eq. 4-122b:

$$V_1 = \frac{9.2504}{6.1417}V_2^{1.75} \quad (4\text{-}126b)$$

and

$$V_3 = \frac{6.1417}{7.1842}V_1$$

The subchannel velocities and associated relative velocities are determined from Eqs. 4-126a and b for successively increasing bundle flow rates until the next subchannel type, the interior subchannel, enters turbulent flow. This occurs at an average bundle velocity of 0.544 m/s. Note that between an average bundle velocity of 0.340 and 0.544 m/s, the relative velocity changes continuously because of the mixed friction characteristics of the subchannels as expressed by the transcendental Eq. 4-126a.

The procedure for determining the relative velocity for higher bundle flows is identical to that described above using values of g_1, g_2, g_3 from Table 4-4 in equations of the type 4-122b and c until the flow is calculated as turbulent in all subchannel types. This occurs at a bundle Reynolds number of approximately 2,388. Note that when $\text{Re}_n = 2,100$, there are two possible values for $g_n(V_n)$, namely: f_{nL} and f_{nT}. By using f_{nL} the velocities are computed at $\text{Re} = 2,100^-$, and by using f_{nT} they are computed at $\text{Re}_n = 2,100^+$. For bundle Reynolds numbers that are higher than 2,388, the relative velocity remains constant. The subchannel relative velocities for bundle flow rates between laminar and turbulent are plotted in Figure 4-18. Figure 4-19 illustrates the associated subchannel Reynolds numbers over this range.

This example is illustrative of the principles involved in calculating velocity using classic tube friction characteristics for subchannel geometries. In actual power reactor hexagonal assemblies, the transition is not sharp, and the flow redistributes itself among the subchannels in a continuous fashion.

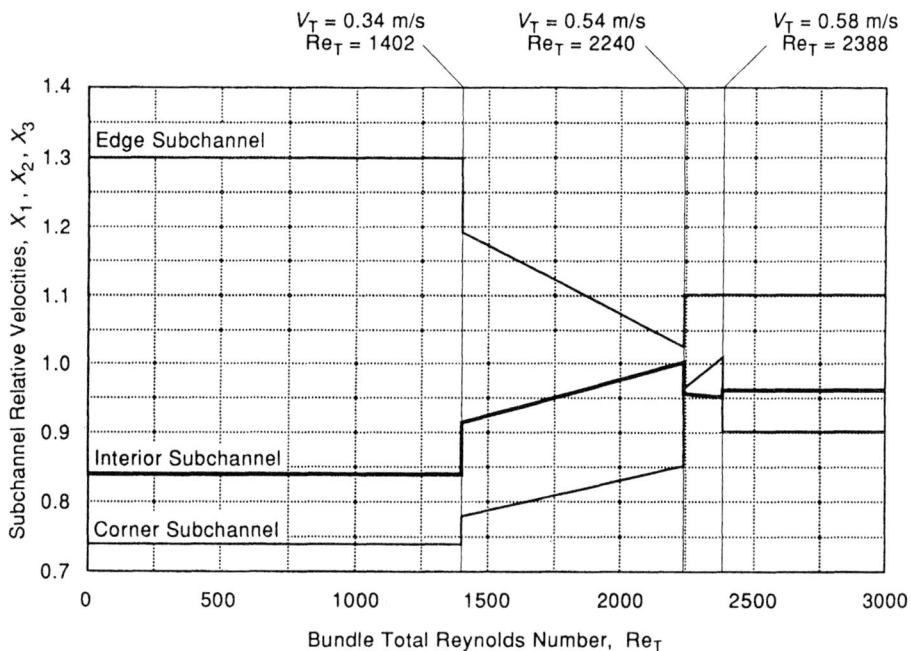

Figure 4-18 Subchannel relative velocities for the hexagonal bundle, Example 4-4.

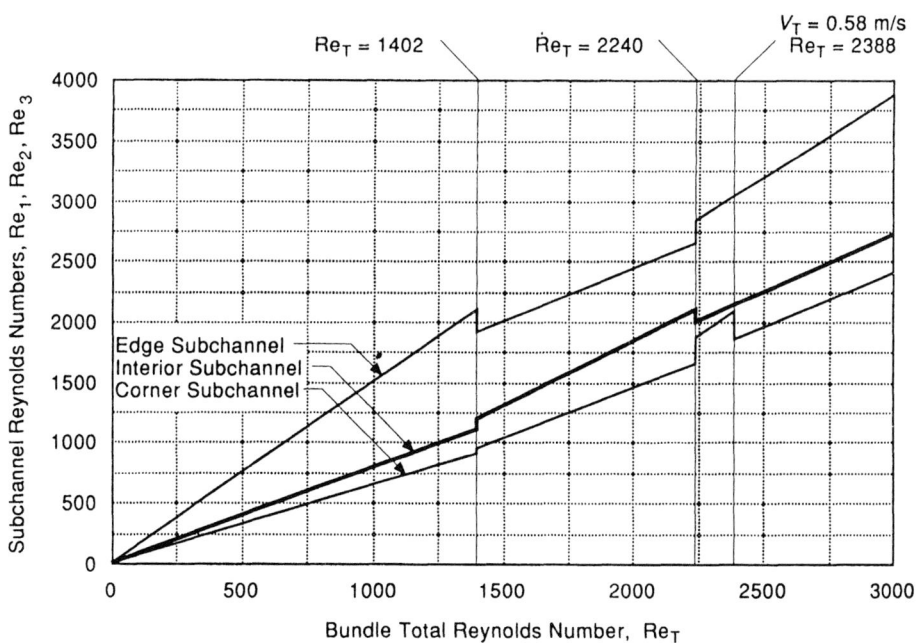

Figure 4-19 Subchannel Reynolds numbers for the hexagonal bundle, Example 4-4.

3 Flow split considering manufacturing tolerance in hexagonal bundles. The actual flow distribution within a bundle is also affected by the manufacturing tolerances existing within a bundle. In this section the effect of these tolerances on the analysis of Sections VIIIB1 and 2 is presented.

This effect is presented in terms of the parameters T and F where:

1. T is defined as the magnitude of the as-fabricated clearance or tolerance in an assembly along the flat-to-flat direction of nominal value D_{ft}, i.e.:

$$T = D_{ft} - 2\left[\left(\frac{\sqrt{3}}{2} N_{rings}\right)(D + D_s)_{NOM} + \frac{D}{2} + D_s\right] \quad (4\text{-}127)$$

where N_{rings} is defined in Table I,1-5 as the number of rings of rods surrounding the central rod. D_{ft} and other geometric parameters are illustrated in Figure 4-20 which is a 19-rod bundle for which $N_{rings} = 2$.

2. F is defined as the fraction of the assembly tolerance T which is distributed around the interior rods of the assembly. The value of F ranges from 0 to 1 where $F = 0$ represents the situation in which all the tolerance is in the edge of the bundle, i.e., the rods are packed together toward the center of the bundle, and $F = 1$ represents the situation in which all the tolerance is

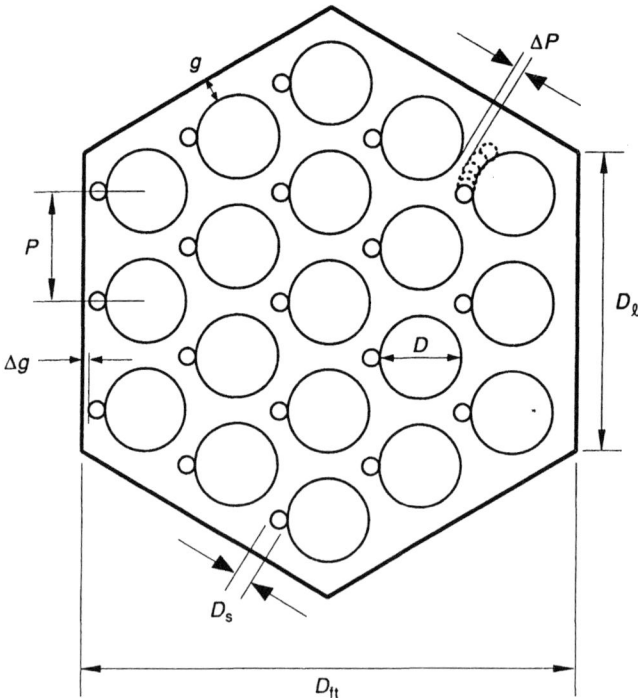

Figure 4-20 Distribution of manufacturing tolerances in the hexagonal bundle.

within the interior subchannels, i.e., the edge and corner rods are packed against the duct walls.

An expression for F is constructed by equating two expressions for the clearance per interior pin, ΔP. Taking P as shown in Figure 4-20 as the actual interior rod-to-rod centerline distance, the first expression for ΔP is based on rod-to-rod parameters, i.e.:

$$\Delta P \equiv P - D - D_s \tag{4-128}$$

Table 4-5 Hexagonal bundle subchannel geometric conditions assuming bundle tolerance

Interior subchannel

$$A_{f_1} = \frac{\sqrt{3}}{4} P^2 - \frac{\pi D^2}{8} - \frac{\pi D_s^2}{8} = \frac{\sqrt{3}}{4}\left(D + D_s + \frac{FT}{\sqrt{3}\,N_{rings}}\right)^2 - \frac{\pi}{8}(D^2 + D_s^2)$$

$$P_{w_1} = \frac{\pi}{2}(D + D_s)$$

$$D_{e_1} = \frac{4A_{f_1}}{P_{w_1}} = \frac{\sqrt{3}\left(D + D_s + \dfrac{FT}{\sqrt{3}\,N_{rings}}\right)^2 - \dfrac{\pi}{2}(D^2 + D_s^2)}{\dfrac{\pi}{2}(D + D_s)}$$

Edge subchannel

$$A_{f_2} = \left(D + D_s + \frac{FT}{\sqrt{3}\,N_{rings}}\right)\left(\frac{D}{2} + D_s + (1 - F)\frac{T}{2}\right) - \frac{\pi}{8}(D^2 + D_s^2)$$

$$P_{w_2} = \frac{\pi}{2}(D + D_s) + \left(D + D_s + \frac{FT}{\sqrt{3}\,N_{rings}}\right)$$

$$D_{e_2} = \frac{4A_{f_2}}{P_{w_2}} = \frac{4\left(D + D_s + \dfrac{FT}{\sqrt{3}\,N_{rings}}\right)\left(\dfrac{D}{2} + D_s + (1 - F)\dfrac{T}{2}\right) - \dfrac{\pi}{2}(D^2 + D_s^2)}{\dfrac{\pi}{2}(D + D_s) + \left(D + D_s + \dfrac{FT}{\sqrt{3}\,N_{rings}}\right)}$$

Corner subchannel

$$A_{f_3} = \frac{1}{\sqrt{3}}\left(\frac{D}{2} + D_s + (1 - F)\frac{T}{2}\right)^2 - \frac{\pi}{24}(D^2 + D_s^2)$$

$$P_{w_3} = \frac{2}{\sqrt{3}}\left(\frac{D}{2} + D_s + (1 - F)\frac{T}{2}\right) + \frac{\pi}{6}(D + D_s)$$

$$D_{e_3} = \frac{\dfrac{4}{\sqrt{3}}\left(\dfrac{D}{2} + D_s + (1 - F)\dfrac{T}{2}\right)^2 - \dfrac{\pi}{6}(D^2 + D_s^2)}{\dfrac{2}{\sqrt{3}}\left(\dfrac{D}{2} + D_s + (1 - F)\dfrac{T}{2}\right) + \dfrac{\pi}{6}(D + D_s)}$$

The second expression is based on total assembly geometric parameters, T, and N_{rings}. The portion of the assembly tolerance within the bundle interior along the rod-to-rod centerline connection path is $F(T/(\sqrt{3}/2))$, and the number of interior pitches along which this tolerance is distributed is $2N_{rings}$. Hence ΔP also equals

$$\Delta P \equiv F \frac{T}{\sqrt{3}N_{rings}} \tag{4-129}$$

We find F from Eqs. 4-128 and 4-129 as

$$F \equiv \left(\frac{P - D - D_s}{T}\right) \sqrt{3}N_{rings} \tag{4-130}$$

For the peripheral rods, the associated parameters are the actual rod centerline-to-wall spacing, $g + R$, and the clearance per peripheral rod Δg. Hence $(g + R)$ is related to F as follows:

$$g + R \equiv \frac{D}{2} + D_s + \Delta g = \frac{D}{2} + D_s + (1 - F)\frac{T}{2} \tag{4-131}$$

To find the flow split as a function of F, we must first find the corresponding areas and wetted perimeters of the individual subchannels. We assume T is given and fixed, and let F vary over the possible range 0.0 to 1.0.

The calculation of flow split and velocity ratios for laminar and turbulent flow utilizes Eqs. 4-112 through 4-116 but with the use of flow areas and hydraulic diameters that are now functions of F. These geometric parameters are expressed in terms of F and T using Eqs. 4-128 through 4-131 in Table 4-5.

In practice it is difficult to determine the appropriate value of F. Further, pin bowing patterns can affect the distribution of flow. For the 61-pin hexagonal bundle of Table 4-3, if a tolerance T of 0.0569 cm (0.0221 inches) is assumed, the subchannel mass flow and velocity ratios change at least 10% from nominal values over the possible range of accommodation of tolerance. This provides an indication of the uncertainty inherent in any analytical assessment of flow distribution within reactor bundles.

REFERENCES

1. Chato, J. C. Natural convection flows in parallel-channel systems, *J. Heat Transfer* 85:339–345, 1963.
2. Iannello, V. Mixed Convection in Parallel Channels Connected Only at Upper and Lower Plena. Ph.D. thesis, MIT, Dept. of Nuclear Engineering, Feb. 1986.
3. Singh, B., and Griffith, P., *Gravity Dominated Two-Phase Flow in Vertical Rod Bundles*. NUREG/CR-1218, Jan. 1980.
4. Yahalom, R., and Bein, M. Boiling thermal hydraulic analysis of multichannel low flow. *Nucl. Eng. Des.* 53:29–38, 1979.
5. Zhukov, A. V., Kudryatseva, L. K., Sviridenko, Ye. Ya, Subbotin, V. T., Talanov, V. D., and Ushakov, P. A. in *Experimental Study of Temperature Fields of Fuel Elements, Using Models*

in Liquid Metals. Kirillov, P. L., Subbotin, V. T., and Ushakov, P. A., Eds. NASA Technical Translation, NASA-TT-F-522, 1969.

PROBLEMS

Problem 4-1 Flow characteristics of an adiabatic channel in a multichannel array (Section VII)

A laboratory experiment has been run to collect Δp^+ versus Re data for the three parallel channel system shown in Figure 4-1. System parameters, operating conditions, and water properties are tabulated below. The work has proceeded smoothly except for a problem when the flow reverses in the adiabatic channel. It seems that a significantly higher Δp^+ is observed when the upflow is decreased to zero than when the downflow is decreased to zero.

Set up the equations for Δp^+ in the adiabatic channel based on the system parameters and properties given below. Plot the curve of Δp^+ versus Re for the adiabatic channel in very low flow. Be careful to identify all important slopes and intercepts. Can you give an explanation for the differences in Δp^+ at zero flow?

System parameters	Operating conditions	Water properties (for a reference temperature at 25 °C)
$L = 2$ m	$\dot{q}_1 = 500$ W	$\beta = 4.75 \times 10^{-4}/°C$
$D_e = 6.35 \times 10^{-3}$ m	$\dot{q}_2 = 0$	$c_p = 4186$ J/kg°C
$A = 1.58 \times 10^{-4}$ m²	$\dot{q}_3 = -200$ W	$\mu = 8.62 \times 10^{-4}$ Pa-s
$S = 1.73$		$\rho^* = 993.1$ kg/m³
$\sum_i K_i = 0$		$p^* = 1.01325 \times 10^5$ Pa
T(upper) $= 30$ °C		
T(lower) $= 20$ °C		
Friction factor correlation (laminar flow)		
$f = \dfrac{95}{\text{Re}}$		

Answer: $\Delta p^+ = 2.375 \times 10^{-3} + 1.426 \times 10^{-5}\,\text{Re}_n$; upflow

$\Delta p^+ = -2.375 \times 10^{-3} + 1.426 \times 10^{-5}\,\text{Re}_n$; downflow

Problem 4-2 Flow characteristics of heated channels in a multichannel array (Section VII)

Consider the three-channel system of Example 4-2. Compute the Reynolds number and Δp^+ values at which freezing occurs in the cooled channel (3) and boiling occurs in the heated channel (1) for both upflow and downflow. Comment on the implications of your results for flow reversal from upflow to downflow in channel 3 and for flow reversal from downflow to upflow in channel 1. Do both transitions behave similarly?

Answer: upflow: $\Delta p_1^+ = -18.60 \times 10^{-3}$ and $\text{Re}_1 = 69.6$ for boiling

$\Delta p_3^+ = +9.45 \times 10^{-3}$ and $\text{Re}_3 = 111.4$ for freezing

downflow: $\Delta p_1^+ = -14.65 \times 10^{-3}$ and $\text{Re}_1 = -69.6$ for boiling

$\Delta p_3^+ = +4.79 \times 10^{-3}$ and $\text{Re}_3 = -111.4$ for freezing

Problem 4-3 Channel preference for downflow (Section VII)

Consider a model LMR core consisting of a fuel, blanket, and poison channel with the following operating conditions and geometry.

	Fuel channel	Blanket channel	Poison channel
Full power, MW	5	2	0.5
Number of pins, N_p	169	61	·7
Pin diameter D, inches	0.3	0.5	1.0
Distance across flats of hexagon D_{ft}, inches	4	4	4
Channel area for flow A_T, square inches	1.90	1.87	8.35
Channel total wetted perimeter P_{wT}, inches	173	109.6	35.8

Order these channels relative to their preference for downflow in a decay heat condition after shutdown from full power.

Answer: The preference for downflow is greatest for the poison channel and least for the fuel channel

Problem 4-4 Flow split among tubes in a U-tube steam generator (Section VIII)

Consider a U-tube steam generator in a typical PWR in which the primary water flows through the U-tubes connected at common plena. The tubes are of equal diameters but of various lengths.

1. Determine the ratio of the water flow rates $\dot{m}_1$ in the short tube and $\dot{m}_2$ in the long tube. The dimensions are given in Figure 4-21. You may make reasonable approximations, but clearly state your reasons for them. Assume the diameter of tube is 20 mm, and the flow is fully turbulent.

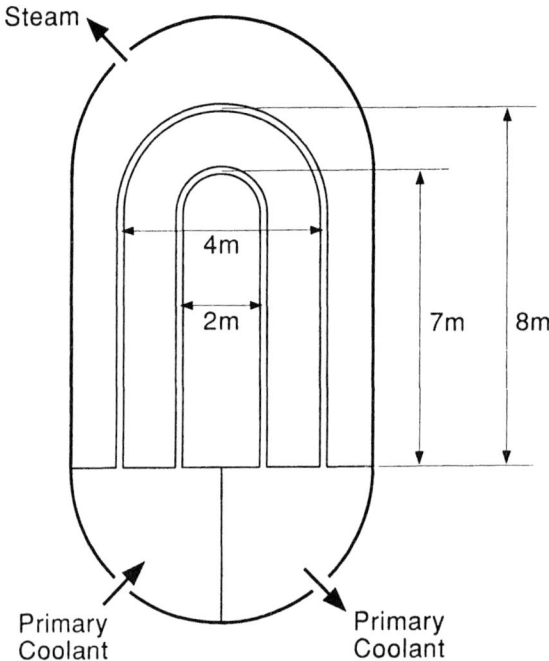

Figure 4-21 U-tube steam generator schematic.

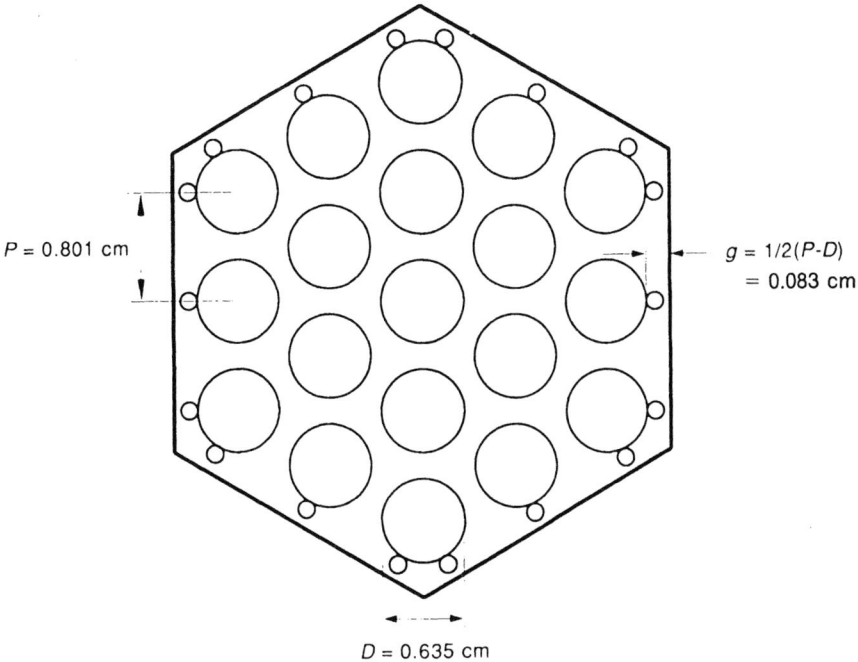

Figure 4-22 Hexagonal bundle for Problem 4-5.

2. What is the pressure drop in the steam generator if there are 10,000 tubes of each kind and the flow rate is 2,000 kg/s. Would the answer to questions a and b above be changed if the steam generator were horizontal and not vertical? Explain your answer.

Water properties	Form loss coefficient due to U-bend
Density $\rho = 690$ kg/m^3	$K = 1.0$
Dynamic viscosity $\mu = 9 \times 10^{-5}$ kg/m·s	

Answer: $\dfrac{\dot{m}_1}{\dot{m}_2} = 1.11$

$$\Delta p = 1.27 \times 10^{-3} \text{ MPa}$$

Problem 4-5 Flow split among subchannels in a hexagonal bundle (Section VIII)
It is proposed to make the coolant temperature rise equal for edge and center channels of the 19-pin bundle in Figure 4-22 by attaching vertical wires to the edge pins (one wire per subchannel). Find the required wire diameter assuming the coolant is single phase at 27 °C.

Answer: $D_s = 0.062$ cm

Problem 4-6 Flow split including manufacturing tolerance (Section VIII)
Recalculate the subchannel velocities and subchannel Reynolds numbers for Example 4-4 assuming the same coolant conditions but now include a tolerance T equal to 0.5245 mm, and assume that this tolerance is distributed with $F = 1$.

ANALYSIS OF INTERACTING CHANNELS
BY THE POROUS MEDIA APPROACH

I INTRODUCTION

This chapter presents the porous media procedure in which the region of interest is divided into a network of volumes or lumped regions, each of which will be characterized by volume-averaged parameters. This approach has been widely used for analysis of large regions containing only fluid (continuum) and for flows in regions with immersed solids. Here the principles of this method will be illustrated for the situation of the behavior of a flow field in which an array of heated rods is immersed.

Typical geometries of interest are a fuel rod array in a nuclear reactor core or the shell side of a steam generator in which an array of tubes carrying the primary coolant is immersed. Normally these arrays are periodic with an anisotropy characterized by the nominal array geometry. In specific circumstances rod distortions or foreign material blockages can exist. The porous media approach is a practical method of analyzing both the reactor core and the steam generator situations. Distributed parameter methods that yield the detailed spatial velocity and temperature fields within the coolant of such arrays are presented in Chapter 7. They are presently effective for only very idealized situations. The subchannel analysis method, which is a special case of the general porous media approach, is presented in Chapter 6.

The presence of solid objects in the flow domain has two significant effects: the geometric effect of displacing fluid, and the physical effect of altering the momentum and the energy exchanges within the flow domain. In the porous media approach solid material equivalent to the real solid material is uniformly dispersed in the flow domain. By equivalent, it is traditionally meant that the same volume of material yields the same overall shear and heat transfer effects between the fluid and solid surface. In this way the characteristics of porosity, flow resistance, and heat source (sink) of each cell replicate the volume-averaged characteristics of their real domain counterparts.

The initial step in the application of the porous media approach to the analysis of a rod array is to select the size of the volume-averaged region. Both upper and lower size limits exist. The volume should be small relative to the large-scale phenomena of interest but large relative to the scale of local phenomena present. To be more specific take the example of a PWR recirculating steam generator in which the primary coolant flows through a square array of U-tubes. Heat is transferred to the secondary coolant that flows over these tubes in upflow within the shell side of the steam generator. This secondary coolant enters the shell region from an annular downcomer. The temperature drop in the primary coolant along the tube length creates hot and cold sides of the steam generator which have markedly different characteristics. Determination of the spatial distribution of velocity, enthalpy, and void fraction within the secondary coolant is desired. A typical modern steam generator has a diameter of 4.5 meters in which about 11,000 coolant tubes are located. If detection of regions of secondary flow recirculation or stagnation is of interest, at least 6 to 8 control volumes should be placed radially across the steam generator diameter. The maximum number of nodes will be dictated on practical grounds by the required computation time but will always include many coolant tubes. Hence for this example the smallest practical computational region does not approach the scale of local phenomena.

On the other hand for some rod bundles of limited size, on the order of 100 or less pins, it becomes computationally practical to use a large number of regions, each of a size that approaches the scale of the local phenomena. However, even in such a case the region-averaged parameters cannot adequately represent local gradients and fine flow structure.

II APPROACHES TO OBTAINING
THE RELEVANT EQUATIONS

There are two approaches to the development of the relevant porous media equations. Mastery of both approaches will be well rewarded by an enhanced ability to survey the existing literature. The approaches are:

- Integration of the differential, fluid conservation equations of Chapters 4 and 5 of Volume I over a volume containing the distributed solids
- Direct application of the conservation principles to a volume containing the distributed solids.

Because most practical applications of porous body modeling are numerical, any set of governing equations must be differenced. The first approach is a conventional, analytical procedure. The second approach yields difference equations directly, but since it is accomplished by observation, terms can be forgotten by the inexperienced analyst. Both approaches will be demonstrated for the mass conservation equations. For the linear momentum and energy equations, only the first approach will be developed. The differential forms of the relevant conservation equations used as the starting point for these derivations are those appearing in Table 4-6 of Volume I with gravity as the only body force.

The conservation equations that are derived in this chapter are instantaneous, volume-averaged relations applicable to single-phase conditions. The formal development of an analogous equation set for two-phase conditions is not yet reported. The interested reader can consult Sha et al. [4] for work in this area. Time averaging of these relations for turbulent conditions will introduce turbulent parameters. The time averaging of the single-phase relations is done in Chapter 6 as part of the procedure of specializing these results to a subchannel-sized control volume.

III FUNDAMENTAL RELATIONS

Consider a domain consisting of a single-phase fluid and distributed solids. Initially the solids will be assumed deformable but stationary in space so that the resulting equations may be also applied to describe a two-phase situation. Heat may be generated or absorbed by the solid structure. About an arbitrary point in the domain, we associate a closed surface A_T enclosing a volume, V_T. The portion of V_T which contains the fluid is V_f. The total fluid–solid interface within the volume V_T is A_{fs}. The portion of A_T through which the fluid may flow is A_f. A schematic of the control volume is illustrated in Figure 5-1. The fluid within the closed volume is bounded by an area A_{fb} which equals the sum of A_{fs} and A_f, i.e.:

$$A_{fb} \equiv A_{fs} + A_f$$

Note that A_{fb} is not identical to A_T.

Two sets of relations are needed for the development of the relevant porous media equations: a set of definitions for the porosity of the media, and a set of theorems. These will be presented next.

A Porosity Definitions

The ratio of fluid volume V_f to the total volume V_T is defined as the volume porosity γ_V. Thus:

$$\gamma_V \equiv \frac{V_f}{V_T} \tag{5-1}$$

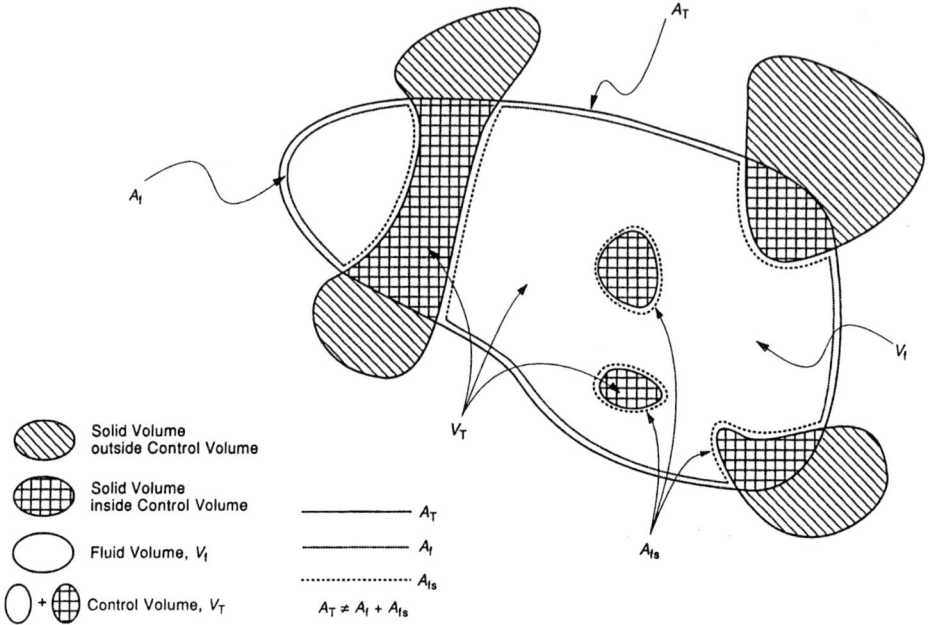

Figure 5-1 Region consisting of a single-phase fluid with stationary solids.

Utilizing the phase density function α_k defined by Eq. I,5-1, the fluid volume V_f can be expressed as:

$$V_f = \int_{V_T} \alpha_\ell dV \tag{5-2}$$

so that γ_V can also be written as:

$$\gamma_V = \frac{1}{V_T} \int_{V_T} \alpha_\ell dV \tag{5-3}$$

In some porous body formulations, only a volume porosity is utilized. Some formulations have introduced the additional concept of an area porosity or percentage area for flow associated with the surface enclosing the volume. As Figure 5-2 shows, two grid selections can have the same volume porosity but different surface porosities. The mathematical definition of the surface porosity γ_A associated with any surface (not necessarily closed) is:

$$\gamma_A \equiv \frac{A_f}{A_T} = \frac{1}{A_T} \int_{A_T} \alpha_\ell dA \tag{5-4}$$

where A_f is the portion of A_T which is occupied by the fluid. Some authors use the term *surface permeability* in place of *surface porosity*. Since permeability is conventionally defined in porous media to provide a relationship between su-

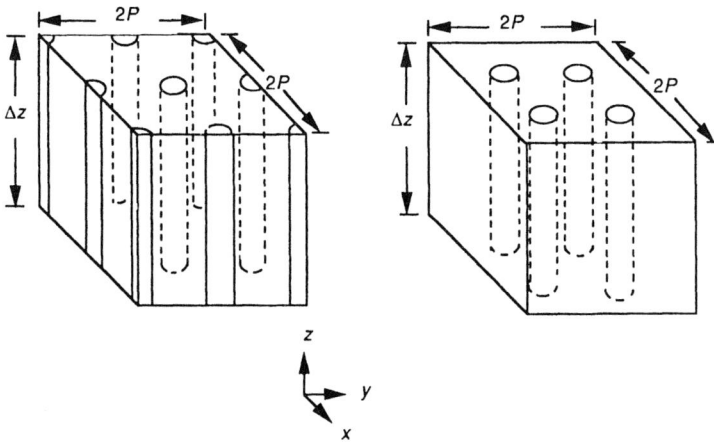

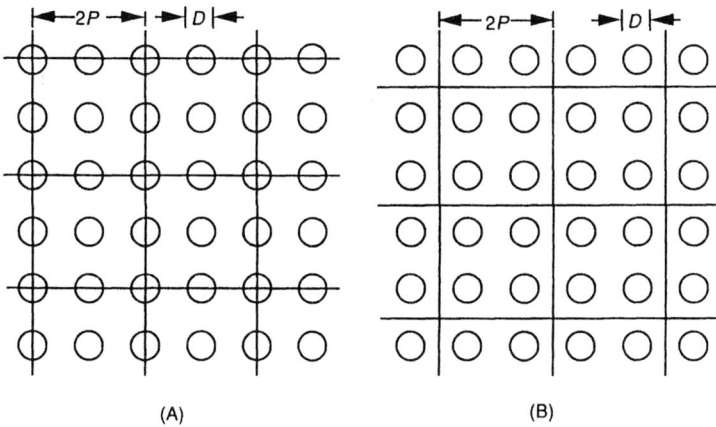

Figure 5-2 Two mesh layouts and associated control volumes for light water reactor fuel.

perficial velocity and pressure gradient, the term *surface porosity* is adopted here for γ_A.

Example 5-1 Computation of control volume characteristics for a PWR square array

PROBLEM The volumes V_T and V_f as well as the areas A_T, A_{fs}, and A_f are defined by Figure 5-1. Compute numerical values for these volumes, areas, volume porosity, and surface porosities for the two control volumes of Figure 5-2 assuming typical PWR characteristics.

SOLUTION Take numerical values of pitch P and rod diameter D as 12.6 mm and 9.5 mm, respectively from Table I,1-3, i.e., $P/D = 1.326$.

The following results are applicable to both arrays:

$$V_T = 4P^2\Delta z = 4(12.6)^2\Delta z = 635.04\Delta z \text{ mm}^3$$

$$V_f = 4P^2\Delta z - 4\left(\frac{\pi D^2\Delta z}{4}\right) = [635.04 - \pi(9.5)^2]\Delta z = 351.1\Delta z \text{ mm}^3$$

$$A_T = 4(2P\Delta z) + 2(2P)^2 = 8(12.6)\Delta z + 8(12.6)^2$$
$$= (100.8\Delta z + 1270.1) \text{ mm}^2$$

$$A_{fs} = 4\pi D\Delta z = 4\pi(9.5)\Delta z = 119.38\Delta z \text{ mm}^2.$$

The fluid area A_f differs for the two grid layouts; for layout A:

$$A_f = 4(2P\Delta z) - 4(2D\Delta z) + 2(2P)^2 - 2\left(4\frac{\pi D^2}{4}\right)$$
$$= 8(12.6)\Delta z - 8(9.5)\Delta z + 8(12.6)^2 - 2\pi(9.5)^2$$
$$= (24.8\Delta z + 703.0)\text{mm}^2$$

and for layout B:

$$A_f = 4(2P\Delta z) + 2(2P)^2 - 2\left(4\frac{\pi D^2}{4}\right)$$
$$= 8(12.6)\Delta z + 8(12.6)^2 - 2\pi(9.5)^2 = (100.8\Delta z + 703.0)\text{mm}^2$$

Next, let us compute the volume porosity γ_V. From Eq. 5-1,

$$\gamma_V \equiv \frac{V_f}{V_T}$$

$$V_T = 4P^2\Delta z$$

and

$$V_f = \left[4P^2 - 4\left(\frac{\pi D^2}{4}\right)\right]\Delta z$$

Hence:

$$\gamma_V = 1 - \frac{\pi}{4}\left(\frac{D}{P}\right)^2$$

$$= 1 - \frac{\pi}{4}\left(\frac{1}{1.326}\right)^2 = 0.554 \quad \text{for both layouts.}$$

Next, consider surface porosities γ_A. From Eq. 5-4:

$$\gamma_A \equiv \frac{A_f}{A_T}$$

By observation: $\gamma_{Az} = \gamma_V = 0.554$ for both layouts.

For layout B, again by observation:

$$\gamma_{Ax} = \gamma_{Ay} = 1$$

However, for layout A:

$$\gamma_{Ax} = \gamma_{Ay} = \frac{2(P - D)}{2P}$$

$$= 1 - D/P$$

$$= 1 - (1/1.326) = 0.246$$

B Theorems

Two theorems dealing with local volume averages are to be introduced. They are derived in Slattery [5] utilizing the general or the Reynolds transport theorem, which have been presented as Eqs. I,4-11 and I,4-12 respectively.

Theorem for local volume averaging of a divergence Let B be interpreted as a spatial vector field or second-order tensor field. This theorem expresses how to average the divergence of such a function $\vec{B}$ over a local volume. The result that is derived in Slattery [5] is:

$$\langle \nabla \cdot \vec{B} \rangle = \nabla \cdot \langle \vec{B} \rangle + \frac{1}{V_T} \int_{A_{fs}} \vec{B} \cdot \vec{n}\, dA \qquad (5\text{-}5)$$

where A_{fs} is the total fluid–solid interface within the volume V_T and $\langle \ \rangle$ designates that the average is associated with volume (as defined by Eq. I,5-2).

Expression for the divergence of an intrinsic local volume average Here let B be any scalar, spatial vector, or second-order tensor associated with the fluid. Now we desire to express the divergence of the intrinsic local volume average of B over the fluid volume only. Slattery [5] obtains this result as a step in the derivation of the theorem for local volume averaging of a divergence as:

$$\nabla \int_{V_f} B\, dV = \int_{A_f} B\, \vec{n}\, dA \qquad \text{for } B, \text{ a scalar} \qquad (5\text{-}6a)$$

and

$$\nabla \cdot \int_{V_f} \vec{B}\, dV = \int_{A_f} \vec{B} \cdot \vec{n}\, dA \qquad \text{for } \vec{B}, \text{ a vector} \qquad (5\text{-}6b)$$

where V_f is the fluid volume and A_f is the portion of the total surface area through which fluid may flow.

IV DERIVATION OF THE VOLUME-AVERAGED MASS CONSERVATION EQUATION

Before integrating the differential form of the mass conservation equation, some useful definitions of the averages will be given.

A Some Useful Definitions of Averages

Let us introduce these definitions by performing the volume integration of the governing differential equation for mass conservation of the fluid noted below:

$$\frac{\partial \rho}{\partial t} + \nabla \cdot (\rho \vec{v}) = 0 \qquad (I,4\text{-}73)$$

Integrating Eq. I,4-73 over the volume V_T and dividing by V_T yields:

$$\frac{1}{V_T} \int_{V_T} \frac{\partial \rho}{\partial t}\, dV + \frac{1}{V_T} \int_{V_T} (\nabla \cdot \rho \vec{v})\, dV = 0 \qquad (5\text{-}7)$$

Since fluid only exists over volume V_f, the integrals reduce to:

$$\frac{1}{V_T} \int_{V_f} \frac{\partial \rho}{\partial t}\, dV + \frac{1}{V_T} \int_{V_f} (\nabla \cdot \rho \vec{v})\, dV = 0 \qquad (5\text{-}8)$$

These volume integrals are commonly written in the following shorthand form as:

$$\left\langle \frac{\partial \rho}{\partial t} \right\rangle + \langle \nabla \cdot \rho \vec{v} \rangle = 0 \qquad (5\text{-}9)$$

Eqs. 5-8 and equivalently 5-9 illustrate volume averages of interest. The formal definition of a volume average operator acting on any parameter c over the entire volume has been given by Eq. I,5-2 as:

$$\langle c \rangle \equiv \frac{1}{V_T} \int\!\!\int\!\!\int_{V_T} c\, dV \qquad (I,5\text{-}2)$$

Now if we specialize the parameter c to be associated with phase k then

$$\langle c_k \rangle \equiv \frac{1}{V_T} \int_{V_T} c\alpha_k\, dV \qquad (5\text{-}10a)$$

where the phase density function α_k has been defined by Eq. I,5-1.

Let us focus attention on intensive properties associated with the fluid which can be scalar, vector, or second-order tensor and define $c_{k=\ell}$ specifically as ψ. Hence Eq. 5-10a can be written as:

$$\langle c_{k=\ell} \rangle \equiv \langle \psi \rangle \equiv \frac{1}{V_T} \int_{V_T} \psi\alpha_{k=\ell}\, dV \qquad (5\text{-}10b)$$

Since ψ equals zero outside the fluid, Eq. 5-10b reduces to:

$$\langle \psi \rangle = \frac{1}{V_T} \int_{V_f} \psi\, dV \qquad (5\text{-}11)$$

This volume averaging of properties is similar to the process for obtaining the volumetric phase averaged properties in Volume I, Section 5III. Recall that the bracket $\langle\ \rangle$ designates the average within a volume, and the bracket $\{\ \}$ designates area averages.

The local volume average defined by Eqs. 5-10a and b and 5-11 is for the total volume V_T. If the average is taken over the fluid volume V_f, a different average is obtained which is appropriately called the intrinsic local volume average of ψ, i.e.:

$$i\langle\psi\rangle \equiv \frac{1}{V_f} \int_{V_f} \psi \, dV \tag{5-12}$$

Utilizing the definition of volume porosity, i.e., Eq. 5-1, note that

$$\langle\psi\rangle = \gamma_V \, i\langle\psi\rangle \tag{5-13}$$

Similarly, the local area average of ψ using the phase density function α_k is:

$$(\vec{r})\{\psi\} \equiv \frac{1}{A_T} \int_{A_T} \psi\alpha_{k=\ell} \, dA = \frac{1}{A_T} \int_{A_f} \psi \, dA \tag{5-14}$$

The associated intrinsic area average is:

$$i(\vec{r})\{\psi\} \equiv \frac{1}{A_f} \int_{A_f} \psi \, dA \tag{5-15}$$

The superscript $\vec{r}$ in the definition $i(\vec{r})\{\psi\}$ is necessary to define the area over which the averaging process is being performed. As an illustration, consider the Cartesian coordinate system of Figure 5-3 in which the averaging volume is the parallelpiped $\Delta x \Delta y \Delta z$. The average mass flux through the surface ΔA_x whose

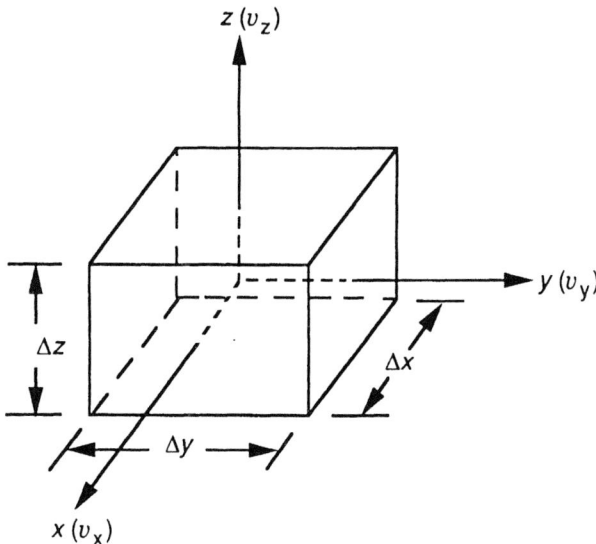

Note: The Centroid of Volume V_T is located at the origin.

Figure 5-3 Local averaging volume V_T in the Cartesian coordinate system.

normal points in the direction of the positive x-axis is:

$$^{(x)}\{\rho v_x\} = \frac{1}{\Delta A_x} \int_{\Delta A_x} \rho v_x \alpha_{k=\ell}\, dA$$

$$= \frac{1}{\Delta A_x} \int_{\Delta A_{x,f}} \rho v_x\, dA \tag{5-16}$$

where $\Delta A_x = \Delta y \Delta z$, $\Delta A_{x,f}$ denotes the fluid portion of the surface ΔA_x, and v_x is the velocity component in the positive x-direction.

The corresponding intrinsic area average is:

$$^{i(x)}\{\rho v_x\rangle = \frac{1}{\Delta A_{x,f}} \int_{\Delta A_{x,f}} \rho v_x\, dA \tag{5-17}$$

B Derivation of the Mass Conservation Equation: Method of Integration over a Control Volume

Integrating the differential mass continuity Eq. I,4-73 over the volume V, yields Eq. 5-9. The final desired form of the volume-averaged continuity equation is found by transforming each term of Eq. 5-9 by use of the established theorems. These transformations are performed next.

To transform the first term, $\langle \partial \rho / \partial t \rangle$, we must utilize a result obtained by specializing the general transport theorem, Eq. I,4-11, to the case of fluid density. Since the fluid density is zero in the solid, Eq. I,4-11 becomes:

$$\frac{d}{dt} \int_{V_f} \rho\, dV = \int_{V_f} \frac{\partial \rho}{\partial t}\, dV + \int_{A_f + A_{fs}} \rho \vec{v}_s \cdot \vec{n}\, dA \tag{5-18}$$

Now taking the volume stationary, the velocity $\vec{v}_s$ on the surface A_f is zero since $\vec{v}_s$ has been defined identically as the velocity of the bounding surface (*not* the fluid velocity) and the total derivative d/dt can be written as a partial derivative $\partial / \partial t$. Utilizing Eq. 5-11 which defines a volume average, Eq. 5-18 becomes:

$$\frac{\partial}{\partial t} \langle \rho \rangle = \left\langle \frac{\partial \rho}{\partial t} \right\rangle + \frac{1}{V_T} \int_{A_{fs}} \rho \vec{v}_{fs} \cdot \vec{n}\, dA \tag{5-19}$$

where $\vec{v}_s$ at the fluid–solid interface is written as $\vec{v}_{fs}$.

Eq. 5-19, valid for deformable solids, is a useful result for two-phase applications. Considering only nondeformable solids like fuel rods, $\vec{v}_{fs} = 0$ since the surface is completely fixed in the space and time domains so that:

$$\frac{1}{V_T} \int_{A_{fs}} \rho \vec{v}_{fs} \cdot \vec{n}\, dA = 0 \tag{5-20}$$

Eq. 5-19 then reduces to the desired result:

$$\left\langle \frac{\partial \rho}{\partial t} \right\rangle = \frac{\partial \langle \rho \rangle}{\partial t} \tag{5-21}$$

Applying Eq. 5-13, the left side term of Eq. 5-21 can be further transformed into an intrinsic local volume average yielding:

$$\left\langle \frac{\partial \rho}{\partial t} \right\rangle = \frac{\partial \langle \rho \rangle}{\partial t} = \gamma_v \frac{\partial^i \langle \rho \rangle}{\partial t} \tag{5-22}$$

Next transform the second term of Eq. 5-9, i.e., $\langle \nabla \cdot \rho \vec{v} \rangle$. Applying Eq. 5-5 where $\vec{B}$ is taken as $\rho \vec{v}$, gives:

$$\langle \nabla \cdot \rho \vec{v} \rangle = \nabla \cdot \langle \rho \vec{v} \rangle \tag{5-23}$$

since the area integral term is zero because $\vec{v} = 0$ everywhere on A_{fs}.

Rewriting Eq. 5-23 in terms of the definition of volume average gives:

$$\nabla \cdot \langle \rho \vec{v} \rangle = \frac{1}{V_T} \nabla \cdot \int_{V_f} \rho \vec{v} \, dV \tag{5-24}$$

which, utilizing Eq. 5-6b, can be expressed as

$$\frac{1}{V_T} \nabla \cdot \int_{V_f} \rho \vec{v} \, dV = \frac{1}{V_T} \int_{A_f} \rho \vec{v} \cdot \vec{n} \, dA \tag{5-25}$$

Finally substituting the results of Eqs. 5-22 and 5-25 into Eq. 5-9 yields the desired result:

$$\gamma_v \frac{\partial^i \langle \rho \rangle}{\partial t} + \frac{1}{V_T} \int_{A_f} \rho \vec{v} \cdot \vec{n} \, dA = 0 \tag{5-26}$$

Let us now consider the special case of a Cartesian coordinate system in which the volume V_T is taken to be $\Delta x \Delta y \Delta z$ as shown in Figure 5-3. The associated coordinate axes are x, y, and z, and their corresponding velocity components are v_x, v_y, and v_z. Evaluating the integral of Eq. 5-26 over the surfaces of the volume V_T yields:

$$\int_{A_f} \rho \vec{v} \cdot \vec{n} \, dA = \int_{A_f \big|_{x + \frac{\Delta x}{2}}} \rho v_x dA_x - \int_{A_f \big|_{x - \frac{\Delta x}{2}}} \rho v_x dA_x + \int_{A_f \big|_{y + \frac{\Delta y}{2}}} \rho v_y dA_y$$

$$\tag{5-27}$$

$$- \int_{A_f \big|_{y - \frac{\Delta y}{2}}} \rho v_y dA_y + \int_{A_f \big|_{z + \frac{\Delta z}{2}}} \rho v_z dA_z - \int_{A_f \big|_{z - \frac{\Delta z}{2}}} \rho v_z dA_z$$

where $A_f\big|_{x + \Delta x/2}$ denotes the free flow area normal to the x-axis located at the position $x + \Delta x/2$. It is equal to:

$$(\Delta y \Delta z) \gamma_{Ax} \big|_{x + \frac{\Delta x}{2}}$$

where:

$$\gamma_{Ax} \big|_{x + \frac{\Delta x}{2}}$$

is the surface permeability. Likewise:

$$A_f\Big|_{x - \frac{\Delta x}{2}} = (\Delta y \Delta z) \gamma_{Ax}\Big|_{x - \frac{\Delta x}{2}}$$

Corresponding terms in the coordinates y and z are similarly evaluated. From the definition of the intrinsic area average, Eq. 5-15, the surface integrals can be written as:

$$\int_{A_f}\Big|_{x + \frac{\Delta x}{2}} \rho v_x dA_x = (\Delta y \Delta z) \ \gamma_{Ax}\Big|_{x + \frac{\Delta x}{2}} {}^{i(x)}\{\rho v_x\}\Big|_{x + \frac{\Delta x}{2}}$$

$$\int_{A_f}\Big|_{x - \frac{\Delta x}{2}} \rho v_x dA_x = (\Delta y \Delta z) \ \gamma_{Ax}\Big|_{x - \frac{\Delta x}{2}} {}^{i(x)}\{\rho v_x\}\Big|_{x - \frac{\Delta x}{2}}$$

$$\int_{A_f}\Big|_{y + \frac{\Delta y}{2}} \rho v_y dA_y = (\Delta z \Delta x) \ \gamma_{Ay}\Big|_{y + \frac{\Delta y}{2}} {}^{i(y)}\{\rho v_y\}\Big|_{y + \frac{\Delta y}{2}}$$

and so on. Consequently, Eq. 5-26 becomes:

$$\gamma_v \frac{\partial^i \langle \rho \rangle}{\partial t} + \frac{\Delta_x(\gamma_{Ax}{}^{i(x)}\{\rho v_x\})}{\Delta x} + \frac{\Delta_y(\gamma_{Ay}{}^{i(y)}\{\rho v_y\})}{\Delta y} + \frac{\Delta_z(\gamma_{Az}{}^{i(z)}\{\rho v_z\})}{\Delta z} = 0 \quad (5\text{-}28)$$

where:

$$\Delta_x(\ ) \equiv (\ )_{x + \frac{\Delta x}{2}} - (\ )_{x - \frac{\Delta x}{2}}$$

$$\Delta_y(\ ) \equiv (\ )_{y + \frac{\Delta y}{2}} - (\ )_{y - \frac{\Delta y}{2}} \quad (5\text{-}29)$$

$$\Delta_z(\ ) \equiv (\ )_{z + \frac{\Delta z}{2}} - (\ )_{z - \frac{\Delta z}{2}}$$

Eq. 5-28 is the volume-averaged mass conservation equation for the volume element shown in Fig. 5-3. Table 5-1 compiles the single-phase porous media equations, i.e., the mass conservation equation which has already been derived, and the linear momentum and energy conservation equations which are derived later. Both the general integral equation form and the specific form for a control volume of size $\Delta x \Delta y \Delta z$ in the Cartesian coordinate system are presented. Eq. 5-28 contains the spatial average of the product of density and various velocities, unknowns that have been introduced by the averaging process. They are commonly dealt with by the approximation of neglecting certain spatially varying cross-products as detailed next.

Analogous to the treatment of instantaneous temporal variations in turbulent flow in Volume I, Section 4VI, the relevant densities and velocities can be expressed as the sum of an average and fluctuating component. Here, however,

Table 5-1 Generalized conservation equations of a single-phase fluid in porous media

	General integral form	Form for control volume $\Delta x\Delta y\Delta z$ in a Cartesian coordinate system. Product of averages of velocities and densities has been approximated as average of products; hence temporal and spatial variations have been neglected.
Continuity equation	$\gamma_V \dfrac{\partial^i\langle\rho\rangle}{\partial t} + \dfrac{1}{V_T}\displaystyle\int_{A_f} \rho\vec{v}\cdot\vec{n}\,dA = 0$ (5-26)	(5-33)
Linear momentum equation	$\gamma_V \dfrac{\partial}{\partial t}{}^i\langle\rho\vec{v}\rangle + \dfrac{1}{V_T}\displaystyle\int_{A_f}\rho\vec{v}(\vec{v}\cdot\vec{n})\,dA = \gamma_V{}^i\langle\rho\rangle\vec{g} + \dfrac{1}{V_T}\displaystyle\int_{A_f}(-p\vec{n}+\bar{\bar{\tau}}\cdot\vec{n})\,dA$ $\qquad\qquad\qquad\qquad\qquad\qquad + \gamma_V\langle\vec{R}\rangle$ (5-43)	z-direction (5-46)
Energy equations in terms of internal energy and enthalpy	$\gamma_V\dfrac{\partial}{\partial t}{}^i\langle\rho u\rangle + \dfrac{1}{V_T}\displaystyle\int_{A_f}\rho u\vec{v}\cdot\vec{n}\,dA = -\gamma_V{}^i\langle p\nabla\cdot\vec{v}\rangle + \dfrac{1}{V_T}\displaystyle\int_{A_f}k_e\vec{n}\cdot\nabla T\,dA$ $\qquad\qquad\qquad + \gamma_V[{}^i\langle q_{rb}'''\rangle + {}^i\langle q'''\rangle + {}^i\langle\phi\rangle]$ (5-49) $\gamma_V\dfrac{\partial}{\partial t}{}^i\langle\rho h\rangle + \dfrac{1}{V_T}\displaystyle\int_{A_f}\rho h\vec{v}\cdot\vec{n}\,dA = \gamma_V\left\langle\dfrac{{}^iDp}{Dt}\right\rangle + \dfrac{1}{V_T}\displaystyle\int_{A_f}k_e\vec{n}\cdot\nabla T\,dA$ $\qquad\qquad\qquad + \gamma_V[{}^i\langle q_{rb}'''\rangle + {}^i\langle q'''\rangle + {}^i\langle\phi\rangle]$ (5-56)	Internal energy (5-55) Enthalpy (5-59)

the average is a spatial average, and the fluctuating component is a local spatial variation. Thus

$$\rho \equiv \{\rho\} + \rho' \quad \text{where } \{\rho'\} \equiv 0 \tag{5-30}$$

and, for any velocity component v:

$$v \equiv \{v\} + v' \quad \text{where } \{v'\} \equiv 0 \tag{5-31}$$

Now by the above definitions, the following averages are zero:

$$\{\{\rho\}v'\} = 0$$

$$\{\rho'\{v\}\} = 0$$

and

$$\{\{\rho\}\{v\}\} = \{\rho\}\{v\}$$

Utilizing the above results the area average value of a typical density–velocity product becomes:

$$\{\rho v\} = \{(\{\rho\} + \rho')(\{v\} + v')\} = \{\{\rho\}\{v\} + \{\rho\}v' + \rho'\{v\} + \{\rho'v'\}\}$$
$$= \{\rho\}\{v\} + \{\rho'v'\} \tag{5-32a}$$

The result is the introduction of the term $\{\rho'v'\}$ stemming from nonuniform spatial differences of v and ρ as the fluid moves through the void space between solids. This term is nonzero even if the local flow is laminar. However, it is usually neglected compared with other terms in the equation. Similar spatially varying cross-products exist but are also neglected in the momentum and energy equations. Rarely is an adequate assessment done of all terms in the equations to determine their comparative magnitudes and thereby justify the neglect of these spatially varying cross-product terms.

Additionally the time averaging of the density–velocity product introduces an additional term, i.e., time averaging Eq. 5-32a yields:

$$\overline{\{\rho v\}} = \{\bar{\rho}\}\{\bar{v}\} + \{\bar{\rho}'\bar{v}'\} + \overline{\{\rho^t\}\{v^t\}} + \overline{\{\rho^t\}\{v^t\}} \tag{5-32b}$$

where superscript prime refers to spatial variation, and superscript t refers to time fluctuation.

The second, third and fourth terms on the right side physically represent time averages of spatial and temporal variations, and should be included where significant. The second term is usually ignored and the third and fourth terms are usually referred to as the turbulent interchange as will be discussed in Chapter 6.

Finally applying these conclusions to Eq. 5-28, the desired volume-averaged equation is obtained:

$$\gamma_v \frac{\partial^i \langle \rho \rangle}{\partial t} + \frac{\Delta_x(\gamma_{Ax}{}^{i(x)}\{\rho\}^{i(x)}\{v_x\})}{\Delta x}$$
$$+ \frac{\Delta_y(\gamma_{Ay}{}^{i(y)}\{\rho\}^{i(y)}\{v_y\})}{\Delta y} + \frac{\Delta_z(\gamma_{Az}{}^{i(z)}\{\rho\}^{i(z)}\{v_z\})}{\Delta z} = 0 \tag{5-33}$$

or in the usual shorthand notation:

$$\gamma_V \frac{\partial \rho}{\partial t} + \frac{\Delta_x(\gamma_{Ax}\{\rho\}\{v_x\})}{\Delta x} + \frac{\Delta_y(\gamma_{Ay}\{\rho\}\{v_y\})}{\Delta y} + \frac{\Delta_z(\gamma_{Az}\{\rho\}\{v_z\})}{\Delta z} = 0 \qquad (5\text{-}34)$$

C Derivation of the Mass Conservation Equation: Application of Conservation Principles to a Volume Containing Distributed Solids

In this derivation a specific coordinate system is adopted. Let us take the Cartesian coordinate system and volume of Figure 5-3 which is considered as before to be fixed in space. Applying the mass conservation principle directly to the volume of Figure 5-3, we obtain:

$$\underbrace{\frac{\partial}{\partial t}(\rho\gamma_V \Delta x \Delta y \Delta z)}_{\text{rate of mass increase}} + \underbrace{\{\rho v_x\gamma_{Ax}\Delta y \Delta z\}_{x+(\Delta x/2)} - \{\rho v_x\gamma_{Ax}\Delta y \Delta z\}_{x-(\Delta x/2)}}_{\text{net outflux in } x \text{ direction}}$$

$$+ \underbrace{\{\rho v_y\gamma_{Ay}\Delta x \Delta z\}_{y+(\Delta y/2)} - \{\rho v_y\gamma_{Ay}\Delta x \Delta z\}_{y-(\Delta y/2)}}_{\text{net outflux in } y \text{ direction}} \qquad (5\text{-}35)$$

$$+ \underbrace{\{\rho v_z\gamma_{Az}\Delta x \Delta y\}_{z+(\Delta z/2)} - \{\rho v_z\gamma_{Az}\Delta x \Delta y\}_{z-(\Delta z/2)}}_{\text{net outflux in } z \text{ direction}} = 0$$

Now recognize that for nondeformable, spatially fixed solids, γ_V is not a function of time. Further, denoting the densities as intrinsic densities over the volume or its surrounding surfaces, Eq. 5-35 becomes identical to Eq. 5-28.

Subsequent treatment of the density–velocity cross-products as presented in Section IVB is directly applicable here.

V DERIVATION OF THE VOLUMETRIC AVERAGED LINEAR MOMENTUM EQUATION

We begin with the dynamic equation of fluid motion with gravity as the only body force:

$$\frac{\partial(\rho\vec{v})}{\partial t} + \nabla \cdot \rho\vec{v}\vec{v} = -\nabla p + \nabla \cdot \bar{\bar{\tau}} + \rho\vec{g} \qquad (\text{I},4\text{-}80)$$

Performing the local volume averaging of Eq. I,4-80 gives

$$\left\langle \frac{\partial(\rho\vec{v})}{\partial t} \right\rangle + \langle \nabla \cdot (\rho\vec{v}\vec{v}) \rangle = \langle\rho\rangle\vec{g} - \langle\nabla p\rangle + \langle\nabla \cdot \bar{\bar{\tau}}\rangle \qquad (5\text{-}36)$$

where $\vec{g}$ is taken as constant.

Further manipulation of Eq. 5-36 will proceed analogously to the treatment of the continuity Eq. 5-9. Specifically:

- The first term $\langle \partial(\rho\vec{v})/\partial t \rangle$ is transformed as the term $\langle \partial\rho/\partial t \rangle$ was transformed in Eqs. 5-18 through 5-21 yielding:

$$\left\langle \frac{\partial(\rho\vec{v})}{\partial t} \right\rangle = \frac{\partial}{\partial t} \langle \rho\vec{v} \rangle \tag{5-37a}$$

- All additional terms other than the gravitational term are transformed using the theorem for local averaging of a divergence, i.e., Eq. 5-5, yielding:

$$\langle \nabla \cdot (\rho\vec{v}\vec{v}) \rangle = \nabla \cdot \langle \rho\vec{v}\vec{v} \rangle \text{ since } \vec{v} \text{ equals zero on } A_{fs} \tag{5-37b}$$

$$\langle \nabla p \rangle = \nabla \langle p \rangle + \frac{1}{V_T} \int_{A_{fs}} p\vec{n} \, dA \tag{5-37c}$$

$$\langle \nabla \cdot \overline{\overline{\tau}} \rangle = \nabla \cdot \langle \overline{\overline{\tau}} \rangle + \frac{1}{V_T} \int_{A_{fs}} \overline{\overline{\tau}} \cdot \vec{n} \, dA \tag{5-37d}$$

Now the divergence terms in Eqs. 5-37a through d can be rewritten using the results for the divergence of an intrinsic local volume average (Eq. 5-6a and b) noting that the velocity $\vec{v}$, pressure p, and shear stress $\overline{\overline{\tau}}$ over the solid region are zero. We obtain:

$$\nabla \cdot \langle \rho\vec{v}\vec{v} \rangle = \nabla \cdot \frac{1}{V_T} \int_{V_f} \rho\vec{v}\vec{v} \, dV = \frac{1}{V_T} \int_{A_f} \rho\vec{v}(\vec{v} \cdot \vec{n}) dA \tag{5-38}$$

$$\nabla \langle p \rangle = \nabla \left(\frac{1}{V_T} \int_{V_f} p \, dV \right) = \frac{1}{V_T} \int_{A_f} p\vec{n} \, dA \tag{5-39}$$

$$\nabla \cdot \langle \overline{\overline{\tau}} \rangle = \nabla \cdot \left(\frac{1}{V_T} \int_{V_f} \overline{\overline{\tau}} \, dV \right) = \frac{1}{V_T} \int_{A_f} \overline{\overline{\tau}} \cdot \vec{n} \, dA \tag{5-40}$$

The distributed resistance $\vec{R}$, a key concept associated with the porous media approach, is the resistance force per unit volume of fluid exerted on the fluid by the dispersed solid. It is defined by the following relation:

$$\int_{V_f} \vec{R} \, dV \equiv \int_{A_{fs}} (-p\vec{n} + \overline{\overline{\tau}} \cdot \vec{n}) dA \tag{5-41}$$

An equivalent but oppositely directed force is exerted on the dispersed solid by the fluid and is an effective drag force per unit volume of fluid. Applying Eqs. 5-37a through 5-37d and 5-38 through 5-41 to Eq. 5-36 yields:

$$\frac{\partial}{\partial t} \langle \rho\vec{v} \rangle + \frac{1}{V_T} \int_{A_f} \rho\vec{v}(\vec{v} \cdot \vec{n}) dA = \langle \rho \rangle \vec{g} + \frac{1}{V_T} \int_{A_f} (-p\vec{n} + \overline{\overline{\tau}} \cdot \vec{n}) dA$$

$$+ \frac{1}{V_T} \int_{V_f} \vec{R} \, dV \tag{5-42}$$

Finally rewriting our result in terms of intrinsic local volume averages utilizing Eqs. 5-12 and 5-13 yields:

$$\gamma_V \frac{\partial}{\partial t} {}^i\langle \rho \vec{v} \rangle + \frac{1}{V_T} \int_{A_f} \rho \vec{v}(\vec{v} \cdot \vec{n}) dA = \gamma_V {}^i\langle \rho \rangle \vec{g} + \frac{1}{V_T} \int_{A_f} (-p\vec{n} + \bar{\bar{\tau}} \cdot \vec{n}) dA$$
$$+ \gamma_V {}^i\langle \vec{R} \rangle \tag{5-43}$$

We next specialize this result to the Cartesian coordinate system of Figure 5-3. By a procedure analogous to that employed for the mass conservation equation, we can write the result for the z-component of the linear momentum equation as:

$$\gamma_V \frac{\partial}{\partial t} {}^i\langle \rho v_z \rangle + \frac{\Delta_x(\gamma_{Ax}{}^{i(x)}\{\rho v_z v_x\})}{\Delta x} + \frac{\Delta_y(\gamma_{Ay}{}^{i(y)}\{\rho v_z v_y\})}{\Delta y} + \frac{\Delta_z(\gamma_{Az}{}^{i(z)}\{\rho v_z^2\})}{\Delta z}$$

$$= -\gamma_V {}^i\langle \rho \rangle g_z - \frac{\Delta_z(\gamma_{Az}{}^{i(z)}\{p\})}{\Delta z} + \frac{\Delta_x(\gamma_{Ax}{}^{i(x)}\{\tau_{xz}\})}{\Delta x} \tag{5-44}$$

$$+ \frac{\Delta_y(\gamma_{Ay}{}^{i(y)}\{\tau_{yz}\})}{\Delta y} + \frac{\Delta_z(\gamma_{Az}{}^{i(z)}\{\tau_{zz}\})}{\Delta z} + \gamma_V {}^i\langle R_z \rangle$$

where gravity is assumed acting along the negative z-axis. Eq. 5-44 is the volume-averaged linear momentum conservation equation for the volume element shown in Figure 5-3. This equation contains the spatial average of the product of density and various velocities, unknowns introduced by the averaging process. These terms are dealt with in a manner analogous to their treatment in the mass conservation equation by neglecting certain spatially varying cross-products.

Expressing the relevant densities and velocities as the sum of a spatially averaged component and a local spatially varying component and proceeding as in Section IVB, the average value of a typical density–velocity–velocity product becomes:

$$\{\rho v_x v_y\} = \{\rho\}\{v_x\}\{v_y\} + \{\rho\}(\{v_x' v_y'\} + \{v_x\}\{\rho' v_y'\} + \{v_y\}\{\rho' v_x'\} + \{\rho' v_x' v_y'\} \tag{5-45}$$

The result is the introduction of terms involving spatial averages of cross-products of local spatially varying components. These terms stem from the nonuniform spatial differences of ρ, v_x, v_y, and v_z as the fluid moves through the void space between solids. They are nonzero even if the local flow is laminar. They are usually neglected compared with other terms.

Finally, making the approximation of equating the average of the product of ρ, v_x, v_y, and v_z to the product of the averages of ρ, v_x, v_y, and v_z, Eq. 5-44 becomes:

$$\gamma_V \overset{(1)}{\frac{\partial}{\partial t} ({}^i\langle\rho\rangle {}^i\langle v_z\rangle)} + \overset{(2)}{\frac{\Delta_x(\gamma_{Ax}{}^{i(x)}\{\rho\}^{i(x)}\{v_z\}^{i(x)}\{v_x\})}{\Delta x}}$$

$$+ \frac{\overset{(3)}{\Delta_y(\gamma_{Ay}{}^{i(y)}\{\rho\}^{i(y)}\{v_z\}^{i(y)}\{v_y\})}}{\Delta y}$$

$$+ \frac{\overset{(4)}{\Delta_z(\gamma_{Az}{}^{i(z)}\{\rho\}^{i(z)}\{v_z\}^{i(z)}\{v_z\})}}{\Delta z} \tag{5-46}$$

$$= \overset{(5)}{-\gamma_v{}^i\langle\rho\rangle g_z} - \frac{\overset{(6)}{\Delta_z(\gamma_{Az}{}^{i(z)}\{p\})}}{\Delta z} + \frac{\overset{(7)}{\Delta_x(\gamma_{Ax}{}^{i(x)}\{\tau_{xz}\})}}{\Delta x}$$

$$+ \frac{\overset{(8)}{\Delta_y(\gamma_{Ay}{}^{i(y)}\{\tau_{yz}\})}}{\Delta y} + \frac{\overset{(9)}{\Delta_z(\gamma_{Az}{}^{i(z)}\{\tau_{zz}\})}}{\Delta z} + \overset{(10)}{\gamma_v{}^i\langle R_z\rangle}$$

The physical meanings of the terms of this equation are:

(1)	(2, 3, and 4)	(5)
rate of increase of linear momentum of the fluid mass in V_T	+ net linear momentum outflux through the surfaces enclosing V_T	= the body force due to gravity acting on the fluid mass
(6)	(7, 8, and 9)	(10)
+ the surface force due to normal fluid stress (pressure) acting on the fluid mass	+ the surface force due to the fluid shear stress acting on the fluid mass	+ the surface force exerted on the fluid by the dispersed solid

Generally the shear stress component τ_{zz} is small compared with τ_{xz} and τ_{yz}. Also for rod bundles R_z is relatively large compared with the shear stress components τ, while for continuum regions like an open plenum, R_z is relatively small compared with the shear stress components. These shear stress components arise from both molecular and turbulent phenomena in turbulent flow.

Example 5-2 Components of the distributed resistance vector for a PWR fuel array

PROBLEM The distributed resistance $^i\langle\vec{R}\rangle$ in Cartesian coordinates consists of three components $^i\langle R_x\rangle$, $^i\langle R_y\rangle$, $^i\langle R_z\rangle$. Compute these components assuming a flow in each direction characterized by a Reynolds number of 10^5 based on the volume-averaged direction velocity and length scales D_e for axial flow and D_V for transverse flow where:

$$D_V \equiv \text{volumetric hydraulic diameter} = \frac{4(\text{net free volume})}{\text{friction surface}}$$

Obtain these resistance coefficients for mesh layouts A and B of Example 5-1 which are based on PWR fuel geometry. Utilize properties based on nominal PWR operating conditions, i.e.:

$$\rho = 726 \text{ kg/m}^3$$

$$\mu = 9.63 \times 10^{-5} \text{ kg/m} \cdot \text{s}$$

SOLUTION The definition of $^i\langle \vec{R} \rangle$ from Eqs. 5-41 and 5-12 is

$$^i\langle \vec{R} \rangle = \frac{1}{V_f} \int_{A_{fs}} (-p\vec{n} + \overline{\overline{\tau}} \cdot \vec{n}) dA$$

The term $-\int_{A_{fs}} -p\vec{n} \, dA$ is the form drag, and the term $-\int_{A_{fs}} \overline{\overline{\tau}} \cdot \vec{n} \, dA$ is the friction drag. The minus sign in each term has been introduced since the drag force (which is the force exerted on the solid by the liquid) is directed opposite to the resistance force (which is the force exerted on the liquid by the solid). For pure axial flow (flow along the rod axis) without any spacers or baffles, there is no form drag. In transverse flows, due to the presence of boundary layer separation phenomena, both form and friction drag forces are important. For flow at an arbitrary direction with respect to the rod axis, separation effects, which are responsible for the transverse form drag, will also affect the axial friction drag. However, due to lack of general constitutive relations accounting for simultaneous action of form drag and friction phenomena for flow at an arbitrary direction, simple combinations of axial and transverse flow resistance correlations are typically used.

For the square array, the transverse directions x and y are symmetrical. We associate the z-coordinate with axial or longitudinal flow. Hence the resistance vector $^i\langle \vec{R} \rangle$ is composed of the transverse components:

$$^i\langle R_x \rangle \text{ and } ^i\langle R_y \rangle \text{ which are equal}$$

and the axial component:

$$^i\langle R_z \rangle$$

For axial flow, recognizing there is no form drag and applying a force balance, we obtain:

$$^i\langle R_z \rangle = \frac{1}{V_f} \vec{k} \cdot \int_{A_{fs}} \overline{\overline{\tau}} \cdot \vec{n} \, dA = -\frac{\Delta p_{fric} A_{flow}}{V_f} = -\frac{\Delta p_{fric}}{\Delta z}$$

since:

$$A_{flow} = \frac{V_f}{\Delta z} \text{ and } \tau_{fs} \equiv \tau_w \text{ which is proportional to } -\Delta p_{fric}$$

Now the axial pressure loss Δp_{fric} is available in terms of the empirical friction factor correlations, i.e., for turbulent flow in a square array with P/D equal to 1.326 at a Re $= 10^5$, using Eq. 7-60, we obtain:

$$\frac{f}{f_{c.t.}} = 1.04 + 0.06(P/D - 1) = 1.06$$

where the circular tube friction factor $f_{c.t.}$ can be obtained from the correlation of Eq. 7-59:

$$\frac{1}{(f_{c.t.})^{1/2}} = 2.035 \log_{10}[\text{Re}_{De}(f_{c.t.})^{1/2}] - 0.989$$

$$f_{c.t.} = 0.0182$$

Hence:

$$f = 1.06 f_{c.t.} = 1.06(0.0182)$$

$$f = 0.0193$$

By definition, in terms of our nomenclature:

$$\Delta p_{fric} \equiv f \frac{\Delta z}{D_e} \frac{{}^i\langle\rho\rangle^i\langle v_z\rangle^2}{2}$$

where for both mesh layouts, A and B:

$$D_e = \frac{4A_f}{P_w} = \frac{4\left[4\left(P^2 - \frac{\pi D^2}{4}\right)\right]}{4\pi D}$$

$$D_e = 11.8 \times 10^{-3} \text{ m}$$

where $P = 12.6$ mm and $D = 9.5$ mm from Table I,1-3.
Combining the previous results gives:

$$^i\langle R_z\rangle = -\frac{f}{D_e}\left[\frac{{}^i\langle\rho\rangle^i\langle v_z\rangle^2}{2}\right]; \frac{\frac{\text{kg}}{\text{m}^3}\frac{\text{m}^2}{\text{s}^2}}{\text{m}}$$

$$= -\frac{0.0193}{11.8 \times 10^{-3}}\left[\frac{{}^i\langle\rho\rangle^i\langle v_z\rangle^2}{2}\right] = -1.63\left[\frac{{}^i\langle\rho\rangle^i\langle v_z\rangle^2}{2}\right]$$

From the definition of Re_{De}:

$$^i\langle v_z\rangle = \frac{\text{Re}_{De}{}^i\langle\mu\rangle}{{}^i\langle\rho\rangle D_e} = \frac{10^5(9.63 \times 10^{-5})}{726(11.8 \times 10^{-3})} = 1.12 \text{ m/s}$$

Hence for both mesh layouts:

$$^i\langle R_z\rangle = -1.63\left[\frac{726(1.12)^2}{2}\right] = -7.44 \times 10^2 \frac{\text{kg(m/s)}^2}{\text{m}^3} \quad \text{or} \quad \frac{\text{Pa}}{\text{m}}$$

For transverse flow in the y direction:

$$^i\langle R_y\rangle = -\frac{\Delta p_{tr} A_{fy}}{V_f}$$

where for both mesh layouts:

$$V_f = \gamma_v 4P^2 \Delta z$$

while for mesh layout A:

$$A_{fy} = 2(P - D)\Delta z$$

and for mesh layout B:

$$A_{fy} = 2P\Delta z$$

since A_{fy} is the flow area in the planes across which Δp_{tr} is taken. Hence:

$$^i\langle R_y\rangle_A = -\frac{\Delta p_{tr} 2(P - D)\Delta z}{\gamma_V 4P^2 \Delta z} = -\frac{\Delta p_{tr}}{\gamma_V 2P}\left(1 - \frac{D}{P}\right)$$

and

$$^i\langle R_y\rangle_B = -\frac{\Delta p_{tr} 2P\Delta z}{\gamma_V 4P^2 \Delta z} = -\frac{\Delta p_{tr}}{\gamma_V 2P}$$

since Δp_{tr} is the same for each mesh layout because each has the same number of transverse flow cells, i.e., two.

A correlation is needed for Δp_{tr}. Use the correlation of Gunter and Shaw [1] which expresses Δp_{tr} in terms of a friction factor based on a Reynolds number, Re_{D_V}, which is defined in terms of a volumetric hydraulic diameter, D_V, and the velocity at the gap between rods, $\{v_y\}_{gap}$, i.e.:

$$\text{Re}_{D_V} \equiv \frac{D_V G}{\mu} = \frac{D_V \{\rho\}_{gap}\{v_y\}_{gap}}{\{\mu\}}$$

The correlation is expressed in the literature as:

$$\frac{f_{tr}}{2} \equiv \frac{\Delta p_{tr} D_V \rho}{G^2 L}\left(\frac{\mu}{\mu_w}\right)^{0.14}\left(\frac{D_V}{S_T}\right)^{-0.4}\left(\frac{S_L}{S_T}\right)^{-0.6}$$

$$= \frac{0.96}{(\text{Re}_{D_V})^{0.145}} \quad \text{for } \text{Re}_{D_V} > 200 \qquad (5\text{-}47a)$$

$$= \frac{90}{\text{Re}_{D_V}} \quad \text{for } \text{Re}_{D_V} < 200 \qquad (5\text{-}47b)$$

Figure 5-4 defines the parameters in the Gunter-Shaw correlation. This correlation is applicable to a variety of rod array shapes provided appropriate definitions of S_L and S_T are employed as discussed in Volume I, Chapter 9. Reformulating the Gunter-Shaw correlation in the nomenclature of this chapter for a PWR square array of pitch P yields Δp_{tr} as:

$$\Delta p_{tr} = \frac{f_{tr}}{2}\frac{L}{D_V}\{\rho\}_{gap}\{v\}_{gap}^2 \left(\frac{^i\langle\mu\rangle}{^i\langle\mu_w\rangle}\right)^{-0.14}\left(\frac{D_V}{P}\right)^{0.4}$$

The desired Δp_{tr} for the two mesh layouts is given by this expression when L is taken as $2P$ since each layout has two transverse unit flow cells.

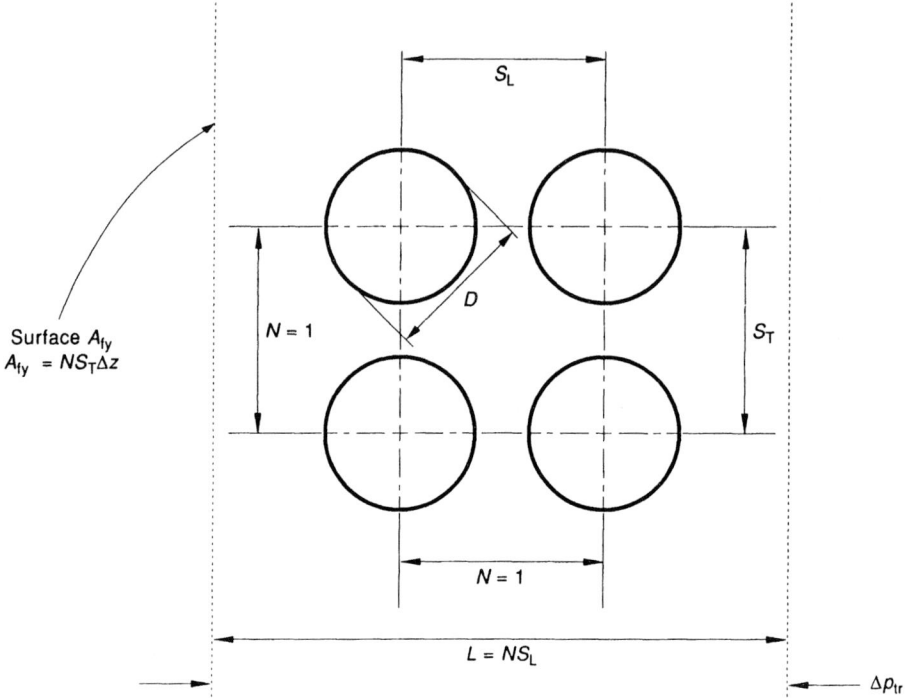

Figure 5-4 Geometric parameters characterizing an in-line square array.

Next the parameters $\{v_y\}_{gap}$ and D_V are expressed in terms of known parameters. Referring to Figure 5-5, from continuity:

$$\{\rho\}_{gap}\{v\}_{gap}A_{gap} = {}^i\langle\rho\rangle^i\langle v_y\rangle^i\langle A_{fy}\rangle$$

For the transverse unit flow cells in both layouts A and B:

$$A_{gap} = (P - D)\Delta z$$

$${}^i\langle A_{fy}\rangle = \frac{V_f}{P} = \left(\frac{V_T}{P}\right)\left(\frac{V_f}{V_T}\right) = P\Delta z\gamma_V$$

note that while the flow areas on the y-faces of the volumes of layouts A and B are not equal, i.e., $A_{fy}|_A \neq A_{fy}|_B$, the volume-averaged y-areas of these layouts are equal, i.e., $\langle A_{fy}\rangle_A = \langle A_{fy}\rangle_B$.

Assume:

$$\{\rho\}_{gap} = {}^i\langle\rho\rangle$$

Then obtain the gap velocity as:

$$\{v_y\}_{gap} = {}^i\langle v_y\rangle\gamma_V\left(\frac{P}{P - D}\right)$$

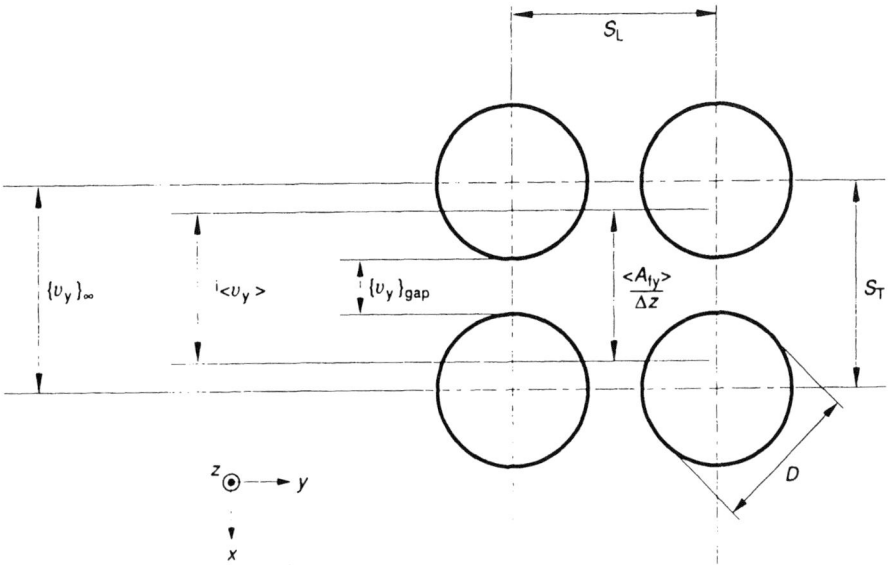

Figure 5-5 Velocities characterizing an in-line square array.

To evaluate f_{tr}, the relevant flow regime must be determined utilizing the given Re. The Re defined in the problem statement is based on the volume-averaged transverse velocity, i.e.:

$$\text{Re} \equiv \frac{{}^i\langle\rho\rangle{}^i\langle v_y\rangle D_V}{{}^i\langle\mu\rangle} = 10^5$$

On the other hand the Gunter-Shaw laminar-turbulent transition is defined in terms of Re_{D_V}. Combining previous results yields Re_{D_V} in terms of known parameters:

$$\text{Re}_{D_V} = \frac{\text{Re}\{v_y\}_{\text{gap}}}{{}^i\langle v_y\rangle} = \text{Re}\,\gamma_V\,\frac{P}{P-D}$$

since D_V is equal to D_e, i.e.:

$$D_V = \frac{4(\text{net free volume})}{\text{friction surface}} = \frac{4[P^2 - \pi D^2/4]\Delta z}{\pi D \Delta z} \equiv D_e$$

From Example 5-1, for both mesh layouts and, consequently, for the transverse unit flow cells of both layouts:

$$\gamma_V = 0.554$$

$$\frac{P}{P-D} = 4.06$$

Hence:

$$\text{Re}_{D_V} = 10^5(0.554)(4.06) = 2.25 \times 10^5$$

Since the flow is turbulent, the friction factor is given by Eq. 5-47a as:

$$\frac{f_{tr}}{2} = \frac{0.96}{(Re_{D_v})^{0.145}}$$

Returning to the expression for Δp_{tr} since $L = 2P$ and $\langle \mu \rangle$ is taken equal to $^i\langle \mu_w \rangle$ obtain for both mesh layouts:

$$\Delta p_{tr} = \frac{0.96}{(Re_{D_v})^{0.145}} \left(\frac{P}{D_V}\right)^{0.6} 2\rho\{v_y\}^2_{gap}$$

Next evaluate D_V and $\{v_y\}_{gap}$ as:

$$D_V = \frac{4(\text{net free volume})}{\text{friction surface}} = \frac{4[P^2 - \pi D^2/4]\Delta z}{\pi D \Delta z}$$

$$= \frac{D[4(P/D)^2 - \pi]}{\pi} = \frac{9.5 \times 10^{-3}[4(1.326)^2 - \pi]}{\pi}$$

$$= 11.8 \times 10^{-3} \text{ m}$$

and

$$\{v_y\}_{gap} \equiv \frac{Re_{D_v}\{\mu\}_{gap}}{D_V\{\rho\}_{gap}} = \frac{2.25 \times 10^5(9.63 \times 10^{-5})}{11.8 \times 10^{-3}(726)}$$

$$= 2.53 \text{ m/s}$$

Applying these values obtain:

$$\Delta p_{tr} = \frac{0.96}{(2.25 \times 10^5)^{0.145}} \left(\frac{12.6}{11.8}\right)^{0.6} 2(726)(2.53)^2$$

$$= 1554.2 \frac{\text{kg}}{\text{m s}^2} \text{ or } P_a$$

Finally returning to the expressions for distributed resistance, for mesh layout A:

$$^i\langle R_y \rangle_A = -\frac{1554.2}{0.554(2)(12.6 \times 10^{-3})} \left(1 - \frac{1}{1.326}\right)$$

$$= -2.74 \times 10^4 \frac{\text{kg}}{\text{m}^2 \text{ s}^2} \text{ or } \frac{\text{Pa}}{\text{m}}$$

while for mesh layout B:

$$^i\langle R_y \rangle_B = -\frac{1554.2}{0.554(2)(12.6 \times 10^{-3})}$$

$$= -1.11 \times 10^5 \frac{\text{kg}}{\text{m}^2 \text{ s}^2} \text{ or } \frac{\text{Pa}}{\text{m}}$$

In summary the following distributed resistances have been computed:

$$\text{Based on } \text{Re}_{D_e}, \qquad {}^i\langle R_z\rangle = -7.44 \times 10^2 \, \frac{\text{Pa}}{\text{m}}$$

$$\text{Based on } \text{Re}_{D_v}, \qquad {}^i\langle R_y\rangle_{\text{layout A}} = -2.74 \times 10^4 \, \frac{\text{Pa}}{\text{m}}$$

$${}^i\langle R_y\rangle_{\text{layout B}} = -1.11 \times 10^5 \, \frac{\text{Pa}}{\text{m}}$$

Hence cross-flow resistance is 50 to 200 times the longitudinal flow resistance for a Reynolds number of 10^5 based on the volume-averaged direction velocity, i.e.:

$$\text{Re}_{D_e} \equiv \frac{{}^i\langle \rho \rangle {}^i\langle v_z\rangle D_e}{{}^i\langle \mu \rangle} = 10^5$$

$$\text{Re}_{D_v} \equiv \frac{{}^i\langle \rho \rangle \{v_y\}_{\text{gap}} D_V}{{}^i\langle \mu \rangle} = 2.25 \times 10^5$$

In practice the transverse Reynolds numbers encountered are generally less than 10^5 so that typical $|{}^i\langle R_y\rangle|$ values are less than the magnitudes displayed in this example.

VI DERIVATION OF THE VOLUMETRIC AVERAGED EQUATIONS OF ENERGY CONSERVATION

A Energy Equation in Terms of Internal Energy

The differential transport equation for internal energy is:

$$\frac{\partial(\rho u)}{\partial t} + \nabla \cdot (\rho u \vec{v}) = -\nabla \cdot \vec{q}'' + q''' - p\nabla \cdot \vec{v} + \phi \qquad (\text{I},4\text{-}109)$$

where ϕ is the dissipation function defined by Eq. I,4-107.

The local volume average of Eq. I,4-109 is:

$$\left\langle \frac{\partial(\rho u)}{\partial t} \right\rangle + \langle \nabla \cdot (\rho u \vec{v}) \rangle = -\langle \nabla \cdot \vec{q}'' \rangle + \langle q''' \rangle - \langle p\nabla \cdot \vec{v} \rangle + \langle \phi \rangle \qquad (5\text{-}48)$$

Following a procedure analogous to that employed with the linear momentum equation by (a) transforming the volume average of a time derivative to the time derivative of a volume average; (b) using the theorem of local volume averaging of a divergence; and (c) using the result for divergence of an intrinsic local volume average and the concept of an intrinsic local average volume, we obtain the volume-averaged internal energy equation:

$$\gamma_V \frac{\partial}{\partial t} {}^i\langle \rho u \rangle + \frac{1}{V_T} \int_{A_f} \rho u \vec{v} \cdot \vec{n} \, dA = -\gamma_V {}^i\langle p \nabla \cdot \vec{v} \rangle$$

$$+ \frac{1}{V_T} \int_{A_f} k_e \vec{n} \cdot \nabla T \, dA + \gamma_V ({}^i\langle q_{rb}''' \rangle + {}^i\langle q''' \rangle + {}^i\langle \phi \rangle)) \quad (5\text{-}49)$$

where k_e is the effective fluid thermal conductivity including both molecular and turbulent effects.

The treatment of the term $\langle \nabla \cdot \vec{q}'' \rangle$ merits elaboration to show the origin of the second and third terms on the right side of Eq. 5-49. Applying the theorem of local volume averaging of a divergence Eq. 5-5 to this term and then expressing the divergence of an intrinsic local volume average we obtain:

$$-\langle \nabla \cdot \vec{q}'' \rangle = -\nabla \cdot \langle \vec{q}'' \rangle - \frac{1}{V_T} \int_{A_{fs}} \vec{q}'' \cdot \vec{n} \, dA$$

$$= -\frac{1}{V_T} \int_{A_f} \vec{q}'' \cdot \vec{n} \, dA - \frac{1}{V_T} \int_{A_{fs}} \vec{q}'' \cdot \vec{n} \, dA \quad (5\text{-}50)$$

The first surface integral in Eq. 5-50 represents conduction heat transfer across the fluid surface A_f, and the second represents conduction heat transfer across all fluid–solid interfaces A_{fs}.

Considering each term individually:

$$-\frac{1}{V_T} \int_{A_f} q'' \cdot \vec{n} \, dA = +\frac{1}{V_T} \int_{A_f} k_e \vec{n} \cdot \nabla T \, dA \quad (5\text{-}51)$$

where Fourier's conduction law is used. The term

$$-\frac{1}{V_T} \int_{A_{fs}} \vec{q}'' \cdot \vec{n} \, dA \quad (5\text{-}52)$$

represents the rate of heat release at all fluid–solid interfaces inside V_T since $\vec{n}$ is the unit normal vector drawn outward from the control volume, i.e., from the fluid into the solid. In reactor cores or heat exchangers with rod bundles, the integral denotes all of the heat released or removed at the immersed rod bundle surface. Hence, it is convenient to define an equivalent, dispersed heat source (or sink) per unit volume of the fluid, q_{rb}''', such that:

$$\int_{V_f} q_{rb}''' \, dV = -\int_{A_{fs}} \vec{q}'' \cdot \vec{n} \, dA \quad (5\text{-}53)$$

Hence rewriting our result in terms of intrinsic local volume averages utilizing Eqs. 5-12 and 5-13 yields:

$$-\frac{1}{V_T} \int_{A_{fs}} \vec{q}'' \cdot \vec{n} \, dA = \frac{1}{V_T} \int_{V_f} q_{rb}''' \, dV \equiv \gamma_V {}^i\langle q_{rb}''' \rangle \quad (5\text{-}54)$$

Example 5-3 Computation of the equivalent dispersed heat source for a PWR fuel assembly

PROBLEM For the PWR square fuel array of Example 5-2 and the grid layouts of Figure 5-2 compute the rate of energy addition expressed as:

$$^i\langle q_{rb}''' \rangle \text{ and } ^i\langle q''' \rangle$$

SOLUTION From Section I,3II about 87 percent of the total energy per fission is produced in the fuel itself. About 3 percent is produced in the moderator. This total energy per fission is expressible in terms of the core average, fuel element linear heat rate $\bar{q}'$ which per Table I,2-3 is 17.8 kW/m.

Now per Eq. 5-54:

$$^i\langle q_{rb}''' \rangle = -\frac{1}{\gamma_v}\frac{1}{V_T}\int_{A_{fs}} \vec{q}'' \cdot \vec{n}\, dA$$

Since:

$$-\int_{A_{fs}} \vec{q}'' \cdot \vec{n}\, dA = N\bar{q}'\Delta z$$

where N is the number of rods each generating a linear power $\bar{q}'$ averaged over Δz,

then:

$$^i\langle q_{rb}''' \rangle = \frac{N\bar{q}'\Delta z}{V_f}$$

For both grid layouts:

$$^i\langle q_{rb}''' \rangle = \frac{4(0.87)\bar{q}'}{0.95 A_f}$$

since only 87 percent of the total energy per fission is recovered in the fuel and 5 percent of the total energy per fission is lost to the neutrinos.

$$A_f = 4\left(P^2 - \frac{\pi D^2}{4}\right)$$

where $P = 12.6$ mm and $D = 9.5$ mm from Table I,1-3.

Hence:

$$^i\langle q_{rb}''' \rangle = \frac{(0.87)\bar{q}'}{0.95 P^2 \left[1 - \frac{\pi}{4}\left(\frac{D}{P}\right)^2\right]}$$

$$= \frac{(0.87)(17.8)}{0.95(12.6 \times 10^{-3})^2 \left[1 - \frac{\pi}{4}\left(\frac{1}{1.326}\right)^2\right]}$$

$$= 1.86 \times 10^5 \text{ kW/m}^3$$

Considering the second term:

$$\langle q''' \rangle \equiv \frac{1}{V_T} \int_{V_f} q''' \, dV \equiv \gamma_V{}^i\langle q''' \rangle$$

and assuming that an infinite lattice q''' represents 3 percent of the total energy produced in fission, we obtain:

$$^i\langle q''' \rangle = \left(\frac{0.03}{0.87} \right)^i\langle q'''_{rb} \rangle = 6.40 \times 10^3 \text{ kW/m}^3$$

We next specialize Eq. 5-49 to the Cartesian coordinate system of Figure 5-3 including the approximation of equating the average of products to the product of the averages. Both steps are performed analogously to the procedures employed for the mass conservation equation. The result is:

$$\overset{(1)}{\gamma_V \frac{\partial}{\partial t} \left(^i\langle\rho\rangle^i\langle u\rangle \right)} + \overset{(2)}{\frac{\Delta_x(\gamma_{Ax}{}^{i(x)}\{\rho\}^{i(x)}\{u\}^{i(x)}\{v_x\})}{\Delta x}}$$

$$+ \overset{(3)}{\frac{\Delta_y(\gamma_{Ay}{}^{i(y)}\{\rho\}^{i(y)}\{u\}^{i(y)}\{v_y\})}{\Delta y}}$$

$$+ \overset{(4)}{\frac{\Delta_z(\gamma_{Az}{}^{i(z)}\{\rho\}^{i(z)}\{u\}^{i(z)}\{v_z\})}{\Delta z}}$$

$$\overset{(5)}{= -\gamma_V {}^i\!\left\langle p \left(\frac{\partial v_x}{\partial x} + \frac{\partial v_y}{\partial y} + \frac{\partial v_z}{\partial z} \right) \right\rangle} \qquad (5\text{-}55)$$

$$+ \overset{(6)}{\frac{\Delta_x\left(\gamma_{Ax}{}^{i(x)}\left\{k_e \frac{\partial T}{\partial x}\right\}\right)}{\Delta x}} + \overset{(7)}{\frac{\Delta_y\left(\gamma_{Ay}{}^{i(y)}\left\{k_e \frac{\partial T}{\partial y}\right\}\right)}{\Delta y}}$$

$$+ \overset{(8)}{\frac{\Delta_z\left(\gamma_{Az}{}^{i(z)}\left\{k_e \frac{\partial T}{\partial z}\right\}\right)}{\Delta z}} + \overset{(9)}{\gamma_V({}^i\langle q'''_{rb}\rangle + {}^i\langle q'''\rangle + {}^i\langle\phi\rangle)}$$

The terms in this equation have the following physical meaning:

(1)	(2, 3, and 4)
rate of increase of internal energy of the fluid mass in V	+ net outflux of internal energy through A

(5)
= reversible rate of pressure work when the density is not constant

(6, 7, and 8)
+ heat conduction through the portion of A that is free to fluid flow

(9)
+ sum of heat liberated (or absorbed) due to immersed solid, extraneous internal sources and viscous dissipative effects.

B Energy Equation in Terms of Enthalpy

The differential equation for the transport of static enthalpy h $(= u + p/\rho)$ is:

$$\frac{\partial(\rho h)}{\partial t} + \nabla \cdot (\rho h \vec{v}) = -\nabla \cdot \vec{q}'' + q''' + \frac{Dp}{Dt} + \phi \qquad (\text{I},4\text{-}108)$$

where ϕ is dissipation function defined by Eq. I,4-107, and the volume-averaged enthalpy equation is:

$$\gamma_V \frac{\partial}{\partial t} {}^i\langle \rho h \rangle + \frac{1}{V_T} \int_{A_f} \rho h \vec{v} \cdot \vec{n} \, dA = + \frac{1}{V_T} \int_{A_f} k_e \vec{n} \cdot \nabla T \, dA + \gamma_V {}^i\!\left\langle \frac{Dp}{Dt} \right\rangle$$
$$+ \gamma_V ({}^i\langle q'''_{rb} \rangle + {}^i\langle q''' \rangle + {}^i\langle \phi \rangle) \qquad (5\text{-}56)$$

Since:

$$\frac{Dp}{Dt} = \frac{\partial p}{\partial t} + \nabla \cdot (p\vec{v}) - p\nabla \cdot \vec{v} \qquad (5\text{-}57)$$

and utilizing Eq. 5-23, the term $\langle Dp/Dt \rangle$ can be expressed alternately as:

$$\left\langle \frac{Dp}{Dt} \right\rangle = \frac{\partial \langle p \rangle}{\partial t} + \nabla \cdot \langle p\vec{v} \rangle - \langle p\nabla \cdot \vec{v} \rangle \qquad (5\text{-}58)$$

Again expressing this result in Cartesian coordinates and approximating the average of products as the product of averages yields:

$$\gamma_V \frac{\partial}{\partial t} ({}^i\langle \rho \rangle {}^i\langle h \rangle) + \frac{\Delta_x(\gamma_{Ax}{}^{i(x)}\{\rho\}^{i(x)}\{h\}^{i(x)}\{v_x\})}{\Delta x}$$

$$+ \frac{\Delta_y(\gamma_{Ay}{}^{i(y)}\{\rho\}^{i(y)}\{h\}^{i(y)}\{v_y\})}{\Delta y} + \frac{\Delta_z(\gamma_{Az}{}^{i(z)}\{\rho\}^{i(z)}\{h\}^{i(z)}\{v_z\})}{\Delta z}$$

$$= \gamma_V {}^i\!\left\langle \frac{Dp}{Dt} \right\rangle + \frac{\Delta_x\left(\gamma_{Ax}{}^{i(x)}\left\{k_e \dfrac{\partial T}{\partial x}\right\}\right)}{\Delta x} + \frac{\Delta_y\left(\gamma_{Ay}{}^{i(y)}\left\{k_e \dfrac{\partial T}{\partial y}\right\}\right)}{\Delta y} \qquad (5\text{-}59)$$

$$+ \frac{\Delta_z\left(\gamma_{Az}{}^{i(z)}\left\{k_e \dfrac{\partial T}{\partial z}\right\}\right)}{\Delta z} + \gamma_V({}^i\langle q'''_{rb} \rangle + {}^i\langle q''' \rangle + {}^i\langle \phi \rangle)$$

VII CONSTITUTIVE RELATIONS

The foregoing derivation of the relevant conservation equations has led to the introduction of certain parameters and the neglect of others. Notably the distributed resistance and heat source (sink) parameters, $\vec{R}$ and q'''_{rb}, respectively, have been introduced. The cross-product terms involving local spatial variations of density and velocity have been neglected.

Although the constitutive relations are derived from a fixed set of physical considerations and correlations, they must be formulated anew for each grid layout. This procedure is laborious but relatively straightforward for the parameters $\vec{R}$ and q'''_{rb}. However, for other parameters like those arising from considering a turbulent flow field, as introduced in Eq. 5-32b, the procedure is complex. An example is the parameter that characterizes turbulent mixing between regions.

VIII CONCLUSION

This chapter has been developed principally from the work of Sha and Chao [3]. As has been noted, most practical applications of these porous body equations are numerical. It is therefore necessary to relate volume and surface averages of the same variables. Such relations with respect to the staggered mesh system commonly employed in numerical models are extensively treated in Sha and Chao [3].

In the example that follows, a simple two-subchannel case is analyzed utilizing the porous medium approach with the distributed resistance values computed in Example 5-2.

Example 5-4 Porous body analysis of two adjacent, interacting subchannels

PROBLEM Consider two adjacent, unheated, interacting subchannels of PWR geometry at nominal PWR fluid conditions as illustrated in Figure 5-6. The total flow rate is 0.58 kg/s. Take the inlet subchannel velocities as unequal where:

$$\frac{v_{zi}(0)}{v_{zj}(0)} = 2$$

Provide all the needed geometric and constitutive relations necessary, and compute the axial distribution of axial mass flux and pressure in each channel and the transverse mass flow rate between the channels assuming

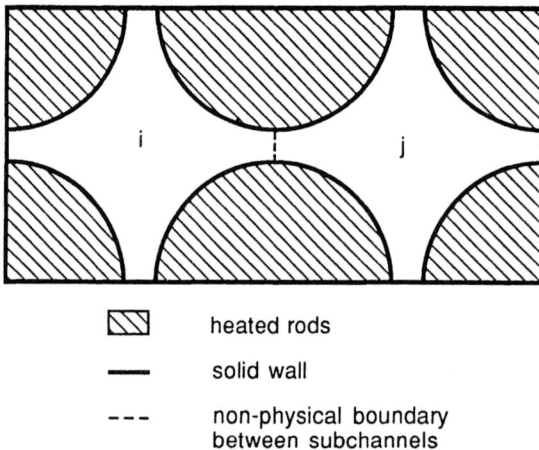

      heated rods

      solid wall

      non-physical boundary between subchannels

Figure 5-6 Test section of Example 5-4.

the exit subchannel pressures are equal at 15.5 MPa, i.e.:

$$p_i(L) = p_j(L) = 15.5 \text{ MPa}$$

Applicable PWR conditions:

$$P/D = 1.326, \ D = 9.5 \text{ mm}, \ L = 3.66 \text{ m}$$

$$T_{in} = 292 \text{ °C (565.15 K)}$$

SOLUTION This example requires simultaneous solution of the continuity, axial momentum, and transverse momentum equations since cross-flow is induced by the inlet velocity upset conditions. This solution can be accomplished by using a numerical code such as THERMIT-2 [2] subject to boundary condition set 3 of Chapter 1. In this case the subchannel inlet axial velocities are used in place of the inlet radial pressure gradient. The resulting boundary conditions are:

exit pressure $\quad\quad\quad\quad p_i(L) = p_j(L) = 15.5 \text{ MPa}$

inlet axial velocity $\quad\quad v_{zi}(0) = 5.94 \text{ m/s}; \ v_{zj}(0) = 2.97 \text{ m/s}$

inlet transverse velocity $v_{yi}(0) = v_{yj}(0) = 0$

which are obtained from the given conditions as follows:

$$\dot{m}_T = \dot{m}_i(0) + \dot{m}_j(0)$$

$$\dot{m}_i(0) = \frac{\dot{m}_T}{\left[1 + \dfrac{v_{zj}(0)}{v_{zi}(0)}\right]}$$

since for identical channels at the same inlet density:

$$\frac{\dot{m}_j(0)}{\dot{m}_i(0)} = \frac{\rho v_{zj}(0) A_{fj}}{\rho v_{zi}(0) A_{fi}} = \frac{v_{zj}(0)}{v_{zi}(0)}$$

Expressing $\dot{m}_i(0)$ in terms of $\rho v_{zi}(0) A_{fi}$ obtain the desired result:

$$v_{zi}(0) = \frac{\dot{m}_T}{\rho A_{fi}} \left[\frac{1}{1 + \dfrac{v_{zj}(0)}{v_{zi}(0)}}\right]$$

$$= \frac{0.58}{724(8.78 \times 10^{-5})} \left[\frac{1}{1 + 0.5}\right] = 5.94 \text{ m/s} \quad\quad (5\text{-}60)$$

$$v_{zj}(0) = \frac{v_{zi}(0)}{2} = \frac{5.94}{2} = 2.97 \text{ m/s}$$

since:

$$A_{fi} = A_{fj} = D^2 \left[\left(\frac{P}{D}\right)^2 - \frac{\pi}{4}\right]$$

$$= (9.5 \times 10^{-3})^2 \left[(1.326)^2 - \frac{\pi}{4}\right] = 8.78 \times 10^{-5} \text{ m}^2$$

In addition to boundary conditions, geometry and constitutive relations must be provided as follows:

1. *Geometry.* The geometry is similar to that of grid layout A, Example 5-1. Hence:

$$\gamma_V = \gamma_{Az} = 1 - \frac{\pi}{4}\left(\frac{D}{P}\right)^2 = 1 - \frac{\pi}{4}\left(\frac{1}{1.326}\right)^2 = 0.554$$

However, these channels are bounded by solid walls except for their interconnection so that:

$\gamma_{Ax} = 0$

$\gamma_{Ay} = 0$ at all boundaries except the gap region

$$\gamma_{Ay} = \frac{P - D}{P} = 1 - \frac{D}{P} = 1 - \frac{1}{1.326} = 0.246 \text{ for boundary that}$$
$$\text{includes the gap region}$$

$P_w = \pi D + 3(P - D) = \pi(9.5) + 3(12.6 - 9.5) = 39.14$ mm

$A_f = 8.78 \times 10^{-5}$ m^2

$$D_V \equiv \frac{4(\text{net free volume})}{\text{friction surface}} = \frac{4A_f}{P_w} = \frac{4(8.78 \times 10^{-5})}{(3.914 \times 10^{-2})} = 8.97 \times 10^{-3} \text{ m}$$

$D_e = D_V = 8.97 \times 10^{-3}$ m

2. *Constitutive laws.*
   a. *Axial distributed resistance, $^i\langle R_z\rangle$.* Following Example 5-2:

   $$^i\langle R_z\rangle = -\frac{f}{D_e}\left[\frac{^i\langle\rho\rangle^i\langle v_z\rangle^2}{2}\right]\frac{\text{Pa}}{\text{m}}$$

   where:

   $$f = f_{c.t.}[1.04 + 0.06(P/D - 1)]$$

   and from Eq. 7-59:

   $$\frac{1}{(f_{c.t.})^{1/2}} = 2.035 \log_{10}[\text{Re}_{De}(f_{c.t.})^{1/2}] - 0.989$$

   (Note $D_e = 8.97 \times 10^{-3}$ m which is different than the D_e of Example 5-2 since the channels here are enclosed by bounding walls.)
   b. *Transverse distributed resistance, $^i\langle R_y\rangle$.* The geometry of the test section here corresponds to mesh layout A of Example 5-2. However, the value of A_{fy} is $\frac{1}{2}$ that of A_{fy} in mesh layout A, while the value of V_f is $\frac{1}{4}$ that in mesh layout A. Hence:

   $$^i\langle R_y\rangle = -\frac{\Delta p_{tr} A_{fy}}{V_f} = -\frac{\Delta p_{tr}}{\gamma_V P}\left(1 - \frac{D}{P}\right)$$

Here $L = P$ so that:

$$\Delta p_{tr} = \frac{0.96}{(\mathrm{Re}_{D_V})^{0.145}} \left(\frac{P}{D_V}\right) \{\rho\}_{gap}\{v_y\}^2_{gap} \left(\frac{D_V}{P}\right)^{0.4}$$

and

$$\mathrm{Re}_{D_V} = \frac{\{\rho\}_{gap}\{v\}_{gap} D_V}{\{\mu\}_{gap}}$$

Hence dropping average symbols on the properties:

$$^i\langle R_y\rangle = -\frac{1}{\gamma_V}\frac{0.96}{(D_V)^{0.745}}\frac{\left(1 - \dfrac{D}{P}\right)}{P^{0.4}}[\rho^{0.855}\mu^{0.145}\{v_y\}^{1.855}_{gap}]$$

$$= -\frac{1}{(0.554)}\frac{0.96}{(8.97 \times 10^{-3})^{0.745}}\frac{\left(1 - \dfrac{1}{1.326}\right)}{(12.6 \times 10^{-3})^{0.4}}[\rho^{0.855}\mu^{0.145}\{v_y\}^{1.855}_{gap}]$$

$$= -82.12^i_i[\rho^{0.855}\mu^{0.145}\{v_y\}^{1.855}_{gap}]\frac{\mathrm{Pa}}{\mathrm{m}}$$

for ρ in $\dfrac{\mathrm{kg}}{\mathrm{m}^3}$, μ in $\dfrac{\mathrm{kg}}{\mathrm{m}\cdot\mathrm{s}}$, v in $\dfrac{\mathrm{m}}{\mathrm{s}}$

c. *Transverse turbulent momentum transfer*, $V_T(\Delta(\gamma_{Ay}\{\tau_{zy}\})/\Delta y)$. The shear stress $\{\tau_{zy}\}$ will be discussed in Chapter 6 as part of the subchannel analysis process. While the THERMIT-2 code is a porous

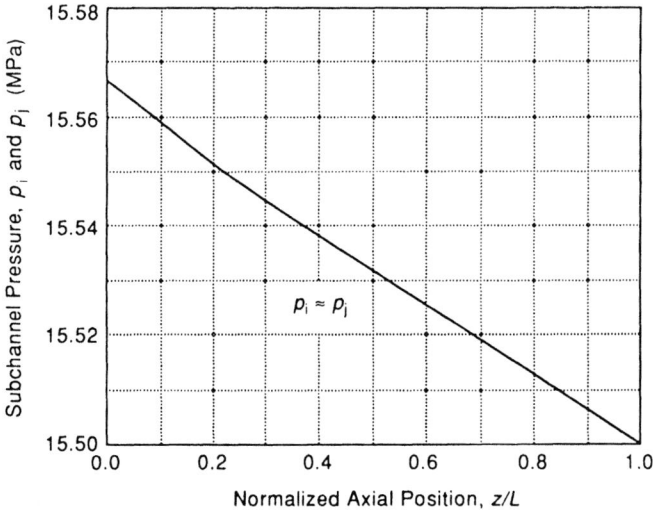

Figure 5-7 Axial pressure profile of Example 5-4.

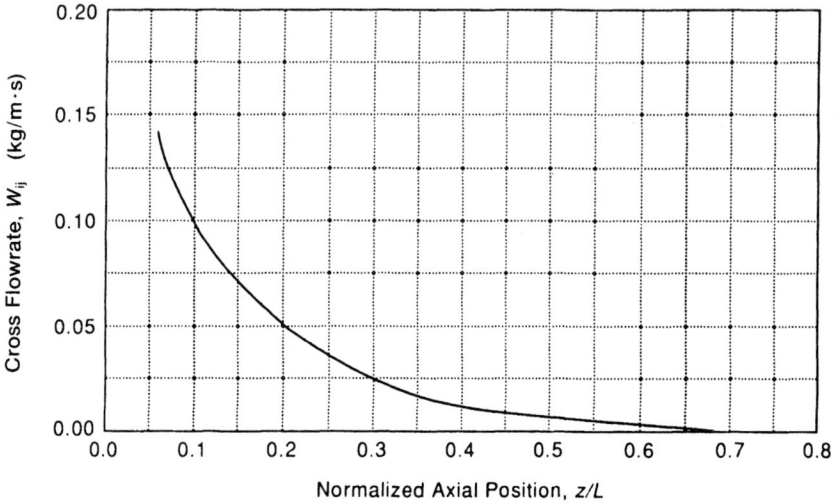

Figure 5-8 Axial distribution of the cross-flow rate between channels i and j of Example 5-4.

body formulation, it has been expanded to include correlations for processes like turbulent momentum transfer which are important predominantly for interactions between subchannel scale regions. The subchannel equivalent parameter for turbulent momentum transfer in a single-phase flow W_{ij}^{*M} which is defined by Eq. 6-15 is

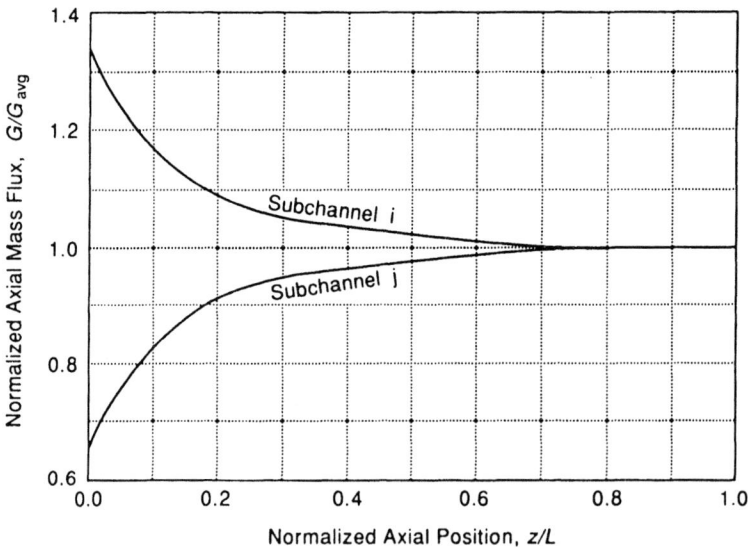

Figure 5-9 Axial distribution of the subchannel axial mass fluxes of Example 5-4.

Table 5-2 Subchannel parameters of Example 5-4

Z/L	p_i(MPa)	p_j(MPa)	W_{ij}(kg/m·s)	Z/L	G_i(kg/m²s)	G_j(kg/m²s)
0.00	15.566764	15.564359		0.0	4403.8	2201.9
0.05	15.562153	15.562147	1.383×10^{-1}	0.1	3827.8	2778.8
0.15	15.555151	15.555153	6.237×10^{-2}	0.2	3561.9	3045.0
0.25	15.548380	15.548380	3.013×10^{-2}	0.3	3435.3	3172.1
0.35	15.541796	15.541796	1.414×10^{-2} †	0.4	3371.4	3236.3
0.45	15.535306	15.535306	1.117×10^{-2}	0.5	3339.9	3267.9
0.55	15.528863	15.528863	2.517×10^{-3}	0.6	3323.5	3284.2
0.65	15.522448	15.522448	1.589×10^{-3}	0.7	3315.1	3292.4
0.75	15.516048	15.516048	1.068×10^{-3}	0.8	3310.2	3297.4
0.85	15.509634	15.509634	-3.52×10^{-3}	0.9	3306.1	3301.2
0.95	15.503232	15.503232	-6.05×10^{-4}	1.0	3306.6	3300.7
1.00	15.500000	15.500000				

† All values of W_{ij} for $Z/L \geq 0.25$ are due to numerical roundoff.

taken from the result of Example 6-4, i.e.:

$$W_{ij}^{*M} = 0.0363 \ \frac{\text{kg}}{\text{m} \cdot \text{s}}$$

Figures 5-7 through 5-9 present the calculated results. Notice that the cross-flow is sufficient to cause the subchannel axial velocities to equilibrate well before the channel exit. The subchannel pressure differences are too small to appear on Figure 5-7. These pressures, accurate to about 10 Pa, are tabulated in Table 5-2.

REFERENCES

1. Gunter, A. Y., and Shaw, W. A. A general correlation of friction factors for various types of surfaces in cross flow. *Trans. ASME* 67:643–660, 1945.
2. Kelly, J. E., Kao, S. P., and Kazimi, M. S. *THERMIT-2: A Two-Fluid Model for Light Water Reactor Subchannel Transient Analysis.* MIT Energy Laboratory Electric Utility Program, Report MIT-EL-81-014, April 1981.
3. Sha, W. T., and Chao, B. T. *Local Volume-Averaged Transport Equations for Single-Phase Flow in Regions Containing Fixed, Dispersed Heat Generating (or Absorbing) Solids.* NUREG/CR-1969, ANL-80-124, April 1981.
4. Sha, W. T., Chao, B. T., and Soo, S. L., *Time- and Volume-Averaged Conservation Equations for Multiphase Flow. Part One: System without Internal Solid Structures.* NUREG/CR-3989, ANL-84-66, December 1984.
5. Slattery, J. C. *Momentum, Energy and Mass Transfer in Continua.* New York: McGraw-Hill, 1972.

PROBLEMS

Problem 5-1 Approach for analysis of the shell side of a U-tube steam generator (Sections I and VII)
Define the thermal hydraulic analysis approach appropriate to determine if flow stagnation areas will exist near the bottom tube sheet of a typical PWR U-tube steam generator under steady-state conditions. This definition should include:

1. The degree of spatial averaging, i.e., field or distributed parameter, porous body or subchannel approach
2. The number of spatial dimensions and number of nodes
3. The boundary conditions
4. Identification of the required constitutive equations.

Problem 5-2 Control volume characteristics of a hexagonal array (Section III)
Compute the surface and volume porosities of the interior subchannels of a 19-pin hexagonal array with the geometry $P/D = 1.24$, $D = 8.65$ mm, and $H/D = 35$. Take the axial length of each subchannel as H, the wire lead length.
   Answer: $\gamma_{AL1} = 0.175$; $\gamma_{V1} = 0.376$
   $\gamma_{TB1} = 0.376$, where $TB1$ means top and bottom surfaces of subchannel type 1, and $AL1$ means lateral surface of subchannel type 1.

Problem 5-3 Distributed resistance for a triangular array (Section V)
Compute the distributed resistances for axial and cross-flow for an equilateral triangular array with geometry typical of a tight pitch light water high converter lattice, $P = 9$ mm and $D = 8.2$ mm,

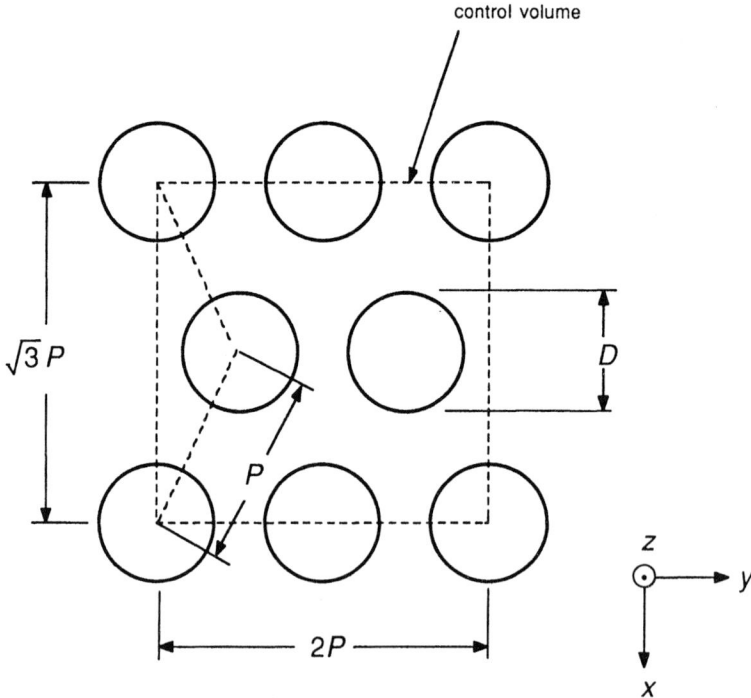

Figure 5-10

for a flow in each direction characterized by a Reynolds number of 10^5 where:

$$\text{for axial flow } Re \equiv \frac{^i\langle \rho \rangle ^i\langle v_z \rangle D_e}{^i\langle \mu \rangle}$$

$$\text{and for transverse flow } Re \equiv \frac{^i\langle \rho \rangle ^i\langle v_y \rangle D_V}{^i\langle \mu \rangle}$$

The array is illustrated in Figure 5-10 which specifies the coordinate system and dimensions. Use water properties and other necessary assumptions from Example 5-2.

Evaluate the ratio $f/f_{c.t.}$ for this triangular array by Eq. I,9-86 and $f_{c.t.}$ from 7-59:

    Answer: axial $^i\langle R_z \rangle = -6.20 \times 10^4$ Pa/m

        transverse $^i\langle R_y \rangle = -4.13 \times 10^6$ Pa/m

                  $^i\langle R_x \rangle = -1.76 \times 10^6$ Pa/m

Problem 5-4 Distributed heat sources for the shell side of U-tube steam generator (Section VI)

    Compute the distributed heat sources, $^i\langle q'''_{rb} \rangle$ and $^i\langle q''' \rangle$ for a control volume located in the hot leg side of the bottom region of a U-tube steam generator. In this steam generator primary coolant flows in tubes and transfers energy to secondary coolant in the shell side. Relevant geometry and operation conditions are given below.

Geometry:

Tube inner diameter	2.05×10^{-2} m
Tube outer diameter	2.22×10^{-2} m
Pitch in square array	3.67×10^{-2} m
Tube material	Inconel
Total number of tubes	3240
Inner diameter of steam	
generator inner shell	3.35 m

Primary coolant operating conditions:

Inlet temperature	324 °C
Outlet temperature	284 °C
Flow rate	4.218×10^3 kg/s

Secondary coolant operating conditions:

Inlet temperature	263.3 °C
Outlet temperature	273.0 °C
Flow rate	1850.58 kg/s
Recirculation ratio	4.1

    Answer: $^i\langle q''' \rangle = 0$

          $^i\langle q'''_{rb} \rangle = 34.9$ MW/m^3

Problem 5-5 Distributed heat source for a pebble bed reactor (Sections I,2IV and VI in this volume)

    The thermal/hydraulic design of a pebble bed gas (helium) cooled reactor is to be considered here. This reactor is fueled by spheres of about 1-inch diameter containing coated fuel particles dispersed within a graphite matrix.

1. What would you recommend for the core thermal design limits for this design? Identify and justify the functional types of limits: exact numerical values are not necessary.

2. In operation assume (for simplicity) that the spheres are levitated by the helium flow (upflow) to space the spheres that operate at an average power of 5 kW/sphere in a square array of $P/D = 1.3$. Take the core as a cube, 12-feet per side.

If you were asked to calculate the coolant temperature field to determine the mean temperature in each region between spheres, identify and justify the solution method you would use among the following:

Porous body, 3D
Subchannel, 2D
Tube type geometry, 1D

For each method identify the heat source term and calculate its numerical value.
    Answer: Part 2
            $\langle q_{rb}''' \rangle = 5155$ kW/ft^3
            $q' = 46.2$ kW/ft
            $Q = 6796/n$ MW/node, where n equals the number of axial nodes by which the core is
                represented.

SUBCHANNEL ANALYSIS

I INTRODUCTION

The subchannel approach for rod bundle analysis was introduced in Section VIII of Chapter 5 by standardizing the porous body control volume equations to a specific nodal layout. This nodal layout defines volumes of a size equivalent to a single fuel rod and its associated fluid. There are two approaches to the definition of this subchannel volume: the coolant-centered and the rod-centered subchannels, as illustrated in Figure 6-1. The traditional approach for rod bundle analysis has been coolant-centered subchannels. That approach is treated in this chapter. However, in two-phase flow, particularly in the annular regime, the liquid flow around the rod is difficult to accommodate by this approach. Hence it makes sense to consider rod-centered subchannels as suggested by Gaspari et al. [11, 12], who demonstrated good results using this concept in the prediction of high-quality critical heat flux. However, little additional systematic work has been accomplished on the development of the constitutive relations required for this rod-centered nodal layout.

Subchannel properties, like axial velocity and density, are represented by single, averaged values. Constitutive equations are required for input parameters like friction factors and lateral momentum and energy exchange rates between adjacent subchannels. This choice of nodal layout in terms of subchannels does not yield simpler constitutive equations than would other choices. Rather, it offers a convenient arrangement for which a specific set of constitutive equations can be formulated. In a practical sense this leads to a focusing of effort on a single set of constitutive relations.

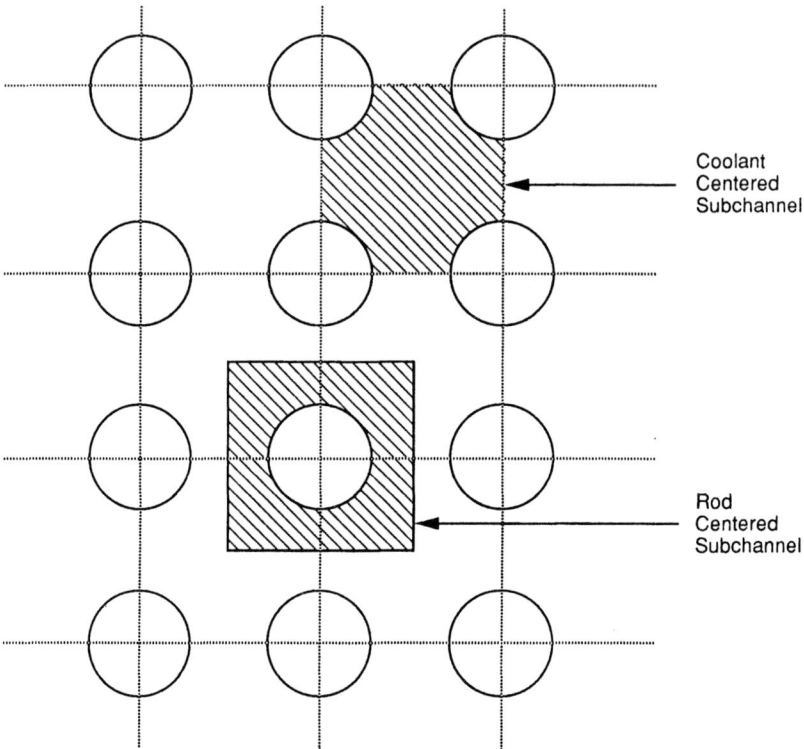

Figure 6-1 Options for subchannel definition.

Further, in the subchannel approach a major simplification is the treatment of lateral exchanges between adjacent subchannels. It is assumed that any lateral flow through the gap region between subchannels loses its sense of direction after leaving the gap region. This allows subchannels to be connected arbitrarily since no fixed lateral coordinate is required. A fully three-dimensional physical situation can be represented simply by connecting the channels in a three-dimensional array. This leads to simplifications in the lateral convective terms of the linear momentum balance equation, that makes the method most appropriate for predominantly axial flow situations. The subchannel analysis approach and its associated boundary conditions are therefore not a fully three-dimensional representation of the flow.

II CONTROL VOLUME SELECTION

Consider the coolant-centered subchannel approach. The actual subchannel control volume encompasses only the coolant, not the fuel rod. It is illustrated in Figure 6-2 relative to the entire reactor core. However, the control volume of

Figure 6-2 is used only for the mass, axial momentum, and energy conservation balances. For the transverse momentum balance, a separate control volume between adjacent subchannels is employed. Typical transverse control volumes are illustrated by the *dashed lines* in Figure 6-3. The dimension $\Delta x'$ must be specified by the analyst. Staggered control volumes are also used in porous body approaches, but they generally differ in three respects from the subchan-

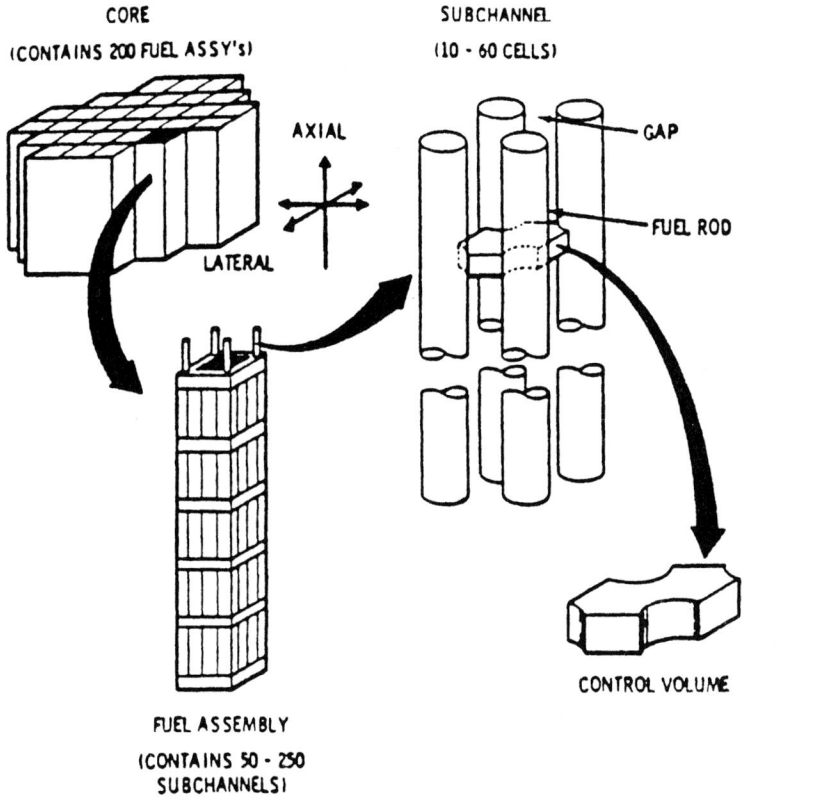

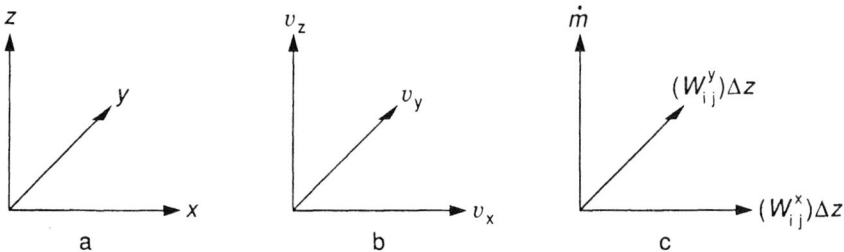

Figure 6-2 Relation of subchannel control volume to reactor core. *(After Stewart et al. [27].)*

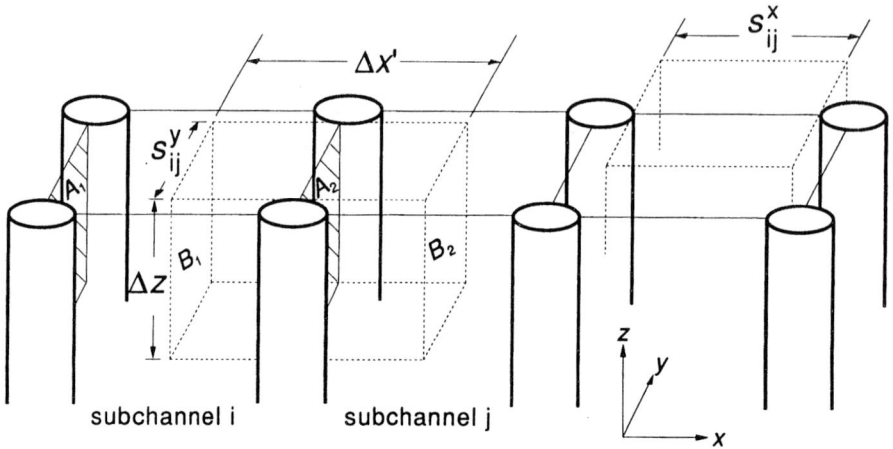

Coordinate Origin in Plane A₂ at Δz/2 and s_{ij}^y /2.

Figure 6-3 Control volume for transverse momentum equation.

nel control volumes, as illustrated in Figure 6-4. First, all porous body control volumes include fueled regions, not just coolant regions. Second, the transverse momentum control volume is of the same dimensions, although staggered, as the control volume for mass and energy. Finally, the axial momentum control volume, like the transverse momentum control volume, is also staggered axially with respect to the mass and energy control volume.

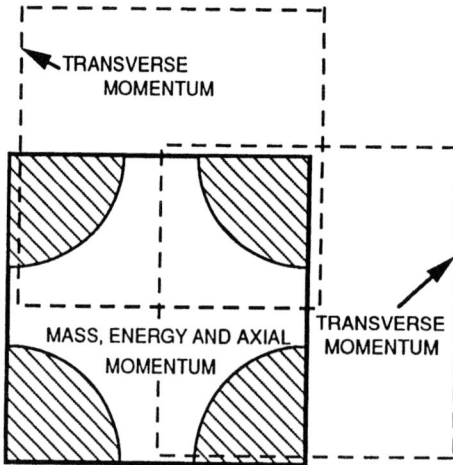

$\boxed{\diagdown}$——FUEL RODS

Figure 6-4 Control volumes for porous body analysis.

III DEFINITIONS OF TERMS IN THE SUBCHANNEL APPROACH

Some important parameters that appear in the subchannel formulations are presented next. Since the subchannel control volume includes only fluid regions, the superscript i for intrinsic volume averages is redundant and will be dropped after it is first introduced.

A Geometry

The key parameter is the spacing or gap size between adjacent fuel rods. The gap size between the adjacent fuel rods, s_{ij}, in conjunction with the axial mesh size, Δz, represents the minimum flow area, $S_{ij} = s_{ij}\Delta z$, available to the fluid in the transverse direction. As Figure 6-3 illustrates, the gaps along the x- and y-directions, s_{ij}^x and s_{ij}^y, respectively, can be different. The fuel rods can also be bowed axially leading to axially-varying gap widths. Most numerical subchannel codes allow for such variations, although in this chapter we consider axially-independent gap widths.

B Mass Flow Rates

There are both axial and transverse mass flow rates associated with the control volume.

C Axial Mass Flow Rate, $\dot{m}_i$

This term represents the predominant axial flow rate of the selected subchannel control volume, i. It has the dimensions of mass/time and is written as:

$$\dot{m}_i = \int_{A_{fi}} \rho v_z dA \tag{6-1}$$

where A_{fi} represents the total axial cross-sectional area of the subchannel which is coincident by definition to the area for coolant flow. Using the area average definition of Eq. 5-15, Eq. 6-1 is written as:

$$\dot{m}_i \equiv {}^{i(z)}\{\rho v_z\} A_{fi} = \{\rho v_z\} A_{fi} \text{ mass/time} \tag{6-2}$$

D Transverse Mass Flow Rate Per Unit Length, W_{ij} and $W_{ij}'^D$

Two mechanisms create transverse mass flows: transverse pressure gradients that drive diversion cross-flow, and turbulent fluctuations in the axial flow that drive turbulent mass interchanges.

1 Diversion cross-flow rate, W_{ij}. In a reactor, transverse pressure gradients can be established by either of two types of phenomena: geometry variations or nonuniform changes in fluid density. Geometry variations include such things as fuel rod bowing and swelling, whereas density changes encompass the small differences across a bundle due to radial heat flux variations and the large local difference due to the onset of boiling. In a tracer experiment, diversion cross-flow will be present if the test section geometry varies along its axis due to fabrication tolerances. The magnitude of diversion cross-flow rate is small compared with axial flow under reactor operational conditions except in cases such as flow blockage or fuel rod bowing. It is written as follows for flow along the x- and y-directions, respectively. Note the dimensions are mass flow rate per unit length:

$$W_{ij}^x = \frac{1}{\Delta z} \int_{\Delta z} \int_{s_{ij}^y} \rho v_x ds \, dz = {}^{i(x)}\{\rho v_x\} s_{ij}^y$$

$$= \{\rho v_x\} s_{ij}^y \text{ (mass/length time)}$$

(6-3)

and

$$W_{ij}^y = \frac{1}{\Delta z} \int_{\Delta z} \int_{s_{ij}^x} \rho v_y ds \, dz = {}^{i(y)}\{\rho v_y\} s_{ij}^x$$

$$= \{\rho v_y\} s_{ij}^x \text{ (mass/length time)}$$

(6-4)

1 Turbulent interchange, $W_{ij}'^D$. This exchange is postulated to involve equal volumes of eddies which cross a transverse subchannel boundary. If these eddies are also of equal density, as they effectively are for single-phase flow conditions, then no net mass exchange results. However, in two-phase flow, a net mass exchange can occur as described in Section VII where the constitutive relations are presented. In single-phase flow although no net mass exchange occurs, both momentum and energy are exchanged between subchannels, and their rates of exchange are characterized in terms of hypothetical turbulent interchange flow rates. These flow rates are defined separately for momentum and energy as $W_{ij}'^M$ and $W_{ij}'^H$. Hence the superscript D is utilized for turbulent mass interchange $W_{ij}'^D$.

Both flow rates W_{ij} and W_{ij}' reflect the nomenclature convention that:

subscript ij   represents flow from subchannel i to j

whereas:

subscript i↔j will be used for net flow between subchannels i and j.

Hence in single-phase flow:

$$W_{ij}'^D = W_{ji}'^D$$

(6-5)

therefore:

$$W_{i \leftrightarrow j}'^D \equiv W_{ij}'^D - W_{ji}'^D = 0$$

(6-6)

E Momentum and Energy Transfer Rates

Three types of transverse momentum and energy transfer exist: transport by diversion cross-flow; transport by turbulent interchange; and viscous transfer due to transverse gradients of axial velocity and temperature.

The momentum and energy transfer by diversion cross-flow is written as area averages of the product of the transverse mass flux, ρv_x or ρv_y, and the transported axial velocity and enthalpy, respectively, i.e., $\{\rho v_x v_z\}$ or $\{\rho v_y v_z\}$ and $\{\rho v_x h\}$ or $\{\rho v_y h\}$. Although the diversion cross-flow rate is defined, the characteristic enthalpy and axial velocity transported cannot be well characterized in terms of defined subchannel properties. For convenience, they are labeled as starred quantities $\{v_z^*\}$ and $\{h^*\}$ for later definition by subchannel analysis computer code users. No distinction is made between the v_x and v_y velocity components with respect to the definition of $\{v_z^*\}$, i.e.:

$$\{v_z^*\} \equiv \frac{\{\rho v_x v_z\}}{\{\rho v_x\}} = \frac{\{\rho v_y v_z\}}{\{\rho v_y\}}, \quad \text{the effective velocity transported by diversion cross-flow rate, length/time}$$

$$\{h_x^*\} \equiv \frac{\{\rho v_x h\}}{\{\rho v_x\}}; \{h_y^*\} \equiv \frac{\{\rho v_y h\}}{\{\rho v_y\}}, \quad \text{the effective enthalpy transported by the diversion cross-flow rate, energy/mass}$$

The turbulent interchange flow rates are defined by the following time-averaged balance equations:

$$\overline{(\tau s)'_{i \leftrightarrow j}} \equiv \overline{W_{ij}'^D v_{zi}} - \overline{W_{ji}'^D v_{zj}} \tag{6-7}$$

$$\overline{(q''s)'_{i \leftrightarrow j}} \equiv \overline{W_{ij}'^D h_i} - \overline{W_{ji}'^D h_j} \tag{6-8}$$

where $\overline{(\tau s)'_{i \leftrightarrow j}}$ and $\overline{(q''s)'_{i \leftrightarrow j}}$ represent, respectively, the net turbulence driven momentum and energy per unit time crossing the gap per unit length. These momentum and energy transfer processes occur for single-phase flow when v_{zi} and h_i are not equal to v_{zj} and h_j. In single phase, $\overline{W_{ij}'^D} = \overline{W_{ji}'^D}$. Eqs. 6-7 and 6-8 can be written for single-phase flow as:

$$\overline{(\tau s)'_{i \leftrightarrow j}} = \overline{W_{ij}'^M}(\overline{v_{zi}} - \overline{v_{zj}}) \tag{6-9}$$

where:

$$\overline{W_{ij}'^M} \equiv \frac{\overline{W_{ij}'^D v_{zi}} - \overline{W_{ji}'^D v_{zj}}}{\overline{v_{zi}} - \overline{v_{zj}}} \tag{6-10}$$

i.e., $\overline{W_{ij}'^M}$ is a hypothetical flow rate for momentum transfer and

$$\overline{(q''s)'_{i \leftrightarrow j}} = \overline{W_{ij}'^H}(\overline{h_i} - \overline{h_j}) \tag{6-11}$$

where:

$$\overline{W_{ij}'^H} \equiv \frac{\overline{W_{ij}'^D h_i} - \overline{W_{ji}'^D h_j}}{\overline{h_i} - \overline{h_j}} \tag{6-12}$$

i.e., $\overline{W_{ij}'^H}$ is a hypothetical flow rate for energy transfer.

However in two-phase flow $\overline{W_{ij}'^D} \neq \overline{W_{ji}'^D}$, so the terms such as $\overline{W_{ij}'^D h_i}$ must be evaluated individually rather than expressed in terms of the product of the hypothetical flow rate $\overline{W_{ij}'^H}$ and the adjacent subchannel enthalpy difference. This is done for the two-phase situation in Section VIIE3 where the product $\overline{W_{i \leftrightarrow j}'^D h_{\text{eff}}}$ is developed as the two-phase analogue to the product $\overline{W_{ij}'^H}(\overline{h_i} - \overline{h_j})$ in single phase.

Table 6-1 Instantaneous transverse flow rates used in subchannel analysis†

	Mass transfer	Momentum transfer	Energy transfer
Flow from subchannels i and j			
Diversion cross-flow	W_{ij}		
Turbulent interchange	$W_{ij}'^D$	$W_{ij}'^M$	$W_{ij}'^H$
Turbulent plus viscous interchange	W_{ij}^{*D}	W_{ij}^{*M}	W_{ij}^{*H}
Net flow between subchannels i and j			
Turbulent interchange	$W_{i\leftrightarrow j}^{*D}$		

† All flow rates are expressed in units of mass/time-length.

Molecular or viscous effects also cause energy and momentum transfer across the gap. These effects are proportional to temperature and velocity gradients at the gap. For compactness and convenience the parameters $W_{ij}'^M$ and $W_{ij}'^H$ can be redefined as W_{ij}^{*M} and W_{ij}^{*H} to include these viscous effects. In this case Eqs. 6-7 and 6-8 are rewritten as:

$$(\tau s)_{i\leftrightarrow j}^* \equiv W_{ij}^{*D}v_{zi} - W_{ji}^{*D}v_{zj} \tag{6-13}$$

$$(q''s)_{i\leftrightarrow j}^* \equiv W_{ij}^{*D}h_i - W_{ji}^{*D}h_j \tag{6-14}$$

and for single-phase Eqs. 6-9 and 6-11 are rewritten as:

$$(\tau s)_{i\leftrightarrow j}^* = W_{ij}^{*M}(v_{zi} - v_{zj}) \tag{6-15}$$

$$(q''s)_{i\leftrightarrow j}^* = W_{ij}^{*H}(h_i - h_j) \tag{6-16}$$

For simplicity the time-averaged overbars are dropped starting with the paragraph preceding Eq. 6-13 whenever these parameters are utilized in the remainder of the chapter. The terms W_{ij}^{*M} and W_{ij}^{*H} will be used in expressing the subchannel conservation equations. The associated turbulent and viscous components of W_{ij}^{*H} and W_{ij}^{*M} are discussed later in this chapter. Table 6-1 summarizes the transverse flow rates that have been defined in this section.

IV DERIVATION OF THE SUBCHANNEL CONSERVATION EQUATIONS: METHOD OF SPECIALIZATION OF THE POROUS MEDIA EQUATIONS

We proceed to derive the subchannel relations by applying the volume-averaged porous body equations of Table 5-1 to the subchannel geometry of Figure 6-5. Use is made of the subchannel parameters defined in Section III. This procedure will make explicit the approximations characteristic of the subchannel approach. Since the porous body equations of Chapter 5 are single-phase

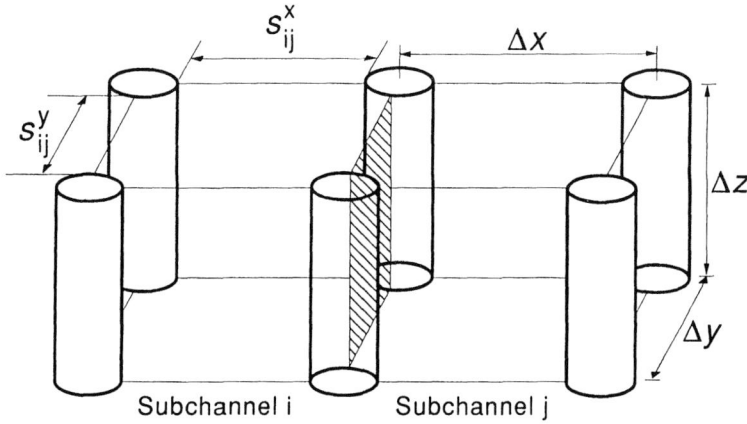

Figure 6-5 Subchannel control volumes.

equations the resulting subchannel equations will be for single-phase conditions. The additional terms to describe two-phase conditions are introduced as appropriate utilizing the definitions of Section III.

A Geometric Relations

The total volume of the porous media unit cell that encompasses a subchannel control volume of Figure 6-5 is:

$$V_T = \Delta x \Delta y \Delta z = A_T \Delta z \tag{6-17}$$

where A_T is the total axial cross-sectional area. This area is taken axially constant in the following derivations. The volume porosity following Eq. 5-1 is:

$$\gamma_V = \frac{V_f}{V_T} \tag{6-18}$$

Also, according to Eq. 5-4 for this control volume, the surface porosity in the z-direction is expressed as:

$$\gamma_{Az} = \frac{A_f}{A_T} \tag{6-19a}$$

Likewise, the surface porosities in the x-direction and the y-direction can be expressed as:

$$\gamma_{Ax} = \frac{s_{ij}^y \Delta z}{\Delta y \Delta z} \tag{6-19b}$$

$$\gamma_{Ay} = \frac{s_{ij}^x \Delta z}{\Delta x \Delta z} \tag{6-19c}$$

Multiplying Eqs. 6-18 and 6-19a, b and c by V_T expressed as Eq. 6-17, we can write:

$$V_T\gamma_V = V_T\frac{V_f}{V_T} = V_f \tag{6-20}$$

$$V_T\gamma_{Az} = (A_T\Delta z)\frac{A_f}{A_T} = A_f\Delta z \tag{6-21a}$$

$$V_T\gamma_{Ax} = (\Delta x\Delta y\Delta z)\frac{s_{ij}^y\Delta z}{\Delta y\Delta z} = s_{ij}^y\Delta x\Delta z \tag{6-22a}$$

$$V_T\gamma_{Ay} = (\Delta x\Delta y\Delta z)\frac{s_{ij}^x\Delta z}{\Delta x\Delta z} = s_{ij}^x\Delta z\Delta y \tag{6-23}$$

These are the geometry relations needed for application to the conservation relations.

B Continuity Equation

By multiplying Eq. 5-28† by V_T and making use of the definitions of Sections III and IVA, we have:

$$V_f\frac{\partial\langle\rho\rangle}{\partial t} + \Delta_x(\{\rho v_x\}s_{ij}^y\Delta z) + \Delta_y(\{\rho v_y\}s_{ij}^x\Delta z)$$
$$+ \Delta_z(\{\rho v_z\}A_f) = 0 \tag{6-24}$$

Applying Eqs. 6-2, 6-3, and 6-4, dividing through by Δz, collecting the transverse direction terms (the second and third terms of Eq. 6-24), and including mass transfer by turbulent interchange, we get the continuity equation for subchannel i:

$$A_{fi}\frac{\partial}{\partial t}\langle\rho_i\rangle + \frac{\Delta\dot{m}_i}{\Delta z} = -\sum_{j=1}^{J}[W_{ij} + W_{i\leftrightarrow j}^{'D}] \tag{6-25}$$

where J is the number of neighboring subchannels. Figure 6-6 illustrates the mass flow rate and diversion cross-flow rate terms of Eq. 6-25 which are crossing the subchannel control volume surfaces.

For single-phase flow, application of Eq. 6-5 simplifies Eq. 6-25 to:

$$A_{fi}\frac{\partial}{\partial t}\langle\rho_i\rangle + \frac{\Delta\dot{m}_i}{\Delta z} = -\sum_{j=1}^{J}W_{ij} \tag{6-26}$$

† We actually use the shorthand version of this equation in which the superscript i is dropped throughout. The lateral turbulent interchange terms must be inserted directly because the time-averaging procedure has not been formally accomplished.

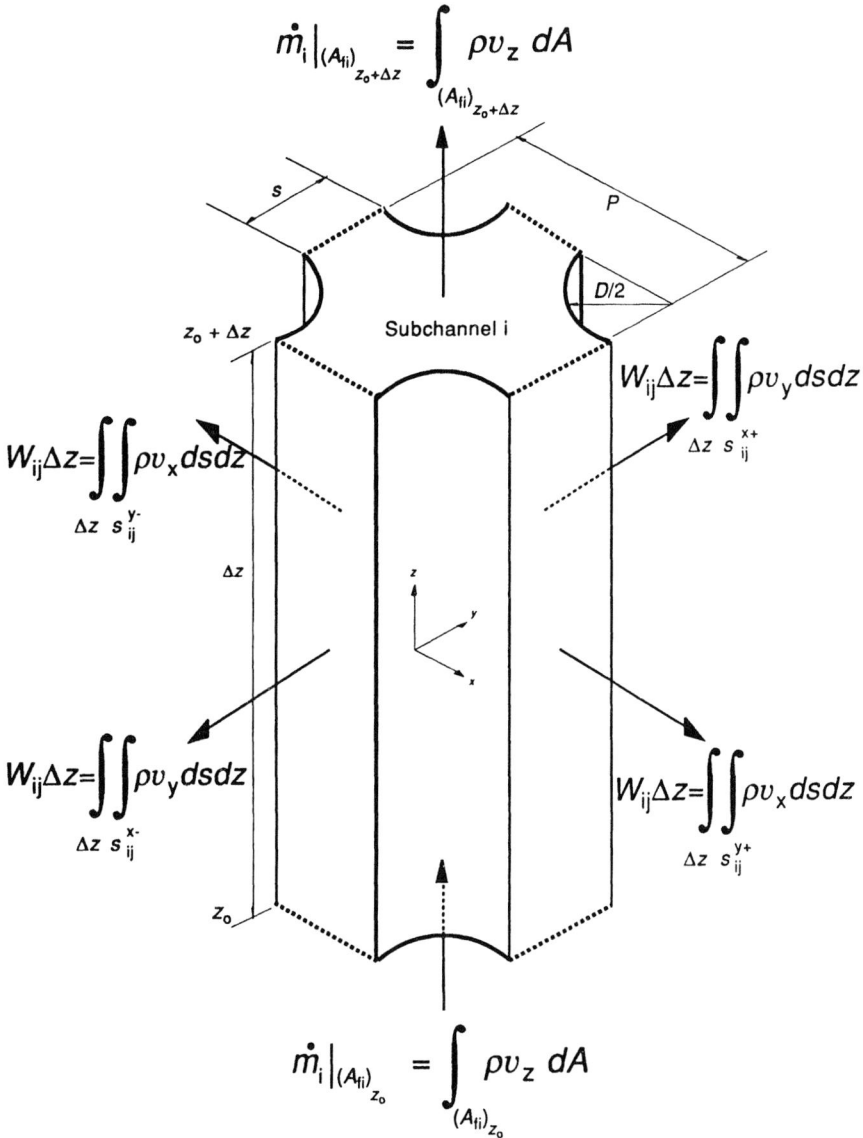

$$\dot{m}_i\big|_{(A_{fi})_{z_0+\Delta z}} = \int_{(A_{fi})_{z_0+\Delta z}} \rho v_z \, dA$$

$$W_{ij}\Delta z = \int\int_{\Delta z \; s_{ij}^{x+}} \rho v_y \, dsdz$$

$$W_{ij}\Delta z = \int\int_{\Delta z \; s_{ij}^{y-}} \rho v_x \, dsdz$$

$$W_{ij}\Delta z = \int\int_{\Delta z \; s_{ij}^{x-}} \rho v_y \, dsdz$$

$$W_{ij}\Delta z = \int\int_{\Delta z \; s_{ij}^{y+}} \rho v_x \, dsdz$$

$$\dot{m}_i\big|_{(A_{fi})_{z_0}} = \int_{(A_{fi})_{z_0}} \rho v_z \, dA$$

Figure 6-6 Subchannel control volume for continuity.

C Energy Equation

Eq. 5-59 represents the energy equation in terms of enthalpy in the porous body approach. Consider the shorthand version of this equation without the superscript *i* and without the approximation of treating the average of products as the product of the averages. Multiplying each term of the equation by V_T and using the definitions of Sections III and IVA, we obtain for subchannel i:

$$V_T \gamma_v \frac{\partial}{\partial t} \langle \rho h \rangle = A_{fi} \frac{\partial}{\partial t} \langle \rho h \rangle_i \Delta z \tag{6-27}$$

$$V_T \frac{\Delta_x}{\Delta x} [\gamma_{Ax} \{\rho h v_x\}] = \Delta_x (W_{ij}^x \{h_x^*\} \Delta z) \tag{6-28}$$

where $\{h_x^*\}$, the enthalpy carried by diversion cross-flow, has been defined as:

$$\{h_x^*\} = \frac{\{\rho v_x h\}}{\{\rho v_x\}} \tag{6-29}$$

The area-averaged character of this diversion transported enthalpy is specifically retained in this definition, in contrast to the definition of axially transported enthalpy presented subsequently in Eq. 6-31:

$$V_T \frac{\Delta_y}{\Delta y} [\gamma_{Ay} \{\rho h v_y\}] = \Delta_y (W_{ij}^y \{h_y^*\} \Delta z) \tag{6-30}$$

where $\{h_y^*\}$ is defined analogously to $\{h_x^*\}$. For the axial transport of enthalpy:

$$V_T \frac{\Delta_z}{\Delta z} [\gamma_{Az} \{\rho h v_z\}] = \Delta_z (\dot{m}_i h_i) \tag{6-31}$$

where for axial-averaged enthalpies it is simply assumed that $\{\rho h v_z\} = h_i \{\rho v_z\}$, i.e., no distinction is made between the area and volume-averaged enthalpy in the subchannel i. Expanding other terms:

$$V_T \gamma_v \left\langle \frac{Dp_i}{Dt} \right\rangle = A_{fi} \left\langle \frac{Dp_i}{Dt} \right\rangle \Delta z \tag{6-32}$$

$$V_T \left[\frac{\Delta_x}{\Delta x} \left[\gamma_{Ax} \left\{ k_e \frac{\partial T}{\partial x} \right\} \right] + \frac{\Delta_y}{\Delta y} \left[\gamma_{Ay} \left\{ k_e \frac{\partial T}{\partial y} \right\} \right] \right]$$
$$= - \sum_j [W_{ij}^{*D} h_i - W_{ji}^{*D} h_j] \Delta z \tag{6-33}$$

where $k_e = k_{\text{molecular}} + k_{\text{eddy}}$, and axial temperature gradients have been neglected. Further, $\{W_{ij}^{*D} h_i\}$ and $\{W_{ji}^{*D} h_j\}$ have been taken equal to $W_{ij}^{*D} h_i$ and $W_{ji}^{*D} h_j$, respectively, consistent with the convention for expressing subchannel enthalpy. For the volumetric energy generation terms:

$$V_T \gamma_v [\langle q_{rb}''' \rangle + \langle q''' \rangle + \langle \phi \rangle] = [\langle q_i' \rangle_{rb} + \langle q_i' \rangle] \Delta z + V_f \langle \phi \rangle \tag{6-34}$$

where $\langle q_{rb}''' \rangle$ is the equivalent dispersed heat source (or sink) per unit volume of the fluid due to immersed solids, and $\langle q''' \rangle$ is the extraneous heat source (or sink) per unit volume of the fluid. Accordingly, $\langle q_i' \rangle_{rb}$ and $\langle q_i' \rangle$ represent the subchannel linear heat generation (or absorption) rates of the aforementioned values. Dividing each term by Δz and regrouping Eq. 5-59, we have for single-phase flow:

$$A_{fi} \frac{\partial}{\partial t} [\langle \rho h \rangle_i] + \frac{\Delta}{\Delta z} [\dot{m}_i h_i] = \langle q_i' \rangle_{rb} - \sum_{j=1}^{J} W_{ij}^{*H}[h_i - h_j]$$

$$- \sum_{j=1}^{J} W_{ij}\{h^*\} + A_{fi} \left\langle \frac{Dp_i}{Dt} \right\rangle$$

(6-35)

where Eq. 6-35 reflects the following simplifying assumptions:

- For single-phase flow: $W_{ij}^{*D}h_i - W_{ji}^{*D}h_j = W_{ij}^{*H}[h_i - h_j]$
- Axial heat conduction in fluid is negligible: $\{k_e(\partial T/\partial z)\} = 0$
- No heat generation in the fluid: $\langle q''' \rangle = 0$
- Heat dissipation due to viscous effects is negligible: $\langle \phi \rangle = 0$
- The enthalpy carried by diversion cross-flow $\{h_x^*\}$ and $\{h_y^*\}$ is represented as $\{h^*\}$
- No distinction between the area and volume averaged enthalpy in channel i: $\{\rho h v_z\}/\{\rho v_z\} \simeq \langle \rho h v_z \rangle/\langle \rho v_z \rangle = h_i$
- Δ_z is written as Δ.

Figure 6-7 illustrates the heat generation rate and convective energy transport rates that cross the subchannel control volume surfaces.

D Axial Linear Momentum Equation

Applying the same procedure and parallel assumptions to those used in derivation of the continuity and energy equations, the terms of Eq. 5-46 (shorthand version without the superscript i) become:

$$V_T \gamma_v \frac{\partial}{\partial t} \langle \rho v_z \rangle = \left(\frac{\partial}{\partial t} \langle \dot{m}_i \rangle \right) \Delta z$$

(6-36)

$$V_T \frac{\Delta_x(\gamma_{Ax}\{\rho v_z v_x\})}{\Delta x} = \Delta_x(W_{ij}^x\{v_z^*\}\Delta z)$$

(6-37)

where the axial velocity associated with the diversion cross-flow has been taken as: $\{v_z^*\} \equiv \{\rho v_x v_z\}/\{v_x\}$.

$$V_T \frac{\Delta_y(\gamma_{Ay}\{\rho v_z v_y\})}{\Delta y} = \Delta_y(W_{ij}^y\{v_z^*\}\Delta z)$$

(6-38)

$$V_T \frac{\Delta_z(\gamma_{Az}\{\rho v_z^2\})}{\Delta z} = \Delta_z(\dot{m}_i\{v_z\}) = \Delta_z(\dot{m}_i v_{zi})$$

(6-39)

$$V_T \gamma_v \langle \rho \rangle g_z = A_{fi} \langle \rho \rangle g_z \Delta z$$

(6-40)

$$V_T \frac{\Delta_z(\gamma_{Az}\{p\})}{\Delta z} = A_{fi} \Delta_z(\{p\})$$

(6-41)

$$\dot{m}_i h_i \big|_{(A_{fi})_{z_0+\Delta z}} = \int_{(A_{fi})_{z_0+\Delta z}} \rho v_z h \, dA$$

$$\frac{1}{\Delta z} \iint_{\Delta z \, P_h} q'' \, dP_h \, dz = (q_i')_{rb} \, \Delta z$$

$$W_{ij}\{h^*\}\Delta z = \frac{1}{\Delta z} \iint_{\Delta z \; s_{ij}^{y+}} \rho v_x h \, ds \, dz$$

$$W_{ij}^{*H} h_i \, \Delta z$$

$$W_{ij}^{*H} h_j \, \Delta z$$

$$\dot{m}_i h_i \big|_{(A_{fi})_{z_0}} = \int_{(A_{fi})_{z_0}} \rho v_z h \, dA$$

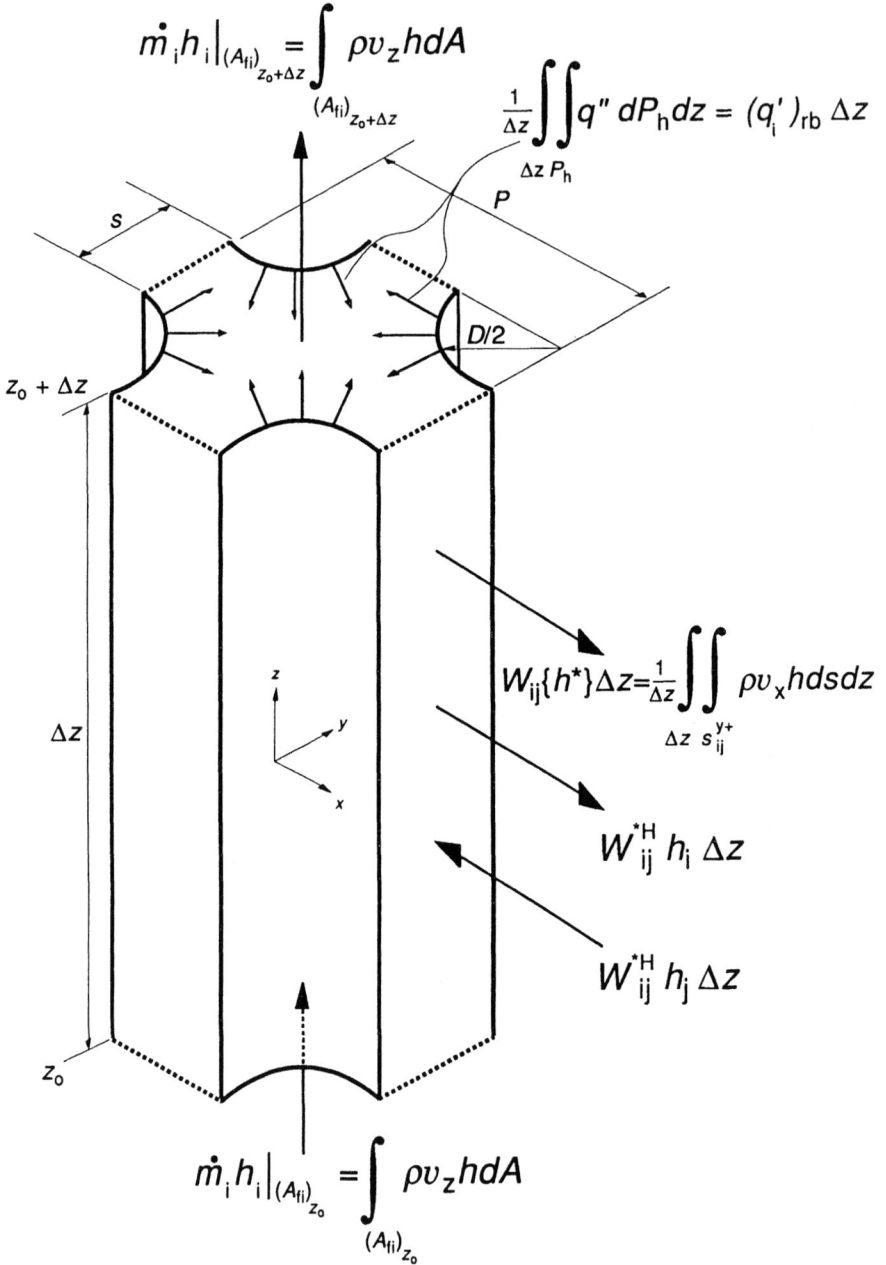

Figure 6-7 Subchannel control volume for energy (enthalpy) balance.

$$V_T \frac{\Delta_x(\gamma_{Ax}\{\tau_{xz}\})}{\Delta x} + V_T \frac{\Delta_y(\gamma_{Ay}\{\tau_{yz}\})}{\Delta y} = -\sum_j [W_{ij}^{*D}v_{zi} - W_{ji}^{*D}v_{zj}]\Delta z \quad (6\text{-}42a)$$

where for single-phase flow utilizing Eq. 6-15 becomes:

$$= -\sum_j W_{ij}^{*M}(v_{zi} - v_{zj})\Delta z \quad (6\text{-}42b)$$

where in this case for velocity, again no distinction is made between area and volume-averaged properties of subchannel i. The effect of the axial component $\{\tau_{zz}\}$ has been neglected:

$$V_T \gamma_V \langle R_z \rangle = V_f \langle R_z \rangle = -\left\{\frac{F_{iz}}{\Delta z}\right\} \Delta z \quad (6\text{-}43)$$

where $\{F_{iz}/\Delta z\}$ is the subchannel circumferentially averaged force per unit length of the fluid on the solid for vertical flow over the solid surfaces in the control volume. Note that the minus sign must be introduced in Eq. 6-43 because $\langle R_z \rangle$ and $\{F_{iz}/\Delta z\}$ are defined as the force of the solid on the fluid and the fluid on the solid, respectively. This difference arises because in subchannel analysis the force that is conventionally expressed and correlated is the drag force, i.e., the force of the fluid on the solid. Dividing each term by Δz and regrouping Eq. 5-46, we obtain:

$$\frac{\partial}{\partial t}\langle \dot{m}_i \rangle + \sum_{j=1}^{J} W_{ij}\{v_z^*\} + \frac{\Delta(\dot{m}_i v_{zi})}{\Delta z} = -A_{fi}\langle \rho \rangle g_z$$

$$- A_{fi}\frac{\Delta\{p\}}{\Delta z} - \sum_{j=1}^{J} W_{ij}^{*M}(v_{zi} - v_{zj}) - \left\{\frac{F_{iz}}{\Delta z}\right\} \quad (6\text{-}44)$$

where again Δ_z has been written as Δ.

Figure 6-8 illustrates the axial momentum that crosses and the axial forces that act on the subchannel control volume surfaces.

In some thermal-hydraulic computer codes such as TORC, LYNX, and various versions of COBRA, different symbols are used in the conservation equations as follows:

symbol for axial direction: x
symbol for the effective axial velocity associated with the diversion cross-flow: v_x^*.

For example, a change of z to x and v_z^* to v_x^* will transform Eq. 6-29 into the form of the COBRA code axial momentum equation, as will be shown in Section VI.

E Transverse Linear Momentum Equation

For each transverse momentum control volume, transverse flow is allowed only along one dimension. Adjacent subchannels (upon which mass, energy,

$$\dot{m}_i v_{zi}\big|_{(A_{fi})_{z_0+\Delta z}} = \int_{(A_{fi})_{z_0+\Delta z}} \rho v_z^2 \, dA \quad p_i\big|_{(A_{fi})_{z_0+\Delta z}}$$

$z_0 + \Delta z$

P

s

$D/2$

$A_{fi}\rho g_z$

z

y

x

Δz

$$W_{ij}^x\{v_z^*\}\Delta z = \iint_{\Delta z \ s_{ij}^{y+}} \rho v_x v_z \, ds \, dz$$

$$W_{ij}^{\bullet M} v_{zi}\Delta z$$

$$\left\{\dfrac{F_{iz}}{\Delta z}\right\}$$

$$W_{ij}^{\bullet M} v_{zj}\,\Delta z$$

z_0

$p_i\big|_{(A_{fi})_{z_0}}$

$$\dot{m}_i v_{zi}\big|_{(A_{fi})_{z_0}} = \int_{(A_{fi})_{z_0}} \rho v_z^2 \, dA$$

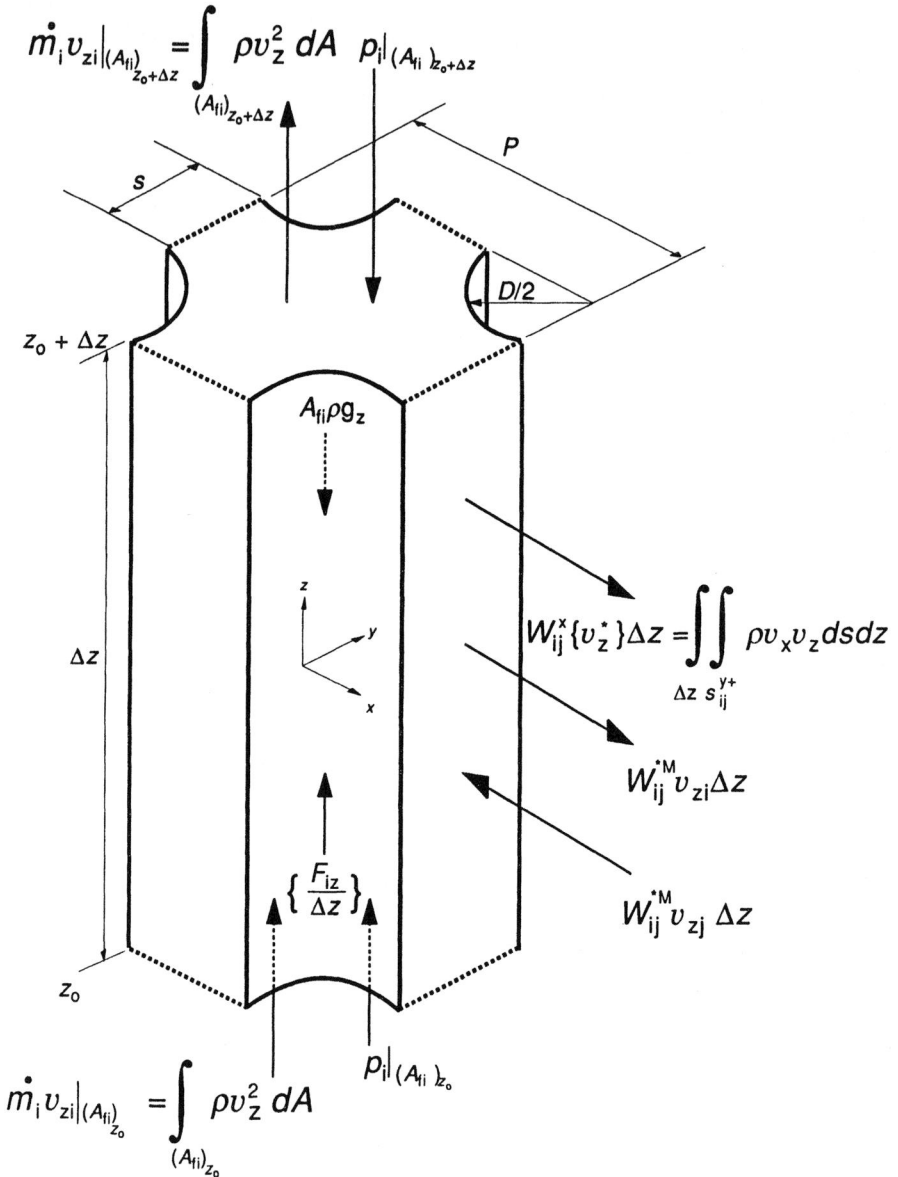

Figure 6-8 Subchannel control volume for axial momentum balance.

and axial momentum balances have been taken) are thereby coupled by similar transverse control volumes that each admit only one-dimensional transverse flow rates. Specifically, typical subchannels i and j are coupled by a transverse flow rate W_{ij} across the gap. The x-direction momentum equation analogous to equation 5-46 (shorthand version without the superscript i) may now be applied to the control volumes shown in Figure 6-3. Taking the control volume lying in the x-direction between the rows of rods as an example:

$$V_T = A_f \Delta x' \tag{6-45}$$

and

$$A_f = \Delta z \Delta y = \Delta z s_{ij}^y \tag{6-46}$$

where $\Delta x'$ may also be replaced by l, the transverse length, which is approximately the distance between adjacent channel centroids. Note that γ_{Az} and γ_{Ax} for this case are unity (γ_{Ay} is also unity, but the term involving the y-direction equals zero because v_y equals zero) so that Eqs. 6-21a and 6-22a become identical, i.e.:

$$V_T \gamma_{Az} = \Delta z s_{ij}^y \Delta x' \tag{6-21b}$$

$$V_T \gamma_{Ax} = s_{ij}^y \Delta x' \Delta z \tag{6-22b}$$

By ignoring shear stresses and gravity terms, Eq. 5-46 (written for x-direction) reduces to:

$$\frac{\partial \langle \rho v_x \rangle}{\partial t} + \frac{\Delta_{x'} \{\rho v_x^2\}}{\Delta x'} + \frac{\Delta_y \{\rho v_x v_y\}}{\Delta y} + \frac{\Delta_z \{\rho v_x v_z\}}{\Delta z} = - \frac{\Delta_{x'} \{p\}}{\Delta x'} + \langle R_x \rangle \tag{6-47}$$

By neglecting the shear stresses, no turbulent mixing term will appear in the final results. Multiplying each term of Eq. 6-47 by V_T and making use of definitions and parallel assumptions to those used previously, we can show that:

$$\frac{\partial}{\partial t} \langle \rho v_x \rangle V_f = \frac{\partial}{\partial t} (\langle \rho v_x \rangle s_{ij}^y) \Delta x' \Delta z = \frac{\partial}{\partial t} (W_{ij}^x) \Delta x' \Delta z \tag{6-48}$$

where $\langle \rho v_x \rangle$ has been taken equal to $\{\rho v_x\}$ at plane A_2 of Figure 6-3;

$$V_T \cdot \frac{\Delta_{x'}}{\Delta x'} \{\rho v_x^2\} = \Delta_{x'} (W_{ij}^x \{v_x\}) \Delta z \tag{6-49}$$

where the v_x velocity on faces B_1 and B_2 equals the v_x velocity on face A_2;

$$V_T \cdot \frac{\Delta_y}{\Delta y} \{\rho v_x v_y\} = 0, \text{ because } v_y = 0 \tag{6-50}$$

$$V_T \cdot \frac{\Delta_z}{\Delta z} \{\rho v_x v_z\} = \Delta_z (W_{ij}^x \{v_z\}) \Delta x' \tag{6-51}$$

where the v_x velocity on xy plane equals the v_x velocity on face A_2;

$$V_T \cdot \frac{\Delta_{x'}}{\Delta x'} \{p\} = \Delta_{x'} (s_{ij}^y \{p\}) \Delta z \tag{6-52}$$

$$V_T \cdot \langle R_x \rangle = - \left\{ \frac{F_{ix}}{2\Delta x' \Delta z} \right\} 2\Delta x' \Delta z \tag{6-53}$$

where $\{F_{ix}/2\Delta x' \Delta z\}$ represents the frictional and form loss or the total drag force per unit area where the area is taken in the xz plane.

Because this $\{F_{ix}/2\Delta x'\Delta z\}$ force is a combination of friction and form effects, it is defined in a way fundamentally different from $\{F_{iz}/\Delta z\}$ in Eq. 6-44, which is a force per axial length. Additionally, Eqs. 6-48 and 6-49 reflect the major assumption that the v_x velocity that exists on the faces B_1 and B_2 of the transverse momentum control volume is equivalent to the v_x velocity on face A_2 of the axial momentum control volume (see Figure 6-3). Similarly, Eq. 6-51 embodies an assumed equivalence of the v_x velocity in the control volume faces in the xy plane to that on face A_2. Dividing Eqs. 6-48 through 6-53 by $\Delta x'\Delta z$, the resulting equation is:

$$\frac{\partial}{\partial t}(W_{ij}^x) + \frac{\Delta}{\Delta x'}(W_{ij}^x\{v_x\}) + \frac{\Delta}{\Delta z}(W_{ij}^x\{v_z\}) = -(s_{ij}^y\frac{\Delta}{\Delta x'}\{p\}) - \left\{\frac{F_{ix}}{\Delta x'\Delta z}\right\} \quad (6\text{-}54)$$

where Δ_x, and Δ_z are written simply as Δ.

Figure 6-9 illustrates the momentum that crosses and the force that acts on the control volume surface. Now if the control volume of Figure 6-3 is selected so that there is no transverse flow through the lateral faces, then we must select a value of $\Delta x' = l$ to allow either $\Delta_x(\{v_x\}) = 0$ or $v_x = 0$ at planes B_1 and B_2. Upon such a control volume selection, Eq. 6-54 reduces to:

$$\frac{\partial}{\partial t}(W_{ij}^x) + \frac{\Delta}{\Delta z}(W_{ij}^x\{v_z\}) = -\frac{s_{ij}^y}{l}(\Delta\{p\}) - \left\{\frac{F_{ix}}{l\Delta z}\right\} \quad (6\text{-}55)$$

The steady-state form of the transverse momentum equation can be found by taking $\partial(W_{ij}^x)/\partial t = 0$, then:

$$\frac{\Delta}{\Delta z}(W_{ij}^x\{v_z\}) = -\frac{s_{ij}^y}{l}(\Delta\{p\}) - \left\{\frac{F_{ix}}{l\Delta z}\right\} \quad (6\text{-}56)$$

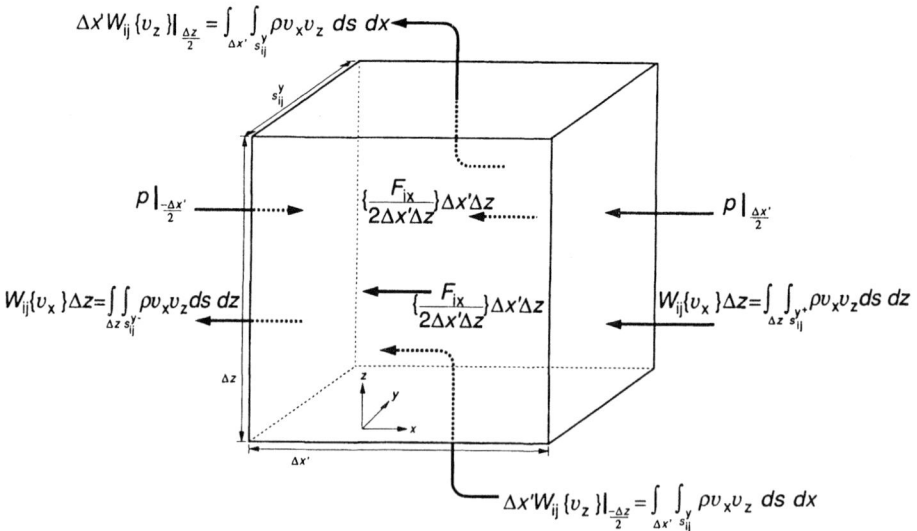

Figure 6-9 Subchannel control volume for transverse momentum.

The simplest form of the transverse momentum equation is applicable in assembly-wise analysis where the term $\Delta(W_{ij}^x\{v_z\})/\Delta z$ is negligible, yielding:

$$\frac{s_{ij}}{l}(\Delta\{p\}) + \left\{\frac{F_{ij}}{l\Delta z}\right\} = 0 \qquad (6\text{-}57)$$

Here, $F_{ix}/l\Delta z$ has been written as $F_{ij}/l\Delta z$ and s_{ij}^y as s_{ij} since Eq. 6-57 is a generalized result that represents the transverse control volume between any subchannel j adjacent to subchannel i.

The above-mentioned cases, in the forms employed in various thermal-hydraulic codes, are presented in Table 6-2. Table 6-3 summarizes the subchannel continuity, axial momentum and enthalpy conservation equations derived in this section.

Example 6-1 Computation of mixing flow rate for a test assembly

PROBLEM Find the mixing flow rate, $W_{ij}^{\prime D}$, between two adjacent channels i and j as shown in Figure 6-10. The channels are geometrically identical so that no diversion cross-flow occurs. Channel i is seeded with 400 ppm of a dye at the channel inlet. The dye concentration at the exit of channel j is 121 ppm. The channel's length is 3.66 m, and flow rate in each channel is 0.29 kg/s.

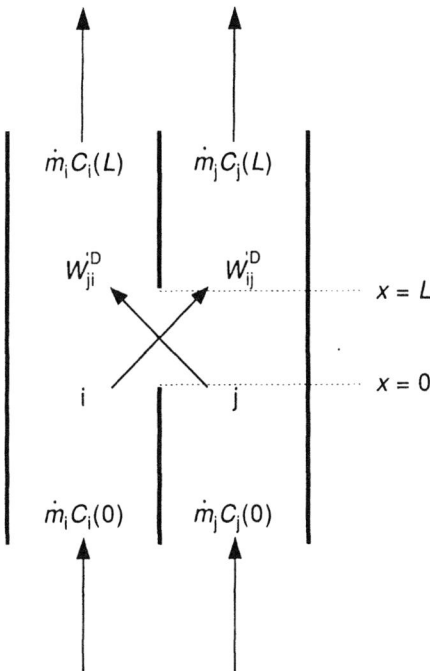

Figure 6-10 Mixing schematic.

Table 6-2 Various forms of the transverse momentum equation

Case	Code	Form of equation in code manual	Reference†
Subchannel-to-subchannel	1. COBRA-IIIC	$\dfrac{\partial W_{ij}}{\partial t} + \dfrac{\partial(UW_{ij})}{\partial x} = \dfrac{s}{l}(P_i - P_j) - F_{ij}$	Rowe [25]
	2. THINC-IV	$\dfrac{\partial(\bar{\rho}\bar{u}v)}{\partial z} + g_c\left(\dfrac{\partial p}{\partial x}\right) + F_x\bar{\rho}vv = 0$	Chelemer et al. [5]
	3. LYNX-2	$\dfrac{d(UW_{ij})}{dZ} = \dfrac{s}{l}(P_i - P_j) - F_{ij}$	LYNX-2, Babcock and Wilcox [19]
	4. TORC	$\dfrac{d(UW_{ij})}{dZ} = \dfrac{s}{l}(P_i - P_j) - F_{ij}$	TORC, Combustion Engineering [28]
Assembly-to-assembly	1. COBRA-IIIC	$\dfrac{\partial W_{ij}}{\partial t} + \dfrac{\partial(UW_{ij})}{\partial x} = \dfrac{s}{l}(P_i - P_j) - F_{ij}$	Rowe [25]
	2. THINC-I	$W_{ij} = \dfrac{2g_c a_{ij}\Delta z \Delta P_{ij}}{\sqrt{K_\infty(2g_c\Delta p_{ij}/\rho_i + \gamma\bar{v}_i^2)}}$	Chelemer et al. [4]
	3. LYNX-1	$\dfrac{d(UW_{ij})}{dZ} = \dfrac{s}{l}(P_i - P_j) - F_{ij}$	Hao and Alcorn [13]
	4. TORC	$(P_i - P_j) - \dfrac{R_{ij}}{s\Delta Z} = 0$	TORC, Combustion Engineering [28]

† See References for definitions of terms.

Table 6-3 Subchannel conservation equations: Single phase

Continuity

$$A_{fi}\frac{\partial}{\partial t}\langle\rho_i\rangle + \frac{\Delta\dot{m}_i}{\Delta z} = -\sum_{j=1}^{J} W_{ij} \tag{6-26}$$

Energy

$$A_{fi}\frac{\partial}{\partial t}[\langle\rho h\rangle_i] + \frac{\Delta}{\Delta z}[\dot{m}_i h_i] = \langle\dot{q}_i'\rangle_{rb} - \sum_{j=1}^{J} W_{ij}^{*H}[h_i - h_j] - \sum_{j=1}^{J} W_{ij}\{h^*\} + A_{fi}\left\langle\frac{Dp_i}{Dt}\right\rangle \tag{6-35}$$

Axial momentum

$$\frac{\partial}{\partial t}\langle\dot{m}_i\rangle + \sum_{j=1}^{J} W_{ij}\{v_z^*\} + \frac{\Delta(\dot{m}_i v_{zi})}{\Delta z} = -A_{fi}\langle\rho\rangle g_z - A_{fi}\frac{\Delta\{p\}}{\Delta z} - \sum_{j=1}^{J} W_{ij}^{*M}(v_{zi} - v_{zj}) - \left\{\frac{F_{iz}}{\Delta z}\right\} \tag{6-44}$$

Transverse momentum
(general form)

$$\frac{\partial}{\partial t}(W_{ij}) + \frac{\Delta}{\Delta x'}(W_{ij}\{v_x\}) + \frac{\Delta}{\Delta x}(W_{ij}\{v_z\}) = -\left(s_{ij}^y\frac{\Delta}{\Delta x'}\{p\}\right) - \left\{\frac{F_{ix}}{\Delta x'\Delta z}\right\} \tag{6-54}$$

SOLUTION Since the adjacent channels are geometrically identical, no diversion cross-flow occurs. Further, for dyes in water, molecular diffusion is normally many times smaller than turbulent effects. Hence $W_{ij}^{*D} = W_{ij}^{'D}$. In this case the subchannel energy equation (Eq. 6-35) can be written in terms of tracer concentration simply as:

$$\frac{d(\dot{m}_i C_i)}{dz} = -W_{ij}^{'D}(C_i - C_j) \tag{6-58}$$

Figure 6-10 illustrates the mixing process. The flow rate $\dot{m}_i$ is a constant and equal to $\dot{m}_j$. Also the mixing flow rates are equal. Hence:

$$\dot{m}_i = \dot{m}_j = \dot{m}$$
$$W_{ij}^{'D} = W_{ji}^{'D} = W'$$

The subchannel equations are:

$$\frac{dC_i}{dz} + \frac{W'}{\dot{m}}(C_i - C_j) = 0 \tag{6-59}$$

$$\frac{dC_j}{dz} + \frac{W'}{\dot{m}}(C_j - C_i) = 0 \tag{6-60}$$

For this two-channel case:

$$C_i(z) = C_0 - C_j(z)$$

where $C_0 \equiv C_i(0)$.

Substituting this result into Eq. 6-60 yields:

$$\frac{dC_j}{dz} + \frac{W'}{\dot{m}}(2C_j - C_0) = 0 \tag{6-61}$$

The solution of this equation, taking W' as a constant, is:

$$C_j(z) = \frac{C_0}{2} + \frac{C_2}{2}\exp\left[-2\frac{W'}{\dot{m}}z\right] \tag{6-62}$$

The initial condition for the untraced subchannel, j, is:

$$C_j(z = 0) = 0$$

Using this result, Eq. 6-62 becomes:

$$C_j(z) = \frac{C_0}{2}\left[1 - \exp\left(-2\frac{W'}{\dot{m}}z\right)\right] \tag{6-63}$$

which upon inversion is

$$W' = -\frac{\dot{m}}{2z}\ell n\left(1 - 2\frac{C_j(z)}{C_0}\right) \tag{6-64}$$

Utilizing the experimentally determined dye concentration at the channel exit, $z = L$, yields:

$$W' = \frac{-0.290}{2(3.66)} \ell n \left(1 - 2 \left(\frac{121}{400} \right) \right)$$

$$W' = 0.0368 \frac{kg}{m \cdot s}$$

V APPROXIMATIONS INHERENT IN THE SUBCHANNEL APPROACH

The subchannel approach is based on two fundamental premises:

• The control volume is fixed—generally as the fluid volume between rods. The crucial point is that a definite selection is made from the myriad possibilities of specific control volumes. This selected volume is used for all conservation equations except the transverse momentum equation.
• Any lateral flow is directed by the rod-to-rod gap through which it flows and loses its sense of direction after leaving the gap region. For this reason transverse momentum flux contributions in a three-dimensional application are not represented completely. Further, a separate control volume is employed for the transverse momentum equation.

The specification of a standard control volume is not an approximation. Rather, it is a step that narrows the user's flexibility in applying the method by dictating the region for which hydrodynamic parameters describing the array can be determined. The advantage is the concurrent focusing of experimental and correlating effort to a fixed configuration for which semiempirical predictions for constitutive relations are needed. These relations are discussed in Section VII.

The treatment of transverse momentum fluxes is an approximation. A fully three-dimensional physical situation can be represented by simply connecting channels in a three-dimensional array. However, because the transverse momentum formulation is incomplete, situations that depart from that of a predominantly axial flow field cannot be well represented. An example of such a situation is flow under conditions of mixed natural and forced convection in rod bundles. Table 6-4 contrasts the conservation equations developed in the previous chapter for the porous body with those of the subchannel formulation. The porous body equations contain several more terms than do the corresponding subchannel equations. However, only the omission in the subchannel transverse momentum equation (Eq. 6-54) of terms 2b and 6b which exist in the porous body formulation (Eq. 5-46) are inherent approximations.

All other deletions could be directly reintroduced into the subchannel equations. The criteria for the reintroduction is only the numerical significance of

Table 6-4 Comparison of terms between the porous body and subchannel forms of the conservation equations

Continuity	
Porous body (Eq. 5-33)	$$\overset{(1)}{\gamma_v \frac{\partial^i(\rho)}{\partial t}} + \overset{(2a)}{\frac{\Delta_x(\gamma_{Ax}{}^{i(x)}\{\rho\}^{i(x)}\{v_x\})}{\Delta x}} + \overset{(2b)}{\frac{\Delta_y(\gamma_{Ay}{}^{i(y)}\{\rho\}^{i(y)}\{v_y\})}{\Delta y}} + \overset{(3)}{\frac{\Delta_z(\gamma_{Az}{}^{i(z)}\{\rho\}^{i(z)}\{v_z\})}{\Delta z}} = 0$$
Subchannel (Eq. 6-26)	$$\overset{(1)}{A_{fi}\frac{\partial}{\partial t}\langle\rho_i\rangle} + \overset{(3)}{\frac{\Delta\dot{m}_i}{\Delta z}} = \overset{(2a+b)}{-\sum_{j=1}^{J} W_{ij}}$$
Energy (enthalpy)	
Porous body (Eq. 5-59)	$$\overset{(1)}{\gamma_v \frac{\partial}{\partial t}{}^i(\rho)^i\langle h\rangle} + \overset{(2a)}{\frac{\Delta_x(\gamma_{Ax}{}^{i(x)}\{\rho\}^{i(x)}\{h\}^{i(x)}\{v_x\})}{\Delta x}} + \overset{(2b)}{\frac{\Delta_y(\gamma_{Ay}{}^{i(y)}\{\rho\}^{i(y)}\{h\}^{i(y)}\{v_y\})}{\Delta y}} + \overset{(3)}{\frac{\Delta_z(\gamma_{Az}{}^{i(z)}\{\rho\}^{i(z)}\{h\}^{i(z)}\{v_z\})}{\Delta z}}$$ $$= \overset{(4)}{\gamma_v {}^i\left\langle\frac{Dp}{Dt}\right\rangle} + \overset{(5a)}{\frac{\Delta_x\left(\gamma_{Ax}{}^{i(x)}\left\{k_e\frac{\partial T}{\partial x}\right\}\right)}{\Delta x}} + \overset{(5b)}{\frac{\Delta_y\left(\gamma_{Ay}{}^{i(y)}\left\{k_e\frac{\partial T}{\partial y}\right\}\right)}{\Delta y}} + \overset{(5c)}{\frac{\Delta_z\left(\gamma_{Az}{}^{i(z)}\left\{k_e\frac{\partial T}{\partial z}\right\}\right)}{\Delta z}}$$ $$\overset{(6a)\qquad(6b)\qquad(6c)}{+\ \gamma_v{}^i(\langle q_{rb}''' \rangle + {}^i\langle q''' \rangle + {}^i\langle\phi\rangle)}$$
Subchannel (Eq. 6-35)	$$\overset{(1)}{A_{fi}\frac{\partial}{\partial t}\langle\rho h\rangle_i} + \overset{(3)}{\frac{\Delta}{\Delta z}[\dot{m}_i h_i]} = \overset{(4)}{A_{fi}\left\langle\frac{Dp_i}{Dt}\right\rangle} + \overset{(6a)}{\langle q_i'\rangle_{rb}} - \overset{(5a+b)}{\sum_{j=1}^{J} W_{ij}^{*H}[h_i - h_{jl}]} - \overset{(2a+b)}{\sum_{j=1}^{J} W_{ij}\{h^*\}}$$

Axial momentum (z-direction)	Porous body (z-component from Eq. 5-46)	$$\overset{(1)}{\gamma_V \frac{\partial}{\partial t}({}^i\langle\rho\rangle{}^i\langle v_z\rangle)} + \overset{(2a)}{\frac{\Delta_x(\gamma_{Ax}{}^{i(x)}\{\rho\}^{i(x)}\{v_z\}^{i(x)}\{v_x\})}{\Delta x}} + \overset{(2b)}{\frac{\Delta_y(\gamma_{Ay}{}^{i(y)}\{\rho\}^{i(y)}\{v_z\}^{i(y)}\{v_y\})}{\Delta y}} + \overset{(3)}{\frac{\Delta_z(\gamma_{Az}{}^{i(z)}\{\rho\}^{i(z)}\{v_z\}^{i(z)}\{v_z\})}{\Delta z}}$$ $$= \overset{(4)}{-\gamma_V{}^i\langle\rho\rangle g_z} - \overset{(5)}{\frac{\Delta_z(\gamma_{Az}{}^{i(z)}\{p\})}{\Delta z}} + \overset{(6a)}{\frac{\Delta_z(\gamma_{Az}{}^{i(z)}\{\tau_{zz}\})}{\Delta z}} + \overset{(6b)}{\frac{\Delta_y(\gamma_{Ay}{}^{i(y)}\{\tau_{yz}\})}{\Delta y}} + \overset{(6c)}{\frac{\Delta_x(\gamma_{Ax}{}^{i(x)}\{\tau_{xz}\})}{\Delta x}} + \overset{(7)}{\gamma_V{}^i\langle R_z\rangle}$$
	Subchannel (Eq. 6-44)	$$\overset{(1)}{\frac{\partial}{\partial t}\langle\dot m_i\rangle} + \overset{(2a+b)}{\sum_{j=1}^{J} W_{ij}\{v_z^*\}} + \overset{(3)}{\frac{\Delta(\dot m_i v_{zi})}{\Delta z}} = \overset{(4)}{-A_{fi}\langle\rho\rangle g_z} - \overset{(5)}{A_{fi}\frac{\Delta\{p\}}{\Delta z}} - \overset{(6b+c)}{\sum_{j=1}^{J} W_{ij}^{*M}(v_{zi} - v_{zj})} - \overset{(7)}{\left\{\frac{F_{iz}}{\Delta z}\right\}}$$
Transverse momentum (x-direction)	Porous body (x-component analogy to Eq. 5-46)	$$\overset{(1)}{\gamma_V \frac{\partial}{\partial t}({}^i\langle\rho\rangle{}^i\langle v_x\rangle)} + \overset{(2a)}{\frac{\Delta_x(\gamma_{Ax}{}^{i(x)}\{\rho\}^{i(x)}\{v_x\}^{i(x)}\{v_x\})}{\Delta x}} + \overset{(2b)}{\frac{\Delta_y(\gamma_{Ay}{}^{i(y)}\{\rho\}^{i(y)}\{v_x\}^{i(y)}\{v_y\})}{\Delta y}} + \overset{(3)}{\frac{\Delta_z(\gamma_{Az}{}^{i(z)}\{\rho\}^{i(z)}\{v_x\}^{i(z)}\{v_z\})}{\Delta z}}$$ $$= \overset{(4)}{-\gamma_V{}^i\langle\rho\rangle g_x} - \overset{(5)}{\frac{\Delta_x(\gamma_{Ax}{}^{i(x)}\{p\})}{\Delta x}} + \overset{(6a)}{\frac{\Delta_x(\gamma_{Ax}{}^{i(x)}\{\tau_{xx}\})}{\Delta x}} + \overset{(6b)}{\frac{\Delta_y(\gamma_{Ay}{}^{i(y)}\{\tau_{yx}\})}{\Delta y}} + \overset{(6c)}{\frac{\Delta_z(\gamma_{Az}{}^{i(z)}\{\tau_{zx}\})}{\Delta z}} + \overset{(7)}{\gamma_V{}^i\langle R_x\rangle}$$
	Subchannel (Eq. 6-54)	$$\overset{(1)}{\frac{\partial}{\partial t}(W_{ij})} + \overset{(2a)}{\frac{\Delta}{\Delta x'}(W_{ij}\{v_x\})} + \overset{(3)}{\frac{\Delta}{\Delta z}(W_{ij}\{v_z\})} = - \overset{(5)}{\left(s_{ij}^{y}\frac{\Delta}{\Delta x'}\{p\}\right)} - \overset{(7)}{\left\{\frac{F_{ix}}{\Delta x'\Delta z}\right\}}$$

TRL

Figure 6-5 Subchannel control volumes.

the term with respect to other terms in the equation. These deleted terms are summarized below.

Energy equation (Eq. 5-59)
   Term 6b. Energy directly deposited in the coolant volume
   Term 6c. Energy from viscous dissipation
   Term 5c. Molecular and turbulent energy exchange in the axial direction
Axial momentum equation (z component from Eq. 5-46)
   Term 6a. Molecular and turbulent momentum flux in the axial direction
Transverse momentum equation (Eq. 5-46)
   Term 4. Transverse gravity or other body force
   Term 6a and 6c. Molecular and turbulent momentum flux in the transverse
      direction.

Finally, note that although the subchannel equations are written with the notation W_{ij}^{*M} and W_{ij}^{*H}, which per Eqs. 6-15 and 6-16 includes both molecular and turbulent transfer of momentum and energy, in practice for high Prandtl number fluids the molecular effect is small.

Example 6-2 Computation of transverse molecular momentum flux and distributed resistance for a PWR rod array

PROBLEM Show that the transverse molecular momentum flux is several orders of magnitude less than the distributed resistance for flow in a typical PWR rod array.

SOLUTION Referring to the porous body axial momentum equation in Table 6-4, we will compare the following terms:

- Molecular portion of term 6b—transverse molecular momentum flux:
$$\frac{\Delta_y(\gamma_{Ay}^{i(y)}\{\tau_{yz}\})}{\Delta y}$$
- Distributed resistance: $\gamma_V^i\langle R_z \rangle$

   Take PWR water properties as:

$\mu \approx 9.63 \times 10^{-5}$ kg/m·s
$\rho \approx 726$ kg/m^3
and an axial velocity v_z of
$v_z \approx 6$ m/s.

   For a typical PWR fuel bundle array:
$$\Delta y \equiv P = (P/D)D = (1.33)\ 9.5\ \text{mm} = 12.6\ \text{mm}$$

From Example 5-1, we obtain:

$$\gamma_V = 0.554$$

$$\gamma_{Ay} = 0.246$$

Taking the molecular portion of the term $^{i(y)}\{\tau_{yz}\}$ as $\mu(\partial v_z/\partial y)$, the transverse molecular momentum flux becomes:

$$\frac{\Delta_y(\gamma_{Ay}{}^{i(y)}\{\tau_{yz}\})}{\Delta y} \simeq \gamma_{Ay}\mu\left(\frac{\Delta^2 v_z}{\Delta y^2}\right)$$

Now assuming that Δv_z is normally, at most, of order v_z, take it as $2v_z$:

$$\gamma_{Ay}\mu\left[\frac{\Delta^2 v_z}{\Delta y^2}\right] \simeq -\frac{0.246(9.63 \times 10^{-5})(2)(6)}{(0.0126)^2}$$

$$\simeq -1.79\,\frac{\text{Pa}}{\text{m}}\,(-1.14 \times 10^{-2}\,\text{lb}_\text{f}/\text{ft}^3)$$

From Example 5-2, the distributed resistance is:

$$\gamma_V{}^i\langle R_z\rangle = -\gamma_V 1.63\left[\frac{{}^i\langle\rho\rangle^i\langle v_z\rangle^2}{2}\right]\frac{\text{Pa}}{\text{m}}$$

with $^i\langle\rho\rangle$ in kg/m^3 and $^i\langle v_z\rangle$ in m/s. Substituting the numerical values:

$$\gamma_V{}^i\langle R_z\rangle \simeq -0.554(1.63)\left[\frac{726(6)^2}{2}\right]$$

$$\simeq -11{,}801\,\frac{\text{Pa}}{\text{m}}\,(-75.12\,\text{lb}_\text{f}/\text{ft}^3)$$

Summarizing, for this PWR example:

$$\text{Transverse molecular momentum flux} \simeq -1.8\,\frac{\text{Pa}}{\text{m}}$$

$$\text{Distributed resistance} \simeq -11{,}801\,\frac{\text{Pa}}{\text{m}}.$$

VI COMMONLY USED FORMS OF THE SUBCHANNEL CONSERVATION EQUATIONS

Since the introduction of numerical subchannel analysis methods in the 1960s, a great number of computer tools have been developed for utilizing this approach. In this section the conservation equations of a typical method for PWR analysis, the COBRA family, will be presented and be shown to be equivalent to the set derived in Section IV (see Table 6-3). The homogeneous equilibrium set of conservation equations of COBRA IV will be used, which were presented by Stewart et al. [27]. The COBRA subchannel control volumes and

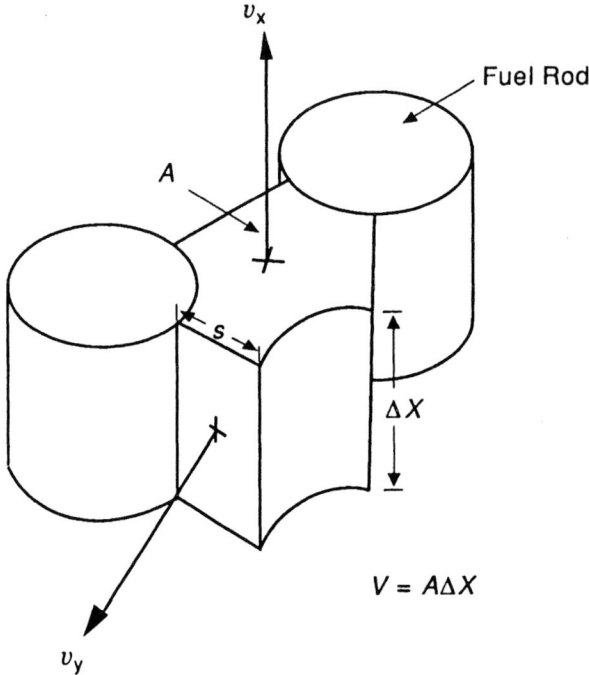

$$V = A\Delta X$$

Figure 6-11 Subchannel control volume for mass, energy, and axial momentum balances. *(From Stewart et al. [27].)*

their associated nomenclature are illustrated in Figures 6-11 and 6-12 for the mass, energy, axial momentum, and the transverse momentum balances, respectively. Note that the fuel rod array is triangular, which will allow demonstration of these conservation equations for a subchannel with nonorthogonal fluid (rod-to-rod gaps) boundaries.

A Definitions

The nomenclature conventions in this text and COBRA are related as follows:

COBRA	This text		
$\langle\langle\phi\rangle\rangle_V$	$\equiv \langle\phi\rangle$	Volume average	(6-65)
$\langle\phi\rangle_A$	$\equiv \{\phi\}$	Surface average	(6-66)
$\{\phi\}$	$\equiv \vec{\phi}$	Vector definition	(6-67)
$[\ ]$	$\equiv [\ ]$	Matrix definition	(6-68)
S	$\equiv s$	Gap width	(6-69)
$X;v_x$	$\equiv z;v_z$	Coordinate system and associated velocity component definitions	(6-70)
w'	$\equiv W_{ij}^{'H}$	Fluctuating cross-flow per unit length for turbulent enthalpy exchange	(6-71)

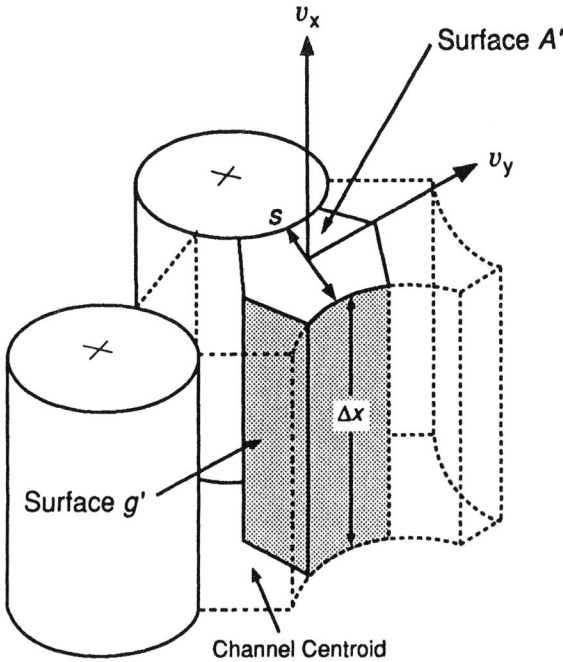

Figure 6-12 Subchannel control volume for transverse momentum balances. *(From Stewart et al. [27].)*

The COBRA equations also make extensive use of vector and matrix notations to define the connections among subchannels, rods, and walls as follows:

- Connections between subchannels use the matrix $[D_c]$ and its transpose $[D_c^T]$
- Connections between rods and subchannels use the matrix $[D_r]$ and its transpose $[D_r^T]$
- Connections between walls and subchannels use the matrix $[D_w]$ and its transpose $[D_w^T]$.

As an example, consider the net lateral heat conduction into channel j for the three channel array of Figure 6-13. For this configuration, we can write:

$$Q_j = C_l(T_i - T_j) - C_m(T_j - T_k) \tag{6-72}$$

where:

C_l, C_m = constant functions of fluid conductivity and gap dimensions

An equivalent compact general form in terms of connection matrix operators $[D_c]$ and $[D_c^T]$ and the diagonal matrix of conduction coefficients $[C]_{kl}$ is:

$$\{Q_j\} = [D_c^T]_{jk}[C]_{kl}[D_c]_{lm}\{T_m\} \tag{6-73}$$

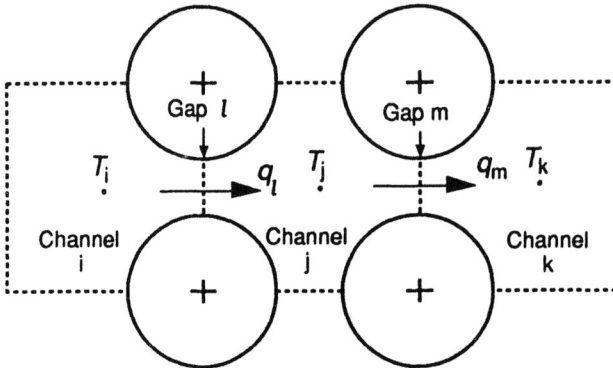

Figure 6-13 Three-channel array for lateral heat conduction problem. *(From Stewart et al. [27].)*

defined as:

$$[D_c] = \begin{bmatrix} 1 & -1 & 0 \\ 0 & 1 & -1 \end{bmatrix}; [D_c^T] = \begin{bmatrix} 1 & 0 \\ -1 & 1 \\ 0 & -1 \end{bmatrix} \quad \text{(6-74 and 6-75)}$$

For those not yet skilled in matrix algebra, it is sufficient to think of these operators as simply providing a shorthand way of managing the sign conventions for exchanges among rod array regions.

B The COBRA Continuity Equation

The COBRA continuity equation is:

$$A \frac{\partial}{\partial t} \langle\langle\rho\rangle\rangle_V + \frac{\partial}{\partial X} \langle\rho v_x\rangle_A A + \{D_c^T\}\langle\rho v_y\rangle_S S = 0 \quad (6\text{-}76)$$

In this case, since the sum of cross-flows around only a single subchannel is represented, the vector $\{D_c^T\}$ has been used. The term-by-term equivalence between the COBRA continuity form (6-76) and the form presented in Eq. 6-26 i.e.:

$$A_{fi} \frac{\partial}{\partial t} \langle\rho_i\rangle + \frac{\Delta \dot{m}_i}{\Delta z} + \sum_{j=1}^{J} W_{ij} = 0 \quad (6\text{-}77)$$

can be confirmed directly when the definitions presented in Section VIII are employed.

C The COBRA Energy Equation

The COBRA subchannel energy equation is:

$$A \frac{\partial}{\partial t} \langle\langle \rho h \rangle\rangle_{\mathrm{v}} + \frac{\partial}{\partial X} \langle \rho v_x h \rangle_{\mathrm{A}} A + \{D_{\mathrm{c}}^{\mathrm{T}}\}\{\langle \rho v_y h \rangle_{\mathrm{S}} S\} =$$

$$\{D_{\mathrm{r}}^{\mathrm{T}}\}[P\phi H][D_{\mathrm{r}}]\{T\} + \{D_{\mathrm{w}}^{\mathrm{T}}\}[LH][D_{\mathrm{w}}]\{T\} +$$

$$\frac{\partial}{\partial X} A \left\langle k \frac{\partial T}{\partial X} \right\rangle_{\mathrm{A}} - \{D_{\mathrm{c}}^{\mathrm{T}}\} \left[\frac{SC\langle k \rangle}{L_{\mathrm{c}}} \right] [D_{\mathrm{c}}]\{T\} - \tag{6-78}$$

$$\{D_{\mathrm{c}}^{\mathrm{T}}\}[w'][D_{\mathrm{c}}]\{h'\}$$

This equation and our general Eq. 6-35 are equivalent when it is recognized that COBRA Eq. 6-78 reflects the following differences from our general Eq. 6-35. The COBRA equation:

- Represents the molecular and turbulent or eddy contributions to energy exchange separately
- Includes heat transfer from walls in addition to rods into the subchannel
- Includes axial heat conduction in the fluid
- Excludes the pressure work term.

D The COBRA Axial Momentum Equation

The COBRA axial momentum equation is:

$$\frac{\partial}{\partial t} \langle\langle \rho v_x \rangle\rangle_{\mathrm{v}} A + \frac{\partial}{\partial X} \langle \rho v_x^2 \rangle_{\mathrm{A}} A + \{D_{\mathrm{c}}^{\mathrm{T}}\}\{\langle \rho v_x v_y \rangle_{\mathrm{S}} S\} =$$

$$-A \frac{\partial}{\partial X} \langle p \rangle_{\mathrm{A}} - \frac{1}{2} \left(\frac{f'}{D_{\mathrm{h}}} + \frac{K}{\Delta X} \right) \langle \rho v_x^2 \rangle_{\mathrm{A}} A - A\langle\langle \rho \rangle\rangle_{\mathrm{v}} g_z \cos\theta - \tag{6-79}$$

$$C_{\mathrm{T}}\{D_{\mathrm{c}}^{\mathrm{T}}\}[w'][D_{\mathrm{c}}]\{v_x'\}$$

where θ is the channel axis orientation angle measured from the vertical.

The COBRA equation neglects only the fluid-to-fluid viscous shear stress relative to the general subchannel axial momentum equation, Eq. 6-44. Specifically, only turbulent momentum transfer across the fluid gap is considered, i.e., molecular effects that are included in our terms $W_{\mathrm{ij}}^{*\mathrm{H}}$ are neglected in the COBRA formulation. The correspondence of the remaining terms† follows directly.

---

† These terms are expressed as $\dfrac{\text{force}}{\text{distance}}$ or $\dfrac{\text{mass}}{\text{time}^2}$.

E The COBRA Transverse Momentum Equation

For a triangular array the COBRA transverse control volume is shown in Figure 6-12. This control volume, V', is bounded laterally by the fuel rod surfaces and by planes joining the adjacent channel centroids and the fuel rod centerlines. The upper and lower surfaces of V' are closed by the flow area A', and the pseudolength, l, is prescribed as:

$$l \equiv \frac{A'}{S} \tag{6-80}$$

Note that l is approximately but not exactly equal to the distance between adjacent channel centroids.

For this control volume the COBRA transverse momentum equation is:

$$\frac{\partial}{\partial t} \langle\langle \rho v_y \rangle\rangle_{V'} S + \frac{\partial}{\partial X} \langle \rho v_y v_x \rangle_{A'} S + C_S \{D_c\}[D_c^T] \left\{ N \frac{S}{l} \langle \rho v_y^2 \rangle_S \cos\Delta\beta \right\} =$$
$$\frac{S}{l} \{D_c\}\{\langle p \rangle_A\} - \frac{1}{2} \frac{S}{l} K_G \langle \rho v_y^2 \rangle_S - \langle\langle \rho \rangle\rangle_{V'} g_z S \sin\theta \cos\beta \tag{6-81}$$

This formulation represents the following approximations:

1 Net lateral momentum flux term This net flux is expressed in terms of lateral momentum carried by cross-flows through the adjacent gaps, i.e.:

$$\langle \rho v_y^2 \rangle_S S \Delta x$$

Each gap is assigned a direction angle, β, from some arbitrary reference direction. The angle $\Delta\beta$ is the difference between the reference angle of a communicating gap and the gap of interest. For the transverse control volume of Figure 6-14, the gap of interest has a direction angle β_1, and the four communicating gaps have reference angles β_2, β_3, β_4, and β_5. A factor C_s, typically taken as unity, is provided to allow for the fact that the lateral momentum flux through a gap may be affected by the adjacent upstream gap conditions. Finally, utilizing the binary operator N to indicate the direction of flow, i.e., $+1$ for cross-flow into subchannels i or j and -1 for cross-flow out of subchannels i or j and the connection matrices defined previously, the net lateral momentum flux out of V' is written as:

$$C_s \{D_c\}[D_c^T]\{N \langle \rho v_y^2 \rangle_S S \Delta X \cos\Delta\beta\}$$

2 Pressure surface force The total lateral pressure force properly given by the difference in forces on the control volume lateral surfaces is approximated by the subchannel area averaged pressure differences, i.e.:

$$S \Delta X \{D_c\}\{\langle p \rangle_A\}$$

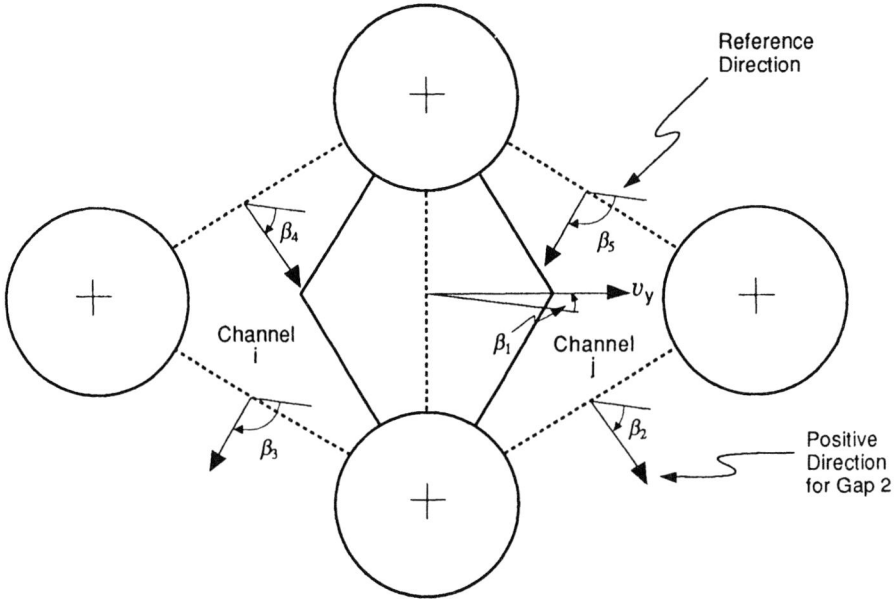

Figure 6-14 Rod orientation from vertical and gap reference angles in plane A. *(After Stewart et al. [27].)*

3 Lateral gravity force Provision for inclined bundles is allowed if the reference direction for β is chosen parallel to the plane described by the bundle axis, and the vertical and θ is the angular orientation from vertical. Then this force is:

$$-\langle\langle\rho\rangle\rangle_{v'} g_z Sl\Delta X \sin\theta \cos\beta$$

The correspondence between the COBRA formulation of Eq. 6-81 and the general subchannel transverse momentum equation, Eq. 6-54, is not as direct as for the other conservation equations because these two forms include different levels of approximation. The two levels of difference are:

1. The general transverse formation has replaced all terms involving ρv_y and a lateral area by the cross-flow W_{ij} by invoking the necessary approximations as described in Section IVE.
2. The COBRA equation has replaced all lateral length differences $\Delta x'$ by the pseudolength l defined on a geometric basis by Eq. 6-80.

With these differences and the differences in symbol definitions in mind, the remaining terms correspond.

Table 6-5 summarizes the COBRA subchannel conservation equations.

Table 6-5 The COBRA subchannel conservation equations

Continuity

$$\left\{ A\frac{\partial}{\partial t}\langle\langle\rho\rangle\rangle_v + \frac{\partial}{\partial X}\langle\rho v_x\rangle_A A + \{D_c^T\}\langle\rho v_y\rangle_S S = 0 \right. \tag{6-76}$$

Energy (enthalpy)

$$A\frac{\partial}{\partial t}\langle\langle\rho h\rangle\rangle_v + \frac{\partial}{\partial X}\langle\rho v_x h\rangle_A A + \{D_c^T\}\langle\rho v_y h\rangle_S S\} = \{D^T\}[P\phi H][D_r]\{T\} + \{D_w^T\}[LH][D_w]\{T\} + \frac{\partial}{\partial X}A\left\langle k\left\langle\frac{\partial T}{\partial X}\right\rangle\right\rangle_A \tag{6-78}$$

$$-\{D_c^T\}\left[\frac{SC\langle k\rangle}{L_c}\right][D_c]\{T\} - \{D_c^T\}\{w'\}[D_c]\{h'\}$$

Axial momentum

$$\frac{\partial}{\partial t}\langle\langle\rho v_x\rangle\rangle_v A + \frac{\partial}{\partial X}\langle\rho v_x^2\rangle_A A + \{D_c^T\}\langle\rho v_x v_y\rangle_S S = -A\frac{\partial}{\partial X}\langle p\rangle_A - \frac{1}{2}\left(\frac{f'}{D_h} + \frac{K}{\Delta X}\right)\langle\rho v_x^2\rangle_A A \tag{6-79}$$

$$- A\langle\langle\rho\rangle\rangle_v g_z\cos\theta - C_T\{D_c^T\}\{w'\}[D_c]\{v_x\}$$

Transverse momentum

$$\left\{\frac{\partial}{\partial t}\langle\langle\rho v_x\rangle\rangle_{v\cdot}S + \frac{\partial}{\partial X}\langle\rho v_y v_x\rangle_{A\cdot}S + C_s\{D_c\}[D_c^T]\{N\frac{S}{l}\langle\rho v_y^2\rangle_S\cos\Delta\beta\} = \frac{S}{l}\{D_c\}\{\langle p\rangle_A\} - \frac{1}{2}\frac{S}{l}K_G\langle\rho v_y^2\rangle_S\right. \tag{6-81}$$

$$- \langle\langle\rho\rangle\rangle_{v\cdot}g_z S\sin\theta\cos\beta$$

VII CONSTITUTIVE EQUATIONS

In addition to the conservation equations, it is necessary to specify fluid properties and constitutive equations to form a closed set of equations for solution. The fluid properties are obtained from an equation of state. The parameters for which constitutive relations are needed can be identified by inspecting the subchannel conservation equations.

The generalized form of the equations of state is:

$$\rho = \rho(h, p) \qquad \text{(analogous to Eq. 2-1)}$$

However, consistent with our assumption in development of the axial momentum equation of neglecting sonic effects, we assume that density may be evaluated as a function of enthalpy only:

$$\rho = \rho(h, p^*) \qquad \text{(analogous to Eq. 2-14)}$$

where p^* is a reference pressure considered constant for the problem. The equation of state, $\rho = \rho(h, p^*)$, is for an incompressible but thermally expanding fluid. This approach was introduced by Meyer [20] as the momentum integral model and is judged most appropriate for a wide class of intermediate speed reactor coolant channel transients. The complete spectrum of models studied by Meyer for the treatment of transient fluid flow through reactor coolant channels is presented in Chapter 2.

The parameters for which constitutive relationships are necessary are listed in Table 6-6. Generally, steady-state formulations are used for both steady and transient situations. Differences between formulations under forced and mixed convection conditions are recognized although forced convection formulations are generally used due to the lack of mixed convection correlations. The parameters are discussed individually in the following sections.

A Surface Heat Transfer Coefficients (Parameter 1) and Axial Friction and Drag (Parameter 4)

Correlations for these parameters have been presented for single-phase flow in circular tubes in Chapters 10 and 9 of Volume I and for two phase flows in circular tubes in Chapters 12 and 11 of Volume I. Adjustments for rod arrays having pitch-to-diameter ratios below which the equivalent diameter approach is adequate are presented in Chapter 9 of Volume I and Chapter 7 of this volume.

B Enthalpy (Parameter 3) and Axial Velocity (Parameter 6) Transported by Pressure-Driven Cross-flow

In reality, enthalpy and axial velocity gradients exist in the vicinity of the gap across which cross-flow occurs. Therefore, the value of enthalpy and axial velocity which is laterally transported across the gap is not possible to specify *a*

Table 6-6 Parameters for which constitutive relations are necessary

Conservation equation	Parameter	General formulation Symbol	General formulation Equation	Equivalent COBRA formulation Symbol	Equivalent COBRA formulation Equation
Continuity	None		6-26		6-76
Energy	1. Linear heat generation rate or surface heat transfer coefficient (COBRA)	$\langle q_1'\rangle_{rb}$		H	6-78
	2. Effective cross-flow rate per unit length associated with transverse energy transport	W_{ij}^{*H}	6-35	v_z (turbulent) C (molecular)	6-78
	3. Enthalpy transported by pressure-driven cross-flow	h^*	6-35	h'	6-78
Axial momentum	4. Friction and drag forces or factors (COBRA)	F_{iz}	6-44	f', K	6-79
	5. Effective cross-flow rate per unit length associated with transverse momentum transport	W_{ij}^{*M}	6-44	$C_T v_z'$ (turbulent)	6-79
	6. Axial velocity transported by pressure-driven cross-flow	v_z^*	6-44	v_x'	6-79
Transverse momentum	7. Transverse friction and form drag forces or coefficient	F_{ix}	6-54	K_G	6-81
	8. Aspect ratio of control volume: Width/length	$s_{ij}^y/\Delta x'$	6-54	S/l	6-81
	9. Flow factor for triangular arrays: Accounts for the portion of flow through the communicating gaps which does not pass through the lateral control volume boundaries	Not included		C_s	6-81

priori. The normal approximation is to use the donor approach in which the transported enthalpy or axial velocity is simply taken equal to the subchannel value from which the cross-flow occurs, i.e.:

if $\qquad\qquad W_{ij} > 0, h^* = h_i \qquad\qquad$ (6-82)

$\qquad\qquad W_{ij} < 0, h^* = h_j \qquad\qquad$ (6-83)

The donor cell method is physically based, easy to compute, and enhances computational stability but at the expense of high numerical diffusion that tends to degrade sharp gradients across the computation mesh.

C Transverse Friction and Form Drag Coefficient (Parameter 7)

The transverse friction and form force are to be specified. Many formulations have been proposed but since they each are empirically reduced from data, it is essential that the form of the transverse momentum equation used to perform this correlation be kept in mind. For example, if the data were interpreted using a form that did not include lateral inertia, then the resulting net lateral force or coefficient must include skin, drag, and lateral inertia effects. As Table 6-2 illustrates, the lateral inertia has not always been included in transverse momentum equation formulations.

Rouhani [26] reviewed the historical development of formulations for this transverse force. For typical rod arrays the coefficient K_G in Eq. 6-81 can be assigned a value of 0.5 or deduced from transverse flow correlations for laminar or turbulent flow as applicable.

D Transverse Control Volume Aspect Ratio (Parameter 8)

The value of the parameter $s^y_{ij}/\Delta x'$ or, equivalently, s/l also depends on the form of the transverse momentum equation to be used. In practice, the control volume is sized so that the term:

$$\frac{\Delta}{\Delta x'} (W_{ij}\{v_x\})$$

is made zero, leading to the use of equations of the form of Eq. 6-55. Physically, this is equivalent to selecting $\Delta x'$ or equivalently l, long enough so that all transverse flow between the adjacent channels is within the transverse control volume, and no flow enters or leaves the lateral faces, i.e., $v_x = 0$ at planes B_1 and B_2 of Figure 6-3. Equivalently, the control volume could be sized to yield $\Delta_{x'}(\{v_x\}) = 0$, but this implies strong enough cross-flow to pass through the control volume undiminished.

Since the gap width is established by the rod array, only $\Delta x'$ or l is to be selected. Based on the physical interpretation above, the value of $\Delta x'$ should be directly proportional to the magnitude of the cross-flow. For typical rod arrays with predominantly axial flow, experience has shown that a value of l equal to twice s, i.e., $s/l = 0.5$, yields satisfactory results. As the cross-flow rate increases, the value of l would be expected to increase, causing s/l to decrease. This trend has been confirmed by Brown et al. [3] who analyzed two-channel data of varying shapes and blockages. However, care must be taken in analyzing blockage situations to ensure both that the numerical tool used can handle flow recirculation situations and that use of an axially variable s/l is allowed to account for diminishing cross-flow downstream of the blockage.

E Effective Cross-Flow Rate for Molecular and Turbulent Momentum and Energy Transport (Parameters 2 and 5)

The evaluation of the terms W_{ij}^{*M} and W_{ij}^{*H} will include:

- Molecular and turbulent contributions
- Single- and two-phase effects.

The molecular contribution is directly proportional to the axial velocity and temperature gradients across the gap separating the subchannels. The turbulent contribution must be modeled phenomenologically. Two options exist. One is to assume that momentum and energy are transferred across the gap by the exchange of globs of fluids of equal mass. The other is to assume that the globs of fluid being exchanged are of equal volume. Either of these models should be satisfactory when the resulting differences are compensated for by the diversion cross-flow rate. The specific procedure of partitioning the transverse mass flow rate between the mixing and diversion cross-flow phenomena is not of concern as long as the applications are within the range of experimental observations.

Let us develop the equi-volume model since it is reducible to the equi-mass model when applied to constant density, single-phase flow situations. The conceptual hypothesis of equal fluid volumes V being exchanged between subchannels across their separating gap is sketched in Figure 6-15. Expressing this hypothesis mathematically:

$$V_{ij}' = V_{ji}' \qquad (6-84)$$

where the following superscript notation will be employed:

fluctuating (in time): $'$
time average: $-$ (note that $\bar{q}$ and $\bar{G}$ are spatial averages).

Since $V' = v_x'A$, where A is an effective cross-sectional area, we take the volume-for-volume exchange model as implying equal transverse fluctuating fluid velocities, i.e.:

$$v_{ij}' = v_{ji}' \qquad (6-85)$$

The fluctuating transverse mass flow rate per unit length across the gap between subchannels i and j is:

$$W_{ij}'^{D} = \frac{\rho_i v_{ij}' s_{ij} \Delta z}{\Delta z} \qquad (6-86)$$

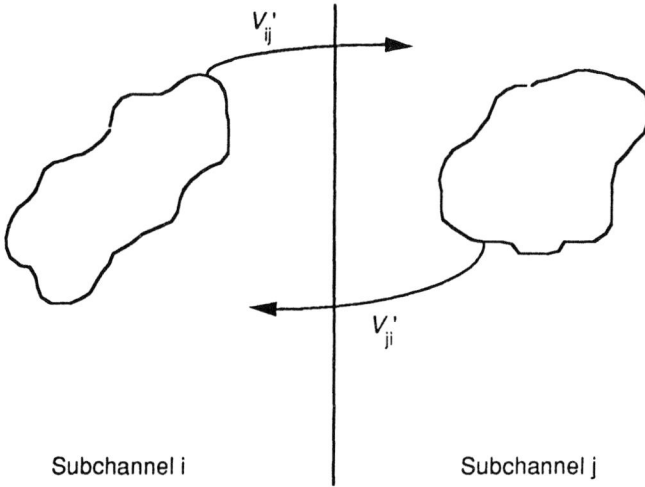

V_{ij}'

V_{ji}'

Subchannel i   Subchannel j

Figure 6-15 Hypothesis of equal fluid volume mixing between subchannels.

Rewriting Eq. 6-86 in terms of mixing length theory, obtain:

$$W_{ij}'^D \text{ is proportional to } \rho_i \left[l \frac{d\bar{v}_z}{dy} \right] s_{ij} = \rho_i \left(\frac{l^2 \dfrac{d\bar{v}_z}{dy}}{l} \right) s_{ij}$$

and therefore is proportional to $\rho_i \dfrac{\varepsilon}{l} s_{ij}$ (6-87)

where l is the turbulent mixing length, and ε is the eddy diffusivity.

Defining l divided by the constant of proportionality of Eq. 6-87 as z_{ij}^T, Eq. 6-87 becomes:

$$W_{ij}'^D = \rho_i \left(\frac{\varepsilon}{z_{ij}^T} \right) s_{ij}$$ (6-88)

Analogously, the fluctuating transverse mass flow rate per unit length from subchannel j to i is:

$$W_{ji}'^D = \rho_j \left(\frac{\varepsilon}{z_{ij}^T} \right) s_{ij}$$ (6-89)

since the values of the geometric parameters z and s are independent of the sequence of the subscripts. The *net* fluctuating transverse mass flow rate per unit length between subchannels i and j, $W_{i\leftrightarrow j}'^D$, is equal to:

$$W_{i\leftrightarrow j}'^D = W_{ij}'^D - W_{ji}'^D = \left(\frac{\varepsilon}{z_{ij}^T} \right) s_{ij}(\rho_i - \rho_j)$$ (6-90)

The physical character of transverse mass flow rates for single- and two-phase flows can be seen directly from Eq. 6-90. In the single-phase case in which the subchannel densities are effectively equal, Eq. 6-90 yields:

- $W_{i\leftrightarrow j}^{'D} = 0$ i.e., the net fluctuating mass flow rate is zero.
- $W_{ij}^{'D} = W_{ji}^{'D}$ i.e., the volume-for-volume exchange model reduces to the mass-for-mass exchange model.

In two-phase flow, since the subchannel densities are not equal, the net fluctuating mass flow rate is nonzero. For example, if $\rho_i > \rho_j$, the net fluctuating mass flow rate is from subchannel i to subchannel j and given by Eq. 6-90. Utilizing the general two-phase definition of density, Eq. I,5-50b in Eq. 6-90, and taking the phasic densities ρ_ℓ and ρ_v as constant, the net fluctuating mass flow rate equals:

$$W_{i\leftrightarrow j}^{'D} \equiv W_{ij}^{'D} - W_{ji}^{'D} = \frac{\varepsilon s_{ij}}{z_{ij}^T} (\rho_\ell - \rho_v)(\alpha_j - \alpha_i) \qquad (6\text{-}91)$$

where α_j and α_i are the void fractions of channels j and i, respectively. Eq. 6-91 illustrates that the net transverse flow rate is from subchannel i to j if the void fraction is larger in subchannel j than i. This net flow rate is composed of a net mass of liquid proportional to $\rho_\ell(\alpha_j - \alpha_i)$ which is transferred from subchannel i to j and a net mass of vapor proportional to $\rho_v(\alpha_j - \alpha_i)$ which is transferred from subchannel j to i.

Both energy and momentum transfer between subchannels must also be characterized. We treat turbulent energy transfer explicitly next. The energy transferred per unit time across the gap area per unit length has been written as:

$$(q''s)_{i\leftrightarrow j}^{'} = W_{ij}^{'D}h_i - W_{ji}^{'D}h_j \qquad (6\text{-}8)$$

Further development of this equation depends on whether the flow is single or two phase.

1 Single phase Equation 6-11 is the result for single-phase turbulent flow:

$$\overline{(q''s)_{i\leftrightarrow j}^{'}} = \overline{W_{ij}^{'H}(\bar{h}_i - \bar{h}_j)} \qquad (6\text{-}11)$$

The molecular contribution to the energy transfer is:

$$(q''s)_{i\leftrightarrow j}^{\text{conduction}} = \int_{s_{ij}} \left(-k \left.\frac{\partial T}{\partial n}\right|_{\text{gap}}\right) ds \qquad (6\text{-}92)$$

which for a constant temperature gradient along the gap can be written in terms of a laminar subchannel mixing length z_{ij}^L as:

$$(q''s)_{i\leftrightarrow j}^{\text{conduction}} = k \frac{T_i - T_j}{z_{ij}^L} s_{ij} \qquad (6\text{-}93)$$

where:

$$\left.\frac{\partial T}{\partial n}\right|_{\text{gap}} \equiv \frac{T_i - T_j}{z_{ij}^L} \qquad (6\text{-}94)$$

The effective mixing length, z_{ij}^L, is that distance defined by Eq. 6-94 which allows the gradient across the gap to be expressed in terms of the available subchannel average temperatures T_i and T_j. These parameters are illustrated in Figure 6-16 which shows the difference between the subchannel centroid-to-centroid distance η_{ij} and the laminar subchannel mixing length z_{ij}^L.

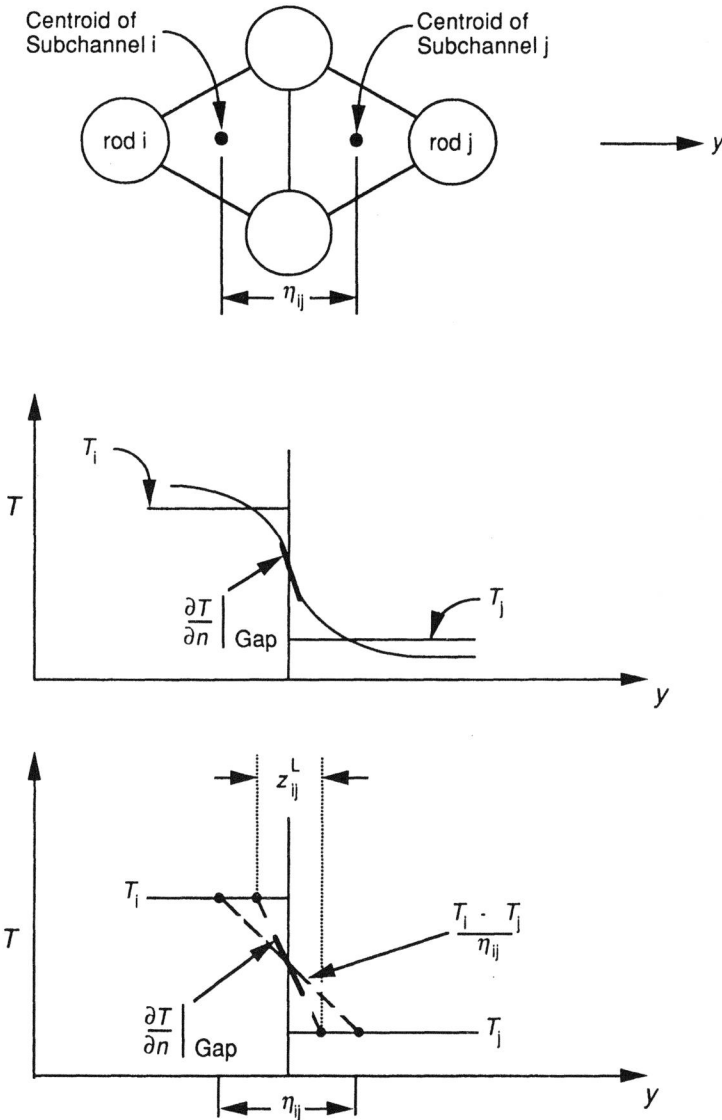

Figure 6-16 Difference between the subchannel centroid-to-centroid distance and the laminar mixing distance.

Summing the turbulent and molecular contributions from Eqs. 6-11 (dropping time-averaged overbars) and 6-93 we obtain:

$$(q''s)_{i\leftrightarrow j} = k \frac{T_i - T_j}{z_{ij}^L} s_{ij} + W_{ij}'^H (h_i - h_j) \tag{6-95}$$

This result has been written compactly in terms of W_{ij}^{*H} as Eq. 6-16:

$$(q''s)_{i\leftrightarrow j}^* = W_{ij}^{*H} (h_i - h_j) \tag{6-16}$$

where W_{ij}^{*H} includes both molecular and turbulent effects. Comparing Eqs. 6-95 and 6-11, it can be seen that W_{ij}^{*H} equals:

$$W_{ij}^{*H} = \frac{k s_{ij}}{C_p z_{ij}^L} + W_{ij}'^H \tag{6-96}$$

which can be expressed in terms of the Prandtl number and eddy diffusivity using Eq. 6-88, i.e.:

$$W_{ij}^{*H} = \mu s_{ij} \left[\frac{1}{z_{ij}^L \mathrm{Pr}} + \frac{\varepsilon}{z_{ij}^T \nu} \right] \tag{6-97}$$

Two principal dimensionless parameters have been introduced to characterize W_{ij}^{*H}. These are the mixing Stanton number M_{ij} and the mixing parameter β. They both are formed by dividing the transverse mass flow rate per unit area, i.e., the transverse mass flux W_{ij}^{*H}/s_{ij} by the axial mass flux G.

$$\text{Dimensionless mixing parameter} \equiv \frac{\text{Transverse mass flux}}{\text{Axial mass flux}} = \frac{W_{ij}^{*H}/s_{ij}}{G} \tag{6-98}$$

The parameters M_{ij} and β differ with respect to the exact axial mass flux employed. The mixing Stanton number employs that of subchannel i, G_i, whereas the mixing parameter β employs that of the interacting subchannels $\bar{G}$. Hence:

$$M_{ij} \equiv \frac{W_{ij}^{*H}}{s_{ij} G_i} \tag{6-99}$$

$$\beta \equiv \frac{W_{ij}^{*H}}{s_{ij} \bar{G}} \tag{6-100}$$

where:

$$\bar{G} = \frac{G_i A_i + G_j A_j}{A_i + A_j} \tag{6-101}$$

Single-phase correlations. Evaluation of W_{ij}^{*H} by Eq. 6-97 requires approximations for the eddy diffusivity and the laminar and turbulent mixing lengths.

The laminar mixing length is an effective constriction factor for conduction across the gap. It can be evaluated directly from the temperature field of a distributed parameter solution. A complexity arises if significant energy transfer between subchannels occurs via the fuel and/or clad. This would be favored if the conductivity and thickness of the clad and the conductivity of the fuel are important relative to the coolant conductivity and gap dimension. The laminar mixing length is obtained by equating Eqs. 6-92 and 6-93 yielding:

$$z_{ij}^L = \frac{k(T_i - T_j)}{\frac{1}{S_{ij}}\left[\int_{S_{ij}} - k\frac{\partial T}{\partial n}\bigg|_{gap} ds\right]} \qquad (6\text{-}102)$$

Eq. 6-102 can be evaluated when the temperature and axial velocity fields are available since the average subchannel temperature equals:

$$T_i = \frac{\iint_{A_i} T v_z dA}{\iint_{A_i} v_z dA} \qquad (6\text{-}103)$$

and the temperature gradient across the gap is directly available. Since the velocity and temperature solutions are dependent on boundary conditions and initial conditions, it should be expected that the mixing lengths similarly depend on these conditions. For example, France and Ginsberg [9] have evaluated mixing lengths for the two-subchannel geometry of Figure 6-16 for a uniform velocity profile and two different developing temperature fields, i.e.:

Case 1. Unheated rods with subchannels *i* and *j* at different inlet temperatures, i.e., an inlet source of heated fluid where the dimensionless temperature T^* boundary conditions are:

$$T_i^*(z = 0) = 0$$
$$T_j^*(z = 0) = 1$$

where:

$$T_n^* = \frac{T_n(z) - T_i(z)}{T_j(0) - T_i(0)}$$

Case 2. Heated rods (rod *i* and rod *j*) at different heat fluxes but identical inlet subchannel temperatures

where:

$$T^* \equiv \frac{T}{(\bar{q}''D/k_{coolant})}$$

and $\bar{q}''$ is the subassembly average fuel element heat flux.

Their results that are shown in Figure 6-17 for a triangular geometry of $P/D = 1.2$ demonstrate the existence of a development length with asymptotic values dependent on boundary and initial conditions. Table 6-7 summarizes the limited investigations to date in which mixing lengths have been determined. All investigations have used a slug, not a laminar velocity profile. In principle, the approach described here can yield turbulent as well as laminar mixing lengths if the corresponding temperature and velocity fields are available.

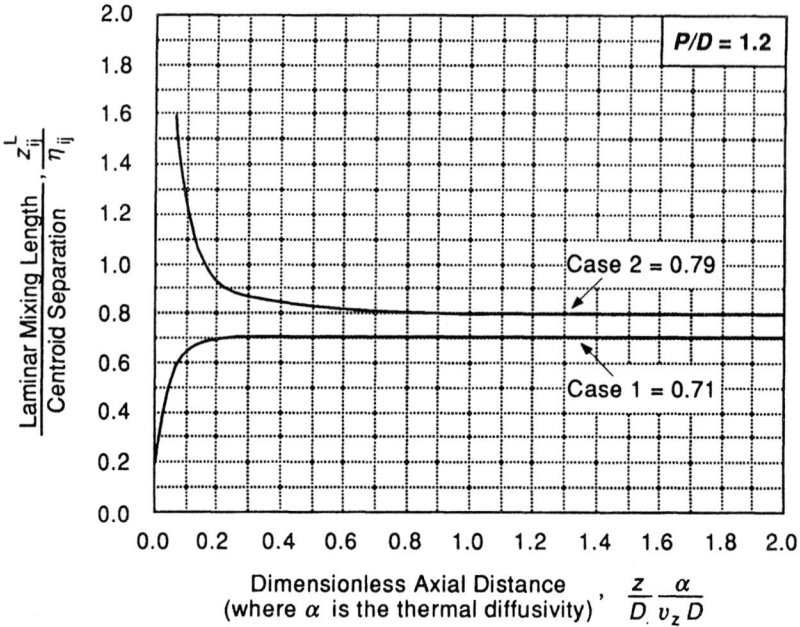

Figure 6-17 Laminar mixing lengths. *(After France and Ginsberg [9].)*

The turbulent mixing flow rate, which from Eq. 6-97 is equal to:

$$W_{ij}'^H = \frac{s_{ij}}{z_{ij}^T} \mu \frac{\varepsilon}{\nu} \tag{6-104}$$

requires introduction of assumptions for the parameters ε/ν and z_{ij}^T. The parameter ε/ν is typically taken proportional to the Reynolds number as:

$$\frac{\varepsilon}{\nu} = K'\text{Re}^b \tag{6-105}$$

The turbulent mixing length has not been consistently represented by investigators. Some have scaled it with fuel rod diameter D, others with gap width, and others with both. The latter approach yields:

$$z_{ij}^T = K_g D \left(\frac{s_{ij}}{D}\right)^r \tag{6-106}$$

where D is the fuel rod diameter. Substituting the results of Eqs. 6-105 and 6-106 into Eq. 6-104 we obtain:

$$\frac{W_{ij}'^H}{\mu} = \frac{K'}{K_g} \text{Re}^b \left(\frac{s_{ij}}{D}\right)^{1-r} \tag{6-107}$$

Table 6-7 Summary of previous work on mixing lengths based on subchannel slug flow

Researcher	Coolant type	Flow distribution	Rod array type	Power distribution
France and Ginsberg [9, 10]	Liquid metal	$\dfrac{U_i}{U_j} = \left[\dfrac{D_{e_i}}{D_{e_j}}\right]^{5/7}$	Finite square and triangular array $1.0 \leq P/D \leq 1.5$	Uniform axially and azimuthally within rods; with and without bundle radial skew
Ramm et al. [23]	Low Pr Fluids Pr: 0.3, 0.003	Uniform	Infinite triangular array P/D: 1.10, 1.30	Unheated rods, nonuniform subchannel temperature distribution at inlet
Yeung and Wolf [31]	Sodium	$\dfrac{U_i}{U_j} = \left[\dfrac{D_{e_i}}{D_{e_j}}\right]^{5/7}$	Finite triangular array	Uniform axially and azimuthally within rods; bundle power skew
Wong and Wolf [30]	Sodium	Uniform	7-Pin triangular array	Uniform axially and radially

Now if subchannels i and j are not of the same shape, an average Reynolds number is used, i.e.:

$$\mathrm{Re}^b = \frac{\mathrm{Re}_i^b + \mathrm{Re}_j^b}{2} = \frac{\mathrm{Re}_i^b}{2}\left[1 + \left(\frac{\mathrm{Re}_j}{\mathrm{Re}_i}\right)^b\right] \tag{6-108}$$

For equal axial pressure drop in the subchannels and equal property values:

$$\frac{V_i}{V_j} = \left(\frac{D_{e_i}}{D_{e_j}}\right)^{(1+n)/(2-n)} \tag{4-115}$$

Hence:

$$\mathrm{Re}^b = \frac{\mathrm{Re}_i^b}{2}\left[1 + \left(\frac{D_{e_j}}{D_{e_i}}\right)^{3b/(2-n)}\right] \tag{6-109}$$

Substituting Eq. 6-109 into 6-107 yields the general form of the correlation for turbulent mixing flow rate, i.e.:

$$\frac{W_{ij}'}{\mu} = \frac{K}{K_g}\mathrm{Re}_i^b\left[1 + \left(\frac{D_{e_j}}{D_{e_i}}\right)^{3b/(2-n)}\right]\left(\frac{s_{ij}}{D}\right)^{1-r} \tag{6-110}$$

where: $K = \dfrac{K'}{2}$ $\tag{6-111}$

Rogers and Tahir [24] have correlated the available literature for the various types of interacting subchannels of design interest. The mixing flow rate in Eq. 6-110 has been intentionally written without superscript H because the correlation of Rogers and Tahir [24] is not explicitly for energy mixing.

Table 6-8 summarizes their recommendations for the parameters K/K_g, b and r of Eq. 6-110.

Example 6-3 Mixing flow rate for square and triangular subchannels

PROBLEM Consider mixing across the gap separating a square and triangular subchannel. Demonstrate that the mixing rate from the square to the triangular subchannel W_{ij}' is equal to that from the triangular to the square subchannel W_{ji}' (as it should be for single-phase flow).

SOLUTION From Eq. 6-110 for equal properties:

$$\frac{W_{ij}'}{W_{ji}'} = \left(\frac{\mathrm{Re}_i}{\mathrm{Re}_j}\right)^b \frac{\left[1 + \left(\frac{D_{e_j}}{D_{e_i}}\right)^{3b/(2-n)}\right]}{\left[1 + \left(\frac{D_{e_i}}{D_{e_j}}\right)^{3b/(2-n)}\right]} \tag{6-112}$$

Table 6-8 Empirical mixing parameters for clean geometries

Array parameter	Triangular-triangular	Square-triangular	Square-square
K/K_g	0.0018/2	0.0027	0.005/2
r	1.4	0.95	0.894
b	0.9	0.9	0.9

Restrictions:

$$Re > 5,000$$

$$\frac{S_{ij}}{D} > 0.032$$

Note: $\left.\dfrac{D_e}{D}\right|_{\text{square array}} = 1.273 \left(\dfrac{P}{D}\right)^2 - 1.$

$\left.\dfrac{D_e}{D}\right|_{\text{triangular array}} = 1.101 \left(\dfrac{P}{D}\right)^2 - 1.$

(After Rogers and Tahir [24] for Eq. 6-110.)

For equal axial pressure drop in the subchannels:

$$\left(\frac{V_i}{V_j}\right) = \left(\frac{D_{e_i}}{D_{e_j}}\right)^{(1+n)/(2-n)} \tag{4-115}$$

Hence:

$$\left(\frac{Re_i}{Re_j}\right)^b \equiv \left(\frac{V_i D_{e_i}}{V_j D_{e_j}}\right)^b = \left(\frac{D_{e_i}}{D_{e_j}}\right)^{3b/(2-n)} \tag{6-113}$$

Therefore we obtain the desired result that:

$$\frac{W'_{ij}}{W'_{ji}} = \left(\frac{D_{e_i}}{D_{e_j}}\right)^{3b/(2-n)} \left[\frac{1 + \left(\frac{D_{e_i}}{D_{e_i}}\right)^{3b/(2-n)}}{1 + \left(\frac{D_{e_i}}{D_{e_j}}\right)^{3b/(2-n)}}\right] = 1 \tag{6-114}$$

The results of Table 6-8 are for subchannel geometries without the presence of spacers, i.e., clean geometries. Also, they have been deduced from experiments in which subchannel interactions have been measured using a variety of tracers. If a radioactive dye tracer is used, then W'_{ij} represents an effective transverse flow for mass exchange, i.e., W'^D_{ij}. Conversely, if a hot water tracer is used, then W'_{ij} represents an effective transverse mass flow rate for energy exchange, i.e., W'^H_{ij}. Finally, in the axial momentum balance, the effective transverse mass flow rate for momentum exchange W'^M_{ij} is needed. In practice, this distinction is not maintained since there have not been sufficient data available for correlations like that of Rogers and Tahir [24] to differentiate between these transverse mass flow rates. Subchannel computer codes like COBRA have recognized the distinction between these factors but crudely relate the momentum and energy terms by a single factor f_T, which in practice is invariably taken as unity:

$$f_T W_{ij}'^H \equiv W_{ij}'^M \tag{6-115}$$

A formal expression for W_{ij}^{*D} starting with the defining balance equations can be developed analogous to Eqs. 6-15 and 6-16 for W_{ij}^{*M} and W_{ij}^{*H}:

$$(js)_{i \leftrightarrow j}^* = W_{ij}^{*D} \frac{(\rho_i - \rho_j)}{\rho_{avg}} \tag{6-116}$$

The analogous parameters for describing molecular and eddy diffusivity of mass, energy, and momentum are summarized in Table 6-9. The corresponding fluxes of mass, momentum, and energy in the x-direction are given by Bird et al. [2]:

$$j_{Ax} = -D_{AB} \frac{d}{dx} (\rho_A)$$

$$\tau_{xz} = -\nu \frac{d}{dx} (\rho v_z)$$

$$q_x'' = -\alpha \frac{d}{dx} (\rho c_p T)$$

where j_{Ax} is the flux of species A diffusing through a binary mixture of A and B because of the concentration gradient of A. Note, although τ_{xz} is shear stress in the z-direction of the fluid surface of constant x, it can be equally interpreted as the viscous flux of z-momentum in the x-direction.

Example 6-4 Computation of mixing flow rate

PROBLEM Compute the mixing flow rate between interior channels of a typical PWR at inlet conditions. Express it as W_{ij}' and β. Take the subchannel inlet flow rate as 0.290 kg/s as in Example 5-4.

Table 6-9 Parameters for describing molecular and eddy diffusivity of mass, energy, and momentum

	Mass	Energy	Momentum
Molecular diffusivity (cm²/s)	D	α	ν
D/ν (dimensionless)	$\dfrac{D}{\nu} = \dfrac{1}{Sc}$	$\dfrac{\alpha}{\nu} = \dfrac{1}{Pr}$	$\dfrac{\nu}{\nu} = 1$
Eddy diffusivity (cm²/s)	ε_D	ε_H	ε_M
Effective mixing flow rate (g/(cm·s))	W_{ij}^{*D}	W_{ij}^{*H}	W_{ij}^{*M}

SOLUTION The PWR array is a square lattice. The fluid viscosity at PWR conditions is 9.63×10^{-5} kg/m s or Pa·s. The correlation of Rogers and Tahir (Eq. 6-107) is appropriate if Re > 5,000 and $s_{ij}/D > 0.032$. For this case:

$$\text{Re} = \frac{\rho v_z D}{\mu} = \frac{\dot{m}D}{\mu A} = \frac{0.290 \frac{\text{kg}}{\text{s}} (9.55 \times 10^{-3} \text{ m})}{9.63 \times 10^{-5} \frac{\text{kg}}{\text{m} \cdot \text{s}} (87.8 \times 10^{-6} \text{ m}^2)}$$

$$= 3 \times 10^5$$

since:

$$A = P^2 - \frac{\pi D^2}{4} = D^2 \left[\left(\frac{P}{D}\right)^2 - \frac{\pi}{r} \right] = (9.5)^2 \left[(1.326)^2 - \frac{\pi}{4} \right]$$

$$= 87.8 \text{ mm}^2$$

Also:

$$\frac{s}{D} = \frac{P - D}{D} = 0.326$$

Hence the parameters of Table 6-8 are applicable, and Eq. 6-107 becomes:

$$\frac{W'_{ij}}{\mu} = \frac{2K}{K_g} \text{Re}^b \left(\frac{s_{ij}}{D}\right)^{1-r}$$

when Eq. 6-111 is utilized.

$$W'_{ij} = \mu(0.005)(\text{Re})^{0.9} \left(\frac{s}{D}\right)^{0.106}$$

$$= 0.0363 \frac{\text{kg}}{\text{m} \cdot \text{s}}$$

from Eq. 6-100:

$$\beta \equiv \frac{W_{ij}^{*H}/s_{ij}}{\bar{G}}$$

For our case of PWR conditions and use of the Rogers-Tahir correlation, take $W_{ij}^{*H} = W'_{ij}$ and $s_{ij} = s$. Now:

$$s = 0.326D = 0.326(9.5) = 3.10 \text{ mm}$$

$$\bar{G} = G_i = \frac{\dot{m}}{A} = \frac{0.290 \text{ kg/s}}{87.8 \times 10^{-6} \text{ m}^2} = 3302 \frac{\text{kg}}{\text{m}^2 \text{ s}}$$

and the transverse mass flux equals:

$$\frac{W'_{ij}}{s} = \frac{0.0363}{3.10 \times 10^{-3}} = 11.71 \frac{kg}{m^2 \, s}$$

Hence:

$$\beta = \frac{W'_{ij}/s}{\bar{G}} = \frac{11.71}{3302} = 0.0035$$

Example 6-5 Computation of the single-phase mixing flow rate for subchannels of a BWR test model

PROBLEM Compute the single-phase mixing flow rate between the side or edge and interior channel of a 3 × 3 rod BWR test model operating at a mass flux of 1356 kg/m² s with the following dimensions.

Rod diameter, D, 14.478 mm
Rod-rod gap, $P - D$, 4.267 mm
Rod-wall gap, g, 3.429 mm
Radius of corner, 10.2 mm
Heated length, L, 1829.0 mm
Assembly face length, D_1, 58.826 mm

Assume that the relevant liquid water properties are:

$\rho = 742.0 \text{ kg/m}^3$

$\mu = 945.5 \times 10^{-7} \text{ kg/ms}$

SOLUTION First compute the assembly Reynolds number:

$$Re = \frac{GD_{eT}}{\mu}$$

Obtain D_e from Table J-2 neglecting spacers (i.e., $\delta = 0$) and rounded corners (i.e., take assembly as square):

$$D_{eT} = \frac{4A_f}{P_w} = \frac{4D_1^2 - N_p \pi D^2}{4D_1 + N_p \pi D}$$

$$= \frac{4(58.826)^2 - 9\pi(14.478)^2}{4(58.826) + 9\pi(14.478)}$$

$$= 12.278 \text{ mm or } 1.23 \times 10^{-2} \text{ m}$$

$$Re = \frac{1.356 \times 10^3 \text{ kg/m}^2s(1.23 \times 10^{-2} \text{ m})}{945.5 \times 10^{-7} \text{ kg/m} \cdot s} = 1.76 \times 10^5$$

Hence the flow is turbulent so that $n = 0.2$.

Also, $s_{ij}/D = 4.267/14.478 = 0.295$.

Both restrictions of Table 6-8 are therefore met. These two channels, although having different equivalent diameters, are on a square array. Hence, select the square-square constants in Eq. 6-110 from Table 6-8 yielding:

$$\frac{W'_{ij}}{\mu} = \frac{0.005}{2} \left[1 + \left(\frac{D_{e_j}}{D_{e_i}}\right)^{1.5} \right] \text{Re}_i^{0.9} \left(\frac{s_{ij}}{D}\right)^{0.106}$$

since:

$$\frac{3b}{2-n} = 1.5$$

Taking i as the side subchannel and j as the interior, from Example 4-3 where $P/D = 1.295$ and $g/D = 0.237$:

$$\frac{D_{e_j}}{D} = 1.273 \left(\frac{P}{D}\right)^2 - 1 = 1.134$$

$$\frac{D_{e_i}}{D} = \frac{\frac{P}{D}\left(2 + \frac{4g}{D}\right) - \frac{\pi}{2}}{P/D + \pi/2} = 0.7836$$

$$s_{ij} = 4.267 \text{ mm}$$

$$D = 14.478 \text{ mm}$$

$$\text{Re}_i \equiv \text{Re}_2 = \frac{GD_{e_2}}{\mu}$$

$$= \frac{1.356 \times 10^3 \frac{\text{kg}}{\text{m}^2\text{s}} [0.7836(14.48) \times 10^{-3} \text{ m}]}{945.5 \times 10^{-7} \frac{\text{kg}}{\text{m} \cdot \text{s}}}$$

$$= 1.63 \times 10^5$$

$$\frac{W'_{ij}}{\mu} = \frac{0.005}{2} \left[1 + \left(\frac{1.134}{0.7836}\right)^{1.5} \right] (1.63 \times 10^5)^{0.9} \left(\frac{4.267}{14.478}\right)^{0.106}$$

$$\frac{W'_{ij}}{\mu} = 295.5$$

The effective transverse mass flow rate (from side to interior) is:

$$W'_{ij} = 295.5(945.5 \times 10^{-7})$$

$$= 0.028 \text{ kg/m} \cdot \text{s}$$

The effective transverse mass flux (from side to interior) is:

$$\frac{W'_{ij}}{s} = \frac{0.028}{4.267 \times 10^{-3}} = 6.56 \text{ kg/m}^2\text{s}$$

Hence:

$$\beta = \frac{W'_{ij}/s}{G} = \frac{6.56}{1356} = 0.00484$$

2 Two Phase In two-phase flow W'_{ij} is not generally equal to W'_{ji}, so the single-phase treatment of Eq. 6-8 is not applicable in this case. Rather, let us expand Eq. 6-8 directly using Eqs. 6-88 and 6-89 which define W'^D_{ij} and W'^D_{ji} and the assumption that the subchannel enthalpy is composed of both saturated liquid and vapor. Hence, for the cross-flow originating in subchannel i:

$$W'^D_{ij} h_i = \frac{\varepsilon}{z^T_{ij}} s_{ij} \rho_i h_i = \frac{\varepsilon s_{ij}}{z^T_{ij}} [\rho_f (1 - \alpha_i) h_f + \rho_g \alpha_i h_g] \qquad (6\text{-}117)$$

Utilizing Eq. 6-8 and the analogous result for cross-flow originating in subchannel j, the net energy transfer across the gap becomes:

$$(q''s)'_{i\leftrightarrow j} = \frac{\varepsilon s_{ij}}{z^T_{ij}} [(\rho_f h_f - \rho_g h_g)(\alpha_j - \alpha_i)] \qquad (6\text{-}118)$$

This result indicates that the net energy transfer is from subchannel i to j when $a_j > \alpha_i$ since:

$$\rho_f h_f > \rho_g h_g \qquad (6\text{-}119)$$

as shown in Figure 6-18. This may seem paradoxical since it indicates energy transfer to the higher void channel. This will be explained shortly. Now the net

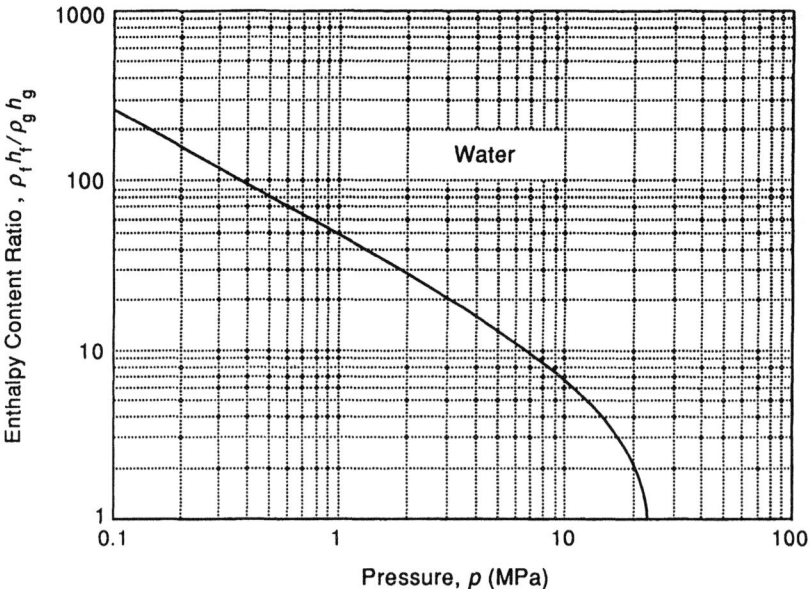

Figure 6-18 Saturated liquid-to-vapor enthalpy content ratio for equal volumes.

energy transferred across the gap per unit length is also equal to the net transverse mass flow rate times the effective enthalpy transported, i.e.:

$$(q''s)'_{i\leftrightarrow j} \equiv W'^{D}_{i\leftrightarrow j} h_{\text{eff}} \tag{6-120}$$

Under two-phase conditions, even when thermal equilibrium does not exist, subchannel temperature differences are small enough so that molecular transport contributions to heat transfer can be neglected.

Substituting the results of Equations 6-118 and 6-91 for saturation conditions into Eq. 6-120, h_{eff} can be obtained as:

$$h_{\text{eff}} = \frac{\rho_f h_f - \rho_g h_g}{\rho_f - \rho_g} = \frac{(\rho_f - \rho_g)h_f + \rho_g(h_f - h_g)}{\rho_f - \rho_g} \tag{6-121}$$

From this result observe that $h_{\text{eff}} \leq h_f$ since $h_g \geq h_f$, i.e.:

$$\frac{\rho_f h_f - \rho_g h_g}{\rho_f - \rho_g} \leq h_f \tag{6-122}$$

The apparently paradoxical implication of Eq. 6-118 can now be explained, i.e., if net energy transfer is into subchannel j with the higher void fraction, does that imply the void fraction is further increased? The component and net exchanges of mass and energy which are expressed by Eqs. 6-91 and 6-118, respectively, are illustrated on Figure 6-19. For $\alpha_j > \alpha_i$, observe that both net mass and energy are transferred to subchannel j. While net vapor is transferred from subchannel j to i, a larger mass of liquid is transferred from subchannel i to j. Now since this net mass carries enthalpy of a value h_{eff} less than saturated liquid, the void fraction of j is not increasing. Rather, the void fraction decreases since mixing decreases the specific energy content of the coolant in the hot subchannel j.

In the two-phase case, $W'^{D}_{i\leftrightarrow j}$ also must be correlated. It is expressed by Eq. 6-91 written for saturated conditions as:

$$W'^{D}_{i\leftrightarrow j} = \varepsilon \frac{s_{ij}}{z^T_{ij}} (\rho_f - \rho_g)(a_j - \alpha_i) \tag{6-123}$$

This equation indicates that the mixing rate is finite whenever a difference in subchannel void fraction exists. This implies that mixing will occur so as to equalize adjacent subchannel void fractions. A body of evidence has been developed, however, which demonstrates that subchannel void fractions do not equalize in rod bundles. In fact, a clear tendency has been demonstrated for void to migrate to that subchannel with the larger cross-sectional area and higher velocity. Figures 6-20 and 6-21 illustrate this behavior for a developing diabatic two-phase case with uniform rod power for a square lattice test bundle. The power-to-flow ratio $\dot{q}/\dot{m}$ for subchannel 3 (corner) is the highest since its area is the smallest as is its mass flux, as Figure 6-20 indicates. Even though the $\dot{q}/\dot{m}$ ratio is highest for subchannel 3, the quality as shown in Figure 6-21 is the lowest of all subchannels. This behavior is also observed in developed adiabatic

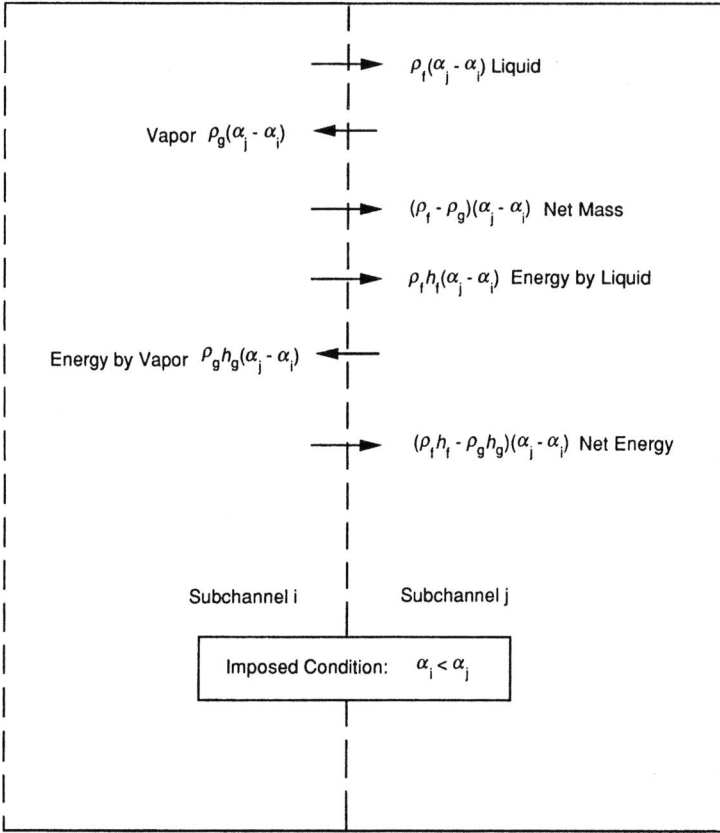

Figure 6-19 Mass and energy exchange between subchannels in two-phase flow.

two-phase flow, although a fundamental understanding of this "void drift" phenomenon is presently unavailable.

Two-phase correlations In practice, the mixing flow rate is reformulated to limit this flow rate as the equilibrium void distribution is approached. Hence Eq. 6-123 is reformulated as:

$$W'^{D}_{i \leftrightarrow j} = \varepsilon \, \frac{s_{ij}}{z^{T}_{ij}} \, (\rho_f - \rho_g)[\alpha_j - \alpha_i - (\alpha_j - \alpha_i)_{equil}] \qquad (6\text{-}124)$$

This result can be rewritten in terms of the single-phase mixing rate expressed by Eq. 6-88 if ρ_i is taken equal to ρ_f. The result is:

$$W'^{D}_{i \leftrightarrow j}|_{TP} = W'^{D}_{ij}|_{SP} \, \theta[\alpha_j - \alpha_i - (\alpha_j - \alpha_i)_{equil}] \qquad (6\text{-}125)$$

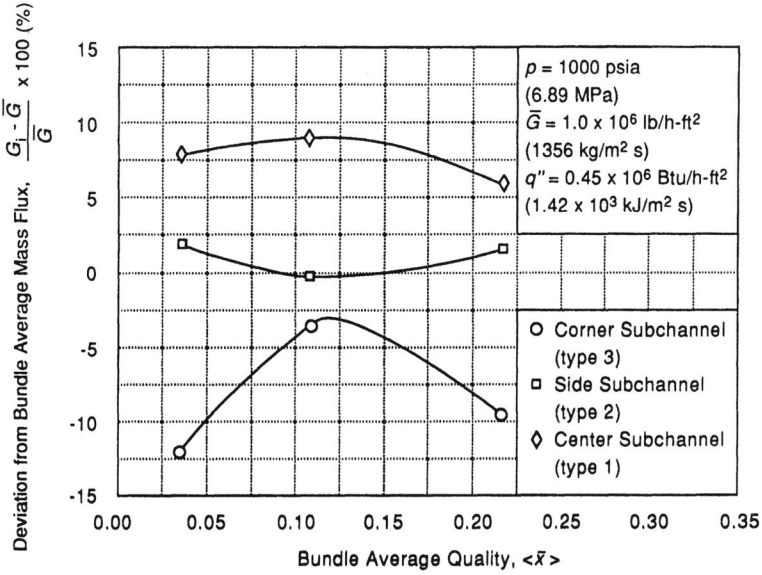

Figure 6-20 Comparison of subchannel flows for the three subchannels. *(After Lahey et al. [18].)*

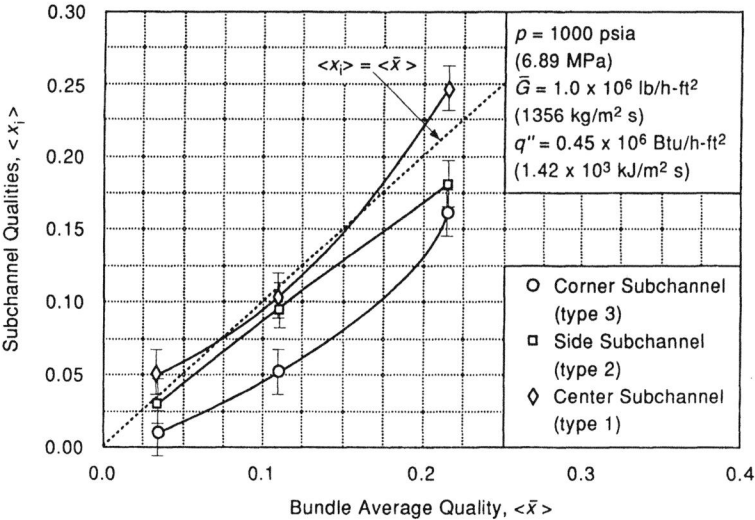

Figure 6-21 Variation of subchannel qualities with average quality for the three subchannels. *(After Lahey et al. [18].)*

where:

$$\theta \equiv \frac{(\varepsilon/z_{ij}^T)_{TP}}{(\varepsilon/z_{ij}^T)_{SP}}\left(1 - \frac{\rho_g}{\rho_f}\right) \tag{6-126}$$

Correlations for θ and $(\alpha_j - \alpha_i)_{equil}$ are required to evaluate $W_{i\leftrightarrow j}^{\prime D}|_{TP}$ from Eq. 6-125. Beus [1] developed a two-phase mixing model that suggested the concept of this two-phase multiplier, θ, and demonstrated that it depends strongly on flow regime. This multiplier has been modeled as increasing linearly with quality x between $x = 0$ and the slug–annular transition point x_M. For qualities greater than x_M, θ is assumed to decrease hyperbolically from its maximum x_M. The functions that describe this behavior are:

$$\theta = 1 + (\theta_M - 1)(x/x_M) \text{ if } x < x_M \tag{6-127a}$$

$$\theta = 1 + (\theta_M - 1)\left(\frac{1 - x_0/x_M}{x/x_M - x_0/x_M}\right) \text{ if } x > x_M \tag{6-127b}$$

where $x_0/x_M = 0.57 \, Re^{0.0417}$ from the work of Beus [1] and $\theta_M = 5$ from the work of Faya [8]. Figure 6-22 illustrates this behavior.

The quality x_M at the slug–annular transition point from Wallis's [29] model is:

$$x_M = \frac{\left[\dfrac{0.4(\rho_f(\rho_f - \rho_g)gD)^{1/2}}{G} + 0.6\right]}{\left[\left(\dfrac{\rho_f}{\rho_g}\right)^{1/2} + 0.6\right]} \tag{6-128}$$

The remaining parameter $(\alpha_j - \alpha_i)_{equil}$ has been developed following the suggestion of Lahey and Moody [17] as proportional to the fully developed (equilib-

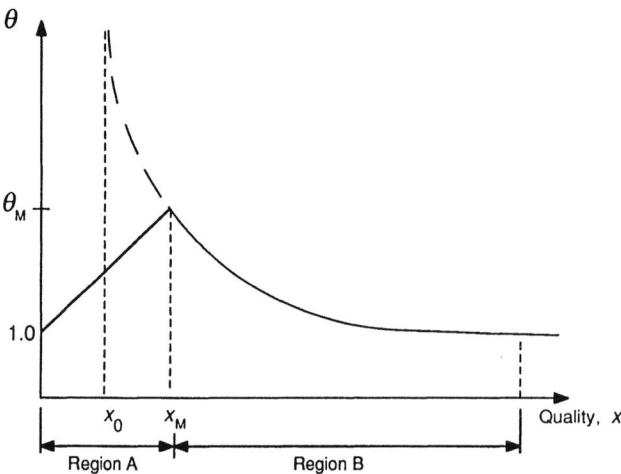

Figure 6-22 Variation of the two-phase mixing parameter with quality.

rium) mass flux difference between channels, i.e.:

$$(\alpha_j - \alpha_i)_{\text{equil}} = K_a \frac{(G_j - G_i)_{\text{equil}}}{G_{\text{avg}}} \qquad (6\text{-}129)$$

where $(G_j - G_i)_{\text{equil}}$ is in practice taken equal to the existing mass flux difference $G_j - G_i$

and

$$K_a = 1.4$$

It is important to recall that for two-phase flow, the assumption adopted here of volume-to-volume exchange caused a net mass flow between channels. As pointed out at the beginning of this section, this is satisfactory since it is to be compensated by a reformulation of the diversion cross-flow rate. The work of Faya et al. [8] is an example of a consistent treatment for diversion, mixing, and void drift effect. This treatment was introduced into the two-fluid formulation of the THERMIT code by Kelly et al. [16]. The effects of mixing and void drift are significant at pressures equal to BWR conditions as illustrated in Figure 6-23. This figure compares additional results from Lahey et al. [18] with THERMIT calculations that both included and excluded the mixing and void drift mechanisms. Kelly et al. [16] also demonstrated that these effects are reduced for higher pressures.

Example 6-6 Computation of two-phase mixing flow rates

PROBLEM For the BWR test assembly of Example 4-3, compute the two-phase mixing flow rate (for enthalpy) and the energy transfer rate from the

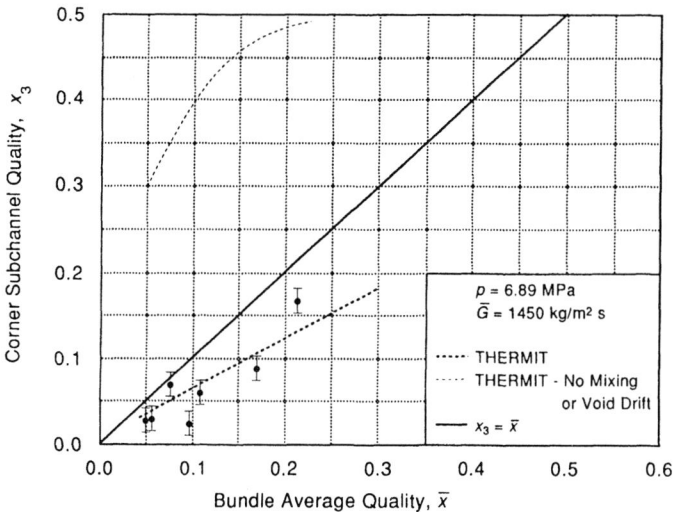

Figure 6-23 Comparison of measured and predicted exit quality in corner subchannels for BWR-type, uniformly heated geometry. *(After Kao and Kazimi [15].)*

edge, type 2, to an interior, type 1, subchannel for a bundle average quality, $\langle \bar{x} \rangle$, equal to 20 percent.

At BWR operating pressure of 70.3×10^5 Pa, the relevant saturation properties for two-phase calculations are:

$\rho_f = 738.2 \text{ kg/m}^3$

$h_f = 1261.4 \text{ kJ/kg}$

$\rho_g = 36.0 \text{ kg/m}^3$

$h_g = 2771.6 \text{ kJ/kg}.$

SOLUTION The relevant equation for calculation of the two-phase mixing flow rate is Eq. 6-125. In evaluating this equation numerically we will need the subchannel mass flux and void fraction distribution for the assembly which are available from Figures 6-20 and 6-21. You should recognize that the mixing flow rate to be calculated, $W_{i \leftrightarrow j}^{'D}|_{TP}$, is an instantaneous rate for a bundle average quality of 0.20 and hence is applicable only at a certain axial position in the bundle. The mixing flow rates above and below this location for higher and lower average bundle qualities are different. A complete axial solution to compare to the experimental results of Figures 6-20 and 6-21 should start at the inlet and progress stepwise to the end of the heated length.

Proceeding with Eq. 6-125 for $i = 2$ (edge) and $j = 1$ (interior):

$$W_{2 \leftrightarrow 1}^{'D}|_{TP} = W_{2 \leftrightarrow 1}^{'D}|_{SP} \, \theta[(\alpha_1 - \alpha_2) - (\alpha_1 - \alpha_2)_{\text{equil}}] \qquad (6\text{-}125)$$

where the relevant parameters are obtained as follows:

$$W_{2 \leftrightarrow 1}^{'D}|_{SP} \text{ from Example 6-5} = 0.028 \text{ kg/m} \cdot \text{s}$$

$$\theta = f(\theta_M, x_0, x_M) \text{ (from 6-127a and b)}$$

where: $\qquad \theta_M = 5$

$$x_0 = x_M 0.57 \, \text{Re}^{0.0417} \text{ (following Eq. 6-127b)}$$

and $\qquad x_M = f(p, G, D) \text{ (from Eq. 6-128)}$

$$\alpha_1, \alpha_2 \text{ (from Figure 6-21)}$$

$$(\alpha_1 - \alpha_2)_{\text{equil}} = f(K_a, G) \text{ (from Eq. 6-129)}$$

First, evaluate θ:

$$x_M = \frac{\left[\dfrac{0.4[\rho_f(\rho_f - \rho_g)gD]^{1/2}}{G} + 0.6 \right]}{\left[\left(\dfrac{\rho_f}{\rho_g} \right)^{1/2} + 0.6 \right]} \qquad (6\text{-}128)$$

Neglecting ρ_g relative to ρ_f and taking their ratio about 20.5 for BWR conditions:

$$x_M = \frac{\dfrac{0.4[(738.2)^2 9.81(1.45 \times 10^{-2})]^{1/2}}{1.356 \times 10^3} + 0.6}{(20.5)^{1/2} + 0.6}$$

$$= \frac{0.682}{5.13} = 0.133$$

Since $x = 0.2$, Eq. 6-127b is applicable for θ:

$$\theta = 1 + (\theta_M - 1) \left[\frac{1 - \dfrac{x_0}{x_M}}{\dfrac{x}{x_M} - \dfrac{x_0}{x_M}} \right]$$

where $x_0/x_M = 0.57 \, Re^{0.0417}$

Take $Re = Re_2 = 1.63 \times 10^5$ from Example 6-5, then:

$$\frac{x_0}{x_M} = 0.57(1.63 \times 10^5)^{0.0417} = 0.94$$

$$\theta = 1 + (5 - 1) \left[\frac{1 - 0.94}{\left(\dfrac{0.2}{0.133}\right) - 0.94} \right] = 1.424$$

Next evaluate $(\alpha_1 - \alpha_2)$ and $(\alpha_1 - \alpha_2)_{equil}$. From Figures 6-20 and 6-21, for $\langle \bar{x} \rangle = 0.2$:

$$\frac{G_1 - \bar{G}}{\bar{G}} = 0.072; \qquad x_1 = 0.223$$

$$\frac{G_2 - \bar{G}}{\bar{G}} = 0.011; \qquad x_2 = 0.168$$

Assuming homogeneous flow:

$$\alpha = \frac{1}{1 + \dfrac{1 - x}{x}\left(\dfrac{\rho_g}{\rho_f}\right)} \tag{I,11-30}$$

yields:

$$\alpha_1 = 0.855$$

$$\alpha_2 = 0.805$$

or

$$\alpha_1 - \alpha_2 = 0.05$$

From Eq. 6-129 and the subsequent discussion:

$$(\alpha_1 - \alpha_2)_{\text{equil}} = K_a \frac{(G_1 - G_2)_{\text{equil}}}{\bar{G}}$$

$$= 1.4 \frac{(G_1 - G_2)_{\text{actual}}}{\bar{G}}$$

Using the results from Figure 6-20:

$$(\alpha_1 - \alpha_2)_{\text{equil}} = 1.4(0.072 - 0.011) = 0.085$$

Substituting these parameter values into Eq. 6-125, obtain the desired result, i.e.:

$$W_{2\leftrightarrow1}'^{D}|_{\text{TP}} = 0.028(1.424)(0.05 - 0.085)$$

$$= -1.40 \times 10^{-3}$$

i.e., the mixing cross-flow rate proceeds from the interior toward the side subchannel.

Next evaluate the energy transfer rate between these subchannels. Recall:

$$(q''s)_{i\leftrightarrow j}' = W_{2\leftrightarrow1}'^{D}|_{\text{TP}} \, h_{\text{eff}} \tag{6-119}$$

where:

$$h_{\text{eff}} = \frac{\rho_f h_f - \rho_g h_g}{\rho_f - \rho_g} \tag{6-120}$$

Using the given properties:

$$h_{\text{eff}} = \frac{738.2(1261.4) - 36.0(2771.6)}{738.2 - 36.0} = 1184.0 \text{ kJ/kg}$$

$$(q''s)_{i\leftrightarrow j}' = -1.40 \times 10^{-3}(1184.0) = -1.657 \text{ kJ/m} \cdot \text{s}$$

Compare this with the energy transferred to the subchannel per meter from the one-half effective fuel rod bordering the subchannel which is operating at conditions noted in Figure 6-21:

$$q' = 0.5q'' \, \pi D$$

$$= 0.5(1.42 \times 10^3 \text{ kJ/m}^2\text{s})[(\pi)(0.01227) \text{ m}]$$

$$= 27.4 \text{ kJ/m} \cdot \text{s}$$

Hence the lateral energy transfer is 6.0 percent of the energy deposited in the channel by the fuel rods, i.e., $1.634/27.4 \times 100 = 6.0$

Example 6-7 Subchannel analysis of two adjacent, interconnecting subchannels

PROBLEM Consider the two adjacent, unheated interacting PWR subchannels of Example 5-4. The problem conditions are:

P/D, 1.326; D, 9.5 mm; L, 3.66 m

Inlet temperature, 292 °C (565.15 K)

Inlet subchannel velocity ratio, $\dfrac{w_i(0)}{w_j(0)} = 2$

Outlet subchannel conditions, $p_i(L) = p_j(L) = 15.5$ MPa

Compute the axial profiles of axial mass flux and pressure in each channel and the transverse mass flow rate between channels.

SOLUTION The solution of this example can be accomplished using a numerical code such as COBRA IIIC/MIT-2 [14] subject to boundary condition set (2), Figure 1.10. The required boundary conditions, geometry, and constitutive relations are presented below (English units are used in COBRA IIIC/MIT-2) which are physically identical to those utilized in Example 5-4.

Boundary conditions:

$h_i(0) = h_j(0) = 1{,}229$ kJ/kg†

$w_i(0) = 6.04$ m/s

$w_j(0) = 3.02$ m/s

$p_i(L) = p_j(L) = 15.5$ MPa

Geometry:

$A_{fi} = A_{fj} = 8.78 \times 10^{-5}$ m²

$P_{wi} = P_{wj} = \pi D + 3(P - D) = \pi(9.5) + 3(12.6 - 9.5) = 39.14$ mm

---

† The numerical values of these enthalpies and velocities are slightly different from those in Example 5-4. This arose because the equation of state in the THERMIT code (Example 5-4) computes subcooled conditions, whereas the COBRA code utilizes an equation of state for saturated conditions only. For example the densities utilized to calculate these velocities were different in the two cases (741.89 kg/m³ (Example 5-4) versus 731.01 kg/m³ (Example 6-7)). The results of these two examples are comparable, however, because the same inlet mass flow rate was used, i.e., the inlet velocities are derived versus input quantities.

Note that the bounding walls provide wetting perimeter in addition to the heated rod walls.

$$D_{ei} = D_{ej} = \frac{4A_f}{P_w} = \frac{4(8.78 \times 10^{-5})}{0.03914} = 8.974 \times 10^{-3} \text{ m}$$

Constitutive laws:

Axial wall friction, f_{axial} (Eq. 7-60):

$$f_{axial} = f_{c.t.}(1.04 + 0.06(P/D - 1)) = 1.06f_{c.t.} \text{ for } P/D = 1.326$$

and $f_{c.t.}$ from Eq. 7-59:

$$\frac{1}{(f_{c.t.})^{1/2}} = 2.035 \log_{10}[\text{Re}_{De}(f_{c.t.})^{1/2}] - 0.989$$

Cross-flow resistance K_G:

$$\Delta p_{tr} \equiv \frac{s}{l} K_G \frac{\rho v_y^2}{2} \qquad \text{(from Eq. 6-81)}$$

To estimate K_G, use the Gunter-Shaw correlation as applied in Example 5-2. For this case taking l as approximately the distance between subchannel centroids:

$$\frac{s}{l} \simeq \frac{P - D}{P} = 1 - \frac{1}{1.326} = 0.25$$

Now from Example 5-4:

$$\Delta p_{tr} = \frac{0.96}{(\text{Re}_{D_V})^{0.145}} \left(\frac{P}{D_V}\right) \{\rho\}_{gap}\{v_y\}_{gap}^2 \left(\frac{D_V}{P}\right)^{0.4}$$

where:

$$\text{Re}_{D_V} = \frac{\{\rho\}_{gap}\{v_y\}_{gap} D_V}{\{\mu\}_{gap}}$$

$$D_V = 8.974 \times 10^{-3} \text{ m}$$

$$P = 12.6 \times 10^{-3} \text{ m}$$

$$\Delta p_{tr} = \frac{0.96}{(\text{Re}_{D_V})^{0.145}} \left(\frac{12.6}{8.974}\right) \rho v_y^2 \left(\frac{8.974}{12.6}\right)^{0.4}$$

which expressed in terms of velocity head $\rho v_y^2/2$ equals:

$$= \frac{2.3535}{(\text{Re}_{D_V})^{0.145}} \frac{\rho v_y^2}{2}$$

To evaluate K from this relation, Re_{D_V} must be estimated. Again from Example 5-4, obtain the maximum cross-flow rate per unit length as:

$$W_{ij} \simeq 0.15 \frac{\text{kg}}{\text{m} \cdot \text{s}}$$

Now:

$$W_{ij} \equiv \frac{\rho v_y A_y}{\Delta z} = \rho v_y (P - D)$$

since:

$$A_y \equiv (P - D)\Delta z$$

for

$$\rho \simeq 742 \text{ kg/m}^3$$

$$\Delta z = 0.366 \text{ m}$$

$$\mu \simeq 9.6 \times 10^{-5} \text{ kg/m·s}$$

$$v_y = \frac{W_{ij}}{\rho(P - D)} = \frac{0.15 \dfrac{\text{kg}}{\text{m·s}}}{742 \text{ kg/m}^3(12.6 - 9.5) \times 10^{-3} \text{ m}} = 6.528 \times 10^{-2} \frac{\text{m}}{\text{s}}$$

$$\text{Re}_{D_V} = \frac{\rho v_y D_V}{\mu} = \frac{742(6.528 \times 10^{-2})(8.974 \times 10^{-3}) \dfrac{\text{kg}}{\text{m}^3} \dfrac{\text{m}}{\text{s}} \text{m}}{9.6 \times 10^{-5}} \dfrac{}{\dfrac{\text{kg}}{\text{m·s}}} = 4,528$$

Hence:

$$\Delta p_{tr} \simeq \frac{2.353}{(4528)^{0.145}} \frac{\rho v_y^2}{2} \simeq 0.694 \frac{\rho v_y^2}{2}$$

Since:

$$\Delta p_{tr} = \frac{s}{l} K_G \frac{\rho v_y^2}{2}$$

and

$$\frac{s}{l} \simeq 0.25$$

then

$$K_G \simeq 2.8$$

Mixing parameters, β:

$$\beta \equiv \frac{W'_{ij}/s}{G}$$

now:

$$W'_{ij} = 0.0363 \frac{\text{kg}}{\text{m·s}}$$

$$s = 12.6 - 9.5 = 3.1 \text{ mm}$$

$$G = \frac{\dot{m}}{A_f} = \frac{0.58 \text{ kg/s}}{2(8.78 \times 10^{-5})\text{m}^2} = 3.30 \times 10^3 \frac{\text{kg}}{\text{m}^2 \text{ s}}$$

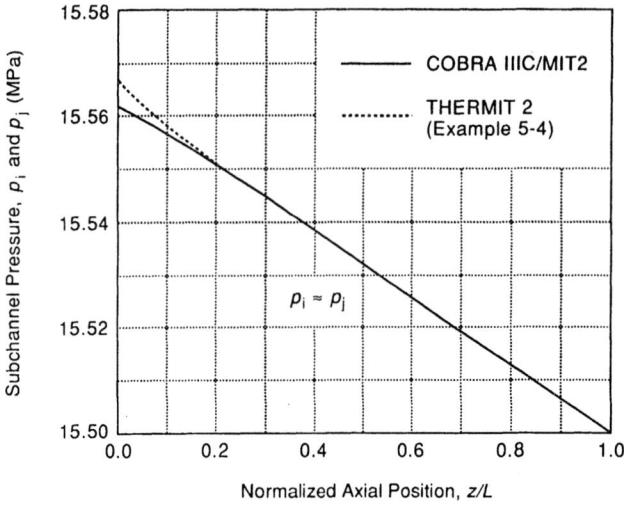

Figure 6-24 Axial pressure profile of Examples 5-4 and 6-7.

Hence:

$$\beta = \frac{0.0363}{3.10 \times 10^{-3}(3.30 \times 10^3)} = 0.0035$$

In summary the following constitutive laws are used:

- Axial friction, $f_{axial} = 1.06 f_{c.t.}$ where $f_{c.t.}$ is obtained from Eq. 7-59
- Cross-flow resistance coefficient, $K_G = 2.8$
- Mixing coefficient, $\beta = 0.0035$

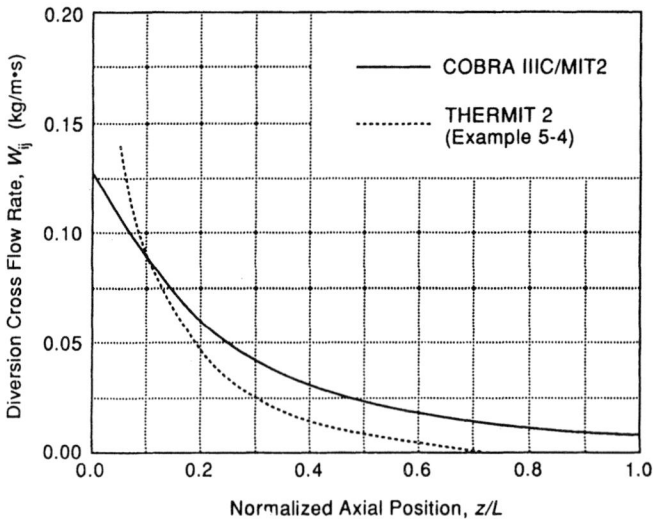

Figure 6-25 Axial distribution of the diversion cross-flow rate of Examples 5-4 and 6-7.

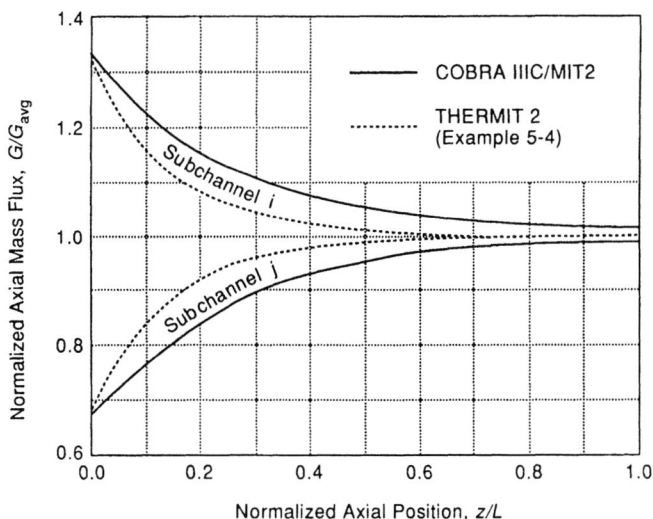

Figure 6-26 Axial distribution of the subchannel axial mass fluxes of Examples 5-4 and 6-7.

RESULTS Figures 6-24 through 6-26 present the results from the COBRA IIIC/MIT-2 [14] subchannel code and compare these results directly with the porous body results of Example 5-4. Differences exist but are not striking. However, it is interesting that such differences exist even though the constitutive relations used in the two examples are derived from the same physical correlations.

VIII APPLICATION OF THE SUBCHANNEL ANALYSIS APPROACH

The ideal way to analyze a reactor core utilizing the lumped subchannel approach is by taking each radial node in the analysis as an actual subchannel. This implies that for a typical PWR core over 30,000 radial nodes should be considered (because of symmetry reasons this number may be reduced to 4,000). These numbers are so large that, currently, numerical solutions are impractical. Therefore, this possibility has historically been ruled out, and two other general approaches have been developed.

One is the chain or cascade (multistage) method, and the other is a one-stage method. The development of both schemes has reflected the limit imposed on the number of radial nodes in any single-pass calculation by previously existing subchannel code formulations. The chain method has maximized the radial mesh representation by performing a multistage analysis of the core. In the first stage the whole core is analyzed on an assembly-to-assembly basis (each radial node represents an actual assembly). From this analysis the hot assembly, i.e., the one with the largest enthalpy, and its boundary conditions can be identified. In the second stage of the two-stage method the hot assembly is analyzed on a subchannel basis (each radial node is an actual subchannel or is created by lumping of a few subchannels) taking advantage of the boundary conditions found in the previous stage. The three-stage method sequentially analyzes the hot assembly as four regions and then the hottest of the four regions by subchannels.

In the simplified one-stage method the core has been analyzed in only one stage using a fine mesh in a zone consisting of those subchannels with the larger radial peaking factors and a coarse mesh outside this zone. However, because available codes require that the boundaries of any mesh node must be connected to another point or be impervious to mass, momentum, and energy exchange, the coarse mesh zone has traditionally been extended to the core boundary. Figure 6-27 illustrates a typical nodalization scheme for simplified one-stage method [22].

It is important to recognize that in both the cascade and one-stage methods, the nodal layout does dictate in some stages and in some locations that several

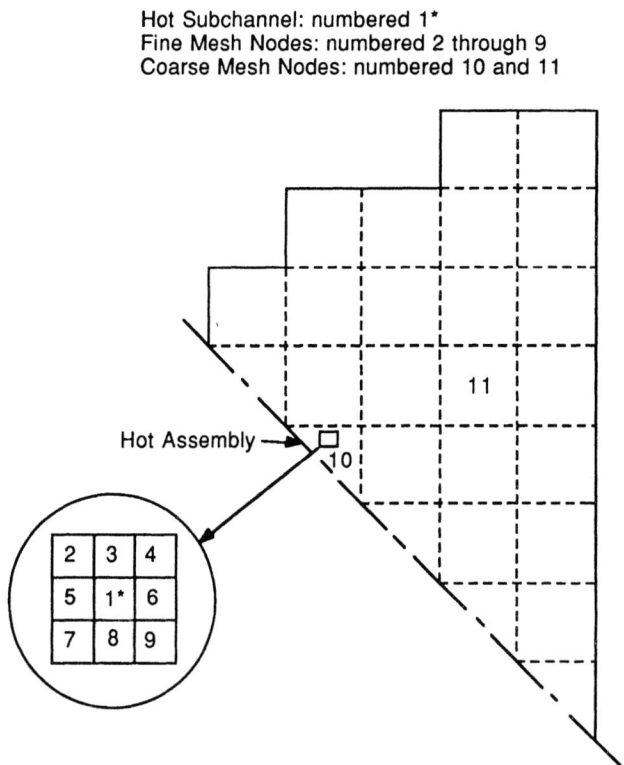

Hot Subchannel: numbered 1*
Fine Mesh Nodes: numbered 2 through 9
Coarse Mesh Nodes: numbered 10 and 11

Figure 6-27 Example of layout of channels used in one-pass method for one hot subchannel in $\frac{1}{8}$ PWR core.

channels be homogenized into one equivalent channel. Now momentum and energy exchanges between nodes are expressed by the subchannel equations as being proportional to differences in nodal properties. In reality, however, such exchanges between nodes are proportional to the gradients of such properties across the boundaries between regions. This difference is analogous to that described in Section VII by Figure 6-16 regarding exchange between single subchannels.

This difference in momentum and energy exchange is illustrated by reference to Figure 6-28 and the subchannel conservation equations. This figure

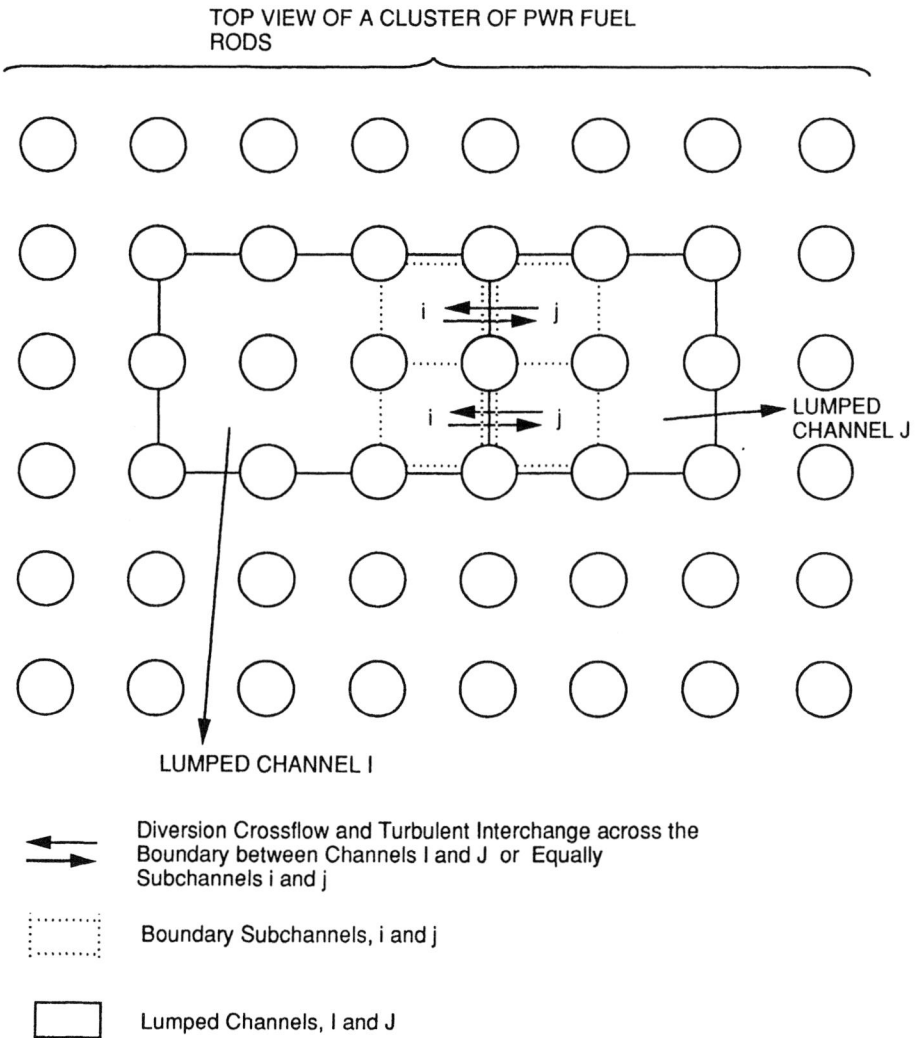

Figure 6-28 Typical lumped channel layout.

illustrates two lumped subchannels I and J composed of six and four subchannels each. The energy and momentum exchange between these lumped subchannels occurs between the boundary subchannels i and j. The relevant terms of the conservation equations for the boundary subchannels and lumped channels are respectively:

$$(h_i - h_j)W_{ij}^{*H} \quad \text{versus} \quad (h_I - h_J)W_{IJ}^{*H} \tag{6-130a}$$

$$(v_{zi} - v_{zj})W_{ij}^{*M} \quad \text{versus} \quad (v_{zI} - v_{zJ})W_{IJ}^{*M} \tag{6-130b}$$

$$p_i - p_j \quad \text{versus} \quad p_I - p_J \tag{6-130c}$$

The values of these terms are different since the enthalpies h, axial velocities v_z, and the pressures p of the boundary subchannels and the lumped channels are not equal. The task is to adjust the lumped channel representation to equal the subchannel representation in a way that is applicable to an arbitrary selection of lumped channel geometry. Hence we introduce coefficients, called transport coefficients, into lumped channel terms so that the relevant terms of Eqs. 6-130a through 130c are equal, i.e.:

$$(h_i - h_j) = \frac{1}{N_H}(h_I - h_J) \tag{6-131}$$

or

$$N_H \equiv \frac{h_I - h_J}{h_i - h_j} \tag{6-132a}$$

$$N_{v_z} \equiv \frac{v_{zI} - v_{zJ}}{v_{zi} - v_{zj}} \tag{6-132b}$$

$$N_p \equiv \frac{p_I - p_J}{p_i - p_j} \tag{6-132c}$$

The task of evaluating these transport coefficients N_H, N_{v_z}, and N_p requires solution of various sized lumped channel regions both as lumped channels and on a detailed subchannel basis. These analyses yield both the numerator and denominator of Eqs. 6-130a through 130c. The task is formidable because the coefficients for any single set of lumped channels are a function of many operating variables. Chiu [6] has performed a sensitivity analysis for PWR conditions and concluded that the calculated values of h_I, v_{zI}, p_I, and W_{IJ} are insensitive to the values used for N_{v_z} and N_p. However, the value of N_H is very important to the predicted channel fluid conditions. Moreno and Leira [21] have developed a correlation for the axially averaged value of the enthalpy transport coefficient, i.e., $\bar{N}_H$, for radially constant inlet flow. The correlation was tested and confirmed accurate for slow (10 percent change per second) power, flow, and pressure transients. It was developed from analyses in which the following variables were studied:

1. *Size of the homogenized channels.* Both fuel assemblies and quarter assemblies were employed in the analysis, i.e., channels identical to those found in a cascade or coupled neutronic–thermohydraulic analysis.
2. *Power axial profile.* All cases were studied with both sinusoidal and uniform axial profile for the heat flux.
3. *Mixing rate.* Values of the mixing parameter β between 0.005 and 0.04 were tested.
4. *Length of the core.* Cores of $L = 3.218$, 3.658, and 4.216 m were analyzed.
5. *Power peaking factors of the channels defining the boundary.* Each of the previous cases was analyzed with the following power ratios for the channels defining the boundary:

   a. $P_R = 1.37/1.35 = 1.01$
   b. $P_R = 1.37/1.30 = 1.05$
   c. $P_R = 1.37/1.20 = 1.14$
   d. $P_R = 1.37/1.00 = 1.37$
   e. $P_R = 1.37/0.65 = 2.10$

   where P_R is defined as the ratio of the larger power peaking factor to the smaller.
6. *Power peaking factors of channels not defining the boundary.* Different configurations were studied. They correspond both to limiting cases, i.e., cases in which the influence of all these channels is added either to increase or to decrease N_H, and intermediate cases.

From the results the two following correlations were suggested:

1. *Problems with $P_R > 1.10$:*

$$\bar{N}_H = 1.0 + a \left(\frac{\beta}{0.005}\right)^{0.6} P_R^{-0.15} \left(\frac{L}{144}\right)^{0.5} \left(\frac{\sqrt{N}}{17}\right)^{0.15} \tag{6-133}$$

where $a = 0.62$ for sinusoidal or quasi-sinusoidal axial power profile, $a = 0.70$ for uniform or quasi-uniform axial power profile, and $N = $ the number of subchannels of each homogenized square channel.

2. *Problems with $P_R < 1.10$:*

$$\bar{N}_H = 1.0 + b \left(\frac{\beta}{0.005}\right)^{0.72} P_R^{-6.6} \left(\frac{L}{144}\right)^{0.5} \tag{6-134}$$

where b is a constant equal to the 1.0 for sinusoidal or quasi-sinusoidal axial power profile, and $b = 1.12$ for uniform or quasi-uniform power axial profile problems.

These equations determine values of $\bar{N}_H$ with an error ranging from 0 to +10%. These results for $\bar{N}_H$ are conservative, i.e., provide values of $\bar{N}_H$ larger

Table 6-10 Expected errors in two-dimensional, homogenized region enthalpy

β	N_p'								
	2	3	4	5	7	9	11	15	23
0.0	0	0	0	0	0	0	0	0	0
0.005	0	−0.66	−0.61	−0.53	−0.40	−0.33	−0.27	−0.206	−0.135
0.02	0	−3.4	−3.6	−3.2	−2.3	−2.0	−1.72	−1.3	−0.86
0.04	0	−7.83	−8.5	−7.56	−5.31	−4.63	−3.9	−2.83	−1.83
0.06	0	−13.3	−14.4	−12.3	−9.34	−7.4	−6.12	−4.53	−2.36

Note: This table is built by using error percent $= \dfrac{1 - 1/\bar{N}_H}{[P_R N_p' A_f/(P_R - 1)L\beta S] - 1/\bar{N}_H}$

where $L = 12$ ft, $A_f = 0.00519$ ft^2, $P_R = 1.5$, $S = 0.22$ in.

$\bar{N}_H = 1 + \ln\left\{1 + \left[353\left(\dfrac{N_p' - 2}{N_p'}\right)^{3.5\beta/(0.015+\beta)}\beta^{1.1}\right]\right\} \pm 5$ percent (a two-dimensional result analogous to the three-dimensional correlations of 6-133 and 6-134.)

$N_p' =$ number of rows of rods
(After Chiu et al. [7].)

than the actual ones. This implies lower interchange of energy due to mixing, which means larger enthalpy for the hot channel.

The importance of the enthalpy transport coefficient depends strongly on the sizes of interacting lumped channels. This is because the coefficient corrects the lateral energy exchange, i.e., the energy exchange across the channel surfaces. The importance of this exchange is measured by its magnitude with respect to internal channel energy addition from the encompassed fuel rods. The larger the channel volume, the larger is the internal energy addition relative to the surface energy exchange. This suggests that N_H is important for smaller channels. However, at the limit of single subchannels, N_H is defined as unity. Therefore, use of the N_H concept becomes most important for intermediate size lumped channels. Table 6-10 from Chiu et al. [7] based on a two-dimensional analysis illustrates this conclusion. It presents the error in subchannel enthalpy by neglect of N_H, i.e., taking $N_H \equiv 1$. This error is largest for intermediate size lumped channels, those composed of four subchannels which for a modern PWR is a quarter subassembly.

Example 6-8 Computation of energy transfer rates between subchannels

PROBLEM For a typical PWR compute the energy transferred across the surface per unit length and the energy internally generated per unit length for (a) single subchannel; and (b) a homogenized group of four subchannels. Make these estimates assuming that the difference in enthalpies between the region of interest and the neighboring subchannels is 5 percent of the nominal enthalpy rise in a channel, i.e., $0.05\bar{q}'L/\dot{m}$. For a typical PWR take:

$\bar{q}'$ = 17.8 kW/m for a fuel rod

$\dot{m}$ = 0.290 kg/s for a subchannel

L = 3.66 m

$W'_{ij} = 0.0363 \, \dfrac{\text{kg}}{\text{m} \cdot \text{s}}$ (Example 6-4)

SOLUTION

1. For a single subchannel:

   The internal generated energy = $\bar{q}'$ = 17.8 kW/m.

   The energy transferred across the subchannel surface (four gaps per subchannel) = $4W'_{ij}(h_i - h_j)$ since W'_{ij} is:

$$\frac{\text{kg across a gap}}{\text{m} \cdot \text{s}}$$

   Now:

$$h_i - h_j = 0.05 \frac{\bar{q}'L}{\dot{m}}$$

$$= 0.05 \frac{17.8(3.66)}{0.290}$$

$$= 11.23 \frac{\text{kJ}}{\text{kg}} \text{ or } \frac{\text{kW}}{\text{kg s}}$$

   Hence, the energy transferred across the subchannel surface

$$= 4(0.0363)(11.23)$$

$$= 1.63 \text{ kW/m}$$

   Therefore:

$$\frac{\text{energy transferred across the subchannel surface}}{\text{energy internally generated}}$$

$$= \frac{1.63}{17.8} = 9.2 \text{ percent.}$$

2. *For a homogenized group of four subchannels:*

   Four equivalent fuel rods exist in this region.

   Eight rod gaps bound this region.

   The internally generated energy = $4\bar{q}'$ = 4(17.8)

$$= 71.2 \text{ kW/m}$$

The energy transferred across the subchannel surface

$$= 8W'_{ij}(h_i - h_j)$$

$$= 8(0.0363)11.23$$

$$= 3.26 \text{ kW/m}.$$

Therefore:

$$\frac{\text{energy transferred across the region surface}}{\text{energy internally generated}}$$

$$= \frac{3.26}{71.2} = 4.6 \text{ percent}.$$

REFERENCES

1. Beus, S. G. *A Two-Phase Turbulent Mixing Model for Flow in Rod Bundles.* WAPD-TM-2438, 1971.
2. Bird, B. R., Stewart, W. E., and Lightfoot, E. N. *Transport Phenomena.* New York: John Wiley and Sons, 1960.
3. Brown, W. D., Khan, E. U., and Todreas, N. E. Production of cross flow due to coolant channel blockages. *Nucl. Sci. Eng.* 57:164–168, 1975.
4. Chelemer, H., Weisman, J., and Tong, L. S. Subchannel thermal analysis of rod bundle cores. *Nucl. Eng. Des.* 21:35–45, 1972.
5. Chelemer, H., Chu, P. T., and Hochreiter, L. E. An improved program for thermal hydraulic analysis of rod bundle cores (THINC-IV). *Nucl. Eng. Des.* 41:219–229, 1977. Also see WCAP-7956.
6. Chiu, C. Three dimensional transport coefficient model and prediction: Correction numerical method for thermal margin analysis of PWR cores. *Nucl. Eng. Des.* 64:103–115, 1981.
7. Chiu, C., Moreno, P., Todreas, N., and Bowring, R. Enthalpy transfer between PWR fuel assemblies in analysis by the lumped subchannel model. *Nucl. Eng. Des.* 53:165–186, 1979.
8. Faya, A., Wolf, L., and Todreas, N. E. *Canal Users Manual.* Energy Laboratory Report MIT-EL-79-028, Nov. 1979.
9. France, D., and Ginsberg, T. Evaluation of lumped parameter heat transfer techniques for nuclear reactor applications. *Nucl. Sci. Eng.* 51:41–51, 1973.
10. France, D., and Ginsberg, T. Comparison of analytic and lumped parameter solutions for steady-state heat transfer in fuel rod assemblies. *Prog. Heat Mass Transfer* 7:167–178, 1973.
11. Gaspari, G. P., Hassid, A., and G. Vanoli. Some considerations on critical heat flux in rod clusters in annular dispersed vertical upward two-phase flow. Paper 4, *Proceeding of the 4th International Heat Transfer Conference,* Session B6, Paris, 1970.
12. Gaspari, G., Hassid, A., and Lucchini, F. A rod-centered subchannel analysis with turbulent (enthalpy) mixing for critical heat flux prediction in rod clusters cooled by boiling water. Paper 12, 5th International Heat Transfer Conference Session B6, Tokyo, 1974.
13. Hao, B. R., and Alcorn, J. M. *LYNX-1: Reactor Fuel Assembly Thermal Hydraulic Analysis Code.* BAW-10129, Oct. 1976.
14. Jackson, J. W., and Todreas, N. E. *COBRA IIIC/MIT-2: A Digital Computer Program for Steady State and Transient Thermal–Hydraulic Analysis of Rod Bundle Nuclear Fuel Elements.* Energy Lab. Report MIT-EL 81-018, June, 1981.
15. Kao, S., and Kazimi, M. Critical heat flux predictions in rod bundles. *Nucl. Technol.* 60:7–13, 1983.

16. Kelly, J. E., Kao, S. P., and Kazimi, M. S. *THERMIT-2: A Two-Fluid Model for Light Water Reactor Subchannel Transient Analysis.* Energy Laboratory Report MIT-EL 81-014, April 1981.
17. Lahey, R. T., Jr., and Moody, F. J. *The Thermal-Hydraulics of a Boiling Water Nuclear Reactor,* Hinsdale, Ill: ANS Monograph, 1977.
18. Lahey, R. T., Jr., Shiralkar, B. S., and Radcliffe, D. W. Mass flux and enthalpy distribution in a rod bundle for single and two-phase flow conditions. *J. Heat Transfer* 93:197–209, 1971.
19. *LYNX-2-Subchannel Thermal Hydraulic Analysis Program.* BAW-10130, October 1976.
20. Meyer, J. Hydrodynamic models for the treatment of reactor thermal transients. *Nucl. Sci. Eng.* 10:269–277, 1961.
21. Moreno, P., and Leira, G. Transport coefficient for enthalpy exchange between PWR homogenized channels. *Nucl. Eng. Des.* 66:269–287, 1981.
22. Moreno, P., Liu, J., Khan, E., and Todreas, N. Steady state thermal analysis of PWRs by a single-pass procedure using a simplified nodal layout. *Nucl. Eng. Des.* 47:35–48, 1978.
23. Ramm, H., Johannsen, K., and Todreas, N. Single phase transport within bare-rod arrays at laminar, transition, and turbulent flow conditions. *Nucl. Eng. Des.* 30:186–204, 1974.
24. Rogers, J. T., and Tahir, A. E. *Turbulent Interchange Mixing in Rod Bundles and the Role of Secondary Flows.* New York: ASME 75-HT-31, 1975.
25. Rowe, D. S. *COBRA IIIC: A Digital Computer Program for Steady-State and Transient Thermal Hydraulic Analysis of Rod Bundle Nuclear Fuel Elements.* BNWL-1965, March 1973.
26. Rouhani, A. Steady-state subchannel analysis. In *Two-Phase Flows and Heat Transfer with Application to Nuclear Reactor Design Problems,* Ginoux, J., Ed. New York: Hemisphere/McGraw Hill, 301–327, 1978.
27. Stewart, C. W., et al. *COBRA IV: The Model and the Method.* BNWL-2214, NRC-4, July 1977.
28. *TORC Code: A Computer Code for Determining the Thermal Margin of a Reactor Core.* Report CENPD-161, Combustion Engineering, Inc., July 1975.
29. Wallis, G. B. *One-Dimensional Two-Phase Flow.* New York: McGraw Hill, 1969.
30. Wong, C.-N., and Wolf, L. A 3-D Slug Flow Heat Transfer Analysis of Coupled Coolant Cells in Finite LMFBR Bundles. M.S. thesis, M.I.T. Department of Nuclear Engineering, Jan., 1978. Also in *Trans. ANS* 30:540–541, 1978.
31. Yeung, M., and Wolf, L. *A Multicell Slug Flow Heat Transfer Analysis of Finite LMFBR Bundles.* COO-2245-68TR, M.I.T. 238–243, December, 1978.

PROBLEMS

Problem 6-1 Evaluation of turbulent interchange flow rate for energy, $W_{ij}'^{H}$ (Section IV)
  Generally, it is postulated that the parameter $W_{ij}'^{M}$ is equal to $W_{ij}'^{H}$. Example 6-1 describes an experiment by which $W_{ij}'^{H}$ was determined. In this problem, you are asked to design an experiment by which $W_{ij}'^{M}$ can be determined using the same two geometrically identical subchannels of Example 6-1.
  Your answer to this question should include:

1. The boundary conditions to be imposed.
2. The parameters to be measured and the locations at which the measurements are to be made.
3. The conservation equation(s) to be used in reducing the data and a description of how the equations are used to obtain $W_{ij}'^{M}$ including:
   a. Exact form of equations to be used.
   b. Description of each necessary parameter in the equation(s) and how you would obtain it.

Be careful to avoid overspecifying the necessary parameters to be measured and thereby creating a redundant situation.

Problem 6-2 The axial momentum equation (Section V)

For the subchannel axial momentum equation (Eq. 6-44) identify:

1. The approximations that are inherent in the equation.
2. The physical quantity represented by v_z^*.
3. The relation of F_{iz} to the Moody friction factor.

Problem 6-3 Subchannel analysis of two interacting test channels (Section VII)

Describe how to apply and solve continuity, momentum, and energy equations to two geometrically identical test channels by answering the following questions:

1. Explain the procedure to solve the equations using the subchannel analysis approach. This should include identifying necessary equations, unknowns, and appropriate boundary conditions.
2. Sketch qualitatively the axial profiles of the identified unknowns.
3. Identify all constitutive inputs necessary to solve these equations.
4. Provide the numerical value for W_{ij}^{*H} under single-phase conditions.

Utilize the following geometry, reference coolant properties, operating conditions, and assumptions.

Geometry:

Typical PWR square array with $P/D = 1.326$, $D = 9.5$ mm, length $= 3.66$ m.

Reference coolant properties:

$\rho^* = 726$ kg/m^3
$\mu^* = 9.63 \times 10^{-4}$ kg/m·s.

Operating conditions:

- Total test section flow rate is 1.452 kg/s
- Same inlet mass flow rate for both channels
- $p_i(L) = p_j(L) = 15.5$ MPa
- $T_i(0) = T_j(0) = 280$ °C
- $(q_1'/q_2') = 1.2$ axially uniform heat input
- Steady-state condition.

Assumption:

Axial friction factor $f = f_{c.t.}[1.04 + 0.06(P/D - 1)]$.

Problem 6-4 Two interacting channels with equilibration of radial pressure difference (Section VII)

Consider two parallel channels of PWR assembly geometry under transient upflow conditions in single-phase flow. The channels differ only by virtue of the energy generation rate differences in the fuel pins bounding the channels due to the radial power profile. For an axial step Δz, derive the set of equations from which the cross-flow rate between the channels can be determined assuming that the channel pressures are equal at the top of the step as a result of the cross-flow over the axial increment Δz. Indicate the manner in which you would ascribe a value to each parameter in this set of equations if you were to calculate the cross-flow rate numerically.

Assumptions:

1. Steady-state conditions exist.
2. Axial friction factor is f.
3. Linear energy generation ratio is $q_1'/q_2' = R$.
4. All inlet conditions to the two-channels are equal.

Problem 6-5 Consequence of two-phase mixing between adjacent, adiabatic interacting channels (Section VII)

Consider turbulent interchange only occurring between two adjacent, adiabatic subchannels, i and j, over an element of length Δz. Inlet conditions to these subchannels are:

Mass flow rate, $\dot{m}_i$ and $\dot{m}_j$
Enthalpy, h_i and h_j
Void fraction, α_i and α_j

If $\alpha_j > \alpha_i$ show that:

1. The exit enthalpy of the element Δz of subchannel j, h_j', is decreased, i.e., $h_j > h_j'$.
2. The exit enthalpy of the element Δz of subchannel i, h_i', is increased, i.e., $h_i' > h_i$.

Problem 6-6 Core-wide, lumped parameter PWR thermal analysis (Section VIII)

Imagine you have the task to use a subchannel code, like COBRA-IIIC in order to make a core-wide analysis of the enthalpy distribution in a PWR. This necessitates, due to computer storage limitations, modeling a full 17×17 rod subassembly simply as one node in which a single rod is surrounded by the appropriate flow area. Therefore, in changing the whole layout you have to specify a new value for the turbulent interchange for energy $W_{ij}'^H$ to link the nodes of the two adjacent "single-rod assemblies." Consider the rod heat flux and flow conditions as being already adjusted.

1. What conditions on enthalpies $h_I(z)$ and $h_J(z)$ of the single-rod assemblies must be satisfied to achieve the desired representation of two full 17×17 subassemblies as illustrated in Figure 6-29?
2. What is the value of $(W_{IJ}^{*H}(z))$ which will achieve the condition listed under 1? Express this in terms of cross-flow between individual subchannels, $(W_{ij}^{*H})(z)$, the subchannel enthalpies within the full assemblies, and the enthalpies of the single-rod assemblies.

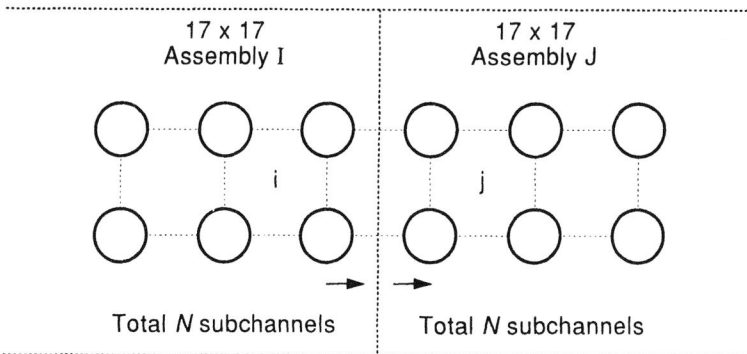

Figure 6-29 Configuration of two 17×17 assemblies.

Answers:

1. $h_I(z) - h_J(z) = \dfrac{1}{N}\left[\displaystyle\sum_{i}^{N} h_i(z) - \sum_{j}^{N} h_j(z)\right]$

2. $W_{IJ}^{*H} = W_{ij}^{*H} \dfrac{\dfrac{1}{2}[h_i(z) - h_j(z)]}{[h_I(z) - h_J(z)]}$

DISTRIBUTED PARAMETER ANALYSIS OF REACTOR FUEL ROD ASSEMBLIES

I INTRODUCTION

In the lumped parameter analysis of Chapter 6, the cross-sectional description of velocity and temperature variations within a subchannel was not considered. This chapter presents the solution of these velocity and temperature fields within a rod array by the distributed parameter method. These fields are considered mainly for fully developed flow leading to the formulation of a two-dimensional problem. As in Chapter 4, the subchannel will be initially viewed as isolated from its neighbors. However, in Section XI the method for treating interacting subchannels will be presented.

At the outset we will make several restricting assumptions allowing solution of some simple problems which will nevertheless illustrate the key features of this method of analysis. Remember that we start initially having made a key assumption that the subchannel is taken as isolated from its neighbors. The other assumptions are:

1. A two-dimensional problem will be considered in r, θ coordinates. (The three-dimensional approaches in which the axial dimension z is included along with interacting subchannels are discussed in Section XI).
2. The subchannel type to be considered will be an interior subchannel. This is the only type subchannel that occurs in an infinite array of rods (analyses of edge and corner subchannels are discussed in Section XI).

The subchannel we wish to analyze is shown in Figure 7-1 for both square and triangular arrays. The domain of interest includes coolant, clad, and fuel. It

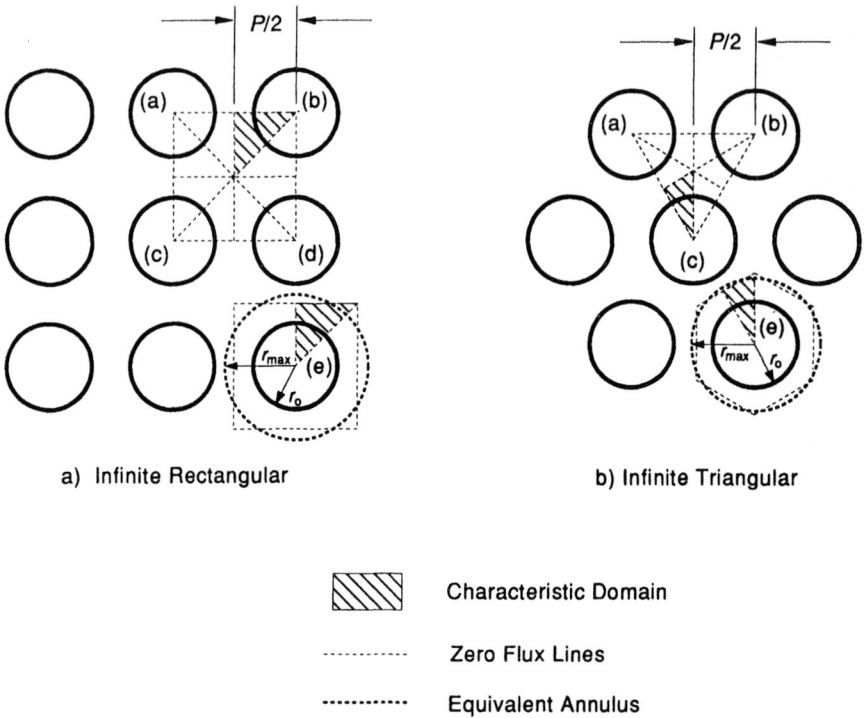

a) Infinite Rectangular   b) Infinite Triangular

Characteristic Domain

Zero Flux Lines

Equivalent Annulus

Figure 7-1 Subchannel and characteristic domains for square and triangular arrays.

is the area within the square or the equilateral triangle formed by con   ng the centers of rods a, b, c, and d, and a, b, and c, respectively. Now if this arrangement has symmetry in the distribution of volumetric energy generation rates, $q'''(r)$, for all rods, a smaller region of symmetry can be defined within the subchannel. This region is cross-hatched in Figure 7-1 and is called the *characteristic domain*.

It is equally possible to view this characteristic domain as the symmetry section of a rod-centered subchannel. Such a subchannel about rod e is also pictured in Figure 7-1 and is the more common viewpoint of distributed parameter analysis. The square or hexagon circumscribed about rod e bounds the cross-sectional area of coolant flow assignable to that particular rod and is assumed to be the locus of points of zero shear stress in the coolant stream. Within a characteristic domain the heat leaving the rod surface will not flow solely radially outward through the coolant stream. This is particularly true for close rod spacings. We will begin by examining whether the rod-centered subchannel can be treated as an equivalent annulus thus neglecting circumferential heat flow and shear stress variations. After this approach is examined, we will turn to the analytical methods of solving the true subchannel geometry considering only the coolant region. The reader should be aware that all our presentations of the distributed parameter analysis of rod bundles, however formidable they appear mathematically, do not include the effect of spacers between rods.

II EQUATIONS FOR MOMENTUM AND HEAT TRANSFER IN THE COOLANT REGION

Only the coolant region equations will be presented here. They will be given in the form necessary for turbulent flow in a noncircular flow region. The forms sufficient for both laminar and turbulent flow in equivalent annuli and laminar flow in noncircular geometries are obtained as simplified forms of these equations.

An incompressible fluid with temperature-independent properties is considered. Under conditions of steady flow and heat transfer, $\partial/\partial t = 0$, the axial (z-direction) momentum and the energy balances for the differential volume element $rd\theta drdz$ of Figure 7-2 are given as follows:

$$\rho \left(v_r \frac{\partial v_z}{\partial r} + \frac{v_\theta}{r} \frac{v_z}{\partial \theta} + v_z \frac{\partial v_z}{\partial z} \right) + \frac{1}{r} \frac{\partial}{\partial r}(r\tau_{rz}) + \frac{1}{r} \frac{\partial \tau_{\theta z}}{\partial \theta} + \frac{\partial \tau_{zz}}{\partial z} = -\frac{\partial \hat{p}}{\partial z} \quad (7\text{-}1)$$

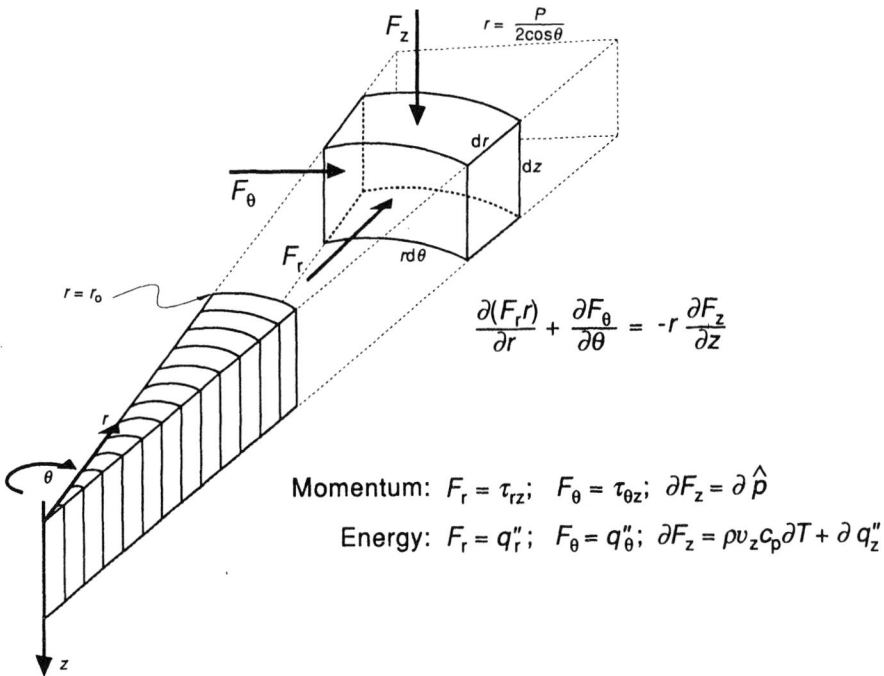

$$\frac{\partial(F_r r)}{\partial r} + \frac{\partial F_\theta}{\partial \theta} = -r \frac{\partial F_z}{\partial z}$$

Momentum: $F_r = \tau_{rz}$; $F_\theta = \tau_{\theta z}$; $\partial F_z = \partial \hat{p}$

Energy: $F_r = q_r''$; $F_\theta = q_\theta''$; $\partial F_z = \rho v_z c_p \partial T + \partial q_z''$

Figure 7-2 Axial momentum and energy balances on a volume element in the coolant region. Momentum is Eq. 7-8, and energy is Eq. 7-9.

where $\hat{p} = p + \rho g z$.

$$\rho c_p \left(v_r \frac{\partial T}{\partial r} + \frac{v_\theta}{r} \frac{\partial T}{\partial \theta} + v_z \frac{\partial T}{\partial z} \right) = - \left[\frac{1}{r} \frac{\partial}{\partial r} (r q_r'') + \frac{1}{r} \frac{\partial q_\theta''}{\partial \theta} + \frac{\partial q_z''}{\partial z} \right] \quad (7\text{-}2)$$

where viscous heating effects are neglected. Figure 7-2 illustrates the momentum and energy components acting on the volume element.

With the further restrictions of fully developed flow, we have:

$$v_\theta = v_r = 0 \quad (7\text{-}3)$$

$$\frac{\partial v_z}{\partial z} = 0 \quad (7\text{-}4)$$

$$\tau_{zz} = 0 \quad (7\text{-}5)$$

$$-\frac{\partial \hat{p}}{\partial z} = 4 \frac{\bar{\tau}_w}{D_e} \quad (7\text{-}6)$$

and for fully developed heat transfer, we have:

$$\frac{\partial}{\partial z} \left(\frac{T_w - T}{T_w - T_b} \right) = 0 \quad (7\text{-}7)$$

where T_b is the bulk fluid temperature.

Note that we cannot evaluate dT/dz yet without specifying the axial surface boundary condition.

Now applying Eqs. 7-3 through 7-5, Eq. 7-1 and 7-2 become:

$$\frac{\partial (\tau_{rz} r)}{\partial r} + \frac{\partial \tau_{\theta z}}{\partial \theta} = -\frac{\partial \hat{p}}{\partial z} r \quad (7\text{-}8)$$

$$\frac{\partial}{\partial r} (q_r'' r) + \frac{\partial q_\theta''}{\partial \theta} + r \frac{\partial q_z''}{\partial z} = -\rho c_p v_z r \frac{\partial T}{\partial z} \quad (7\text{-}9)$$

Now for a constant axial wall heat flux, boundary condition $q_{z-w}'' = $ constant and constant heat transfer coefficient h:

$$\frac{\partial T_w}{\partial z} = \frac{\partial T_b}{\partial z} \quad (7\text{-}10)$$

and, with this result Eq. 7-7 reduces to:

$$\frac{\partial T}{\partial z} = \frac{\partial T_w}{\partial z} = \frac{\partial T_b}{\partial z}, \text{ independent of } r \quad (7\text{-}11)$$

The magnitude of $\partial T/\partial z$ can be determined from an energy balance as:

$$\frac{\partial T}{\partial z} = \frac{4 \bar{q}_{z-w}''}{\{v_z\} c_p \rho D_e} = \frac{q'}{\dot{m} c_p} = \text{constant} \quad (7\text{-}12)$$

where $\{v_z\}$ = average coolant speed;

and also, from Eq. 7-12 we note:

$$-k \frac{\partial^2 T}{\partial z^2} = \frac{\partial q''_{z-w}}{\partial z} = 0 \qquad (7\text{-}13)$$

On the other hand, for a constant axial wall temperature boundary condition, $T_w(z) = $ constant, then:

$$\frac{\partial T_w}{\partial z} = 0 \qquad (7\text{-}14)$$

and Eq. 7-7 reduces to:

$$\frac{\partial T}{\partial z} = \left(\frac{T_w - T}{T_w - T_b} \right) \frac{\partial T_b}{\partial z} = f \left(\frac{r}{r_0} \right) \frac{\partial T_b}{\partial z} \qquad (7\text{-}15)$$

and we note specifically:

$$\frac{\partial q''_{z-w}}{\partial z} \neq 0$$

Therefore to eliminate the axial conduction term of Eq. 7-9, it must be assumed that either axial conduction effects are negligible or a constant axial heat flux boundary condition exists. Under these circumstances Eq. 7-9 simplifies to:

$$\frac{\partial}{\partial r} (q''_r r) + \frac{\partial q''_\theta}{\partial \theta} = -\rho c_p v_z r \frac{\partial T}{\partial z} \qquad (7\text{-}16)$$

For fully developed flow, the r- and θ-direction momentum balances reduce to the condition of constant pressure in these directions. The momentum fluxes τ_{rz} and $\tau_{\theta z}$ incorporate transport contributions due to viscous effects, turbulent velocity fluctuations and, for the geometrical configuration considered here, also that due to secondary flow. Secondary flows are the radial and circumferential velocities that exist in noncircular ducts under turbulent but fully developed flow. They do not exist in laminar flow in noncircular ducts or in annuli under laminar or turbulent flow. The momentum fluxes are expressed by the equations:

$$\tau_{rz} = -\rho v \left[\frac{\partial v_z}{\partial r} + \frac{\partial v_r}{\partial z} \right] + \rho \overline{v'_z v'_r} \qquad (7\text{-}17)$$

and

$$\tau_{\theta z} = -\rho v \left[\frac{\partial v_z}{r \partial \theta} + \frac{\partial v_\theta}{\partial z} \right] + \rho \overline{v'_z v'_\theta} \qquad (7\text{-}18)$$

where ρ and v are the fluid density and kinematic viscosity; v_z, v_r, and v_θ are time-averaged velocity components; and v'_z, v'_r and v'_θ are fluctuating velocity components. The turbulent stress terms $\rho \overline{v'_z v'_r}$ and $\rho \overline{v'_\theta v'_z}$ representing one-point correlations of mutually perpendicular velocity fluctuations, have a value different from zero. In a similar manner, the heat fluxes q_r'' and q_θ'' can be expressed as:

$$q_r'' = -c_p\rho\alpha \frac{\partial T}{\partial r} + c_p\rho\overline{T'v_r'} \tag{7-19}$$

$$q_\theta'' = -c_p\rho\alpha \frac{\partial T}{r\partial\theta} + c_p\rho\overline{T'v_\theta'} \tag{7-20}$$

where T is the time-average temperature and T' its fluctuating component.

A Turbulent Flow Equations for Noncircular Channels

The preceding equations can be developed further only when expressions are available relating the turbulent flux terms to the mean flow properties. These aspects are dealt with in detail by Nijsing and Eifler [35] for the general case which includes anisotropy effect terms and develop the general form for the turbulent flux terms.

Here, however, we will treat the simpler case in which secondary flow can be neglected and the velocity fluctuations can be related to the velocity gradients by turbulent or eddy diffusivities, i.e.:

$$\overline{v_z'v_r'} = -\varepsilon_{Mr} \frac{\partial v_z}{\partial r} \tag{7-21}$$

$$\overline{v_z'v_\theta'} = -\varepsilon_{M\theta} \frac{1}{r} \frac{\partial v_z}{\partial\theta} \tag{7-22}$$

$$\overline{T'v_r'} = -\varepsilon_{Hr} \frac{\partial T}{\partial r} \tag{7-23}$$

$$\overline{T'v_\theta'} = -\varepsilon_{H\theta} \frac{1}{r} \frac{\partial T}{\partial\theta} \tag{7-24}$$

Applying Eqs. 7-17 through 7-24 to Eqs. 7-8 and 7-9, the fully developed turbulent flow equations are obtained as:

$$\frac{1}{r} \frac{\partial}{\partial r}\left[\rho r(\nu + \varepsilon_{Mr})\frac{\partial v_z}{\partial r}\right] + \frac{1}{r^2}\frac{\partial}{\partial\theta}\left[\rho(\nu + \varepsilon_{M\theta})\frac{\partial v_z}{\partial\theta}\right] = \frac{\partial\bar{p}}{\partial z} \tag{7-25}$$

$$\frac{1}{r} \frac{\partial}{\partial r}\left[c_p\rho r(\alpha + \varepsilon_{Hr})\frac{\partial T}{\partial r}\right] + \frac{1}{r^2}\frac{\partial}{\partial\theta}\left[c_p\rho(\alpha + \varepsilon_{H\theta})\frac{\partial T}{\partial\theta}\right] = c_p\rho v_z\frac{\partial T}{\partial z} \tag{7-26}$$

where it should be recalled that the term $\partial^2 T/\partial z^2$ has been eliminated from Eq. 7-26 by Eq. 7-13.

B Laminar Flow Equations for Noncircular Channels

In this case Eqs. 7-25 and 7-26 are simplified only by taking the eddy diffusivities as identically zero. With that step, we obtain:

$$\frac{\partial^2 v_z}{\partial r^2} + \frac{1}{r}\frac{\partial v_z}{\partial r} + \frac{1}{r^2}\frac{\partial^2 v_z}{\partial \theta^2} = \frac{1}{\mu}\frac{\partial \hat{p}}{\partial z} \tag{7-27}$$

$$\frac{\partial^2 T}{\partial r^2} + \frac{1}{r}\frac{\partial T}{\partial r} + \frac{1}{r^2}\frac{\partial^2 T}{\partial \theta^2} = \frac{v_z}{\alpha}\frac{\partial T}{\partial z} \tag{7-28}$$

C Turbulent and Laminar Flow Equations for Annuli

In this case symmetry exists in the θ direction. Upon eliminating the θ-dependent terms and expanding the r-dependent terms, Eqs. 7-25 and 7-26 reduce to the following forms for turbulent flow:

$$\frac{\partial}{\partial r}\left[\rho(\nu + \varepsilon_{\mathrm{Mr}})\frac{\partial v_z}{\partial r}\right] + \frac{1}{r}\frac{\partial v_z}{\partial r}\left[\rho(\nu + \varepsilon_{\mathrm{Mr}})\right] = \frac{\partial \hat{p}}{\partial z} \tag{7-29}$$

$$\frac{\partial}{\partial r}\left[(\alpha + \varepsilon_{\mathrm{Hr}})\frac{\partial T}{\partial r}\right] + \frac{1}{r}\frac{\partial T}{\partial r}(\alpha + \varepsilon_{\mathrm{Hr}}) = v_z\frac{\partial T}{\partial z} \tag{7-30}$$

where v_z, T, ε_M and ε_H only are position dependent. For laminar flow the eddy diffusivities are identically zero, and Eqs. 7-29 and 7-30 become:

$$\frac{\partial^2 v_z}{\partial r^2} + \frac{1}{r}\frac{\partial v_z}{\partial r} = \frac{1}{\mu}\frac{\partial \hat{p}}{\partial z} \tag{7-31}$$

$$\frac{\partial^2 T}{\partial r^2} + \frac{1}{r}\frac{\partial T}{\partial r} = \frac{v_z}{\alpha}\frac{\partial T}{\partial z} \tag{7-32}$$

III THE EQUIVALENT ANNULUS MODEL

The dotted circle about rod "e" in Figure 7-1 together with the rod surface define an annulus having a flow area equal to that bounded by the square or the hexagon and the rod surface but with a zero shear stress condition on the outer boundary. Consequently this approach is called the equivalent annulus model. This equivalent annular flow area is assumed to be fluid-dynamically equivalent to that portion of an annulus which extends from its inner radius to its radius of maximum axial velocity. From the condition of equal flow areas between the equivalent annulus and the actual flow area bounded by the square or the hexagon:

$$\pi(r_{\mathrm{max}}^2 - r_0^2) = 12\left[\frac{P^2}{8}\tan\frac{\pi}{6} - \frac{\pi r_0^2}{12}\right] \tag{7-33}$$

The position of the radius of maximum axial velocity is:

$$r_{\mathrm{max}} = r_0\left(\frac{2\sqrt{3}}{\pi}\right)^{1/2} P/D \text{ (for triangular array)} \tag{7-34a}$$

$$r_{\mathrm{max}} = r_0\left(\frac{4}{\pi}\right)^{1/2} P/D \text{ (for square array)} \tag{7-34b}$$

The boundary conditions at this location at zero gradients of axial velocity and temperature:

$$\frac{\partial v_z}{\partial r}\bigg|_{r_{max}} = \frac{\partial T}{\partial r}\bigg|_{r_{max}} = 0 \qquad \begin{matrix}(7\text{-}35)\\(7\text{-}36)\end{matrix}$$

The boundary conditions at the inner wall are no slip, uniform axial heat flux, and both uniform circumferential heat flux and wall temperature:

$$v_z|_{r_0} = 0 \qquad (7\text{-}37)$$

$$\frac{\partial}{\partial z}\left(\frac{\partial T}{\partial r}\bigg|_{r_0}\right) = \frac{\partial}{\partial \theta}\left(\frac{\partial T}{\partial r}\bigg|_{r_0}\right) = \text{constant}; \ T_w(\theta) = \text{independent of } \theta \equiv T_w \qquad \begin{matrix}(7\text{-}38)\\(7\text{-}39)\\(7\text{-}40)\end{matrix}$$

Note that circumferentially, the conditions of uniform heat flux and wall temperature are compatible for the equivalent annulus by virtue of its symmetrical geometry.

Take the case of fully developed conditions with no temperature dependence of thermal properties. Let us first examine solutions to the momentum equation and then the energy equation for this geometry.

A Momentum Balance Solutions

1 Laminar flow Eq. 7-31 is the applicable equation. Applying the boundary conditions of Eqs. 7-35 and 7-37 the velocity distribution resulting from the solution of Eq. 7-31 is:

$$\frac{v_z}{\dfrac{P^2}{4}\left(-\dfrac{1}{\mu}\dfrac{\partial \hat{p}}{\partial z}\right)} = \frac{\sqrt{3}}{\pi}\ln\frac{r}{r_0} - \frac{1}{4}\left[\left(\frac{r}{P/2}\right)^2 - \left(\frac{r_0}{P/2}\right)^2\right] \qquad (7\text{-}41)$$

We can now calculate the friction factor–Reynolds number product. First note that the definitions of friction factor f and Reynolds number Re_{De} are:

$$-\frac{d\hat{p}}{dz} \equiv \frac{4f'}{D_e}\frac{\rho\{v_z\}^2}{2} = \frac{f}{D_e}\rho\frac{\{v_z\}^2}{2} \qquad (7\text{-}42)$$

and

$$\text{Re}_{De} = \frac{D_e\{v_z\}\rho}{\mu} \qquad (7\text{-}43)$$

where:

$$D_e = \frac{4\pi(r_{max}^2 - r_0^2)}{2\pi r_0} \qquad (7\text{-}44)$$

Hence:

$$f\text{Re}_{De} \equiv -\left[2\left(\frac{1}{\mu}\frac{d\hat{p}}{dz}\right)\frac{D_e^2}{\{v_z\}}\right] \qquad (7\text{-}45)$$

where $\{v_z\}$ is obtained from Eq. 7-41 as:

$$\frac{\{v_z\}}{\left(-\frac{1}{\mu}\frac{d\hat{p}}{dz}\right) r_{max}^2} = \frac{X_m^2 \, \ell n \, X_m}{2(X_m^2 - 1)} + \left(\frac{2}{X_m^2 - 1}\right)\left[\frac{1}{4} - \frac{1}{16X_m^2} - \frac{3}{16} X_m^2\right] \quad (7\text{-}46)$$

or consolidating:

$$f\text{Re}_{De} = \frac{64(X_m^2 - 1)^3}{-3X_m^4 + 4X_m^4 \, \ell n \, X_m - 1 + 4X_m^2} \quad (7\text{-}47)$$

where:

$$X_m \equiv \frac{r_{max}}{r_0} \quad (7\text{-}48)$$

which is related to the P/D per Eq. 7-34 as:

$$X_m = 1.05 P/D$$

Eq. 7-47 represents the equivalent annulus solution plotted in Figure I,9-22. Note that for our case of fully developed laminar flow, Eq. 7-47 is a function of geometry only. This approach is a direct solution of the equivalent annulus problem.

Example 7-1 next illustrates the method for prediction of $f\text{Re}_{De}$ for an equivalent annulus geometry using the existing annular literature. The two geometries are shown in Figure 7-3.

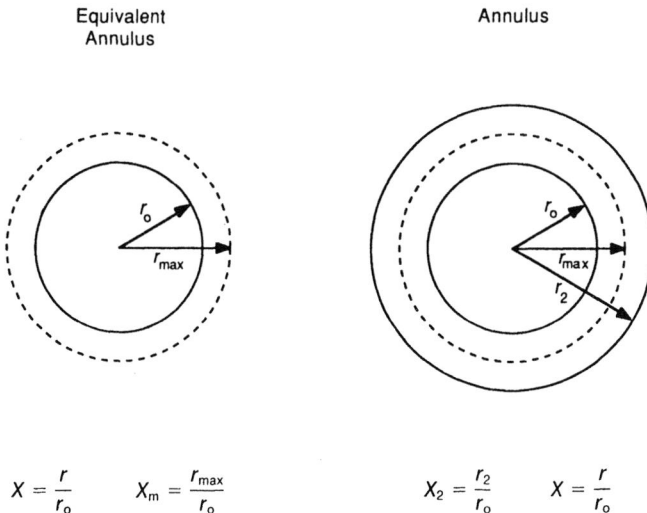

Figure 7-3 Annular geometries.

Example 7-1 Determination of the friction factor–Reynolds product for a triangular array using existing annular literature

PROBLEM Find $f\mathrm{Re}_{D_e}$ for a triangular array of $P/D = 2$ in laminar flow by the equivalent annulus method from existing annular literature.

SOLUTION

A. Given (P/D), find (r_{max}/r_0) from Eq. 7-34 as:

$$\frac{r_{max}}{r_0} = 2 \left(\frac{2\sqrt{3}}{\pi} \right)^{1/2}$$

$$\text{or } r_{max}/r_0 = 2.1$$

B. Given the knowledge that r_{max} is the point of maximum velocity, $dv_z/dr = 0$, find r_2/r_0 for the annulus geometry:

Axial momentum equation:

$$\frac{1}{r}\frac{d}{dr}\left(r \frac{dv_z}{dr} \right) = \frac{1}{\mu}\frac{d\hat{p}}{dz} = K_0$$

integrate:

$$\frac{dv_z}{dr} = \frac{c}{r} + \frac{K_0 r}{2}; \frac{dv_z}{dr} = 0 \text{ at } r = r_{max}$$

Therefore:

$$c = \frac{-K_0 r_{max}^2}{2}$$

integrate:

$$v_z = \frac{-K_0 r_{max}^2}{2} \ell n\, r + \frac{K_0 r^2}{4} + c_1;$$

$v_z = 0$ at $r = r_0$ and r_2. Apply both boundary conditions and then subtract to obtain:

$$r_2^2 - 2r_{max}^2 \ell n\, r_2 = r_0^2 - 2r_{max}^2 \ell n\, r_0$$

which upon rearrangement yields:

$$\left(\frac{r_{max}}{r_0} \right)^2 = \frac{(r_2/r_0)^2 - 1}{2 \ell n(r_2/r_0)} \tag{7-49}$$

Now utilizing the result of part A for r_{max}/r_0 in Eq. 7-49, by trial and error find $r_2/r_0 = 3.458$.

C. Now knowing r_2/r_0 for laminar flow look up the annulus friction–Reynolds number product, i.e., $f\mathrm{Re}]_A$ from Kays and Perkins [24] who also treated turbulent conditions:

$$f\text{Re}]_A = (r_2/r_0 = 3.5) = 94$$

where f is the Moody friction factor.

D. Now relate the equivalent annulus friction–Reynolds number product, i.e., $f\text{R}]_{EA}$ to $f\text{Re}]_A$. First express $f\text{Re}$ utilizing Eq. 7-45 as:

$$f\text{Re} \equiv -\frac{2}{\mu}\frac{d\hat{p}}{dz}\frac{D_e^2}{\{v_z\}}$$

$$\text{or } f\text{Re}]_{EA} = f\text{Re}]_A \frac{\{v_z\}_A}{\{v_z\}_{EA}}\frac{D_e^2]_{EA}}{D_e^2]_A} \tag{7-50}$$

E. Hydraulic diameter:

$$D_e]_{EA} = \frac{4\pi(r_{max}^2 - r_0^2)}{2\pi r_0} = 2\left(\left(\frac{r_{max}}{r_0}\right)^2 - 1\right)r_0 = 2(3.41)r_0 \tag{7-51a}$$

$$D_e]_A = \frac{4\pi(r_2^2 - r_0^2)}{2\pi(r_2 + r_0)} = 2\frac{((r_2/r_0)^2 - 1)}{(r_2/r_0 + 1)}r_0 = 2(2.46)r_0 \tag{7-51b}$$

Find $\{v_z\}$ for the equivalent annulus and the annulus utilizing these definitions:

$$X_m = \frac{r_{max}}{r_0}; \; X_2 = \frac{r_2}{r_0}; \; X = \frac{r}{r_0}$$

$$\{v_z\}_{EA} = \frac{\left(\int_{r_0}^{r_{max}} v_z(r)r\,dr\right)}{\left(\dfrac{r_{max}^2}{2} - \dfrac{r_0^2}{2}\right)} \tag{7-52}$$

$$\{v_z\}_A = \frac{\left(\int_{r_0}^{r_2} v_z(r)r\,dr\right)}{\left(\dfrac{r_2^2}{2} - \dfrac{r_0^2}{2}\right)} \tag{7-53}$$

$$\frac{\{v_z\}_{EA}}{-K_0 r_{max}^2} = \frac{1}{2(X_m^2 - 1)}\left[\frac{X^2}{2X_m^2} - \frac{X^4}{4X_m^2} + 2X^2\left(\frac{\ell n\, X}{2} - \frac{1}{4}\right)\right]_1^{X_m} = 0.133$$

$$\frac{\{v_z\}_A}{-K_0 r_{max}^2} = \frac{1}{2(X_2^2 - 1)}\left[\frac{X^2}{2X_m^2} - \frac{X^4}{4X_m^2} + 2X^2\left(\frac{\ell n\, X}{2} - \frac{1}{4}\right)\right]_1^{X_2} = 0.1164$$

G. Utilize Eq. 7-50 to obtain the desired result:

$$f\text{Re}]_{EA} = 94\left(\frac{0.1164}{0.133}\right)\left(\frac{3.41}{2.46}\right)^2 = 160$$

compared to 155 from Figure 9-22, Vol. I

2 Turbulent flow Eq. 7-29 is the applicable equation for this condition. To solve this equation directly, the radial description of the eddy diffusivity of momentum, $\varepsilon_M(r)$ is the necessary input. The selection of $\varepsilon_M(r)$ is not a straight-forward procedure. Even for the simple case of a circular tube, disagreements exist regarding the proper form. For a discussion of turbulent diffusivities of momentum and heat in tubes, parallel plates, and annuli, the interested reader should consult the review of Nijsing [33].

Rather than select an eddy diffusivity distribution and perform the required integrations, we will quote the result of Maubach [31] for the turbulent friction factor of the equivalent annulus. He obtained this result assuming that the universal velocity profile for circular tubes written below applied equally to the annulus:

$$v_z^+ = 2.5 \, \ell n \, y^+ + 5.5 \tag{7-54}$$

where:

$$v_z^+ = \frac{v_z}{v_z^*} \text{ (dimensionless velocity)} \tag{7-55}$$

$$y^+ = \frac{y v_z^*}{\nu} \text{ (dimensionless distance from the wall)} \tag{7-56}$$

$$v_z^* = (\tau_w/\rho)^{1/2} \text{ (friction or shear velocity)} \tag{7-57}$$

The resulting friction factor expression that appears in Rehme [41] is:

$$\left(\frac{8}{f}\right)^{1/2} = 2.5 \, \ell n \left[\text{Re}_{De} \left(\frac{f}{8}\right)^{1/2} \right] + 5.5 - \frac{3.966 + 1.25X_m}{1 + X_m}$$

$$- 2.5 \, \ell n \, 2(1 + X_m) \tag{7-58}$$

Based on $f_{c.t.}$ (circular tube) given by:

$$\frac{1}{(f_{c.t.})^{1/2}} = 2.035 \log_{10}[\text{Re}_{De}(f_{c.t.})^{1/2}] - 0.989 \tag{7-59}$$

the ratio of $\dfrac{f}{f_{c.t.}}$ from Eqs. 7-58 and 7-59 is:

for a triangular array per Rehme [41]:

$$\text{Re}_{De} = 10^4: \frac{f}{f_{c.t.}} = 1.045 + 0.071(P/D - 1) \tag{I,9-85}$$

$$\text{Re}_{De} = 10^5: \frac{f}{f_{c.t.}} = 1.036 + 0.054(P/D - 1) \tag{I,9-86}$$

for a square array per Marek et al. [28]:

$$\text{Re}_{De} = 10^5: \frac{f}{f_{c.t.}} = 1.04 + 0.06(P/D - 1) \tag{7-60}$$

Eq. 7-60 is the equivalent annulus result which is plotted in Figure I,10-12. Note that from this equation the friction factor ratio continually increases with P/D.

B Energy Balance Solutions

1 Laminar flow Eq. 7-32 is the relevant equation to be solved subject to the boundary conditions of Eqs. 7-36, 7-38, and 7-39 or 7-40. The required velocity profile is given by Eq. 7-41. To solve this problem let us first express $\partial T/\partial z$ which for our case of uniform heat transfer per unit length (Eqs. 7-10 to 7-12) is:

$$\frac{\partial T}{\partial z} = \text{constant} = \frac{q'}{\dot{m}c_p} = \frac{q'/12}{(\dot{m}/12)c_p} \tag{7-61}$$

where $\dot{m}$ and q', respectively, represent the mass flow and the heat transfer per unit length of the equivalent annulus. If we define a dimensionless grouping M as:

$$M \equiv \frac{\dot{m}/12\rho}{\dfrac{P^4}{16}\left(-\dfrac{1}{\mu}\dfrac{d\hat{p}}{dz}\right)} \tag{7-62}$$

then with the velocity profile on Eq. 7-41 the governing differential Eq. 7-32 becomes:

$$\frac{\partial^2 T}{\partial r^2} + \frac{1}{r}\frac{\partial T}{\partial r} = \frac{q'/12k}{MP^2/4}\left\{\frac{\sqrt{3}}{\pi}\,\ell n\,\frac{r}{r_0} - \frac{1}{4}\left[\left(\frac{r}{P/2}\right)^2 - \left(\frac{r_0}{P/2}\right)^2\right]\right\} \tag{7-63}$$

where the flow parameter M in laminar flow is obtained as follows when the velocity v_z is related to the flow rate $\dot{m}$:

$$M = \frac{1}{2\pi}\,\ell n\,\frac{r_{max}}{r_0} - \frac{3}{8\pi} + \frac{\sqrt{3}}{12}\left(\frac{r_0}{P/2}\right)^2 - \frac{\pi}{96}\left(\frac{r_0}{P/2}\right)^4 \tag{7-64}$$

The basis for the selected form of the parameter M is not obvious from this example but is based on convenience in the later analysis of triangular arrays where the symmetry segment will be seen to be $\frac{1}{12}$ of the coolant hexagon surrounding each rod.

Now for boundary conditions Eqs. 7-36 and 7-40, the temperature distribution is:

$$T - T_w = \frac{q'/12k}{MP^2/4}\left\{\left[\frac{\sqrt{3}}{4\pi}(r^2 + r_0^2) - \frac{1}{16}\frac{r_0^4}{P^2/4} - \frac{6MP^2/4}{\pi}\right]\ell n\,\frac{r}{r_0}\right.$$
$$\left. - \frac{1}{64}\frac{(r^4 - r_0^4)}{P^2/4} + \left[-\frac{\sqrt{3}}{4\pi} + \frac{1}{16}\left(\frac{r_0}{P/2}\right)^2\right](r^2 - r_0^2)\right\} \tag{7-65}$$

Finally the desired Nusselt number can be obtained from the definitions below since $T - T_w$, q', and $\dot{m}$ are known, i.e.:

$$\text{Nu}_{D_e} = \frac{hD_e}{k} = \frac{(X_m^2 - 1)q'\dot{m}}{k2\pi^2\rho \left[\int_{r_0}^{r_{max}} (T_w - T)v_z r dr\right]} \tag{7-66}$$

where:

$$h \equiv \frac{q'/\pi 2r_0}{T_w - T_b} \tag{7-67}$$

and

$$T_w - T_b = \frac{2\pi\rho \int_{r_0}^{r_{max}} (T_w - T)v_z r dr}{\dot{m}} \tag{7-68}$$

The results of Nu_{D_e} evaluated from Eq. 7-66 as a function of P/D are presented in Figure I,10-6.

2 Turbulent flow: Transformation of bilateral heated annular results

Eq. 7-30 is the relevant equation. To solve this equation directly the radial description of the eddy diffusivity of heat, $\varepsilon_H(r)$, and the radial velocity distribution are the necessary inputs. Selection of $\varepsilon_H(r)$ is even more complex than the selection of $\varepsilon_M(r)$, and again the interested reader should consult the review of Nijsing [36] for a lucid discussion of the options available. Although the method to solve Eq. 7-52 is direct, the algebraic complexity of the diffusivity and velocity distributions lead to an involved procedure.

Therefore we first examine how to transform the extensive results in the literature of bilaterally heated full annuli to our case of an equivalent annulus. The turbulent and laminar flow bilaterally heated annulus cases have been solved by Kays and Lueng [23] and by Lundberg et al. [27], respectively. Of course the turbulent results are not unique as are the laminar solutions but depend on the assumptions made for velocity profile and eddy diffusivity. These authors solved the two cases of (a) inner surface heated, outer insulated, and (b) inner surface insulated, outer heated. From these cases, it is possible because of the linearity of the energy equation, Eq. 7-30, to use superposition methods to solve any desired case. The equivalent annulus is in fact the intermediate case of the inner surface heated and a zero temperature gradient at the position of maximum axial velocity. Let us examine how to use the literature results for that case.

The Nusselt numbers, Nu_i and Nu_o, for any intermediate case are given by (utilizing the literature nomenclature):

$$\text{Nu}_i = \frac{\text{Nu}_{ii}}{1 - \frac{q_o''}{q_i''} \Theta_i^*} \tag{7-69}$$

$$\text{Nu}_0 = \frac{\text{Nu}_\infty}{1 - \frac{q_i''}{q_o''} \Theta_o^*} \tag{7-70}$$

$$\text{Nu}_i \equiv \frac{h_i D_e}{k} \tag{7-71}$$

$$\text{Nu}_o \equiv \frac{h_o D_e}{k} \tag{7-72}$$

where i and o refer to inner and outer surfaces

$$D_e \equiv D_e|_A = 2(r_o - r_i) \text{ for the full annulus} \tag{7-51b}$$

Nu_{ii} is the Nusselt number for the inner surface with the inner surface heated, outer surface insulated

Nu_∞ is the Nusselt number for the outer surface with the inner surface insulated, outer surface heated and

Θ_i^*, Θ_o^* are influence coefficients obtained from the fundamental solutions.

These results will be transformed into the nomenclature illustrated in Figure 7-3 which is based on the fact that the solid circle in this figure represents a fuel element having an outside radius of r_o. Hence, the literature symbols r_i and r_o are equivalent to our symbols r_0 and r_2, respectively.

$r_i(\text{literature}) = r_0$
$r_o(\text{literature}) = r_2$

In our case we desire the Nusselt number on surface r_0, i.e., Nu_{r_0} defined in terms of the equivalent annulus hydraulic diameter, $D_e|_{EA}$ given by Eq. 7-51a. From the above equations the desired Nusselt number can be expressed in our nomenclature as:

$$\text{Nu}_{r_0} = \text{Nu}_i \frac{D_e]_{EA}}{D_e]_A} = \text{Nu}_i \frac{2(r_{max}^2 - r_0^2)/r_0}{2(r_2 - r_0)} \tag{7-73}$$

$$= \left[\frac{\text{Nu}_{r_0 r_0}}{1 - \frac{q_{r_2}''}{q_{r_0}''} \Theta_i^*} \right] \left[\frac{\left(\frac{r_{max}}{r_0}\right)^2 - 1}{\frac{1}{r^*} - 1} \right] \tag{7-74}$$

where $r^* \equiv r_0/r_2$ for the bilaterally heated annulus.

Evaluation of Nu_{r_0} from Eq. 7-74 first requires determination of q_{r_2}''/q_{r_0}'' and r^*. In the existing literature Nusselt numbers for bilaterally heated annuli are expressed in terms of the bulk annulus temperature. Therefore the first condition for evaluating the above two parameters is that the bulk temperatures of the two annuli are equal, i.e. (if length, inlet temperature, and fluid are the same):

$$\frac{q_{r_0}'' 2\pi r_0}{\{v_z\}\pi(r_{max}^2 - r_0^2)}\Bigg]_{EA} = \frac{q_{r_0}'' 2\pi r_0 + q_{r_2}'' 2\pi r_2}{\pi(r_2^2 - r_0^2)\{v_z\}}\Bigg]_A \qquad (7\text{-}75)$$

which reduces to:

$$\frac{q_{r_2}''}{q_{r_0}''} = \frac{r_0}{r_2}\left[\left(\frac{\{v_z\}_A}{\{v_z\}_{EA}}\right)\frac{\left(\frac{r_2}{r_0}\right)^2 - 1}{\left(\frac{r_{max}}{r_0}\right)^2 - 1} - 1\right] \qquad (7\text{-}76)$$

The second condition is that the radius of minimum temperature (r_t) of the bilaterally heated annulus corresponds to the radius (r_{max}) of minimum temperature (also zero shear stress) of the equivalent annulus. Now since we have made the boundary conditions similar, r_t is also the radius of zero shear stress. We know r_{max} from the lattice dimensions. Thus if we know the relationship between r_{max} and r_2 we have solved the problem. Rehme [43] gives r_{max} in terms of r_2 for turbulent flow as:

$$s^* = r^{*\,0.386} \qquad (7\text{-}77)$$

where:

$$s^* = \frac{r_{max}^* - r^*}{1 - r_{max}^*}$$

$$r^* = \frac{r_0}{r_2}$$

$$r_{max}^* = \frac{r_{max}}{r_2}$$

One should note that in turbulent flow the positions of zero shear stress and maximum velocity are not coincident because of the diffusion of turbulent energy, as discussed by Rehme [44].

Example 7-2 Determination of the Nusselt number for a square array using existing bilaterally heated annuli literature

PROBLEM Find Nu for a square array of $P/D = 1.5$ in turbulent flow by the equivalent annulus method from existing literature for bilaterally heated annuli.

Given:

• square array
• $P/D = 1.5$
• $D = 1.12$ cm
• Re = 10^5; Pr = 1

SOLUTION

A. Given P/D find r_{max} by the same method as in Example 7-1 but for a square array:

$$r_{max}/r_0 = 1.68$$

B. Given knowledge that r_{max} is the point of maximum velocity, $dv_z/dr = 0$, find r_2/r_0 for the annulus geometry. Use Rehme's turbulent profile for r_{max}:

$$\left(\frac{r_0}{r_2}\right)^{0.386} = \frac{(r_{max}/r_2) - (r_0/r_2)}{1 - \frac{r_{max}}{r_2}} \qquad (7\text{-}77)$$

By trial and error:

$$\frac{r_2}{r_0} = 2.67$$

C. Find $\dfrac{q''_{r_2}}{q''_{r_0}}$. Assume that $\dfrac{\{v_z\}_A}{\{v_z\}_{EA}} = 1$. Then Eq. 7-76 gives:

$$\frac{q''_{r_2}}{q''_{r_0}} = 0.888$$

E. Now given $\dfrac{r_2}{r_0}, \dfrac{r_{max}}{r_0}, \dfrac{q''_{r_2}}{q''_{r_0}}$, Re and Pr, we find Θ_i^*, and $Nu_{r_0r_0}$ from Kays and Leung [23] as 0.225 and 227, respectively. Hence:

$$Nu_{r_0} = \left[\frac{Nu_{r_0r_0}}{1 - \frac{q''_{r_2}}{q''_{r_0}}\Theta_i^*}\right]\left[\frac{\left(\frac{r_{max}}{r_0}\right)^2 - 1}{\frac{r_2}{r_0} - 1}\right]$$

$$= \left[\frac{227}{1 - 0.888(0.225)}\right]\left[\frac{(1.68)^2 - 1}{(2.67) - 1}\right] \qquad (7\text{-}74)$$

$$= 309$$

From the well known Dittus–Boelter equation:

$$Nu_{c.t.} = 0.023\ Re^{0.8} \cdot Pr^{0.4}$$

obtain $Nu_{c.t.} = 230$.

Therefore:

$$\frac{Nu_{r_0}}{Nu_{c.t.}} = 1.34$$

The direct literature solutions of this ratio, defined as G, are tabulated below for comparison. A wide range of numerical values exists for this turbulent flow situation.

Investigators	G	
Deissler and Taylor [7]	1.1	equivalent annulus solutions for boundary condition case A
Presser $(f/f_{c.t.})$ [39]	1.06	equivalent annulus solutions
Presser [39]	1.14	square array solution
Weisman [53]	1.69	square array solution
Markoczy [30]	1.2	square array solution

where $f_{c.t.} \equiv f$ for circular tube.

3 Turbulent flow: Direct solution approach Although this approach was developed for application to liquid metals, the general method presented holds for all Prandtl number fluids. Numerical evaluations which are made here are limited to liquid metals. We desire to find the heat transfer coefficient h. It does not vary circumferentially by virtue of our equivalent annulus geometry.

By Newton's cooling law define h as:

$$h \equiv \frac{q'_{r_0}}{2\pi r_0 (T_{r_0} - T_b)} \tag{7-78}$$

The Nusselt number for the equivalent annulus is given by:

$$Nu]_{EA} \equiv \frac{hD_e}{k} \tag{7-79}$$

Now, q'_{r_0}, the linear heat generation rate, is readily expressible by an axial energy balance as:

$$q'_{r_0} = \pi(r_{max}^2 - r_0^2)\{v_z\}\rho c_p \frac{dT_b}{dz} \tag{7-80}$$

which, when substituted into Eq. 7-78, yields:

$$h = \frac{(r_{max}^2 - r_0^2)}{2r_0} \{v_z\}\rho c_p \left(\frac{\dfrac{dT_b}{dz}}{T_{r_0} - T_b}\right) \tag{7-81}$$

Next express the temperature difference between the tube wall and the bulk coolant temperature, $T_{r_0} - T_b$, in terms of known variables. This temperature difference is defined using Eq. 7-68 as:

$$T_{r_0} - T_b \equiv \frac{\displaystyle\int_{r_0}^{r_{max}} (T_{r_0} - T(r))v_z(r)2\pi r\,dr}{\pi(r_{max}^2 - r_0^2)\{v_z\}} \tag{7-82}$$

where:

$$T_{r_0} - T \equiv \int_{r_0}^{r} -\frac{\partial T}{\partial r}\,dr \tag{7-83}$$

Now establish $-\partial T/\partial r$ in terms of Fourier's equation, i.e.:

$$\frac{\partial T}{\partial r} = \frac{-q'}{k_{\text{eff}} 2\pi r} \tag{7-84}$$

where q' is the linear heat flow rate from an annulus with an outer radius r, and k_{eff} is the effective thermal conductivity which is equal to the sum of the molecular thermal conductivity and the eddy (k_e) thermal conductivity of the fluid for the general case of turbulent flow, i.e., $k_{\text{eff}} = k + \rho c_p \varepsilon_H$. This heat transfer rate must be equal to the rate of heat transport by the coolant flowing in the annulus with an inner radius r and outer radius r_{max}:

$$q' = \int_r^{r_{\text{max}}} 2\pi r \rho v_z(r) c_p \frac{\partial T}{\partial z} \, dr \tag{7-85}$$

Combining Eqs. 7-85, 7-84, and 7-83 with 7-82 obtain:

$$T_{r_0} - T_b = \frac{2}{\{v_z\}(r_{\text{max}}^2 - r_0^2)} \int_{r_0}^{r_{\text{max}}} \left[\int_{r_0}^{r} \frac{\left(\int_{r'}^{r_{\text{max}}} r'' v_z(r'') \rho c_p \frac{\partial T}{\partial z} \, dr'' \right)}{r' k_{\text{eff}}} \, dr' \right] v_z(r) r \, dr \tag{7-86}$$

Now for uniform radial and axial heat flux and axially independent heat transfer coefficient:

$$\frac{dT}{dz} = \frac{dT_b}{dz} \neq f(r) \tag{7-87}$$

as was shown in Section II, since the flow and heat transfer conditions have been taken as fully developed and the wall heat flux has been taken as uniform axially as well as radially. Hence in Eq. 7-86 the axial temperature gradient dT/dz is a constant in the integration and when Eq. 7-86 is substituted into Eq. 7-81, the axial temperature gradients dT/dz and dT_b/dz will cancel each other. Similarly, since we assume no changes in physical properties the product ρc_p cancels, yielding:

$$h = \frac{\{v_z\}^2 (r_{\text{max}}^2 - r_0^2)^2}{4 r_0} \left[\int_{r_0}^{r_{\text{max}}} \left(\int_{r_0}^{r} \frac{\int_{r'}^{r_{\text{max}}} r'' v_z(r'') \, dr''}{r' k_{\text{eff}}} \, dr' \right) v_z(r) r \, dr \right]^{-1} \tag{7-88}$$

Now we can express the average axial velocity $\{v_z\}$ more generally in terms of an integral of the local velocity by equating equivalent expressions for the rate of heat flow from the rod per unit length. These expressions are Eqs. 7-80 and 7-85 with $r = r_0$. Applying the equality of temperature gradients expressed by Eq. 7-87:

$$\{v_z\} \equiv \frac{2 \int_{r_0}^{r_{\text{max}}} r v_z(r) \, dr}{r_{\text{max}}^2 - r_0^2} \tag{7-89}$$

We will use this relation together with the following definitions of the equivalent diameter of the annular flow area D_e and dimensionless distance X to obtain our general equation for the Nusselt number:

$$D_e \equiv \frac{4\pi(r_{max}^2 - r_0^2)}{2\pi r_0} = 2r_0(X_{max}^2 - 1) \tag{7-90}$$

where:

$$X_{max} \equiv \frac{r_{max}}{r_0}; \quad X \equiv \frac{r}{r_0} \tag{7-91a}$$
$$\tag{7-91b}$$

Substituting Eqs. 7-89, 7-90, and 7-91 into Eq. 7-88, we obtain the desired general equation for the Nusselt number:

$$\text{Nu]}_{EA} \equiv \frac{hD_e}{k} = \frac{2(X_{max}^2 - 1)\left[\int_1^{X_{max}} v_z(X)XdX\right]^2}{\int_1^{X_{max}}\left[\int_1^x \frac{\int_{X'}^{X_{max}} v_z(X'')X''dX''}{X'\frac{k_{eff}}{k}} dX'\right] v_z(X)XdX} \tag{7-92}$$

This general equation for the Nusselt number for the equivalent annulus model can now be applied to the cases of turbulent flow and laminar flow.

For slug flow, the velocity is taken as uniform with radius while the heat transfer is by conduction only, i.e., $k_{eff} \equiv k$. For these conditions, the integration of Eq. 7-92 can be performed directly, yielding:

$$\text{Nu}_s]_{EA} = \frac{8(X_{max}^2 - 1)^3}{4X_{max}^4 \, \ell n \, X_{max} - (3X_{max}^2 - 1)(X_{max}^2 - 1)} \tag{7-93}$$

For turbulent flow integration of Eq. 7-92 requires specification of the radial variation of both the velocity and the effective conductivity k_{eff}. The velocity correlations of Rothfus et al. [46] (which rely on coincidence of the positions of zero shear stress and maximum velocity) have been used for this purpose by Dwyer and Tu [16]. The problem of expressing k_{eff} explicitly has been circumvented by invoking the analogy between eddy diffusivity of heat and momentum and expressing this analogy in parametric form in terms of the parameter which is the ratio of these respective diffusivities. In our nomenclature, we have:

$$\varepsilon_H \equiv \varepsilon_M \Psi \tag{7-94}$$

$$\frac{k_e}{\rho c_p} \equiv \frac{\mu_e}{\rho} \Psi \tag{7-95}$$

where:

$\varepsilon_H \equiv$ eddy diffusivity of heat, m^2/s (ft^2/s)

$\varepsilon_M \equiv$ eddy diffusivity of momentum, m^2/s (ft^2/s)

and

$\mu_e \equiv$ effective eddy viscosity, $kg/m \cdot s$ ($lbm/ft \cdot s$)

The term μ_e is made available at every radial position in the annulus from the defining equation:

$$\tau(r) \equiv (\mu + \mu_e) \frac{dv_z}{dr} \tag{7-96}$$

where:

$$\tau(r) = \frac{\Delta p}{L} \frac{r_{max}^2 - r^2}{2r} = \frac{f\rho\{v_z\}^2}{2D_e} \left[\frac{r_{max}^2 - r^2}{2r} \right] \tag{7-97}$$

The specification of ε_H becomes complete if we prescribe with Eqs. 7-96 and 7-97 a friction law for flow through our selected array of parallel rods by which the shear stress τ can be evaluated. The velocity correlation cited earlier is also used in Eq. 7-96 to permit evaluation of the local velocity gradient, dv_z/dr. This procedure is applicable for all Prandtl numbers since this effect is included in the specification of the ratio k_{eff}/k. The practical difficulty is the specification of Ψ which is actually a function of Re, Pr, and position.

Results for Pr < 0.01: Liquid metals in turbulent flow Following the above procedure, Dwyer and Tu [16] first solved the equivalent annulus turbulent flow problem for liquid metals graphically, and later Maresca and Dwyer [29] obtained a numerical solution. Both solutions were based on the same fundamental assumptions, and only minor differences were found between the calculations. The latter investigators correlated their results very well by the semiempirical equation:

$$\mathrm{Nu_t]_{EA}} = \alpha + \beta(\bar{\Psi}\mathrm{Pe})^\gamma \tag{7-98}$$

where:

$$\alpha = 6.66 + 3.126(P/D) + 1.184(P/D)^2 \tag{7-99}$$

$$\beta = 0.0155 \tag{7-100}$$

$$\gamma = 0.86 \tag{7-101}$$

for the ranges $1.3 < P/D < 3.0$ and $10^2 < \mathrm{Pe} < 10^4$. Note that in Eq. 7-98 the spatially averaged value $\bar{\Psi}$ was utilized by the original investigators for reasons of convenience. To evaluate the turbulent Nusselt number by Eq. 7-98, an auxiliary relation for $\bar{\Psi}$ is required, which for liquid metals, Dwyer [8] recommends as:

$$\bar{\Psi} = 1 - \frac{1.82}{\mathrm{Pr}(\varepsilon_M/\nu)_{max}^{1.4}} \tag{7-102}$$

This equation should be used with care at small Pe since it yields negative values of Ψ. Values for $(\varepsilon_M/\nu)_{max}$ for rod bundles are to be obtained from Figure 2 of Dwyer [8].

The dependence of Eq. 7-98 on the term $\bar{\Psi}\mathrm{Pe}$ can be traced directly from the term k_{eff}/k in our general equation, Eq. 7-92. The eddy dependent portion of this term, k_e/k, can be expressed by utilizing the definitions of $k_e \equiv \rho c_p \varepsilon_H$, Eq. 7-94 and the Prandtl number; we write:

$$\frac{k_e}{k} \equiv \frac{\rho c_p \varepsilon_M \Psi}{k} \cdot \frac{\mu}{\mu} \equiv \frac{\varepsilon_M Pr \Psi}{\nu} \tag{7-103a}$$

Now since ε_M/ν is practically proportional to Re, and $RePr \equiv Pe$, we obtain the parameter $\bar{\Psi}Pe$ which led the investigators to correlate results by a relationship of the form:

$$Nu = f(P/D, \bar{\Psi}Pe) \tag{7-103b}$$

of which Eq. 7-98 is one form.

The logic of the development leading to Eq. 7-103b is in fact applicable to fluids of all Prandtl number. For a fluid of $Pr \approx 1$, the Nusselt number is expressed as:

$$Nu_t]_{EA} = \beta'(\bar{\Psi}Pe)^{\gamma'} \tag{7-103c}$$

Now since $\bar{\Psi}$ is a function of Pr, this dependence can be included by reformulating Eq. 7-103c as:

$$Nu_t]_{EA} = \beta'(Re^{\gamma'}Pr^{\gamma''})$$

which is the expected Dittus–Boelter form of the forced convection equation for fluid of $Pr \approx 1$. In this case the constant β' is a function of geometry.

IV COMPARISON OF LAMINAR, SLUG, AND TURBULENT FLOW NUSSELT NUMBERS FOR LIQUID METALS IN EQUIVALENT ANNULI

Table 7-1 summarizes these results. Since the effective diffusivity is a variable only for turbulent conditions, only for such conditions does the Nusselt number vary with Peclet number for a fixed P/D. The slug flow condition is one of uniform velocity with molecular conduction only. Therefore, we would expect that slug flow Nusselt numbers will be higher than laminar flow Nusselt numbers for the same geometry because the laminar velocity gradient at the wall is less than that for slug flow. Comparing slug to turbulent flow conditions, the wall velocity gradients and effective thermal conductivities yield opposite trends. For most flow conditions, as Table 7-1 illustrates, the eddy contribution to the thermal conductivity is significant enough to overcome the velocity gradient difference and make turbulent Nusselt numbers greater than comparable slug results. The values of Table 7-1 also demonstrate that for suitably low values of flow rate, i.e., $\Psi Pe = 100$, slug Nusselt numbers can exceed turbulent Nusselt numbers at fixed P/D ratios. It must be emphasized that agreement of slug flow predictions with experimental data or with results of more realistic analyses will occur only for special conditions that cannot be specified *a priori*.

Table 7-1 Nusselt numbers calculated using the equivalent annulus model for longitudinal flow of coolant through infinite unbaffled rod bundles under conditions of uniform heat flux and fully developed velocity and temperature fields

P/D	$Nu_L]_{EA}$	$Nu_s]_{EA}$	$Nu_t]_{EA}$		
1.01	8.58	—			
1.02	8.64	—			
1.03	8.70	—			
1.04	8.76	—			
1.05	8.82	12.66			
1.06	8.88	—			
1.07	8.94	12.71			
1.10	9.12	12.96			
1.20	9.75	13.63			
1.30	10.40	14.34			
1.40	11.07	15.06	$\bar{\Psi}Pe$	$Nu_t]_{EA}$	
1.50	11.75	15.81	10^2	14.8	
1.51	—	—	10^3	19.9	
1.60	12.46	16.57	10^4	56.6	
1.80	13.92	18.15			
1.90	14.68	—			
2.00	15.46	19.81			
2.10	16.21	20.66	10^2	19.5	
2.13	—	—	10^3	24.6	
2.38	18.56	23.16	10^4	61.3	
2.86	22.77	27.70	10^2	27.5	
3.01	—	—	10^3	32.6	
3.81	32.22	37.70	10^4	69.3	

Sources (for reference use only):
$Nu_L]_{EA}$: Table 2 in [11]; Table 2.17 in [9].
$Nu_s]_{EA}$: Table 1 in [13]; Table 2.17 in [9].
$Nu_t]_{EA}$: Table 1 in [10].

It is instructive also to examine such results in graphical form as illustrated in Figure 7-4. Note that Nusselt numbers are plotted against the Peclet number of a fixed P/D of 1.5. The figure legend refers to these as results for rod bundles since for such large P/D ratios, the equivalent annulus represents rod bundle geometry. Curves A-A and B-B represent the turbulent flow solution, Eq. 7-98, with $\bar{\Psi}$ taken as 1 for curve A-A and evaluated by Eq. 7-102 for curve B-B. As the Peclet number decreases, the corresponding Nusselt number decreases reflecting the decreasing contribution of eddy conduction to the heat transfer process. At sufficiently low Peclet numbers, the solution of Eq. 7-92 with a turbulent velocity profile but with eddy effects neglected, i.e., $k_{eff}/k = 1$ would

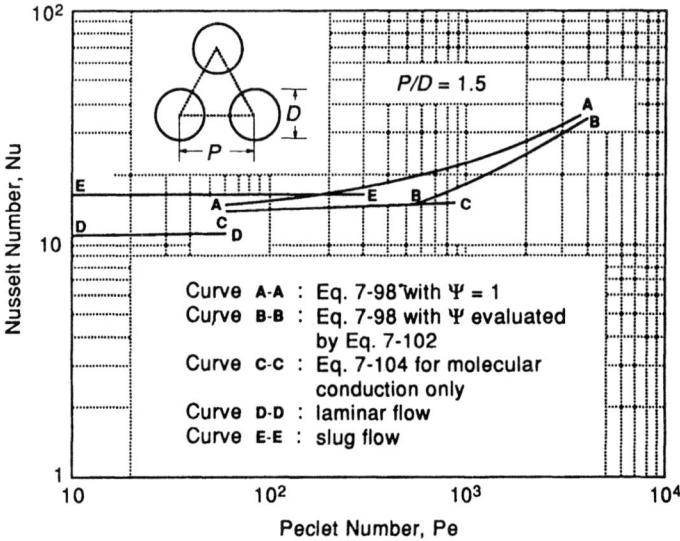

Figure 7-4 Analytical predictions for longitudinal flow of liquid metals through unbaffled rod bundles under conditions of uniform heat flux and fully developed velocity and temperature profiles. *(After Dwyer [10].)*

represent the limit for the turbulent solution. This solution was obtained by Maresca and Dwyer [29] who correlated their results by the empirical equation:

$$Nu_{t-mc} = -2.79 + 3.97(P/D) + 1.025(P/D)^2 + 3.12 \log_{10} Re - 0.265(\log_{10} Re)^2 \qquad (7-104)$$

Eq. 7-104 appears on Figure 7-4 as curve C-C. The laminar flow solution curve D-D differs from curve C-C in that the velocity profile to be introduced in the general equation, Eq. 7-92, must be the laminar annulus flow profile. Finally the slug solution $Nu_s]_{EA} = 15.81$ from Table 7-1 illustrated on the figure as curve E-E is slightly larger than turbulent values at low Peclet numbers. It should be noted that the actual Pe for transition from laminar to turbulent conditions varies with the fluid through its Prandtl number and has only been drawn illustratively in Figure 7-4. In summary then, the predicted Nu versus Pe curve is a composite curve with a dogleg shape of the type C-B-B with the C-B portion representing the molecular conduction regime but with a turbulent velocity profile and the B-B portion representing the molecular plus eddy conduction regime. The exact magnitudes of the Nusselt numbers along these curves are dependent on the relationships used to evaluate velocity profiles and the Ψ function. Relationships other than the illustrative ones presented here can also be employed.

V LAMINAR FLOW BETWEEN LONGITUDINAL HEATED CYLINDERS (FUEL RODS)

Let us now return to the problem of the velocity and temperature fields within a symmetrical portion of an infinite array of rods presented in Section I, but abandon the simplified equivalent annulus approach and consider circumferential heat transfer which can occur. The elemental coolant flow area of interest was introduced in Figure 7-1.

In the next several sections we will address the problem of fully developed laminar flow with boundary conditions applied to those surfaces bounding only the coolant region. This problem is aptly defined as a single-region problem. The more complex but more realistic multiregion problems include the various additional portions of the fuel rod with the coolant region and involve the simultaneous solution of the energy equations in these regions.

The momentum and energy conservation equations can be solved sequentially by virtue of the assumption that physical properties can be taken independent of temperature. However, the momentum equation solution must first be solved to provide the required velocity field for the energy equation. The challenge of this problem comes with the task of fitting the general solutions of the conservation equations to the particular boundary conditions in view of our unusual geometric flow configuration.

VI MOMENTUM TRANSFER IN LAMINAR FLOW BETWEEN FUEL RODS

The applicable form of the conservation of momentum equation is Eq. 7-27. This partial differential equation is frequently called Poisson's equation.

We will follow the solution of Sparrow and Loeffler [49] by expressing Eq. 7-27 in terms of a reduced velocity, v_z^*. Here $\hat{p}$ is utilized for generality versus the use of p in the Sparrow and Loeffler [49] definition of v_z^*.

$$v_z^* \equiv v_z - \frac{r^2}{4}\left(\frac{1}{\mu}\frac{d\hat{p}}{dz}\right) \tag{7-105}$$

Under this transformation Eq. 7-27 becomes:

$$\frac{\partial^2 v_z^*}{\partial r^2} + \frac{1}{r}\frac{\partial v_z^*}{\partial r} + \frac{1}{r^2}\frac{\partial^2 v_z^*}{\partial \theta^2} = 0 \tag{7-106}$$

which is the well-known Laplace equation. The general solution of Eq. 7-106 is:

$$v_z^* = A + B\,\ell n\,r + \sum_{k=1}^{\infty}(C_k r^k + D_k r^{-k})\cdot(E_k\cos k\theta + F_k\sin k\theta) \tag{7-107}$$

where k takes on integral values to ensure that the velocity is single valued, i.e., the velocity computed at a location (r,θ) is identical to that computed at $(r, \theta +$

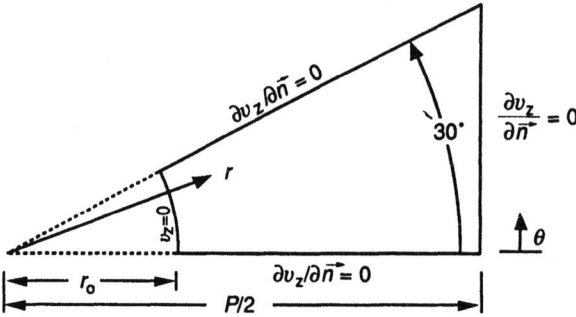

Figure 7-5 Momentum boundary conditions on characteristic coolant element of an infinite triangular array.

2π). The constants A, B through F_k in Eq. 7-107 will be determined from the boundary conditions, as are the number of terms k of the series.

Combining Eqs. 7-105 and 7-107 yields the general velocity solution:

$$v_z^* = A + B \ln r - \frac{r^2}{4}\left(-\frac{1}{\mu}\frac{d\hat{p}}{dz}\right) + \sum_{k=1}^{\infty}(C_k r^k + D_k r^{-k})$$
$$\cdot (E_k \cos k\theta + F_k \sin k\theta) \tag{7-108}$$

This solution is next specialized to flow parallel to cylindrical rods in regular arrays.

The characteristic coolant domain for a triangular array is shown in Figure 7-5 with its applicable momentum boundary conditions. From symmetry $\partial v_z/\partial \vec{n} = 0$ at $\theta = 0$, $\theta = 30$ degrees and along the line $r = (P/2)/\cos\theta$. The no-slip requirement of viscous flow at the inner boundary is $v_z = 0$.

Now apply these boundary conditions to determine the constants of Eq. 7-108. From the symmetry condition at $\theta = 0$:

$$F_k = 0 \tag{7-109}$$

while from this condition at $\theta = 30$ degrees:

$$k = 6, 12, 18 \ldots \tag{7-110}$$

to guarantee that $\sin k\pi/6 = 0$. Next imposing the condition that $v_z = 0$ at $r = r_0$ yields:

$$D_k = -C_k r_0^{2k}$$
$$A = -B \ln r_0 + \frac{r_0^2}{4}\left(-\frac{1}{\mu}\frac{d\hat{p}}{dz}\right) \tag{7-111}$$

Further, the total friction force exerted on the fluid by the solid rod must be balanced by the net pressure force acting over the entire cross-section of the typical element yielding:

$$\int_0^{\pi/6} \mu\left(\frac{\partial v_z}{\partial r}\right)_{r=r_0} r_0 d\theta = \int_0^{\pi/6}\int_{r_0}^{P/(2\cos\theta)}\left(\frac{d\hat{p}}{dz}\right) r\,dr\,d\theta \tag{7-112}$$

Evaluating this overall force balance from Eq. 7-108 yields:

$$B = \frac{\sqrt{3}}{\pi} \left(\frac{P}{2}\right)^2 \left(-\frac{1}{\mu}\frac{d\hat{p}}{dz}\right) \tag{7-113}$$

Utilizing the preceding results, Eq. 7-108 becomes:

$$v_z = \frac{\sqrt{3}}{\pi} \left(\frac{P}{2}\right)^2 \left(-\frac{1}{\mu}\frac{d\hat{p}}{dz}\right) \ell n \frac{r}{r_0} - \frac{1}{4} \left(-\frac{1}{\mu}\frac{d\hat{p}}{dz}\right)(r^2 - r_0^2)$$

$$+ \sum_{j=1}^{\infty} G_j \left(r^{6j} - \frac{r_0^{12j}}{r^{6j}}\right) \cos 6j\theta \tag{7-114}$$

where $G_j = C_j E_j$.

G_j must still be obtained by the condition that $\partial v_z / \partial \vec{n}$ is zero on the boundary where $r = (P/2)/\cos\theta$. It is convenient to make use of the identity:

$$\frac{\partial v_z}{\partial \vec{n}} = \frac{\partial v_z}{\partial r}\cos\theta - \frac{\partial v_z}{\partial \theta}\frac{\sin\theta}{r} \tag{7-115}$$

Substituting Eq. 7-114 into 7-115 and setting $\partial v_z / \partial \vec{n} = 0$ when $r = P/2\cos\theta$, obtain after rearrangement:

$$\sum_{j=1}^{\infty} \Delta_j(\cos\theta)^{1-6j} \left[\cos(6j - 1)\theta + \left(\frac{D\cos\theta}{P}\right)^{12j}\cos(6j + 1)\theta\right]$$

$$+ \frac{\sqrt{3}}{\pi}\cos^2\theta - \frac{1}{2} = 0 \tag{7-116}$$

where:

$$\Delta_j = G_j \frac{6j(P/2)^{6j}}{\left(-\frac{1}{\mu}\frac{d\hat{p}}{dz}\right)\left(\frac{P}{2}\right)^2} \tag{7-117}$$

The Δ_j (that is G_j) can be determined from Eq. 7-116. However, the nature of Eq. 7-116 precludes the use of Fourier analysis to provide an infinite set of Δ_j because the cosine terms in Eq. 7-116 cannot be multiplied by appropriate sine or cosine functions to yield orthogonal functions.

The method used by Sparrow and Loeffler [49] was to apply Eq. 7-116 at a finite number of points along the boundary. This approach of selecting j values of θ and truncating the series to Δ_j terms is known as a point-matching technique. It represents the simplest approach to match equations and unknowns. If the number of values of θ used is greater than the number of coefficients from the series truncation, the system of equations is overdetermined and a boundary least-squares method is used. Another more rigorous approach is to determine Δ_j by having the equation boundary condition agree in an integral sense; this is called the integral boundary least-squares method. The numerical values of Δ_j, computed as outlined above for up to five points between $\theta = 0$

Table 7-2 Listing of Δ_j and δ_j values for various spacing ratios

	Values of Δ_j					
P/D	Δ_1	Δ_2	Δ_3	Δ_4	Δ_5	Δ_6
4.0	−0.0505	−0.0008	0.0000	———————————————→		
2.0	−0.0505	−0.0008	0.0000	———————————————→		
1.5	−0.0502	−0.0007	0.0000	———————————————→		
1.2	−0.0469	0.0007	0.0002	0.0000	——————————→	
1.1	−0.0416	0.0028	0.0004	0.0000	——————————→	
1.05	−0.0368	0.0043	0.0003	−0.0001	0.0000	—————→
1.04	−0.0357	0.0046	0.0002	−0.0001	0.0000	—————→
1.03	−0.0345	0.0049	0.0002	−0.0001	0.0000	—————→
1.02	−0.0332	0.0051	0.0000	−0.0001	0.0000	—————→
1.01	−0.0319	0.0052	−0.0001	−0.0002	0.0000	—————→
1.00	−0.0305	0.0053	−0.0003	−0.0002	0.0000	—————→

	Values of δ_j					
P/D	δ_1	δ_2	δ_3	δ_4	δ_5	δ_6
4.0	−0.1253	−0.0106	−0.0006	0.0000	——————————→	
2.0	−0.1250	−0.0105	−0.0006	0.0000	——————————→	
1.5	−0.1225	−0.0091	−0.0002	0.0000	——————————→	
1.2	−0.1104	−0.0024	−0.0015	0.0003	0.0001	0.0000 —→
1.1	−0.0987	0.0036	0.0029	0.0005	0.0000	—————→
1.05	−0.0904	0.0073	0.0032	0.0002	−0.0001	0.0000 —→

(From Sparrow and Loeffler [49].)

degrees and 30 degrees are listed in Table 7-2. The P/D ratio is a parameter of Table 7-2 since it appears in Eq. 7-116. Since P/D is raised to a large negative power in Eq. 7-116, the Δ_j values are little affected by increases of spacing for large P/D. The tabulation is given to four decimal places because this is sufficient for the shear stress and velocity computations that follow, although additional figures were used by the original authors to satisfy Eq. 7-116 to the desired accuracy. With the determination of G_j through Δ_j, the velocity distribution for the infinite triangular array is now available by Eq. 7-114.

The key results in the application of this same procedure to flow between rods in a square array is outlined next. The characteristic domain for the square array has an opening angle equal to 45 degrees rather than the 60 degrees of the triangular array. Therefore with only this exception, the boundary conditions illustrated in Figure 7-5 for the triangular array apply to the square array. The velocity solution as given by Eq. 7-107 applies, but the constants A, B, through F_k which are appropriate to the square array must be found.

Proceeding as before from the conditions that $\partial v_z/\partial \vec{n} \equiv \partial v_z/\partial \vec{\theta} = 0$ at $\theta = 0$ degrees and $\theta = 45$ degrees, find:

$$F_k \equiv 0$$

and

$$k = 4, 8, 12 \ldots$$

$$D_k = -C_k r_0^{2k}$$

$$A = -B \, \ell n \, r_0 + \frac{r_0^2}{4} \left(-\frac{1}{\mu} \frac{d\hat{p}}{dz} \right) \tag{7-118}$$

The overall force balance between net pressure and wall shear, Eq. 7-112, with $\pi/6$ replaced by $\pi/4$, yields:

$$B = \frac{2}{\pi} \left(\frac{P}{2} \right)^2 \left(-\frac{1}{\mu} \frac{d\hat{p}}{dz} \right) \tag{7-119}$$

The velocity is obtained by substituting these results into Eq. 7-108 yielding:

$$v_z = \frac{2}{\pi} \left(\frac{P}{2} \right)^2 \left(-\frac{1}{\mu} \frac{d\hat{p}}{dz} \right) \ell n \frac{r}{r_0} - \frac{1}{4} \left(-\frac{1}{\mu} \frac{d\hat{p}}{dz} \right) (r^2 - r_0^2)$$

$$+ \sum_{j=1}^{\infty} G_j \left(r^{4j} - \frac{r_0^{8j}}{r^{4j}} \right) \cos 4j\theta \tag{7-120}$$

where the $G_j \, (= C_j E_j)$ still remain to be determined. Again applying the condition that $\partial v_z / \partial \vec{n} = 0$ on the right boundary of Figure 7-5 for which $r = (P/2)/\cos\theta$ yields:

$$\sum_{j=1}^{\infty} \delta_j (\cos\theta)^{1-4j} \left[\cos(4j - 1)\theta + \left(\frac{D\cos\theta}{P} \right)^{8j} \cos(4j + 1)\theta \right]$$

$$+ \frac{2\cos^2\theta}{\pi} - \frac{1}{2} = 0 \quad (7\text{-}121)$$

where:

$$\delta_i = G_j \frac{4j(P/2)^{4j}}{\left(-\frac{1}{\mu} \frac{d\hat{p}}{dz} \right) \left(\frac{P}{2} \right)^2}$$

Numerical values of the δ_i obtained by the point-matching technique are also listed in Table 7-2 as a function of P/D.

Results from the Momentum Solution

Dimensionless velocity contours are presented in Figure 7-6 for the triangular array at spacing ratios of $P/D = 1.1$ to represent close packings and $P/D = 2.0$ to represent open packings. The contours for the large spacing are essentially circular for a sizable region near the rod surface indicating little effect there of the neighboring rods. On the other hand for the small spacing example, the influence of the neighboring rods extends to regions very near the rod surface.

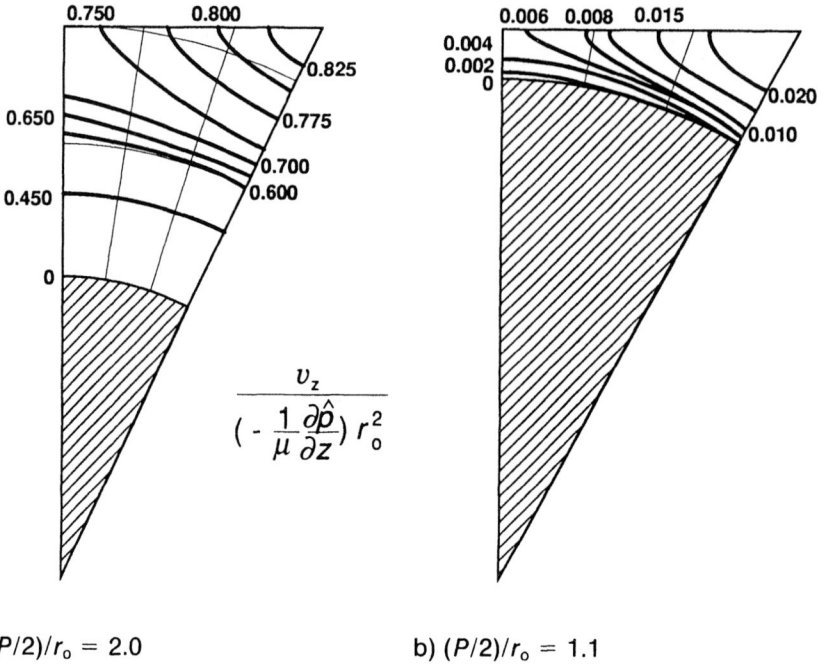

a) $(P/2)/r_0 = 2.0$ b) $(P/2)/r_0 = 1.1$

Figure 7-6 Representative velocity contour lines for an equilateral triangular array. *(After Sparrow and Loeffler [49].)*

The local wall shear stress, exerted by the fluid on the wall τ_w, is given by:

$$\tau_w(\theta) = \mu \left. \frac{dv_z}{dr} \right|_{r=r_0} \tag{7-122}$$

which varies with angular position because the velocity gradient at the wall varies with angular position. For the triangular array the wall shear stress can be evaluated utilizing Eq. 7-114 as:

$$\tau_w(\theta) = \left(-\frac{d\hat{p}}{dz} \right) r_0 \left\{ \frac{\sqrt{3}}{\pi} \left(\frac{P}{2r_0} \right)^2 - \frac{1}{2} + \sum_{j=1}^{\infty} 2\Delta_j \left(\frac{P}{2r_0} \right)^{2-6j} \cos 6j\theta \right\} \tag{7-123}$$

Since we will plot the ratio of the local wall shear stress to the average value, $\tau_w(\theta)/\bar{\tau}_w$, we next evaluate the average wall shear stress. This is most simply done for our case of fully developed flow by performing a force balance on a unit length of our elemental coolant area A_f. The force balance yields:

$$\bar{\tau}_w r_0 \theta_0 dz = -d\hat{p} A_f \tag{7-124}$$

and since A_f is equal to:

$$A_f = \left(\frac{P}{2} \right)^2 \frac{\tan \theta_0}{2} - \frac{r_0^2 \theta_0}{2} \tag{7-125}$$

we obtain the following general expression for the average wall shear stress:

$$\bar{\tau}_w = \left(- \frac{d\hat{p}}{dz} \right) \left[\left(\frac{1}{2} \right) \frac{P^2 \tan\theta_0}{4r_0\theta_0} - \frac{r_0}{2} \right] \tag{7-126}$$

Now for the triangular array:

$$\theta_0 = \frac{\pi}{6} \text{ and } \tan\theta_0 = \frac{\sqrt{3}}{3}$$

For these conditions, the average wall shear stress becomes:

$$\bar{\tau}_w = \left(- \frac{d\hat{p}}{dz} \right) r_0 \left[\frac{\sqrt{3}}{\pi} \left(\frac{P}{2r_0} \right)^2 - \frac{1}{2} \right] \text{ (for triangular array)} \tag{7-127}$$

and finally, the desired ratio of local to average wall shear stress is:

$$\frac{\tau_w(\theta)}{\bar{\tau}_w} = 1 + \frac{\sum\limits_{j=1}^{\infty} 2\Delta_j \left(\frac{P}{2r_0} \right)^{2-6j} \cos 6j\theta}{\frac{\sqrt{3}}{\pi} \left(\frac{P}{2r_0} \right)^2 - \frac{1}{2}} \text{ (for triangular array)} \tag{7-128}$$

Similarly, utilizing the velocity distribution for the square array, the comparable ratio of wall stresses for that array is:

$$\frac{\tau_w(\theta)}{\bar{\tau}_w} = 1 + \frac{\sum\limits_{j=1}^{\infty} 2\delta_j \left(\frac{P}{2r_0} \right)^{2-4j} \cos 4j\theta}{\frac{2}{\pi} \left(\frac{P}{2r_0} \right)^2 - \frac{1}{2}} \text{ (for square array)} \tag{7-129}$$

The manner in which the shear stress ratio varies is shown in Figure 7-7 for the triangular array. These curves illustrate the role of neighboring rods on the flow

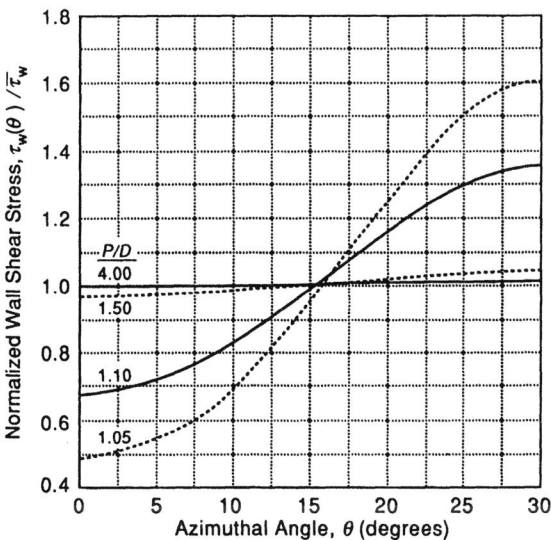

Figure 7-7 Local wall shear stress distribution for an equilateral triangular array. *(From Sparrow and Loeffler [49].)*

pattern around a given rod. For large spacings, for example $P/D \geq 1.5$, the local shear stress is essentially a constant around the periphery of the rod. The angular dependence of the local wall shear stress increases as the spacing decreases. This is a consequence of the increasing asymmetry of the flow due to interference of neighbors.

The highest shear stress is associated with the location of highest velocities ($\theta = 30$ degrees), and the smallest shear stress is at the location of lowest velocity ($\theta = 0$ degrees).

Finally, we will develop the friction factor–Reynolds number relationships for these arrays. First write the definition of the friction factor as:

$$f \equiv \frac{4\bar{\tau}_w}{\dfrac{\rho\{v_z\}^2}{2}} \tag{7-130}$$

Now utilizing our general expression for $\bar{\tau}_w$, Eq. 7-124, the friction factor can be expressed as:

$$f = \frac{8}{\rho\{v_z\}^2}\left(-\frac{d\hat{p}}{dz}\right)\frac{A_f}{r_0\theta_0} \tag{7-131}$$

where for simplicity we have retained the designation A_f rather than its expanded form. Now let us introduce a Reynolds number based on the rod diameter D as:

$$\mathrm{Re}_D \equiv \frac{\rho\{v_z\}D}{\mu}$$

and recognize that the volume flow rate Q is given by:

$$Q = \{v_z\}A_f$$

Utilizing these two relations and Eq. 7-131 the desired product is:

$$f\mathrm{Re}_D = \left[\frac{(D/2)^4}{Q\mu}\left(-\frac{d\hat{p}}{dz}\right)\right]\left[\frac{16A_f^2}{\theta_0(D/2)^4}\right] \tag{7-132}$$

From reference to Eq. 7-125, we see that for a given array type the second term of Eq. 7-132 is only a function of P/D. If we can show also that the first term is only a function of P/D for a specified array type, the product $f\mathrm{Re}_D$ is then only a function of P/D and thus related to the fractional cross-section available for flow.

Let us examine the first term by calculating Q through integration of the velocity over the flow area. This integral is expressed as:

$$Q = \int\int_{A_f} v_z r\,dr\,d\theta = \int_0^{\theta_0}\int_{r_0}^{P/(2\cos\theta)} v_z r\,dr\,d\theta \tag{7-133}$$

Now introducing the dimensionless variable:

$$\zeta \equiv \frac{2r}{D} \tag{7-134}$$

and taking the example case of the square array, we apply the velocity profile of Eq. 7-120 utilizing the defining Eq. 7-121 for the constants G_j in terms of the tabulated (Table 7-2) constants δ_j. Under these conditions, the volumetric rate of flow Q from Eq. 7-133 becomes:

$$Q = \left(-\frac{d\hat{p}}{dz}\right) \frac{r_0^4}{\mu} \left\{ \int_0^{\pi/4} \int_1^{(P/D)/\cos\theta} \left[\frac{2}{\pi}\left(\frac{P}{D}\right)^2 \ell n\zeta - \frac{1}{4}(\zeta^2 - 1) \right. \right.$$

$$\left. \left. + \sum_{j=1}^{\infty} \frac{\delta_j (D/P)^{4j-2}}{4j} \cdot (\zeta^{4j} - \zeta^{-4j})\cos 4j\theta \right] \zeta d\zeta d\theta \right\}$$

(7-135)

where $\theta_0 = \pi/4$ has been introduced for the square array. The integral of Eq. 7-135 is only a function of P/D since ζ is a dummy variable.

The integral appearing in Eq. 7-135 can be directly evaluated giving:

$$\frac{Q\mu}{(-d\hat{p}/dz)(D/2)^4} = \left(\frac{P}{D}\right)^4 \left[\frac{1}{2\pi}(2\,\ell n(P/D) + \ell n 2 - 3) + \frac{1}{6}\right]$$

$$+ \left(\frac{P}{D}\right)^4 \sum_{j=1}^{\infty} \frac{\delta_j}{4j} \left[\frac{\Lambda_j}{4j+2} + \left(\frac{D}{P}\right)^{8j} \frac{B_j}{4j-2}\right] + \frac{1}{4}\left(\frac{P}{D}\right)^2 - \frac{\pi}{64}$$

(7-136)

where:

$$\Lambda_j = \int_0^{\pi/4} \frac{\cos 4j\theta}{(\cos\theta)^{4j+2}} d\theta$$

(7-137)

$$B_j = \int_0^{\pi/4} \frac{\cos 4j\theta}{(\cos\theta)^{2-4j}} d\theta$$

(7-138)

The definite integrals represented by Λ_j and β_j were computed numerically by the Runge–Kutta method by Sparrow and Loeffler [49].

Now note that the left side of Eq. 7-136 is the inverse of the first term of the right side product in Eq. 7-132. As Eq. 7-136 demonstrates, this term is only a function of the ratio P/D. This term is known as the pressure drop/flow relationship. From the foregoing analysis the product $f\mathrm{Re_D}$ can be plotted solely as a function of P/D for a defined array. This has been done in Figure 7-8. In this figure the porosity ε, the fraction of the total cross-sectional area available to flow, is a coordinate along with the P/D ratio.

Utilizing the velocity profile of Eq. 7-114:

$$G_j = \Delta_j \frac{\left(-\frac{d\hat{p}}{dz}\right)\left(\frac{P}{2}\right)^2}{6j\mu\left(\frac{P}{2}\right)^{6j}}$$

(7-139)

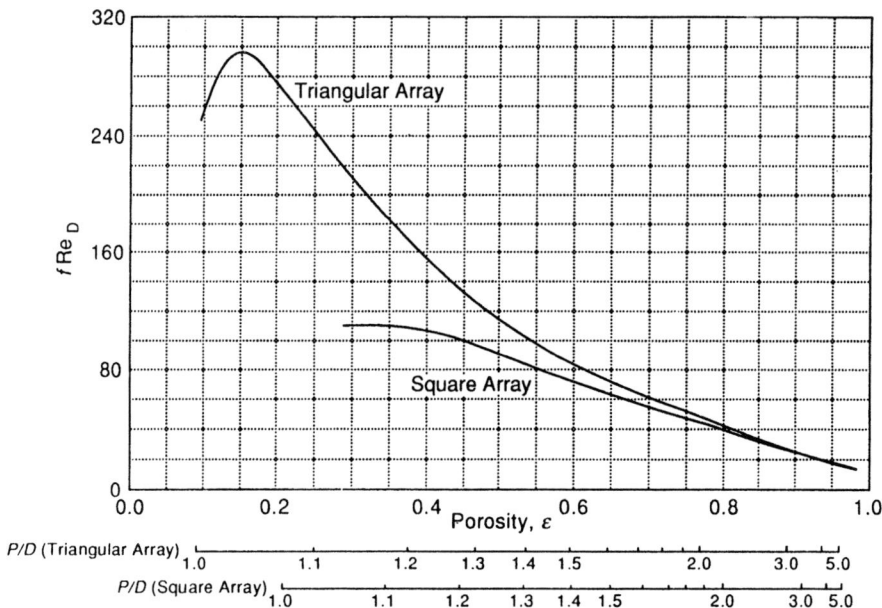

Figure 7-8 Friction factor–Reynolds number as a function of rod spacing. *(From Sparrow and Loeffler [49].)*

the final result for the flow rate for the triangular array is:

$$\frac{Q\mu}{(-d\hat{p}/dz)(D/2)^4} = \left(\frac{P}{D}\right)^4 \left[\frac{1}{2\pi}\left(\ell n\frac{P}{D} - \ell n\cos 30 \text{ degrees} - \frac{3}{2}\right) + \frac{13\sqrt{3}}{216}\right]$$

$$+ \left(\frac{P}{D}\right)^4 \sum_{j=1}^{\infty} \frac{\Delta_j}{6j}\left[\frac{\Omega_j}{6j+2} + \left(\frac{D}{P}\right)^{12j}\frac{\chi_j}{6j-2}\right] + \frac{\sqrt{3}}{12}\left(\frac{P}{D}\right)^2 - \frac{\pi}{96}$$

$$(7\text{-}140)$$

where:

$$\Omega_j = \int_0^{\pi/6} \frac{\cos 6j\theta}{(\cos\theta)^{6j+2}}\, d\theta \tag{7-141}$$

$$\chi_j = \int_0^{\pi/6} \frac{\cos 6j\theta}{(\cos\theta)^{2-6j}}\, d\theta \tag{7-142}$$

On Figure 7-8 the curve representing the triangular array lies above that for the square array. For tight packings, the curves for the two arrays are not similar in shape. This is because at small spacings the flow passages of the two arrays are very different. In the limiting case of rods touching ($P/D = 1.0$), the passages of the square array are curvilinear squares, whereas those of the triangular array are curvilinear triangles giving different flow profiles. On the other hand, at large spacings, the flow passages of the two arrays are almost geometrically similar so the curves merge as we expect.

VII HEAT TRANSFER IN LAMINAR FLOW BETWEEN FUEL RODS

The appropriate conservation of energy equation is Eq. 7-28 where for fully developed flow and heat transfer and uniform axial heat flux boundary condition, the term $\partial T/\partial z$ is a constant, i.e.:

$$\frac{\partial T}{\partial z} = \frac{4\bar{q}''_{\text{w}}}{\{v_z\}\rho c_p D_{\text{e}}} = \frac{q'}{\dot{m} c_p} = \text{constant} \tag{7-143}$$

In the previous section solutions were obtained for the velocity distribution. Taking the triangular array case, rewrite the velocity solution of Eqs. 7-114 and 7-117 in the form:

$$\frac{v_z}{\left(\frac{P}{2}\right)^2\left(-\frac{1}{\mu}\frac{d\hat{p}}{dz}\right)} = \frac{\sqrt{3}}{\pi}\,\ell n\,\frac{r}{r_0} - \frac{1}{4}\left[\left(\frac{2r}{P}\right)^2 - \left(\frac{2r_0}{P}\right)^2\right]$$

$$+ \sum_{n=1}^{\infty}\frac{\Delta_n}{6n}\left(\frac{2r}{P}\right)^{6n}\left[1 - \left(\frac{r_0}{r}\right)^{12n}\right]\cos 6n\theta \tag{7-144}$$

where the index n has been introduced in place of j to avoid later algebraic confusion with this equation in obtaining the temperature field solution.

When the relations 7-143 and 7-144 are introduced into the energy equation, Eq. 7-28, the resulting governing equation for the temperature field is:

$$\frac{\partial^2 T}{\partial r^2} + \frac{1}{r}\frac{\partial T}{\partial r} + \frac{1}{r^2}\frac{\partial^2 T}{\partial \theta^2} = \frac{q'/12k}{M(P/2)^2}\left\{\frac{\sqrt{3}}{\pi}\,\ell n\,\frac{r}{r_0} - \frac{1}{4}\left[\left(\frac{2r}{P}\right)^2 - \left(\frac{2r_0}{P}\right)^2\right]\right.$$

$$\left. + \sum_{n=1}^{\infty}\frac{\Delta_n}{6n}\left(\frac{2r}{P}\right)^{6n}\left[1 - \left(\frac{r_0}{r}\right)^{12n}\right]\cos 6n\theta\right\} \tag{7-145}$$

where for convenience the dimensionless grouping M has been introduced:

$$M \equiv \frac{(\dot{m}/12\rho)}{\left(\frac{P}{2}\right)^4\left(-\frac{1}{\mu}\frac{d\hat{p}}{dz}\right)} \tag{7-146}$$

Numerical values of M can be obtained as a function of spacing ratio from the pressure drop/flow parameter $(-(1/\mu)(d\hat{p}/dz))r_0^4/Q$ plotted in Sparrow and Loeffler [49] by noting the following relation between this parameter and M, i.e.:

$$\left(-\frac{1}{\mu}\frac{d\hat{p}}{dz}\right)\frac{r_0^4}{Q} = \frac{1}{M}\frac{1}{12}\left(\frac{r_0}{P/2}\right)^4 \tag{7-147}$$

since:

$$\left(-\frac{1}{\mu}\frac{d\hat{p}}{dz}\right)\frac{r_0^4}{Q} \times \frac{\rho}{\rho}\frac{12}{12}\frac{(P/2)^4}{(P/2)^4} = \frac{\left(\frac{P}{2}\right)^4\left(-\frac{1}{\mu}\frac{d\hat{p}}{dz}\right)}{\dot{m}/12\rho} \times \frac{1}{12}\left(\frac{r_0}{P/2}\right)^4$$

The general solution of Eq. 7-145 is expressed as the sum of particular and homogeneous solutions:

$$T = T_p + T_h$$

For a particular solution, any function satisfying Eq. 7-145 will suffice. It can be verified by direct substitution that the following expression is a satisfactory particular solution:

$$T_p = \frac{(q'/12k)r^2}{M(P/2)^2} \left\{ -\frac{\sqrt{3}}{4\pi} + \frac{1}{16}\left(\frac{D}{P}\right)^2 - \frac{1}{64}\left(\frac{2r}{P}\right)^2 + \frac{\sqrt{3}}{4\pi}\ell n \frac{r}{r_0} \right.$$

$$\left. + \sum_{n=1}^{\infty} \frac{\Delta_n}{24n} \left(\frac{2r}{P}\right)^{6n} \left[\frac{1}{6n+1} + \frac{(r_0/r)^{12n}}{6n-1} \right] \cos 6n\theta \right\} \quad (7\text{-}148)$$

The homogeneous equation is obtained by setting the right side of Eq. 7-145 to zero yielding:

$$\frac{\partial^2 T_h}{\partial r^2} + \frac{1}{r}\frac{\partial T_h}{\partial r} + \frac{1}{r^2}\frac{\partial^2 T_h}{\partial \theta^2} = 0 \quad (7\text{-}149)$$

Like Eq. 7-106 this is a Laplace equation that has a general solution of the form of Eq. 7-108. To avoid confusion we will write the series portion of this equation for T_h in terms of the index m as:

$$T_h = A + B\,\ell n\,r + \sum_{m=1}^{\infty} (C_m r^m + D_m r^{-m})(E_m \cos m\theta + F_m \sin m\theta) \quad (7\text{-}150)$$

where m takes on integral values to ensure that T is single valued, i.e., $T(\theta) = T(\theta + 2\pi)$. As before, the constants A, B, through F_m remain to be determined from the boundary conditions.

The characteristic coolant domain with its applicable thermal boundary conditions is shown in Figure 7-9. The rod surface condition shown is one of circumferentially uniform surface temperature. An alternate limiting possibility is that of circumferentially uniform heat flux. Both reactor fuel rods and electrically heated rods generally used in low temperature experimental studies have low thermal conductivity ceramic cores encased in metallic claddings. For

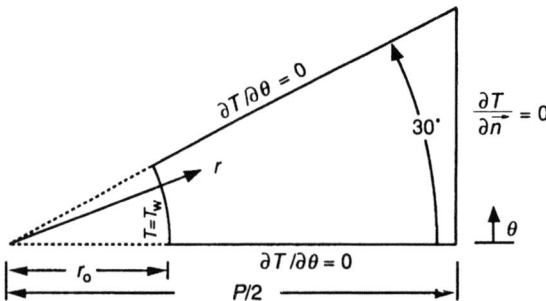

Figure 7-9 Thermal boundary conditions on the characteristic coolant element of an infinite triangular array for the circumferentially uniform rod surface temperature assumption.

these materials, a nearly uniform wall heat flux is achieved on the inner wall of the cladding. Under these circumstances the boundary condition of circumferentially uniform surface temperature is approached for thick and high-conductivity claddings, whereas the condition of circumferentially uniform heat flux is approached for thin and low-conductivity claddings. Numerical criteria for these conditions are given in Section IX.

We proceed here with the solution for the uniform surface temperature condition because it is more instructive to follow an analytical versus numerical solution procedure, and only for this condition has a nearly completely analytical solution been performed [50]. Following closely their procedure, we start the application of the boundary conditions by first imposing the symmetry requirements that $\partial T/\partial \vec{n} = 0$ at $\theta = 0$ degrees and at $\theta = 30$ degrees. The condition at $\theta = 0$ degrees yields:

$$F_m \equiv 0$$

and the condition at $\theta = 30$ degrees yields:

$$m = 6, 12, 18, \ldots$$

to ensure that $\sin (m\pi/6) = 0$. The temperature solution, Eq. 7-150, then becomes:

$$T = T_p + A + B \ln r + \sum_{j=1}^{\infty} (C_j r^{6j} + D_j r^{-6j})\cos 6j\theta \qquad (7\text{-}151)$$

where E_m has been set equal to unity without loss of generality and $j = m/6$.

Next at the surface of the rod, $r = D/2$, the temperature equals T_w independent of angle. Applying this condition to Eq. 7-151 yields:

$$A = T_w - B \ln r_0 - \frac{(q'/12k)r_0^2}{M(P/2)^2} \left\{ -\frac{\sqrt{3}}{4\pi} + \frac{3}{64} \left(\frac{D}{P}\right)^2 \right\} \qquad (7\text{-}152)$$

and

$$D_j = -C_j r_0^{12j} - \frac{(q'/12k)}{M(P/2)^2} \frac{\Delta_n}{2(36n^2 - 1)} \frac{r_0^{(12n+2)}}{(P/2)^{6n}}, \qquad n = j \qquad (7\text{-}153)$$

Note that in Eq. 7-153 and thereafter, the condition $n = j$ applies so that the two subscripts in Eq. 7-153 could be combined. However, following Sparrow et al. [50] for convenience we will retain both.

Further, since a point-matching procedure will be applied for the last boundary condition, an energy conservation requirement is imposed for the typical element of Figure 7-9. This ensures that there is no net energy transport across the other boundaries of the typical element. Equating the heat transfer per unit length $q'/12$ to the heat conducted at the surface of the fuel rod, obtain:

$$q'/12 = \int_0^{\pi/6} -k \left(\frac{\partial T}{\partial r}\right)\bigg|_{r_0} r_0 d\theta \qquad (7\text{-}154)$$

Inserting Eq. 7-151 into the energy conservation condition 7-154 gives:

$$B = \frac{q'/12k}{M(P^2/4)}\left[-\frac{6M}{\pi}\left(\frac{P^2}{4}\right) + \frac{\sqrt{3}}{4\pi}\left(\frac{D^2}{4}\right) - \frac{1}{16}\left(\frac{D^4}{4P^2}\right)\right] \tag{7-155}$$

Introducing Eqs. 7-152, 7-153, and 7-154 into the temperature solution 7-151 and substituting for T_p from Eq. 7-148 gives:

$$\begin{aligned}
T - T_w &= \frac{q'/12k}{M(P^2/4)}\left\{\left[\frac{\sqrt{3}}{4\pi}\left(r^2 + \frac{D^2}{4}\right) - \frac{1}{16}\left(\frac{D^4}{4P^2}\right) - \frac{6M}{\pi}\left(\frac{P^2}{4}\right)\right]\ell n\,\frac{r}{(D/2)}\right.\\
&\quad -\frac{1}{64}\frac{(r^4 - (D^4/16))}{(P^2/4)} + \left[-\frac{\sqrt{3}}{4\pi} + \frac{1}{16}\left(\frac{D}{P}\right)^2\right]\left(r^2 - \left(\frac{D^2}{4}\right)\right)\right\}\\
&\quad + \sum_{j=1}^{\infty} C_j r^{6j}\left[1 - \left(\frac{D}{2r}\right)^{12j}\right]\cos 6j\theta\\
&\quad + \frac{(q'/12k)r^2}{M(P^2/4)}\sum_{n=1}^{\infty}\Delta_n\cos 6n\theta\left\{\frac{(2r/P)^{6n}}{24n}\left[\frac{1}{6n+1} + \frac{(D/2r)^{12n}}{6n-1}\right]\right.\\
&\quad \left.-\frac{(D/2r)^2}{2(36n^2-1)}\frac{(D/2)^{12n}}{r^{6n}(P/2)^{6n}}\right\}
\end{aligned} \tag{7-156}$$

Since Δ_n and M are known numerical constants whose values depend upon P/D, the temperature distribution will be completely determined when the C_j are found. The condition available to determine C_j is that $\partial T/\partial \vec{n} = 0$ on the right boundary of the typical element where $r = (P/2)/\cos\theta$. The following identity is useful in applying this condition:

$$\frac{\partial T}{\partial \vec{n}} = \frac{\partial T}{\partial r}\cos\theta - \frac{\partial T}{\partial\theta}\frac{\sin\theta}{r} \tag{7-157}$$

then utilizing the temperature distribution of 7-156 and setting $\partial T/\partial \vec{n} = 0$ for $r = (P/2)/\cos\theta$, after considerable rearrangement one obtains:

$$\begin{aligned}
\sum_{j=1}^{\infty}&\omega_j(\cos\theta)^{1-6j}\left[\cos(6j-1)\theta + \left(\frac{D\cos\theta}{P}\right)^{12j}\cos(6j+1)\theta\right]\\
&= -\sum_{n=1}^{\infty}\Delta_n\left\{\frac{3n}{36n^2-1}\left(\frac{D}{P}\right)^{12n+2}(\cos\theta)^{6n+1}\cos(6n+1)\theta\right.\\
&\quad + \frac{\cos(6n-1)\theta + (1/3n)\cos\theta\cos 6n\theta}{4(6n+1)(\cos\theta)^{6n+1}}\\
&\quad \left.+ \frac{(D/P)^{12n}(\cos\theta)^{6n-1}}{4(1-6n)}[\cos(6n+1)\theta - (1/3n)\cos\theta\cos 6n\theta]\right\}\\
&\quad -\left[-\frac{6M}{\pi} + \frac{\sqrt{3}}{4\pi}\left(\frac{D}{P}\right)^2 - \frac{1}{16}\left(\frac{D}{P}\right)^4\right]\cos^2\theta\\
&\quad -\frac{1}{8}\left(\frac{D}{P}\right)^2 - \frac{\sqrt{3}}{2\pi}\ell n\left(\frac{P}{D\cos\theta}\right) + \frac{1}{16\cos^2\theta} + \frac{\sqrt{3}}{4\pi}
\end{aligned} \tag{7-158}$$

where:

$$\omega_j = C_j \frac{M6j \left(\frac{P}{2}\right)^{6j}}{(Q/12k)} \tag{7-159}$$

The only unknowns appearing in Eq. 7-158 are the coefficients ω_j (i.e., C_j). Sparrow et al. [50] applied the point-matching technique in which j values of θ were selected and the series was truncated to ω_j terms. This yields j linear equations, each of which contains j unknown values of ω_j. As soon as the P/D ratio is specified, the values of Δ_n and M become available, and the right side of each equation can be reduced to a numerical constant. The numerical results that have been obtained for the ω_j are listed in Table 7-3 as a function of the spacing ratio P/D.

The coefficients C_j can be determined by Eq. 7-159 knowing the ω_j values. The summation involving C_j in the temperature distribution Eq. 7-156 can be expressed in terms of the ω_j values as:

$$\sum_{j=1}^{\infty} C_j r^{6j} \left[1 - \left(\frac{r_0}{r}\right)^{12j} \right] \cos 6j\theta$$

$$= \frac{Q/12k}{M} \sum_{j=1}^{\infty} \frac{\omega_j}{6j} \left(\frac{2r}{P}\right)^{6j} \left[1 - \left(\frac{D}{2r}\right)^{12j} \right] \cos 6j\theta \tag{7-160}$$

The temperature distribution within the coolant region of an array with specified geometry, i.e., r_0 and $P/2$, can be determined from Eq. 7-156 since C_j, Δ_n, and M are now known. Presentation of heat transfer results follows.

Results from the Temperature Distribution Solution

Let us first compute the Nusselt number for the circumferentially uniform rod surface temperature condition just solved. Defining the Nusselt number for the infinite rod bundle in terms of the equivalent diameter, we have:

$$\overline{Nu}_L \equiv \frac{\bar{h}D_e}{k} = \left(\frac{12}{\pi}\right)\left(\frac{D_e}{D}\right)\frac{q'/12k}{(T_w - T_b)} \tag{7-161}$$

Table 7-3 Listing of ω_i values for various spacing ratios

P/D	$\omega_1 \times 10^2$	$\omega_2 \times 10^3$	$\omega_3 \times 10^5$	$\omega_4 \times 10^7$
4.0	2.9649	0.39239	0.579	~0.3
2.0	1.2712	0.12031	0.136	~0.05
1.5	0.71081	0.014556	−0.195	~−1.5
1.2	0.37039	−0.1323	−1.20	~−6.0
1.1	0.25961	−0.2065	−1.12	~7.0

(From Sparrow et al. [50].)

where:

$$\bar{h} \equiv \frac{q'}{\pi D(T_w - T_b)} \tag{7-162}$$

and

$$D_e \equiv D \left[\frac{6}{\sqrt{3}\pi} \left(\frac{P}{D}\right)^2 - 1 \right] \tag{7-163}$$

Regarding notation, we have used the subscript EA for equivalent annulus and unsubscripted Nusselt numbers are taken to be applicable to rod arrays. The bar superscript refers to parameters averaged over the entire rod surface of the characteristic coolant region. The standard definition of the bulk temperature is used to compute the wall-to-bulk temperature difference:

$$T_b \equiv \frac{\displaystyle\iint_{\text{area}} T\rho v_z r dr d\theta}{\displaystyle\iint_{\text{area}} \rho v_z r dr d\theta} \tag{7-164}$$

where the denominator is the mass flow rate. Specializing to the elemental coolant area and subtracting the wall temperature, we obtain:

$$T_w - T_b = \frac{\rho \displaystyle\int_0^{30 \text{ degrees}} \int_{r_0}^{P/(2\cos\theta)} (T_w - T)v_z r dr d\theta}{\dot{m}/12} \tag{7-165}$$

The integrand is evaluated using the velocity distribution from Eq. 7-114 and the temperature distribution from Eq. 7-156 with C_j given by Eq. 7-159. Evaluation of Eq. 7-165 leads to the following relation:

$$\frac{T_w - T_b}{q'/12k} = f(P/D) \tag{7-166}$$

When this is applied to Eq. 7-161 and D_e is evaluated by Eq. 7-163, we find that the resulting laminar Nusselt number is solely a function of P/D for a fixed type array. Numerical values of $[\overline{Nu_L}]$ for this case are presented in Table 7-4 as boundary condition case A.

At this point, observe that since only the quantity $T_w - T$ from Eq. 7-156 entered the Nusselt number computation, we have proceeded so far without the need to establish a wall temperature value numerically. For our case of prescribed radially averaged uniform axial heat flux $\bar{q}''$, the following relationships between T_w, $\bar{q}''$ and $T_w - T_b$ exist. First the bulk coolant temperature is known from an axial energy balance as:

$$T_b(z) = T_{\text{inlet}} + \left(\frac{\bar{q}''\pi D}{\dot{m}c_p}\right) z \tag{7-167}$$

Table 7-4 Heat transfer characteristics for in-line, laminar flow through unbaffled rod bundles and for the thermal boundary conditions of: (A) uniform wall heat flux axially and uniform rod surface temperature circumferentially and (B) uniform wall heat flux in all directions

| | $[\overline{Nu_L}]_{r,b} \equiv \overline{Nu_L}$ | | |
| | Boundary condition | Boundary condition | |
P/D	A	B	$Nu_L]_{EA}$
1.001	1.26	0.149	8.52
1.01	1.52	0.263	8.58
1.02	1.82	0.404	8.64
1.03	2.14	0.580	8.70
1.04	2.48	0.795	8.76
1.05	2.82	1.06	8.82
1.06	3.18	1.36	8.88
1.07	3.54	1.70	8.94
1.10	4.62	2.94	9.12
1.20	7.48	6.90	9.75
1.30	9.19	9.03	10.40
1.40	10.34	10.28	11.07
1.50	11.26	11.22	11.75
1.60	12.08	12.05	12.46
1.75	13.28	13.26	13.55
1.80	13.68	13.66	13.92
1.90	14.47	14.46	14.68
2.00	15.27	15.26	15.46

(From Dwyer and Berry [11].)

Now, using a defining relationship for $T_w(z)$ and Eq. 7-167 we obtain the desired result linking $T_w(z)$ to other known parameters, i.e.:

$$T_w(z) \equiv T_b(z) + (T_w - T_b) = T_{inlet} + \frac{\bar{q}''\pi D}{\dot{m}c_p} z + (T_w - T_b) \qquad (7\text{-}168)$$

where it should be observed that although $T_w(z)$ increases with z (for a heated fuel rod), $T_w - T_b$ is a constant with z as demonstrated by Eq. 7-166 for our case of fully developed laminar flow and heat transfer.

For the boundary condition case of circumferentially uniform heat flux on the rod surface, a fully numerical solution has been performed by Dwyer and Berry [11] for the triangular array using a velocity profile of the form of Eq. 7-114. Values of the rod-averaged Nusselt numbers for this case are presented in

Table 7-4 as boundary condition case B. Also the Nusselt numbers calculated from the equivalent annulus model and already presented in Table 7-1 are tabulated in Table 7-4 for comparison. From Table 7-4 it is seen that the results for boundary condition cases A and B are very different for small spacings, $P/D < 1.2$, but above this value they rapidly approach each other and the limiting equivalent annulus model results. At P/D ratios of 1.20, 1.50, and 2.00, the simple equivalent annulus model gives Nusselt numbers that are too high versus case B by 41.3, 4.7, and 1.3 percent, respectively. For the slug flow case [13] these comparable differences are less because in that situation, only a geometrical difference exists between the annulus and the true flow area assignable to each rod, whereas in the case of laminar velocity profile there is a difference between velocity distributions in these flow areas as well.

The fact that the results for cases A and B approach those of the equivalent annulus model at large spacing follows physically from the observation that in the annulus model all the heat is transferred radially. In this situation boundary conditions A and B coexist. Another way to demonstrate this merging of boundary conditions at large spacings is to examine directly the circumferential variation of rod surface heat flux and temperature for boundary conditions A and B, respectively. To compute the local wall heat flux variation for case A, we use Fourier's law as:

$$q'' = -k \left(\frac{\partial T}{\partial r} \right)_{r_0} \tag{7-169}$$

The temperature distribution is introduced from Eq. 7-156 with C_j coefficients given by Eq. 7-159, and after some rearrangement, it is found that:

$$\frac{q''}{q''}\bigg|_{r_0} = 1 - \frac{\pi}{6M} \left\{ \sum_{j=1}^{\infty} 2\omega_j \left(\frac{D}{P} \right)^{6j} \cos 6j\theta + \sum_{n=1}^{\infty} \frac{\Delta_n \left(\frac{P}{D} \right)^{6n+2}}{2(6n-1)} \cos 6n\theta \right\} \tag{7-170}$$

Utilizing tabulated values of ω_j from Table 7-3, Δ_n from Table 7-2, and M from Sparrow and Loeffler [49], the circumferential variation of q'' can be evaluated. The results are tabulated in Dwyer and Berry [11] and plotted in Figure 7-10 as a function of θ for varying spacing ratios P/D.

The local wall temperature variation for case B has been computed directly [11]. The local wall temperature is of interest from the standpoints of mass transfer corrosion, thermal stress, and the possibility of nucleate boiling incep-

$$\frac{(T_{w,\theta} - T_b)k}{q'' D_e} \tag{7-171}$$

From Eq. 7-161 it can be seen that this ratio is actually the reciprocal of the local Nusselt number. It is seen that for $P/D = 1.07$, the wall temperature at $\theta = 30$ degrees is very slightly less than the bulk temperature of the coolant. As the P/D ratio is decreased below 1.07, a small circumferential region of negative values of $(T_{w,\theta} - T_b)$ occurs. Of course under all conditions heat flows from

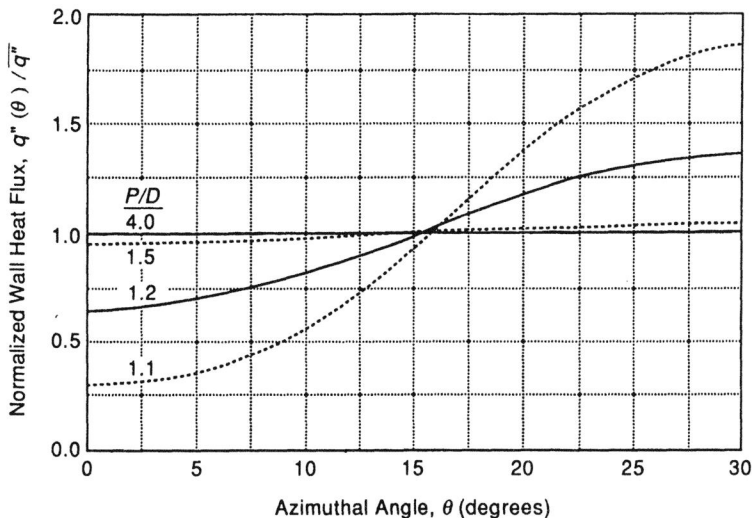

Figure 7-10 Circumferential wall heat flux variation for laminar flow through equilateral triangular array rod bundles under conditions of uniform wall heat flux axially and uniform wall surface temperature circumferentially (case A). *(From Sparrow et al. [50].)*

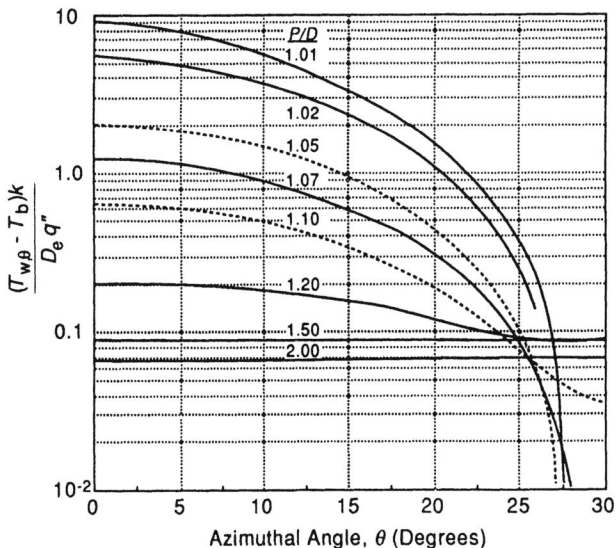

Figure 7-11 Circumferential wall temperature variation for laminar flow through equilateral triangular rod bundles under conditions of uniform wall heat flux in all directions (case B.) *(From Dwyer and Berry [11].)*

the rod to the coolant so that the local radial temperature gradient at the rod surface is always positive. Because the use of T_b introduces negative values in the normal definition of the local heat transfer coefficient, we have not chosen to present any results from the literature in these terms. Now from both Figures 7-10 and 7-11 it is obvious that circumferential variations in the plotted variables become negligible for P/D ratios of 1.5 and higher. This behavior is entirely consistent with the tendency of the rod-averaged Nusselt numbers in Table 7-4 to merge at high P/D ratios and demonstrates that for wide spacings boundary conditions A and B coexist.

The single-region laminar flow problem has now been completed. The physical conditions to which boundary condition cases A and B apply have been qualitatively discussed. More precise numerical criteria require the more complex analysis of simultaneously considering the clad or the fuel and clad with the elemental coolant region. These analyses are called multiregion analyses and represent two- and three-region problems. These analyses will be addressed in Section IX.

VIII TURBULENT LONGITUDINAL FLOW IN ROD BUNDLES

The relevant differential equations that have been derived in Section II are Eqs. 7-25 and 7-26. It should be recalled that these equations are based on (a) the eddy diffusivity approach for modeling the Reynolds shear stresses and turbulent heat fluxes, and (b) the neglect of secondary flows.

The eddy diffusivity approach is only one, albeit the most practical, method for handling the turbulent fluxes. The others are the phenomenological turbulence model approach and the statistical turbulence model approach. The former is centered on the mixing length theories proposed by Prandtl [38] and Taylor [52]. Buleev's [4, 5] extension of these theories falls into this category. The inherent assumptions of this approach are not in many respects consistent with observation [33], and additionally this method cannot provide needed information with respect to turbulent transfer tangential to the channel wall. The latter approach concentrates on a few of the statistical properties of turbulence which are supposed to obey laws of generation, dissipation, and transport. The proposed models consist of a set of differential equations and associated algebraic ones, the solution of which simulate the real behavior of turbulent fluids in important respects. A short review of the different approaches arranged in order of the number of equations may be found in the review by Launder and Spaulding [26]. However, prediction methods are sophisticated and still require a trial-and-error adjustment of the empirical input. Furthermore, they have not yet been used extensively enough to study flow and heat transfer in nonuniform channels to permit their general recommendation. An example of a theoretical and experimental study of the interior subchannel of a triangular array by this method using a two equation model is the work of Bartzis and Todreas [2].

A Eddy Diffusivities and Secondary Flow

Using the eddy diffusivity approach one is still faced with providing ε_{Mr}, $\varepsilon_{M\theta}$, ε_{Hr}, and $\varepsilon_{H\theta}$ distributions. Contrasted to our previous example of the circumferentially symmetrical equivalent annulus, the subchannel is nonsymmetric. Further, experimental evidence indicates that tangential diffusivities can be several times the radial diffusivities, thereby giving rise to so-called anisotropic effects. These effects are important since they are effective mechanisms for lateral transport of momentum and energy. The necessity to provide thermal diffusivities encountered in the equivalent annulus problem also exists here but now is even more complicated since both ε_{Hr} and $\varepsilon_{H\theta}$ are required.

In reality all four diffusivity terms above are spatially varying. Since reliable information on such variations is only available for the radial component of momentum transfer ε_{Mr} in uniform channels, one can appreciate that providing the necessary input for solution of Eqs. 7-25 and 7-26 is a major difficulty. Generally the path taken has been to postulate a distribution of ε_{Mr} from circular tube results based on the observation that the radial distribution of axial velocity for a nonuniform channel also obeys a universal velocity profile if based on the local wall shear. Then a global (infrequently a local) value for the ratio of $\bar{\varepsilon}_{M\theta}/\bar{\varepsilon}_{Mr}$ is postulated from experiment. The resulting global value of circumferential eddy diffusivity $\bar{\varepsilon}_{M\theta}$ is either used directly or used as the base to postulate a local distribution of circumferential eddy diffusivity. Finally either local or global eddy diffusivities of heat are related to those of momentum by factors that stem from semiempirical theories or heat transfer experiments.

The direct solution of the turbulent flow equations including secondary flow for triangular array geometries is a difficult task. An alternate approach useful for assessing the importance of secondary flows is to postulate their distribution function and estimate their amplitude based on allied experimental results. This result is then fed into the analysis for axial velocity distribution and heat transfer behavior. An example of this approach can be found in Nijsing and Eifler [35].

In view of the large variety of options available for specifying the eddy diffusivities and secondary flows and the complexity of the algebraic operations involved in the solution of the relevant differential equations, we will not present a detailed solution of the turbulent problem as we did for laminar flow. Rather we will present results from the Nijsing and Eifler [35] study which shows the importance of the different elements in the analysis. Proceeding from the equivalent annulus solution these new elements are:

1. Channel geometry, isotropic turbulence:

$$\varepsilon_{Mr} = \varepsilon_{M\theta}$$

$$\varepsilon_{Hr} = \varepsilon_{H\theta}$$

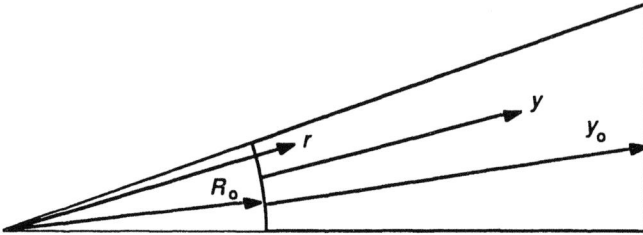

Figure 7-12 Geometric parameter definitions for turbulent flow in the coolant region.

2. Anisotropic turbulence:

$$\varepsilon_{M\theta} \neq \varepsilon_{Mr}$$

$$\varepsilon_{H\theta} \neq \varepsilon_{Hr}$$

3. Secondary flow.

These results are for a triangular array of $P/D = 1.05$ for a fluid of Pr = 0.0045 with uniform heat flux (case B) boundary condition. The geometric parameters in which the results are expressed, R_0 and y_0, are defined in Figure 7-12. The momentum solution is given in Figure 7-13 in terms of the circumferential distribution of the dimensionless velocity v_θ^* which denotes the ratio for the

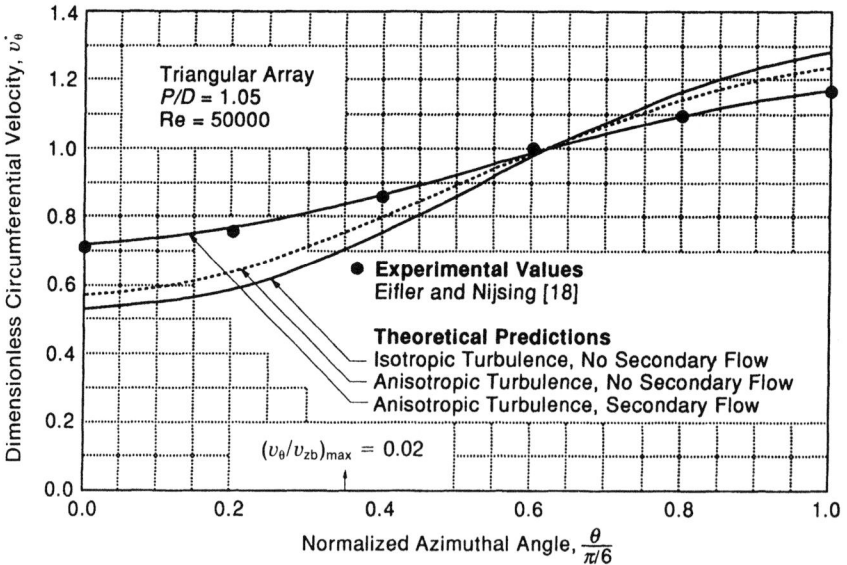

Figure 7-13 Circumferential distribution of dimensionless velocity v_θ^*. *(From Nijsing and Eifler [35].)*

radially averaged (from R_0 to $r_0 + y_0$) coolant velocity at position θ to the bulk coolant velocity v_{zb}.

In evaluating the velocity profile the following eddy diffusivity properties were used:

1. ε_{Mr} given by the following relationship developed by Eifler as reported in [35]

$$\varepsilon_{Mr} = \frac{K}{2(1 + b)} y \left[b + \left(1 - \frac{y}{y_0}\right)^2 \right] \left[2 - \frac{y}{y_0}\right] \frac{\tau_r}{\tau_w} \left(\frac{y_0}{y_0 - y}\right) \left(\frac{\tau_w}{\rho}\right)^{1/2} \quad (7\text{-}172)$$

where K is a coefficient depending weakly on the ratio y_0/r_0 and b is a parameter depending both on Re and on y_0/r_0, given by:

$$K = 0.407 + 0.02 y_0/r_0 \quad (7\text{-}173)$$

$$b = \left[1.82 - 1.3 \frac{y_0}{r_0}\left(1 + 0.7 \frac{(\log_{10} \text{Re}) - 5}{5}\right)\right]^{-1} \quad (7\text{-}174)$$

2. $\bar{\varepsilon}_{M\theta} = 0.185 y_0 (\tau_w/\rho)^{1/2}$ is a radially averaged diffusivity which is two to three times greater than $\bar{\varepsilon}_{Mr}$.

Examining Figure 7-13 we first see that variations of v_θ^* with θ are induced by the subchannel geometry and would not exist for a circularly symmetrical channel like an annulus. Next the curves for zero secondary flow show that the influence of circumferential turbulent diffusion is relatively small since the curves for anisotropic turbulence and isotropic turbulence differ only slightly from each other. Finally the effect of secondary flow is pronounced and tends to smooth the variations of v_θ^* with angle.

The energy solution is given in terms of the dimensionless circumferential clad temperature distribution in Figure 7-14. The dimensionless temperature parameter is:

$$\frac{k_{coolant}[T_{R_0}(\theta) - \bar{T}_{R_0}]}{q'} \quad (7\text{-}175)$$

In evaluating this parameter, the following additional eddy diffusivity properties were used:

1. $\dfrac{\varepsilon_{Hr}}{\varepsilon_{Mr}} = 1.45 \left(\dfrac{v_z}{v_{z\,max}}\right)^{0.5} \left[1 - \exp(-0.62(10^{-4})) \left(\dfrac{\tau_w}{\bar{\tau}_w}\right)^{1.5} \text{RePr}^{1/3}\right] \quad (7\text{-}176)$

2. $\dfrac{\varepsilon_{H\theta}}{\varepsilon_{M\theta}} = 1.45 \left(\dfrac{v_z}{v_{z\,max}}\right)^{0.5} \left[1 - \exp(-1.24(10^{-4})) \left(\dfrac{\tau_w}{\bar{\tau}_w}\right)^{1.5} \text{RePr}^{1/3}\right] \quad (7\text{-}177)$

The results of Figure 7-14 indicate that circumferential turbulent diffusion has a much larger effect on cladding temperature variation than secondary

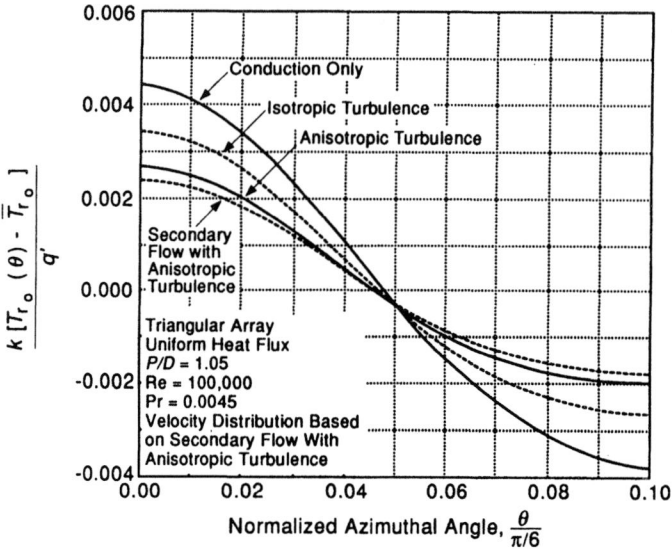

Figure 7-14 The effect of turbulent heat transport on circumferential cladding temperature variation. *(From Nijsing and Eifler [35].)*

flow. This is contrary to the behavior exhibited in Figure 7-13 for the circumferential velocity distribution. The small effect of secondary flow on circumferential heat transport in the coolant is due to the fact that the radial temperature variation in the coolant is very small compared with the circumferential coolant temperature variation.

B Turbulent Friction Factors and Nusselt Numbers

As with laminar flow the analysis can be extended to predict friction factors and Nusselt numbers (for example [15]). These turbulent results are sensitive to the approach for modeling Reynolds shear stresses and turbulent heat fluxes. The range among results encompasses results from finite rod array experiments and analyses which have already been given in Volume I, Chapters 9 and 10, respectively, for friction factors and Nusselt numbers.

IX MULTIREGION ANALYSIS OF LONGITUDINAL LAMINAR FLOW

A The General Solution Procedure

To illustrate the principles involved, we will be content to present a two-region analysis, i.e., clad and coolant. As mentioned in Section VII, however, since the ceramic fuel of power reactors and the core of electrically heated rods have low thermal conductivity, the assumption of uniform circumferential wall heat flux which will be applied to the inner surface of the clad in our two-region example is physically realistic.

The problem is formulated by expressing the conservation of energy equations and boundary conditions for the coolant and clad regions as shown in Figure 7-15. Retaining the assumptions of no physical property variations with temperature and no axial heat conduction, the coolant energy conservation equation is still Eq. 7-28 i.e.:

$$\rho c_p v_z \frac{\partial T}{\partial z} = k_f \left(\frac{\partial^2 T}{\partial r^2} + \frac{1}{r} \frac{\partial T}{\partial r} + \frac{1}{r^2} \frac{\partial^2 T}{\partial \theta^2} \right) \tag{7-28}$$

where the subscript f has been added to distinguish the coolant and clad thermal conductivities. The boundary conditions at $\theta = 0$ degrees, $\theta = 30$ degrees, and $r = P/(2\cos\theta)$ are also the same, i.e.:

$$\frac{\partial T}{\partial \theta} = 0 \text{ at } \theta = 0 \text{ degrees and 30 degrees} \tag{7-178}$$

$$\frac{\partial T}{\partial \vec{n}} = 0 \text{ at } r = P/(2\cos\theta) \tag{7-179}$$

However, the boundary condition on the rod surface $r = r_0$ is now different and reflects continuity of heat flux and temperature at the rod surface; for $r/r_0 = 1$,

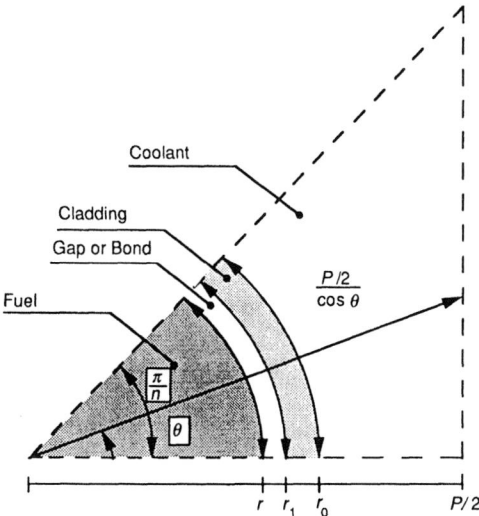

— — — Zero Flux Lines

$n = 6$: triangular array
$n = 4$: rectangular array

Figure 7-15 Characteristic multi-region domain of an infinite rod array.

0 degrees $< \theta <$ 30 degrees:

$$-k_f \left. \frac{\partial T}{\partial r} \right|_{r_0} = -k_w \left. \frac{\partial T}{\partial r} \right|_{r_0} = q''_{r_0}(\theta) \tag{7-180}$$

$$T_f|_{r_0} = T_w|_{r_0} = T_{r_0}(\theta) \tag{7-181}$$

The cladding energy conservation equation, assuming no energy generation in the clad as for the coolant, is:

$$\frac{\partial^2 T}{\partial r^2} + \frac{1}{r} \frac{\partial T}{\partial r} + \frac{1}{r^2} \frac{\partial^2 T}{\partial \theta^2} = 0 \tag{7-182}$$

for which the boundary condition, in addition to Eqs. 7-180 and 7-181, are the symmetry conditions at $\theta = 0$ and 30 degrees, i.e.:

$$\frac{\partial T}{\partial \theta} = 0 \text{ at } \theta = 0 \text{ and 30 degrees} \tag{7-178}$$

and a prescribed heat flux at the inner clad surface, $r = r_1$, i.e.:

$$-k_w \frac{\partial T}{\partial r} = q''_{r_1} = \text{a constant} \tag{7-183}$$

For the laminar flow case that we are examining, an analytical three-region solution assuming the volumetric heat generation rate in the fuel to be constant has been performed by Axford [1] utilizing the method of finite Fourier cosine transforms. For two-region problems only an iterative, numerical solution by Dwyer and Berry [12] exists. It is the Dwyer and Berry results that we will present along with first a summary of their procedure because it is instructive to see how the boundary conditions are adapted for an iterative solution. Similar procedures have been applied to the more complicated problem of coupled solution of adjacent but geometrically dissimilar coolant regions, i.e., edge subchannel to interior subchannel.

For the iterative procedure the differential equations and boundary conditions of Eqs. 7-28 and 7-178 through 7-183 are applied in the following way. The wall flux boundary condition of Eq. 7-180 is temporarily assumed to be a constant with θ, i.e., case B. Now utilizing Eqs. 7-178 and 7-179, Eq. 7-28 is solved to give the wall temperature distribution. Using this distribution, which is the boundary condition given by Eq. 7-181 as well as Eqs. 7-178 and 7-183, Eq. 7-182 is solved to give a clad temperature distribution from which a wall heat flux can be calculated as:

$$-k_w \left. \frac{\partial T}{\partial r} \right|_{r_0} \text{ for all } \theta$$

At this point we now have a circumferentially dependent wall heat flux to compare against our originally assumed constant heat flux and have completed one iteration. The iterative process is continued until the circumferential heat flux distribution at r_0 (boundary condition 7-180) used to obtain the solution of

Eq. 7-28 is in close agreement with the circumferential heat flux distribution at r_0 as obtained from the temperature distribution resulting from the solution of Eq. 7-182. Dwyer and Berry state that their results are quite accurate for the conditions investigated since in the numerical solution a very fine mesh size was used, the series expansion for the velocity distribution was carried to at least five terms, and boundary condition 7-179 was satisfied at nine points along its applicable line of symmetry.

B Results from a Two-Region Analysis of Longitudinal Flow between Fuel Rods

1 Average Nusselt number Values of the average Nusselt number as a function of three independent parameters P/D, $(r_0 - r_1)/r_0$, and k_w/k_f are given in Table 7-5. The parameters $(r_0 - r_1)/r_0$ and k_w/k_f characterize the geometry and thermal characteristics (for steady-state behavior) of the second region, the clad. Table 7-5 also gives for reference the corresponding values of $[\overline{Nu_L}]_{q''}$ and $[\overline{Nu_L}]_T$ for the two limiting cases of boundary conditions B and A, respectively. The blank entries in Table 7-5 represent cases in which more than 16 iterations would have been required to get convergence. These blank entries occur at low values of P/D and high values of k_w/k_f and $(r_0 - r_1)/r_0$. Physically these are for cases in which a relatively large variation in circumferential heat transfer through the coolant and the cladding exists.

The results can be represented by the empirical equation:

$$\overline{Nu_L} = \frac{[\overline{Nu_L}]_{q''}}{1 + A} + \frac{A[\overline{Nu_L}]_T}{1 + A} \tag{7-184}$$

where:

$$A = 2.0(k_w/k_f)[(r_0 - r_1)/r_0]^{3/4} \tag{7-185}$$

Eq. 7-184 represents the results in Table 7-5 with an average deviation of about $\pm 2\%$. From this equation, it is apparent that for the same percentage change, the thermal conductivity of the wall has an appreciably greater influence on the rod-averaged heat transfer coefficient than the cladding thickness. From Eq. 7-184 it is seen that the case of boundary condition B is applicable for the parameter $A \ll 1$ while the case of boundary condition A is applicable for values of the parameter $A \gg 1$. As an example, for sodium-cooled fast breeder reactors where the P/D is about 1.20, typical values of k_w/k_f and $(r_0 - r_1)/r_0$ are 0.20 and 0.10, respectively. The parameter A per Eq. 7-185 is then 0.071, thereby justifying boundary condition B for that application.

From the average Nusselt values given in the table, it is seen that they increase with an increase in either the cladding thickness or the cladding conductivity.

These effects, however, decrease as the P/D ratio increases, until at a $P/D \approx 1.2$, they are practically negligible. However, at the lower P/D ratios,

Table 7-5 Rod-averaged laminar flow Nusselt numbers, illustrating the effects of variations in the independent parameters P/D, $(r_2 - r_1)/r_2$, and k_w/k_f†

P/D	$[\overline{Nu}_L]_q$	$[\overline{Nu}_L]_T$	$\dfrac{r_2 - r_1}{r_2}$	$\overline{Nu}_L$						
				$k_w/k_f = 0.1$	$k_w/k_f = 0.2$	$k_w/k_f = 0.4$	$k_w/k_f = 1.0$	$k_w/k_f = 2.0$	$k_w/k_f = 3.0$	$k_w/k_f = 4.0$
1.05	1.055	2.825	0.025	1.078	1.098	1.139	1.249	1.402	—	—
			0.050	1.097	1.138	1.212	1.399	1.626	—	—
			0.075	1.116	1.173	1.275	1.514	1.778	—	—
			0.100	1.131	1.204	1.327	1.602	1.885	—	—
			0.150	1.158	1.248	1.401	1.716	—	—	—
			0.200	1.173	1.275	1.443	1.779	—	—	—
			0.300	1.186	1.297	1.477	1.822	—	—	—
1.06	1.359	3.183	0.025	1.383	1.404	1.446	1.566	1.728	1.859	1.968
			0.050	1.404	1.446	1.527	1.725	1.964	2.137	2.262
			0.075	1.424	1.485	1.594	1.846	2.125	2.303	—
			0.100	1.441	1.517	1.649	1.938	2.230	—	—
			0.150	1.468	1.565	1.727	2.058	2.362	—	—
			0.200	1.485	1.593	1.772	2.121	2.424	—	—
			0.300	1.498	1.617	1.807	2.168	2.469	—	—
1.07	1.705	3.545	0.025	1.730	1.752	1.797	1.920	2.089	2.224	2.335
			0.050	1.755	1.796	1.880	2.085	2.330	2.503	2.639
			0.075	1.773	1.836	1.950	2.211	2.492	2.673	2.798
			0.100	1.790	1.870	2.007	2.305	2.607	2.781	—
			0.150	1.817	1.920	2.087	2.426	2.731	—	—
			0.200	1.836	1.949	2.134	2.491	2.796	—	—
			0.300	1.850	1.973	2.170	2.540	—	—	—

1.10	2.936	4.620	0.025	2.961	2.986	3.028	3.150	3.313	3.443	3.548
			0.050	2.986	3.027	3.110	3.310	3.543	3.703	3.820
			0.075	3.009	3.064	3.178	3.430	3.693	3.857	3.977
			0.100	3.022	3.100	3.234	3.519	3.793	3.960	4.058
			0.150	3.048	3.149	3.312	3.633	3.911	4.063	4.156
			0.200	3.064	3.178	3.356	3.692	3.968	4.109	4.200
			0.300	3.077	3.201	3.392	3.737	4.014	4.151	—
1.15	5.135	6.222	0.025	5.152	5.169	5.203	5.274	5.388	5.473	5.542
			0.050	5.169	5.202	5.250	5.386	5.539	5.641	5.715
			0.075	5.182	5.223	5.298	5.465	5.635	5.738	5.808
			0.150	5.212	5.273	5.387	5.596	5.772	5.865	5.923
			0.300	5.232	5.314	5.440	5.663	5.833	5.924	5.974
1.20	6.902	7.484	0.025	6.911	6.920	6.937	6.974	7.033	7.077	7.112
			0.050	6.919	6.937	6.962	7.032	7.111	7.164	7.202
			0.075	6.927	6.953	6.983	7.073	7.161	7.214	7.250
			0.150	6.946	6.974	7.033	7.141	7.232	7.280	7.309
			0.300	6.959	6.990	7.060	7.175	7.263	7.306	7.332
1.30	9.033	9.194	0.025	9.033	9.036	9.040	9.052	9.059	9.072	9.080
			0.050	9.036	9.040	9.048	9.059	9.080	9.092	9.101
			0.075	9.038	9.044	9.051	9.071	9.092	9.104	9.112
			0.150	9.042	9.052	9.059	9.087	9.108	9.119	9.126
			0.300	9.045	9.053	9.068	9.095	9.116	9.126	9.132

† $\dfrac{r_2 - r_1}{r_2}$ (original ref.) $\equiv \dfrac{r_0 - r_1}{r_0}$ (this chapter); $\left.\dfrac{k_w}{k_f}\right|_{H_2O} = 30$; $\left.\dfrac{k_w}{k_f}\right|_{sodium} \simeq 0.25$.

(From Dwyer and Berry [12].)

$\overline{Nu}_L$ can be significantly affected by changes in either cladding thickness or cladding thermal conductivity, particularly the latter.

2 Circumferential temperature variation around outer surface of cladding The local clad surface temperature is a maximum at the gap, $\theta = 0$ degrees. Since it is in general a function of θ, $\bar{q}''_w$, P/D, $(r_0 - r_1)/r_0$ and k_w/k_f, it is not feasible to develop an empirical correlation to handle all these variables. However, as in the analogous study on slug flow ([13]), $T_{w,\theta}$ can be very well represented by the empirical equation:

$$\frac{T_{w,\theta} - T_{w,30\text{ degrees}}}{\bar{T}_{w,\theta} - T_b} = \frac{\left[\dfrac{T_{w,\theta} - T_{w,30\text{ degrees}}}{\bar{T}_{w,\theta} - T_b}\right]_{q''}}{1 + a(k_w/k_f)} \tag{7-186}$$

where the coefficient a is a function of P/D and $(r_0 - r_1)/r_0$, and the subscript q'' refers to the case of boundary condition B. Values of a and $[(T_{w,\theta} - T_{w,30\text{ degrees}})/(\bar{T}_{w,\theta} - T_b)]_{q''}$ are given in Figure 7-16 and Table 7-6, respectively. Dwyer and Berry [12] estimated that values of the ratio $[(T_{w,\theta} - T_{w,30\text{ degrees}})/(\bar{T}_{w,\theta} - T_b)]_{q''}$ from Eq. 7-186 were correct to $\pm 1\%$ except possibly those for P/D values equal to 1.05 for $(r_0 - r_1)/r_0$ greater than 0.1 and for k_w/k_f values greater than 2.0. In this very limited range, the accuracy was estimated to be within $\pm 2\%$. Further they found that for $(r_0 - r_1)/r_0$ values less than 0.10, the effect on the circumferential temperature variation was the same as for k_w/k_f. However, above $(r_0 - r_1)/r_0 = 0.1$, changes in cladding thickness had progressively less influence than changes in k_w/k_f.

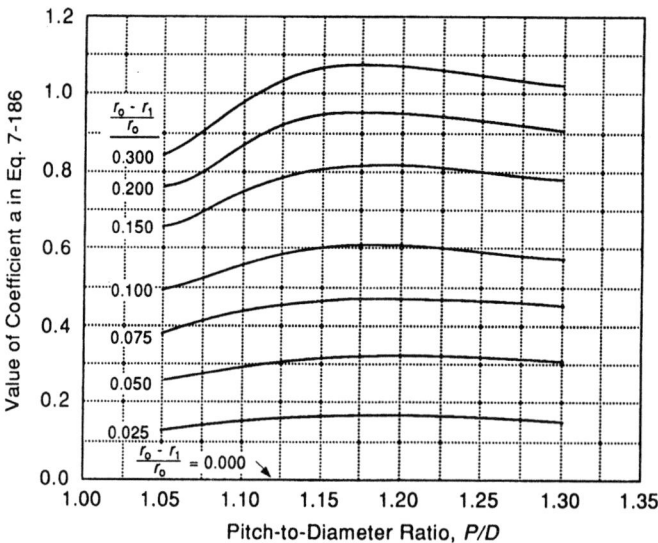

Figure 7-16 Effects of variations in rod spacing and relative cladding thickness on the value of the constant a in Eq. 7-186. (*From Dwyer and Berry [12].*)

Example 7-3 Prediction of local clad surface temperature

PROBLEM Suppose we have the conditions:

$P/D = 1.10$
$(r_0 - r_1)/r_0 = 0.2$
$k_w/k_f = 0.2$
$k_f = 69.24$ W/m °C (40 Btu/hr ft F)
$r_0 = 3.175$ mm (0.125 in)
$\bar{q}''_{r_0} = 3.154 \times 10^6$ W/m² (10^6 BTU/hr/ft)
$T_b = 593.33$ °C (1100 F)

and wish to predict T_w at $\theta = 8$ degrees.

SOLUTION The equivalent diameter D_e for this geometry per Eq. 7-163 is 2.12 mm (0.006963 ft). From Table 7-5 we see that $\overline{Nu}_L = 3.178$, and from this we determine from Eq. 7-161 that:

$$\bar{T}_{w,\theta} - T_b = \frac{12}{\pi} \frac{D_e}{D} \frac{q'/12k_f}{\overline{Nu}_L} = \frac{D_e}{k_f} \frac{\bar{q}''_{r_0}}{\overline{Nu}_L} = \frac{2.12 \times 10^{-3}}{69.24} \left(\frac{3.154 \times 10^6}{3.178} \right)$$

$$= 30.4 \text{ °C} \tag{7-187}$$

$$\bar{T}_{w,\theta} = 593.3 + 30.4 = 623.7 \text{ °C}$$

The next step is to calculate $T_{w,30 \text{ degrees}}$. We begin by recognizing that:

$$\left[\frac{T_{w,\theta} - T_{w,30 \text{ degrees}}}{\bar{T}_{w,\theta} - T_b} \right]_{avg} = \frac{\bar{T}_{w,\theta} - T_{w,30 \text{ degrees}}}{\bar{T}_{w,\theta} - T_b} \tag{7-188}$$

which permits Eq. 7-186 to be transformed to:

$$\frac{\bar{T}_{w,\theta} - T_{w,30 \text{ degrees}}}{\bar{T}_{w,\theta} - T_b} = \frac{\left[\left[\frac{T_{w,\theta} - T_{w,30 \text{ degrees}}}{\bar{T}_{w,\theta} - T_b} \right]_{q''} \right]_{avg}}{1 + a(k_w/k_f)} \tag{7-189}$$

Now the term $\left[\left[\frac{T_{w,\theta} - T_{w,30 \text{ degrees}}}{\bar{T}_{w,\theta} - T_b} \right]_{q''} \right]_{avg}$ is tabulated on the bottom of Table 7-6 for $P/D = 1.10$ as 0.8942. From Figure 7-16, the coefficient a is equal to 0.875. We now solve Eq. 7-189 for $T_{w,30}$:

$$T_{w,30 \text{ degrees}} = 623.7 - \left[\frac{0.8942}{1 + 0.875(0.2)} \right] (30.4) = 600.6 \text{ °C}$$

Next employing Eq. 7-186 directly and taking the value of:

$$\left[\frac{T_{w,\theta} - T_{w,30 \text{ degrees}}}{\bar{T}_{w,\theta} - T_b} \right]_{q''}$$

Table 7-6 Calculated values of $[(T_{w,\theta} - \bar{T}_{w,30}/(\bar{T}_{w,\theta} - T_b)]_{q''}$ for use in Eqs. 7-186 and 7-189

θ, degrees		P/D						
	1.05	1.06	1.07	1.10	1.12	1.15	1.20	1.30
0	2.1352	2.0934	2.0404	1.8046	1.5990	1.2723	0.8090	0.3161
1	2.1282	2.0868	2.0342	1.7994	1.5944	1.2687	0.8068	0.3152
2	2.1075	2.0673	2.0158	1.7839	1.5810	1.2581	0.8001	0.3126
3	2.0736	2.0355	1.9856	1.7585	1.5588	1.2407	0.7891	0.3084
4	2.0268	1.9914	1.9438	1.7231	1.5279	1.2164	0.7738	0.3024
5	1.9684	1.9360	1.8912	1.6786	1.4889	1.1858	0.7545	0.2949
6	1.8995	1.8706	1.8289	1.6254	1.4424	1.1492	0.7313	0.2859
7	1.8206	1.7954	1.7570	1.5640	1.3886	1.1067	0.7045	0.2754
8	1.7336	1.7121	1.6771	1.4954	1.3284	1.0591	0.6744	0.2637
9	1.6399	1.6219	1.5905	1.4205	1.2625	1.0071	0.6415	0.2509
10	1.5400	1.5254	1.4974	1.3397	1.1914	0.9508	0.6058	0.2370
11	1.4359	1.4244	1.3996	1.2543	1.1161	0.8911	0.5680	0.2222
12	1.3290	1.3201	1.2984	1.1655	1.0377	0.8289	0.5285	0.2068
13	1.2196	1.2130	1.1942	1.0737	0.9564	0.7643	0.4875	0.1908
14	1.1097	1.1050	1.0888	0.9803	0.8737	0.6985	0.4456	0.1744
15	1.0003	0.9971	0.9833	0.8865	0.7905	0.6322	0.4035	0.1579
16	0.8918	0.8899	0.8781	0.7927	0.7071	0.5657	0.3611	0.1414
17	0.7859	0.7848	0.7750	0.7003	0.6249	0.5001	0.3194	0.1250
18	0.6836	0.6831	0.6749	0.6105	0.5449	0.4363	0.2786	0.1091
19	0.5851	0.5851	0.5783	0.5235	0.4674	0.3743	0.2391	0.0936
20	0.4920	0.4922	0.4866	0.4408	0.3937	0.3153	0.2015	0.0789
21	0.4049	0.4053	0.4008	0.3633	0.3246	0.2600	0.1661	0.0651
22	0.3244	0.3247	0.3213	0.2913	0.2603	0.2086	0.1333	0.0522
23	0.2515	0.2518	0.2492	0.2260	0.2020	0.1618	0.1034	0.0405
24	0.1869	0.1872	0.1852	0.1681	0.1502	0.1204	0.0770	0.0315
25	0.1309	0.1311	0.1298	0.1178	0.1053	0.0844	0.0539	0.0211
26	0.0844	0.0846	0.0837	0.0760	0.0679	0.0544	0.0348	0.0136
27	0.0478	0.0790	0.0474	0.0430	0.0385	0.0308	0.0197	0.0077
28	0.0213	0.0214	0.0211	0.0192	0.0172	0.0137	0.0088	0.0034
29	0.0054	0.0054	0.0053	0.0048	0.0043	0.0035	0.0022	0.0009
30	0.0000	0.0000	0.0000	0.0000	0.0000	0.0000	0.0000	0.0000
Av	1.0332	1.0226	1.0014	0.8942	0.7949	0.6341	0.4040	0.1580

(From Dwyer and Berry [12].)

for 8 degrees from Table 7-6 as 1.4954, we calculate $T_{w,8\,degrees}$ as:

$$T_{w,8\,degrees} = T_{w,30\,degrees} + \frac{\bar{T}_{w,\theta} - T_b}{1 + a\left(\dfrac{k_w}{k_f}\right)} \left[\frac{T_{w,\theta} - T_{w,30}}{\bar{T}_{w,\theta} - T_b}\right]_{q''}$$

$$= 600.6 + \frac{30.4}{1 + 0.875(0.2)} [1.4954] = 639.3 \, °C$$

It is possible if desired to calculate the local values of the wall heat flux and subsequently the heat transfer coefficient from additional empirical equations and tables given by Dwyer and Berry [12].

X SURVEY OF ISOLATED CELL PROBLEMS SOLVED BY THE DISTRIBUTED PARAMETER METHOD

Having completed the description of a specific case, a two-region analysis of laminar flow in an infinite array of heat-generating fuel rods, we will now return to the general class of problems which this case typifies. A vast amount of literature exists on the solution of these problems which can be made more easily accessible to the reader by introducing a classification system based on common key problem characteristics.

Single Versus Multirod Analyses

In analyzing rod bundles with longitudinal flow, the first step generally is to apply symmetry conditions to divide the bundle into characteristic unit cells. True characteristic cells neither influence nor are influenced by their neighboring cells, i.e., neither mass, momentum, nor heat transfer occurs across the boundaries of the cells. Since each such cell includes a portion of one of the rods of the array, this analysis procedure has been called *single-rod analysis*. In our example case of fully developed laminar and turbulent flow and heat transfer in an infinite array of fuel rods each with axisymmetrical and identical mean radial energy generation rates, this method is correct for the stated boundary conditions, cases A or B. For the general case of arbitrarily prescribed axial and circumferential wall temperature or heat flux distribution, solutions can be obtained using these standard boundary conditions using the superposition technique. This technique is applicable to our case of fully developed steady flow with constant fluid properties since all the governing energy equations of Section II are linear and homogeneous in temperature. Therefore any linear combination of solutions is itself a solution. Sutherland and Kays [51] present an example of the exploitation of this approach for turbulent flow in infinite rod arrays.

If the bundle were to be limited by a fixed wall, if irregularities of rod spacing exist, and/or if the energy generation rate of the rods varies radially across the bundle, the single-rod analysis method can be considered only as an approximation. Exact multirod analyses have been performed to date only for a limited number of problems. For both cases of analysis the full range of flow situations can be addressed, i.e., laminar, slug, and turbulent flow. Additionally the mixed convection condition, a situation that includes the effects of buoyancy or free convection, can also be addressed. Further, cases of both symmetrically located and asymmetrically located rods in an assembly can be defined for each case although the approximation involved in solving the asymmetri-

Table 7-7 Some single-rod analyses of longitudinal flow in rod bundles

	Laminar		Laminar-mixed convection		Slug		Turbulent	
	Interior	Edge/corner	Interior	Edge/corner	Interior	Edge/corner	Interior	Edge/corner
Symmetrically located rods:								
One-region boundary conditions								
$v_{zw} = 0$	49†						18	
T_w = constant	50		13		13		15, 35	
$\dot{q}''_w$ = constant	11		13		13		15, 35	
Unspecified since Pr $\geq$ 0.7							30, 37, 39, 53	30
Multiregion	1, 12	20	25	25	1, 14	54	3, 17, 34	3
Asymmetrically located rods:								
One-region boundary conditions								
T_w = constant $\dot{q}''_w$ = constant	21				19, 22			
Multiregion								

† Reference number.

cally located rod problem with the single-rod method may need investigation depending on the application of the results. Hsu [19] has analyzed the effect of central rod displacement on heat transfer for slug flow.

With the above definitions/characteristics in hand, Table 7-7 identifies some of the existing literature developed using the single-rod method. Notice that for the single-rod method only, it is necessary to identify which portion of the rod array is being considered, i.e., interior subchannels or cells, edge cells or corner cells. Further, for the single-rod method in either the symmetrical or asymmetrical cases, one-region or multiregion analyses can be performed. Recall that this refers to whether the coolant alone is considered or the coolant in combination simultaneously with the clad, clad and bond, or clad, bond, and fuel regions. For the multirod method it is consistent to perform only a multiregion analysis. For single-region and those multiregion analyses that do not extend to the rod centerline, the circumferential boundary condition on the innermost defined surface must be prescribed as indicated in Table 7-7. As can be seen from Table 7-7 for our laminar flow example, we drew our discussion from the papers of Sparrow et al. [49] and Dwyer and Berry [11–15]. We mentioned the work of Axford [1] but did not present it since it involves a three-region analysis. For turbulent flow we drew our examples from the review of Nijsing and Eifler [36]. The paper of Dwyer and Berry [15] includes an interesting comparison of the effect of various assumptions for eddy diffusivities on Nusselt numbers. Multiregion turbulent results were not presented as was done for laminar flow but can be obtained from the papers of Dwyer et al. [17] and Nijsing and Eifler [34]. The work of Bobkov et al. [3] and Markoczy [30] is significant in that prediction methods applicable to edge and corner subchannels are given.

Although Table 7-7 does not give a comprehensive listing of all the relevant literature, if it is taken in combination with the reviews of Nijsing [33] and Dwyer [10] and the latest publications, the interested reader should easily be able to become acquainted with the existing literature in the area of distributed parameter methods in rod bundles.

XI ANALYSIS OF INTERACTING CHANNELS

In the previous section the distributed parameter analysis procedure applied to single rods or single cells was discussed. However, cells can be considered as isolated from their neighbors only in the case of a regular, infinite array of rods in which each rod has an axisymmetrical and equal mean energy generation rate. Such conditions are very restrictive in both a geometric and a thermal sense. In practical situations we must consider geometrically finite rod arrays that can also exhibit local position distortions. Further reactor power profiles usually lead to spatially varying energy generation rates across rod arrays, i.e., so-called *power skews*. Finally, self-shielding effects in rods containing fissionable, fertile or absorber materials can lead to asymmetrical energy generation rates within the rods.

For these reasons most practical situations are multicell problems for which solution attempts by the distributed parameter approaches lead to formidable mathematical difficulties. In fact, this area requires very specialized analytical and numerical techniques. For these reasons the current approaches will be identified but not described in detail as was done for the single-cell approaches. The interested reader can consult Shah and London [48], Chen et al. [6], and Yeung and Wolf [56] for the literature covering the three principal approaches to the multicell problem.

A The Three Methods of Solution

The most complex statement of the problem to be solved is the determination of the velocity and temperature fields in a geometrically arbitrary array of rods with spacers where each rod has a prescribed but asymmetrical energy generation rate and the array has an arbitrary power skew. Conversely a very simple statement would involve a geometrically regular but finite bare rod array with rods having an axisymmetrical but equivalent mean energy generation rate. To address the spectrum of cases suggested by these extremes three general approaches exist:

1. Direct analytical solution of the multicell problem
2. Direct numerical solution of the multicell problem
3. Utilization of single-cell results by iterative or superposition procedures.

As might be anticipated, the analytical approaches are applicable to a much narrower range of cases than are the numerical approaches, as will be discussed next.

B Direct Analytical Solution Methods

One of the earliest solutions [47] is a good starting point because it deals with a simple but physically important problem. The situation of developed laminar flow in a semiinfinite square array limited by a fixed wall was investigated to determine the influence of the wall. An analytical solution of the governing Poisson equation in polar coordinates was employed using point matching of boundary conditions at the cell boundaries. This was necessitated by the non-orthogonality of the coordinate system with portions of the finite array boundary—a difficulty for any analytical approach to the general multicell problem. Figure 7-17 presents a typical result for the influence of the wall on volumetric flow rates for the case of $P/D = 2.0$ and a range of first-cell (cell adjacent to the wall) geometries relative to the remainder of the cells as characterized by the ratios of hydraulic diameters of the cell next to the wall, D_{e_1}, to the other cells, D_{e_i}, i.e., D_{e_1}/D_{e_i}. The abscissa indicates the cell position from the wall. The

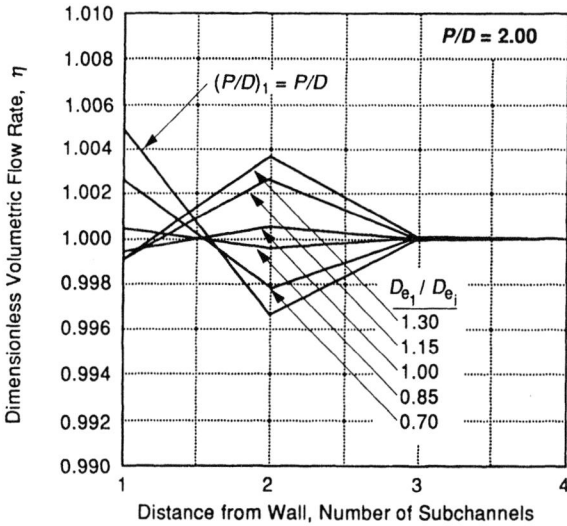

Figure 7-17 Influence of wall on subchannel dimensionless volumetric flow rate. *(After Schmid [47].)*

ordinate η is equal to the ratio of cell volume flow rates in the presence of the wall to that for the infinite array, i.e., no wall influence. The figure demonstrates that the ratio of cell volume flow rates between the case with the wall present and absent is less than 1% for the first and second cells for the conditions considered. It is more negligible for the third and subsequent row of cells.

The other analytical solutions have utilized special mathematical functions but have been able to solve only a limited class of problems. The majority have been limited to laminar, fully developed flow around a triangular array of seven rods located inside a flow tube as illustrated in Figure 7-18. Some solutions for

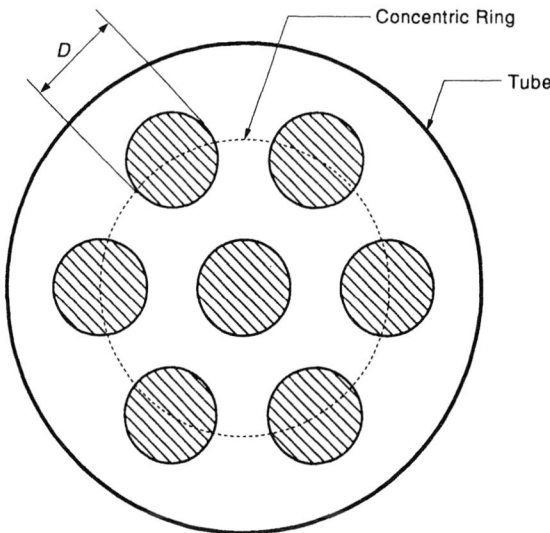

Figure 7-18 One-ring concentric seven-rod bundle in a tube.

this geometry for the thermally developing, hydrodynamically developed case have been made. Mottaghian and Wolf [32] have relaxed this one-ring problem and solved the case of an arbitrary number of rods, with different radii, placed in concentric rings about a central rod. For the finite hexagonal array Wong and Wolf [55] have solved the interesting case of a seven-pin cluster with a radial power tilt under three-dimensional, steady-state conditions.

References and results for these solutions are available in Shah and London [48]. However, as this brief summary demonstrates, the range of cases successfully solved by analytical (or semianalytical) methods is limited.

C Direct Numerical Solution Methods

Accurate numerical solutions are also limited by the lack of coincidence of the numerical mesh and the system boundaries. In the normal finite difference case this lack of coincidence requires interpolation. However, it is particularly important to avoid interpolation with strong curvature or slope discontinuities, both of which are common in physical applications. Consequently an essential part of a general numerical approach for solution of the Navier–Stokes and the energy equations such as the technique of boundary-fitted coordinate systems is the generation of a curvilinear coordinate system with coordinate lines coincident with all boundaries. This procedure does not use conformal transformation and consequently is not limited to two dimensions. The Navier–Stokes and the energy equations are solved in the transformed plane using the finite difference technique once the curvilinear coordinates are generated.

This procedure has been incorporated in a computer code developed for application to a broad range of rod array problems. The interested reader is referred to the work of Chen et al. [6] for details of the formulation and application of this method.

D Utilization of Single-Cell Results

There are two approaches to the utilization of single-cell results: the iterative approach and a superposition approach.

The iterative approach described here was applied by Yeung and Wolf [56] to the solution of a two-dimensional, finite array in slug flow. However, it is applicable to any array for which the unit cell solutions are available for boundary conditions of prescribed temperatures and zero heat flux. Consider the symmetry section of a 19-rod bundle as illustrated in Figure 7-19. Divide the section into cells for which unit cell solutions exist or can be obtained. In Figure 7-19 the boundaries between these cells are labeled A through D. Boundaries E and F are external boundaries of the symmetry section which are also boundaries of cells 3 and 4.

The iterative solution procedure starts with the temperature fields for the unit cells for adiabatic boundary conditions. This not only yields a temperature field within each cell but also along the adiabatic boundary since the rod linear

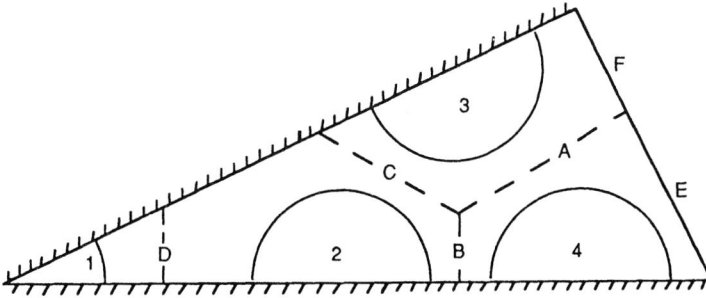

Figure 7-19 Configuration of the symmetry section of a 19-rod bundle.

power generation rate has been prescribed. The resulting temperatures along the common boundary of two cells are generally different. The initial adiabatic boundary condition is then relaxed by assuming that the temperature along the common boundary is the average of the temperatures of the two neighboring cells resulting from the initial calculation. This new boundary condition is used to produce an updated temperature field for one of the unit cells from which an updated heat flux distribution along the common boundary is developed. This heat flux distribution in turn is used as the boundary condition for its neighboring cell, producing a new temperature field for the neighboring cell. The second loop of the iteration utilizes these temperature boundary conditions at the outset, and the procedure continues until an acceptable solution is achieved.

For adiabatic boundaries E and F of Figure 7-19 this calculation sequence is illustrated in Figure 7-20. This figure shows that with average temperatures

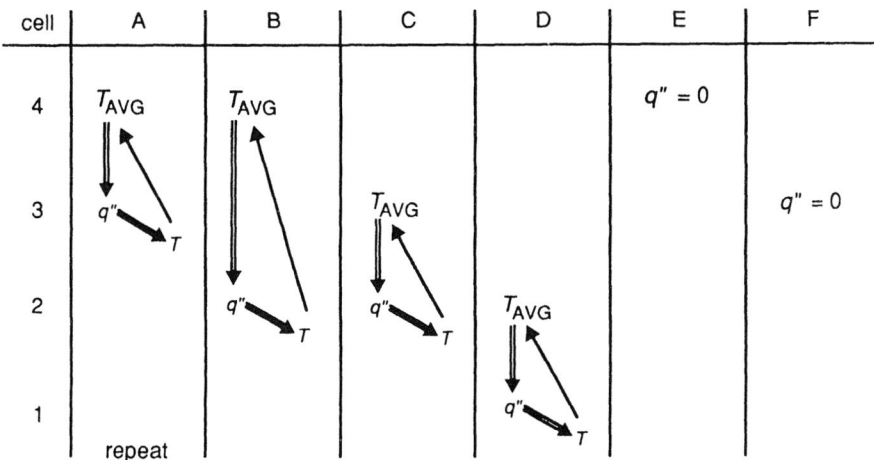

cell	A	B	C	D	E	F
4	T_{AVG}	T_{AVG}			$q'' = 0$	
3	q'' T		T_{AVG}			$q'' = 0$
2		q'' T	q'' T	T_{AVG}		
1				q'' T		
	repeat					

Figure 7-20 The iterative solution procedure for multicell arrays. Start with adiabatic boundaries on each cell, solve for T field, find T_{avg} on each boundary as listed for starting point.

initially available on all interior boundaries, the procedure is to calculate cells in the sequence 4, 3, 2, 1. Such a series of calculations comprises an iteration that yields a set of new average temperatures along the interior boundaries. The coolant temperature along boundary D converges rapidly so that a practical solution is obtained in three iterations.

The range of application of this iterative procedure is limited since it requires availability of single-cell solutions. It could be used to determine the velocity field under laminar or turbulent conditions. However for the temperature field in finite arrays under nonslug conditions, single-cell temperature solutions are not available for the velocity fields that would exist. This is because the velocity fields are slightly distorted due to the presence of the bundle wall. In principle however these single cell temperature solutions could be obtained from an initial solution for the velocity field. For mixed convection conditions, on the other hand, the velocity and temperature fields cannot be sequentially addressed. Therefore, the iterative solution is not applicable under these conditions.

A superposition approach has been developed by Rehme [40] for friction factors for finite arrays in laminar flow using individual subchannel friction factor values which he also developed. He later developed a method to predict subchannel and bundle friction factors for turbulent flow on the basis of these laminar flow results [42]. The friction factors of central subchannels with $P/D > 1.2$ were also shown to be predicted with good approximation from the equivalent annulus solutions [41]. These laminar and turbulent flow results were summarized by Rehme and Trippe [45] and appear in Volume I as friction factor—Reynolds number products for subchannels and for triangular and square array rod bundles (see Section VI, Chapter 9).

REFERENCES

1. Axford, R. A. Two-dimensional multiregion analysis of temperature fields in reactor tube bundles. *Nucl. Eng. Des.* 6:25–42, 1967.
2. Bartzis, J. G., and Todreas, N. E. Turbulence modeling of axial flow in a bare rod bundle. *J. Heat Transfer* 101:628–634, 1979.
3. Bobkov, V., Ibramov, M., and Subbotin, V. Generalized relationships for heat transfer in the fuel assemblies of nuclear reactors with liquid metal cooling. Translated from *Tepl. Vys. Temp.* 10:795–803, 1972, and published in *High Temp.* 10:713–720, 1972.
4. Buleev, N. I. Theoretical model of the mechanism of turbulent exchange in fluid flows. Translated from *Teploperedacha*, 64–98, USSR Academy of Science, Moscow (1962). Published as Report AERE-Trans. 957.
5. Buleev, N. I. Theoretical model for turbulent transfer in three-dimensional fluid flow. In *Proceedings of the Third United Nations International Conference on the Peaceful Uses of Atomic Energy*, Geneva, 1964. Paper 329, United Nations, Geneva, 1964.
6. Chen, B. C.-J., Vanka, S. P., and Sha, W. T. Some recent computations of rod bundle thermal hydraulics using boundary fitted coordinates. *Nucl. Eng. Des.* 62:123–136, 1980.
7. Deissler, R. G., and Taylor, M. F. Analysis of axial turbulent flow and heat transfer through banks of rods or tubes. In TID-7529, *Reactor Heat Transfer Conference*, New York, Part 1, Book 1, 1957, pp. 416–461.

8. Dwyer, O. E. Eddy transport in liquid–metal heat transfer. *A.I.Ch.E.J.* 9:261, 1963.
9. Dwyer, O. E. Liquid metal heat transfer. In *Sodium-Nak Engineering Handbook,* Vol. 2. Foust, O. J., Ed. London: Gordon and Breach Science Publishers, 1968.
10. Dwyer, O. E. Heat transfer to liquid metals flowing in line through unbaffled rod bundles: A review. *Nucl. Eng. Des.* 10:3–20, 1969.
11. Dwyer, O. E., and Berry, H. C. Laminar flow heat transfer for in-line flow through unbaffled rod bundles. *Nucl. Sci. Eng.* 42:81–88, 1970.
12. Dwyer, O. E., and Berry, H. C. Effects of cladding thickness and thermal conductivity on heat transfer for laminar in-line flow through rod bundles. *Nucl. Sci. Eng.* 42:69–80, 1970.
13. Dwyer, O. E., and Berry, H. C. Slug-flow Nusselt numbers for in-line flow through unbaffled rod bundles. *Nucl. Sci. Eng.* 39:143–170, 1970.
14. Dwyer, O. E., and Berry, H. C. Effects of cladding thickness and thermal conductivity on heat transfer to liquid metal flowing in-line through bundles of closely spaced reactor fuel rods. *Nucl. Sci. Eng.* 40:317–330, 1970.
15. Dwyer, O. E., and Berry, H. C. Heat transfer to liquid metals flowing turbulently and longitudinally through closely spaced rod bundles: Part 1. *Nucl. Eng. Des.* 23:273–294, 1972.
16. Dwyer, O. E., and Tu, P. S. Analytical study of heat transfer rates for parallel flow of liquid metals through tube bundles: Part 1. *Chem. Eng. Prog. Symp. Ser.* 56:183–193, 1960.
17. Dwyer, O. E., Berry, H. C., and Hlavac, P. Heat transfer to liquid metals flowing turbulently and longitudinally through closely spaced rod bundles: Part 2. *Nucl. Eng. Des.* 23:295–308, 1972.
18. Eifler, W., and Nijsing, R. Experimental investigation of velocity distribution and flow resistance in a triangular array of parallel rods. *Nucl. Eng. Des.* 5:22–42, 1967; also Rep. EUR 2193-3II, 1965.
19. Hsu, C. J. Multiregion analysis of the effect of rod displacement on heat transfer in slug flow through closely packed rod bundles. *Nucl. Sci. Eng.* 47:380–388, 1972.
20. Hsu, C. J. Laminar- and slug-flow heat transfer characteristics of fuel rods adjacent to fuel subassembly walls. *Nucl. Sci. Eng.* 49:398–404, 1972.
21. Hsu, C. J. Multiregion analysis of the effect of lateral rod displacement on heat transfer in laminar flow through closely packed reactor fuel rods. In *Progress in Heat and Mass Transfer,* Vol. 7. Dwyer, O. E., Ed. Oxford: Pergamon Press, 1973, pp. 219–237. Original presented at the International Seminar on Heat Transfer in Liquid Metals, Trogir, Yugoslavia, 1971.
22. Johannsen, K., Ullrich, R., and Wolf, L. Multiregion heat transfer analysis for displaced central, corner and side elements of unbaffled LMFBR fuel rod bundles. *Nucl. Eng. Des.* 15:29–32, 1971.
23. Kays, W. M., and Leung, E. Y. Heat transfer in annular passages: Hydrodynamically developed turbulent flow with arbitrarily prescribed heat flux. *Int. J. Heat Mass Transfer* 6:537–557, 1963.
24. Kays, W., and Perkins, H. C. Forced convection internal flow in ducts. In *Heat Transfer Handbook.* Hartnet, J. P., and Rohsenow, W. M., Eds. pp. 7.1–7.193. New York: McGraw Hill, 1972.
25. Kim, J., and Wolf, L. Laminar mixed convection heat transfer in finite hexagonal bundles. In *Transactions of the American Nuclear Society,* Vol. 27, San Francisco, November, 1977.
26. Launder, B. E., and Spalding, D. B. *Mathematical Models of Turbulence.* New York: Academic Press, 1972.
27. Lundberg, R. E., McCuen, P. A., and Reynolds, W. C. Heat transfer in annular passages: Hydrodynamically developed laminar flow with arbitrarily prescribed wall temperatures or heat fluxes. *Int. J. Heat Mass Transfer* 6:495–529, 1963.
28. Marek, J., Maubach, K., and Rehme, K. Heat transfer and pressure drop performance of rod bundles arranged in square arrays. *Int. J. Heat Mass Transfer* 16:2215–2228, 1973.
29. Maresca, M. W., and Dwyer, O. E. Heat transfer to mercury flowing in-line through a bundle of circular tubes. *J. Heat Transfer,* 86:180–186, 1964.
30. Markoczy, G. *Konvektive Warmeubertragung in langsangestromten Rohr-oder Stabbundeln.* EIR-Bericht 198, April 1972.

31. Maubach, K. Reibungsgesetze turbulenter stromungen. *Chem. Ing. Tech.* 42:995–1004, 1970.
32. Mottaghian, R., and Wolf, L. A two-dimensional analysis of laminar fluid flow in rod bundles of arbitrary arrangement. *Int. J. Heat Mass Transfer* 17:1121–1128, 1974.
33. Nijsing, R. Heat exchange and heat exchangers with liquid metals. In *AGARD Lecture Series No. 57 on Heat Exchangers*. Ginoux, J. J., Lecture Series Director, AGARD-LS-57-72, 1972.
34. Nijsing, R., and Eifler, W. Analysis of liquid metal heat transfer in assemblies of closely spaced fuel rods. *Nucl. Eng. Des.* 10:21–54, 1969.
35. Nijsing, R., and Eifler, W. Temperature fields in liquid-metal cooled rod assemblies. In *Progress in Heat and Mass Transfer*, Vol. 7. Oxford: Pergamon Press, 1973, pp. 115–149. Original presented at the International Seminar on Heat Transfer in Liquid Metals, Trogir, Yugoslavia, 1971.
36. Nijsing, R., and Eifler, W. Axially varying heat flux effects in tubes, flat ducts and widely spaced bundles cooled by a turbulent flow of liquid metal. *Nucl. Eng. Des.* 23:331–346, 1972.
37. Nijsing, R., Garganti, I., and Eifler, W. Analysis of fluid flow and heat transfer in a triangular array of parallel heat generating rods. *Nucl. Eng. Des.* 4:375–398, 1966.
38. Prandtl, L. Untersuchungen zur ausgebildeten turbulenz. *Z. Angew. Math. Mech.* 5:136–146, 1925.
39. Presser, K. H. *Warmeubergang und Druckverlust an Reaktorbrennelementen in Form langsdurchstromter Rundstabbundel.* Jul-486-RB, KFA, Julich, 1967.
40. Rehme, K. Laminarstromung in stabbundeln. *Chem. Ing. Tech.* 43:962–966, 1971.
41. Rehme, K. Pressure drop performance of rod bundles in hexagonal arrangement. *J. Heat Mass Transfer* 15:2499–2517, 1972.
42. Rehme, K. Simple method of predicting friction factors of turbulent flow in non-circular channels. *Int. J. Heat Mass Transfer* 16:933–950, 1973.
43. Rehme, K. Turbulent flow in smooth concentric annuli with small radius ratios. *J. Fluid Mech.* 64:263–287, 1974.
44. Rehme, K. Turbulence measurements in smooth concentric annuli with small radius ratios. *J. Fluid Mech.* 72:189–206, 1975.
45. Rehme, K., and Trippe, G. Pressure drop and velocity distribution in rod bundles with spacer grids. *Nucl. Eng. Des.* 62:349–359, 1980.
46. Rothfus, R. R., Walker, J. E., and Whan, G. A. Correlation of local velocities in tubes, annuli and parallel plates. *A.I.Ch.E. J.* 4:240, 1958.
47. Schmid, J. Longitudinal laminar flow in an array of circular cylinders. *Int. J. Heat Mass Transfer* 9:925–937, 1966.
48. Shah, R. K., and London, A. L. *Laminar Flow Forced Convection in Ducts.* New York: Academic Press, 1978.
49. Sparrow, E. M., and Loeffler, A. L., Jr. Longitudinal laminar flow between cylinders arranged in regular array. *A.I.Ch.E. J.* 5:325–330, 1959.
50. Sparrow, E. M., Loeffler, A. L., and Hubbard, H. A. Heat transfer to longitudinal laminar flow between cylinders. *Trans. ASME, J. Heat Transfer* 83:415–422, 1961.
51. Sutherland, W., and Kays, W. Heat transfer in parallel rod arrays. *J. Heat Transfer* 88:117–124, 1966.
52. Taylor, G. I. Eddy motion in the atmosphere. *Phil. Trans. R. Soc. Lond.* 215:1–26, 1915.
53. Weisman, J. Heat transfer to water flowing parallel to tube bundles. *Nucl. Sci. Eng.* 6:78–79, 1959.
54. Wolf, L., and Johanssen, K. Two-dimensional multiregion analysis of temperature fields in finite rod bundles cooled by liquid metals. Presented at the *First International Conference on Structural Mechanics in Reactor Technology* Berlin, Germany, September 1971.
55. Wong, C-N., and Wolf, L. A 3-D slug flow heat transfer analysis of coupled coolant cells in finite LMFBR bundles. M.S. thesis, Department of Nuclear Engineering, M.I.T., January 1978. Also in *Trans. Am. Nucl. Soc.* 30:540–541, 1978.
56. Yeung, M. R., and Wolf, L. Multi-cell slug flow heat transfer analysis for finite FMFBR bundles. *Nucl. Eng. Des.* 62:101–122, 1980.

PROBLEMS

Problem 7-1 Friction factor–Reynolds number product for equivalent annulus in laminar and turbulent flow (Section III)

Compute the value of $f\text{Re}_{De}$ for an interior subchannel in a triangular array of $P/D = 1.50$ using the equivalent annulus formulation if you can justify its applicability.

Answer: Laminar $f\text{Re}_{De}]_{EA} = 126.4$

Repeat for a Reynolds number equal to 5×10^4.

Answer: Turbulent $f]_{EA} = 0.023$

Problem 7-2 Nusselt number for an equivalent annulus in laminar and turbulent flow (Sections III and IV)

Compute the value of Nu for an interior subchannel in a triangular array of $P/D = 1.50$ using the equivalent annulus formulation if you can justify its applicability.

Repeat for a fluid at $\text{Re} = 5 \times 10^4$ with a $\text{Pr} = 0.0045$.

Answer: Laminar $\text{Nu}]_{EA} = 11.75$

Turbulent $\text{Nu}]_{EA} = 14.28$

Problem 7-3 Friction factor–Reynolds number product for an interior subchannel in laminar flow (Section VI)

Determine the value of $f\text{Re}_{De}$ for an interior subchannel in a triangular array of $P/D = 1.10$.

Answer: $f\text{Re}_{De} = 80.2$

Problem 7-4 Nusselt number for an interior subchannel in laminar flow (Section VII)

Determine the Nusselt number for an interior subchannel in laminar flow in a triangular array of $P/D = 1.10$ for the boundary condition of uniform wall heat flux.

Answer: $\overline{\text{Nu}} = 2.94$

Problem 7-5 Nusselt number for an interior subchannel in laminar flow including pin geometry and coolant-clad conductivities (Section IX)

Repeat Problem 7-4 for the clad dimensions and material thermal conductivities typical of fast reactor core mixed/oxide fuel. Note that for these practical conditions, the clad boundary condition is neither uniform wall heat flux or temperature.

Answer: $\overline{\text{Nu}} = 6.96$

Problem 7-6 Friction factor–Reynolds number product for a finite triangular array laminar pressure drop (Section XI)

Compute the $f\text{Re}_{De}$ for a seven-rod triangular array bundle with a $P/D = 1.10$. Take the gap g between the pin outside diameter and the wall equal to $(P - D)/2$.

Compare your result with the answer to Problem 7-3 and explain any differences.

Answer: $f\text{Re}_{De} = 52$.

EIGHT

TREATMENT OF UNCERTAINTIES
IN REACTOR THERMAL ANALYSIS

I INTRODUCTION

The designer of a reactor core strives to achieve the goal of safe and reliable economic operation. Uncertainties in core materials and operating conditions affect the achievement of this goal. These uncertainties stem from two main sources:

1. Uncertainties from randomness inherent in a process, e.g., fuel pellet diameter variation from a manufacturing process or reactor power level determination variation arising from temperature variations due to turbulent flow conditions
2. Uncertainties from imperfect modeling or estimation of a parameter, e.g., the degree of overheating caused by rod bowing.

Uncertainties of the first type can be reduced but not eliminated by gathering more data whereas uncertainties of the second type can be effectively eliminated by development and confirmation of modeling procedures. In practice both types of uncertainties are treated, although historically many uncertainties of the second type introduced in early thermal design procedures have been eliminated over the ensuing years. This chapter describes the principles inherent in the methods utilized for treatment of uncertainties in reactor thermal analysis [8].

II RELEVANT STATISTICAL FUNDAMENTALS

A knowledge of some underlying statistical relationships is necessary to understand the methods of treating uncertainties in reactor thermal analysis. This section reviews concepts and definitions relevant to these methods. Because of its usefulness in this problem area, emphasis is placed on the characteristics of the normal distribution. Of interest are:

1. The properties of the normal distribution
2. The means for estimating normal distribution population parameters from sampling data.

A Estimating the Mean and Standard Deviation of Distributions

Let us identify the random variable x_j as that which is of interest in our analysis. As an example, let it be the percentage of U-235 enrichment in the fuel pellets. Hence the population involved in the characterization of x_j for this case is all the fuel pellets in the core. If every pellet were assayed and its percent enrichment reported, we would have a distribution of data points which can be characterized by its mean, μ_j, a measure of its central tendency and its standard deviation, σ_j, a measure of its dispersion. If N is the population size and $_ix_j$ is the value of the ith sample of x_j:

$$\mu_j \equiv \frac{1}{N} \sum_{i=1}^{N} {_ix_j} \tag{8-1}$$

$$\sigma_j \equiv \left\{ \frac{1}{N} \sum_{i=1}^{N} ({_ix_j} - \mu_j)^2 \right\}^{1/2} \tag{8-2}$$

Since it is generally impractical to sample the entire parent population, a smaller sample is employed to estimate μ_j and σ_j of the parent population. Let n be the number of pellets sampled where commonly $n \ll N$. If the enrichment of the ith sampled pellet was reported as $_ix_j$, then the mean and standard deviations of the parent population can be estimated from the data of the smaller sample as:

$$\hat{\mu}_j \equiv \frac{\sum_{i=1}^{n} {_ix_j}}{n} \tag{8-3}$$

and

$$\hat{\sigma}_j \equiv \left\{ \frac{\sum_{i=1}^{n} {_ix_j^2}}{n-1} - \left(\frac{n}{n-1}\right) \hat{\mu}_j^2 \right\}^{1/2} \tag{8-4}$$

where the parameter estimates are signified by a caret (^), and again the subscript j represents the property of interest, and the subscript i represents a sample of the population. Eq. 8-4 uses a denominator of $n - 1$ instead of n to correct for statistical bias. It also is presented in a form somewhat easier to use than Eq. 8-2. Note that all of these equations are independent of the parent population distribution.

Example 8-1 Estimates of mean and standard deviation of a population

PROBLEM Compute the mean and standard deviation estimators from the following sample of the U-235 enrichment percentages in 12 fuel pellets. (This and several subsequent examples of this section are taken from [11].)

	Sample
n	$_ix_j$
1	1.8620
2	1.8535
3	1.8495
4	1.8560
5	1.8595
6	1.8485
7	1.8585
8	1.8715
9	1.8515
10	1.8585
11	1.8515
12	1.8595

$$\sum_{i=1}^{12} {}_ix_j = 22.2800$$

SOLUTION From the given data, $\sum_{i=1}^{n} {}_ix_j = 22.2800$. From Eq. 8-3, $\hat{\mu}_j =$

$\dfrac{22.2800}{12} = 1.85667$; $\hat{\mu}_j^2 = 3.4472$. To compute $\hat{\sigma}_j$, form the following table:

n	$_ix_j^2$
1	3.46704400
2	3.43546225
3	3.42065025
4	3.44473600
5	3.45774025
6	3.41695225
7	3.45402225
8	3.50251225
9	3.42805225
10	3.45402225
11	3.42805225
12	3.45774025

$$\sum_{i=1}^{12} {}_ix_j^2 = 41.3669865$$

From Eq. 8-4, $\hat{\sigma}_j = \left[\dfrac{41.3669865}{11} - \dfrac{12}{11}(3.4472111)\right]^{1/2} = 0.00642.$

B The Normal Distribution

We inherently think of a set of data as coming from some distribution. Among the set of continuous statistical distributions, one of the most useful is the *normal distribution*. This arises from the central limit theorem whose existence allows us to use the normal distribution to characterize the distribution of the means of random samples drawn from any distribution whose mean μ and variance σ^2 are finite. With this characterization, we can proceed to use certain statistics derived from samples even though we do not know the exact form of the distribution from which the samples are drawn.

Suppose the random variable x_j is normally distributed. The probability P_x that x_j will be less than some value, x, is:

$$P_x \equiv P\{x_j \leq x\} = \int_{-\infty}^{x} p(x')dx' \tag{8-5}$$

where $p(x)$ is called the *probability density function* and is for a normal distribution given by:

$$p(x) = \frac{1}{\sigma\sqrt{2\pi}} \exp\left\{-\frac{1}{2}\left[\frac{(x-\mu)}{\sigma}\right]^2\right\} \tag{8-6}$$

which is plotted in Figure 8-1. Therefore, the probability that x_j will lie between x_1 and x_2 is:†

$$P\{x_1 < x_j \leq x_2\} = \int_{x_1}^{x_2} p(x')dx' \tag{8-7}$$

If the random variable x_j is expressed as a number of standard deviations about the mean, we obtain:

$$x = \mu + z\sigma \tag{8-8a}$$

where z can be a real number (e.g., fraction or integer and positive or negative) or rearranging:

$$z = \frac{x - \mu}{\sigma} \tag{8-8b}$$

and

$$dx = \sigma dz \tag{8-9}$$

since μ and σ are constants. Since the probability is unaffected by the variables of measurement i.e. $p(x)dx = p(z)dz$, from Eq. 8-6 we obtain the standardized normal distribution:

$$p(z)dz = \frac{1}{\sqrt{2\pi}} \exp\{-z^2/2\}dz \tag{8-10}$$

† The form of the LHS of Eq. 8-7 explicitly excludes the event $x_j = x_1$ which is dropped through the subtraction of P_{x_1} from P_{x_2}. Practically, this is only limiting if there is a singularity in the probability density function at $x_j = x_1$.

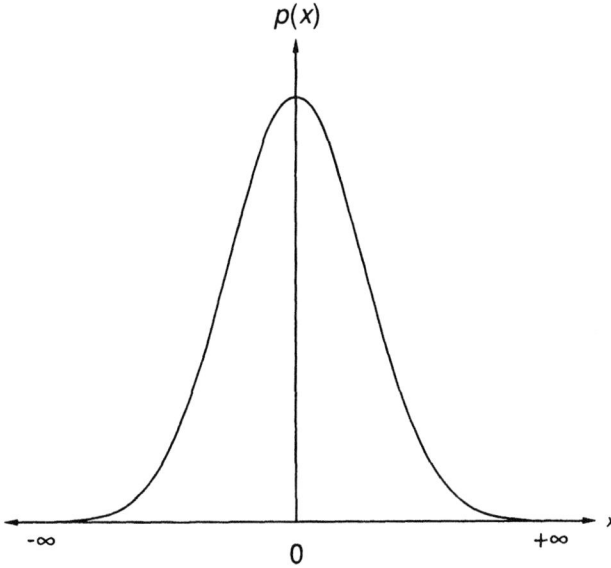

Figure 8-1 The probability density function for the normal distribution.

or

$$P\{z_1 < z \le z_2\} \equiv P_{z2} - P_{z1} = \frac{1}{\sqrt{2\pi}} \int_{z_1}^{z_2} \exp\left\{-\frac{1}{2} z'^2\right\} dz' \qquad (8\text{-}11)$$

P_z values from Eq. 8-11 are compiled in Table 8-1 for $z_1 = -\infty$.

We frequently need to evaluate P_z to determine the probability that a normally distributed variable will lie within certain bounds. For example if z_1 and z_2 are chosen to be ∓ 2 (corresponding to x_j lying within two standard deviations about the mean), then $P\{-2 < z \le 2\} = 0.9545$. This is interpreted as being a probability of 95.45 percent that a sample x_j drawn from a parent population that is normally distributed, will lie within two standard deviations of the mean, μ_j, of the parent population. This result is obtained from Eq. 8-11 and Table 8-1 using the fact that $p(z)$ is an even function around $z = 0$. The specific calculation procedure is:

$$P(-2 < z \le 2) = \left[\frac{1}{\sqrt{2\pi}} \int_{-2.0}^{+2.0} \exp\left\{-\frac{z'^2}{2}\right\} dz'\right]$$

$$= 2\left[\frac{1}{\sqrt{2\pi}} \int_{-2.0}^{0} \exp\left\{-\frac{z'^2}{2}\right\} dz'\right]$$

$$= 2\left[\frac{1}{\sqrt{2\pi}} \int_{-\infty}^{0} \exp\left\{-\frac{z'^2}{2}\right\} dz' - \frac{1}{\sqrt{2\pi}} \int_{-\infty}^{-2.0} \exp\left\{-\frac{z'^2}{2}\right\} dz'\right]$$

$$= 2[0.50000 - 0.02275]$$

$$= 0.95450$$

Table 8-1 Fractional area under the standard normal distribution from $-\infty$ to z,

i.e., $\displaystyle\int_{-\infty}^{z} \frac{1}{\sqrt{2\pi}} e^{-z'^2/2}dz' \equiv P_z$

z	0	1	2	3	4	5	6	7	8	9
-2.9	0.00187	0.00181	0.00175	0.00169	0.00164	0.00159	0.00154	0.00149	0.00144	0.00139
-2.8	0.00256	0.00248	0.00240	0.00233	0.00226	0.00219	0.00212	0.00205	0.00199	0.00193
-2.7	0.00347	0.00336	0.00326	0.00317	0.00307	0.00298	0.00289	0.00280	0.00272	0.00264
-2.6	0.00466	0.00453	0.00440	0.00427	0.00415	0.00402	0.00391	0.00379	0.00368	0.00357
-2.5	0.00621	0.00604	0.00587	0.00570	0.00554	0.00539	0.00523	0.00508	0.00494	0.00480
-2.4	0.00820	0.00798	0.00776	0.00755	0.00734	0.00714	0.00695	0.00676	0.00657	0.00639
-2.3	0.01072	0.01044	0.01017	0.00990	0.00964	0.00939	0.00914	0.00889	0.00866	0.00842
-2.2	0.01390	0.01355	0.01321	0.01287	0.01255	0.01222	0.01191	0.01160	0.01130	0.01101
-2.1	0.01786	0.01743	0.01700	0.01659	0.01618	0.01578	0.01539	0.01500	0.01463	0.01426
-2.0	0.02275	0.02222	0.02169	0.02118	0.02068	0.02018	0.01970	0.01923	0.01876	0.01831

−1.9	0.02872	0.02807	0.02743	0.02680	0.02619	0.02559	0.02500	0.02442	0.02385	0.02330
−1.8	0.03593	0.03515	0.03438	0.03362	0.03288	0.03216	0.03144	0.03074	0.03005	0.02938
−1.7	0.04457	0.04363	0.04272	0.04182	0.04093	0.04006	0.03920	0.03836	0.03754	0.03673
−1.6	0.05480	0.05370	0.05262	0.05155	0.05050	0.04947	0.04846	0.04746	0.04648	0.04551
−1.5	0.06681	0.06552	0.06426	0.06301	0.06178	0.06057	0.05938	0.05821	0.05705	0.05592
−1.4	0.08076	0.07927	0.07780	0.07636	0.07493	0.07353	0.07215	0.07078	0.06944	0.06811
−1.3	0.09680	0.09510	0.09342	0.09176	0.09012	0.08851	0.08692	0.08534	0.08379	0.08226
−1.2	0.11507	0.11314	0.11123	0.10935	0.10749	0.10565	0.10383	0.10204	0.10027	0.09853
−1.1	0.13567	0.13350	0.13136	0.12924	0.12714	0.12507	0.12302	0.12100	0.11900	0.11702
−1.0	0.15866	0.15625	0.15386	0.15151	0.14917	0.14686	0.14457	0.14231	0.14007	0.13786
−0.9	0.18406	0.18141	0.17879	0.17619	0.17361	0.17106	0.16853	0.16602	0.16354	0.16109
−0.8	0.21186	0.20897	0.20611	0.20327	0.20045	0.19766	0.19489	0.19215	0.18943	0.18673
−0.7	0.24196	0.23885	0.23576	0.23270	0.22965	0.22663	0.22363	0.22065	0.21770	0.21476
−0.6	0.27425	0.27093	0.26763	0.26435	0.26109	0.25785	0.25463	0.25143	0.24825	0.24510
−0.5	0.30854	0.30503	0.30153	0.29806	0.29460	0.29116	0.28774	0.28434	0.28096	0.27760
−0.4	0.34458	0.34090	0.33724	0.33360	0.32997	0.32636	0.32276	0.31918	0.31561	0.31207
−0.3	0.38209	0.37828	0.37448	0.37070	0.36693	0.36317	0.35942	0.35569	0.35197	0.34827
−0.2	0.42074	0.41683	0.41294	0.40905	0.40517	0.40129	0.39743	0.39358	0.38974	0.38591
−0.1	0.46017	0.45620	0.45224	0.44828	0.44433	0.44038	0.43644	0.43251	0.42858	0.42465
−0.0	0.50000	0.49601	0.49202	0.48803	0.48405	0.48006	0.47608	0.47210	0.46812	0.46414

Example 8-2 Properties of a sample drawn from a population

PROBLEM Suppose the parent population of values of pellet U-235 enrichment percentage is normally distributed about a mean value of $\mu = 1.8534$ with a standard deviation of $\sigma = 0.00996$. What is the probability that a sample drawn from this distribution will lie within the extremes of those samples reported in Example 8-1?

SOLUTION The extreme values of U-235 reported in Example 8-1 are 1.8715 (high) and 1.8485 (low). Hence:

$$z_1 = \frac{1.8485 - 1.8534}{0.00996} \text{ and } z_2 = \frac{1.8715 - 1.8534}{0.00996}$$

or

$$z_1 = -0.492 \simeq -0.49; \ z_2 = 1.817 \simeq 1.82$$

Therefore:

$$P\{-0.49 < z \le 1.82\} = \frac{1}{\sqrt{2\pi}} \int_{-0.49}^{+1.82} \exp\left\{-\frac{z'^2}{2}\right\} dz'$$

which from Table 8-1 yields:

$$= [0.50000 - 0.03438 + 0.50000 - 0.31207]$$

$$= 0.65355$$

C Confidence Level

Complete characterization of any finite population is usually impractical because of the large number of samples required. Consequently, limited sampling procedures are used to estimate parent population parameters. Scaling down the number of samples, however, conflicts with the requirements for an accurate result. Intuitively we feel that the confidence in the estimation process improves as the number of samples increases. Note also that our confidence that the parent population parameter lies within a certain range increases as we extend that range. It is desirable to quantify such trends by quantifying the level of confidence in our estimates of parent population parameters.

The confidence level expresses the probability that a parent population parameter estimated from a sample is within a stated range. We denote the confidence level as α where α is the probability obtained by integrating a probability density function between appropriate limits as in Eq. 8-7. Now the stated range or interval for the parameter is termed the *confidence interval* and the end points of the interval are called *confidence limits*. Using Eq. 8-7 for the normal distribution:

$x_2 - x_1$ is the confidence interval.
x_2 and x_1 are confidence limits

For example, suppose we wish to find confidence limits for the parent population mean μ using the sample mean $\hat{\mu}$. If the population is normally distributed, the ratio $[(\mu - \hat{\mu})/\hat{\sigma}]$ is proportional to a quantity t which has a known distribution $p(t)$ called the *Student t distribution;* μ is then determined to a prescribed confidence level α by finding the limits of integration of the known distribution so that the required probability α is obtained, i.e.:

$$\int_{t_1}^{t_2} p(t')dt' = \alpha \tag{8-12}$$

or

$$\alpha \equiv P\{t_1 < t \le t_2\} = P_{t2} - P_{t1}$$

If t_1 and t_2 have equal absolute values, the equation $t = f[(\hat{\mu} - \mu)/\hat{\sigma}]$ yields μ_1 and μ_2 for α prescribed. This procedure will be demonstrated in Example 8-3.

When t_1 becomes $-\infty$ and t_2 is finite, a lower single-ended confidence interval is obtained. Conversely, when t_2 tends to $+\infty$ and t_1 is finite, an upper confidence interval is defined. Both the upper and lower confidence intervals are single-ended confidence intervals.

D Estimating the Population Mean, μ

The estimation of a population mean μ depends on a particular relationship that exists between it and the mean of some sample $\hat{\mu}$ which is drawn from the population characterized by μ. The standard procedure for estimating μ, based on the asymptotic normality of the random variable $\hat{\mu}$ (or exact normality if the population distribution is normal) assumes that the difference between μ and $\hat{\mu}$ divided by $\hat{\sigma}_{\hat{\mu}}$, the estimated standard deviation for $\hat{\mu}$, follows the Student t distribution with $n - 1$ degrees of freedom.[†] The ratio t is thus defined as:

$$t = \frac{\hat{\mu} - \mu}{\hat{\sigma}_{\hat{\mu}}} = \frac{\hat{\mu} - \mu}{\hat{\sigma}/\sqrt{n}} \tag{8-13}$$

where t can be positive or negative. Consequently, t increases with sample size n.

The relationship

$$\hat{\sigma}_{\hat{\mu}} = \frac{\hat{\sigma}}{\sqrt{n}} \tag{8-14}$$

used in Eq. 8-13 can be derived by determining the true standard deviation $\sigma_{\hat{\mu}}$ of the sample mean, $\hat{\mu}$ (recall that the $\hat{\mu}$ is a random variable in this case) and

---

† The number of degrees of freedom f of a statistic (t is an example of a statistic) is defined as the number n of independent observations in the sample or equivalently the sample size minus the number k of population parameters which must be estimated from sample observations. Hence $f = n - k$. In the case of the statistic t, the number of independent observations in the sample is n from which $\hat{\mu}$ and $\hat{\sigma}$ can be computed. Since μ must be estimated, k equals unity yielding $f = n - 1$.

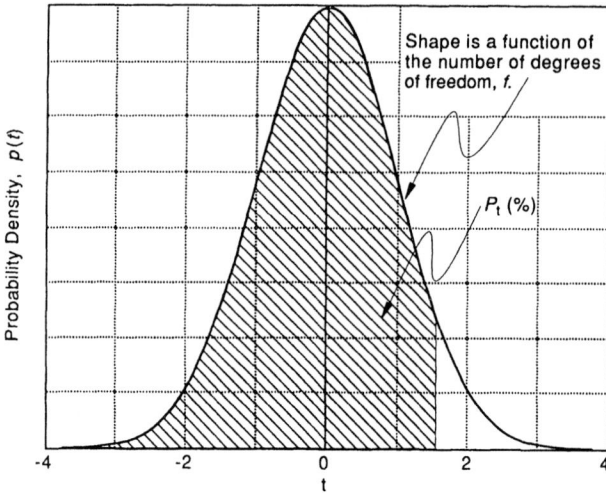

Figure 8-2 Relationship between $P_t(\%)$, f, and t of Table 8-2 for the Student t distribution.

then employing the estimator $\hat{\sigma}$ for the true standard deviation of the population σ.

The distribution of t for some degrees of freedom f are listed in Table 8-2. Values of t and $P_t(\%)$ correspond to those defined in Figure 8-2. Notice that as either $P_t(\%)$ gets smaller for a given number of degrees of freedom or the number of degrees of freedom gets greater for a given $P_t(\%)$, the t values decrease.

As an example of using Table 8-2 and Eq. 8-12, consider a system with 11 degrees of freedom for which it was desired to find the values of t at which $P_t(\%)$ would equal 95 percent. For $f = 11$ and $P_t(\%) = 95$, Table 8-2 lists a t value of 1.796. By definition, the 95 percent includes areas between $+1.796$ and $-\infty$, i.e., in terms of the equation for $P_t(\%)$ given in Table 8-2:

$$\int_{-\infty}^{+1.796} p(t')dt' = 95 \text{ percent}$$

Hence this value of t defines the single-ended lower confidence interval.

The Student t distribution is symmetrical and approaches the normal distribution as n approaches ∞. Hence for an ∞ number of degrees of freedom, the results of Table 8-2 for the Student t distribution are equivalent to those of Table 8-1 for the normal distribution. For example, from Table 8-2 for $P_t(\%) = 97.5$ and $f = \infty$, $t = 1.960$, which is equivalent to the value of $P_z = 0.025$ in Table 8-1 for $z = 1.960$.

Example 8-3 Interval estimate for the population mean value

PROBLEM Derive the symmetrical 95 percent confidence interval estimate for the parent population mean value of U-235 pellet enrichment μ from the sample given in Example 8-1.

SOLUTION From Eq. 8-13 obtain μ as:

$$\hat{\mu} - \mu = \frac{t\hat{\sigma}}{\sqrt{n}}$$

where the following sample parameters were found in Example 8-1:

$$n = 12$$

$$\hat{\sigma} = 0.00642$$

$$\hat{\mu} = 1.8567$$

To obtain the symmetrical 95 percent confidence interval (t_1 and t_2 as in Eq. 8-12), 5 percent of the total area under the t distribution should lie in the two tails or 2.5 percent of that area for each tail (due to symmetry). From Table 8-2 (which only gives upper confidence limits, i.e., P_t) for a probability of 97.5 percent and $f = 11$ degrees of freedom obtain t_2 as 2.201. By symmetry, $t_1 = -2.201$.

Hence:

$$\mu = \hat{\mu} - \frac{t\hat{\sigma}}{\sqrt{n}}$$

$$= 1.8567 - \frac{(\pm 2.201)(0.00642)}{\sqrt{12}}$$

The upper and lower confidence limits are then:

$$\mu_2 = 1.8608$$

$$\mu_1 = 1.8526$$

E Estimating the Population Standard Deviation σ

The Student t distribution was used to estimate the population mean μ based on a sample mean $\hat{\mu}$. Another distribution is used to estimate the population standard deviation σ in terms of the sample standard deviation $\hat{\sigma}$.

Define the statistic χ^2 as:

$$\chi^2 = \frac{\hat{\sigma}^2(n-1)}{\sigma^2} \tag{8-15}$$

This statistic follows the chi-squared (χ^2) distribution with $n - 1$ degrees of freedom.

Values of the χ^2 distribution for various degrees of freedom are listed in Table 8-3. Since χ^2 is always greater than zero, the values of χ^2 are based on the following relationship:

$$\int_0^{\chi^2} g(\chi'^2, f) d\chi'^2 = P_{\chi^2} \tag{8-16}$$

Table 8-2 *t* values for the Student *t* distribution

$$\int_{-\infty}^{t} p(t')dt' = P_t(\%)$$

f / $P_t(\%)$	55	60	70	80	90	95	97.5	99	99.5	99.9
1	0.158	0.325	0.727	1.376	3.078	6.314	12.71	31.82	63.66	318.0
2	0.142	0.289	0.617	1.061	1.886	2.920	4.303	6.965	9.925	22.30
3	0.137	0.277	0.584	0.978	1.638	2.353	3.182	4.541	5.841	10.20
4	0.134	0.271	0.569	0.941	1.533	2.132	2.776	3.747	4.604	7.173
5	0.132	0.267	0.559	0.920	1.476	2.015	2.571	3.365	4.032	5.893
6	0.131	0.265	0.553	0.906	1.440	1.943	2.447	3.143	3.707	5.208
7	0.130	0.263	0.549	0.896	1.415	1.895	2.365	2.998	3.499	4.785
8	0.130	0.262	0.546	0.889	1.397	1.860	2.306	2.896	3.355	4.501
9	0.129	0.261	0.543	0.883	1.383	1.833	2.262	2.821	3.250	4.297
10	0.129	0.260	0.542	0.879	1.372	1.812	2.228	2.764	3.169	4.144
11	0.129	0.260	0.540	0.876	1.363	1.796	2.201	2.718	3.106	4.025
12	0.128	0.259	0.539	0.873	1.356	1.782	2.179	2.681	3.055	3.930

13	0.128	0.259	0.538	0.870	1.350	1.771	2.160	2.650	3.012	3.852
14	0.128	0.258	0.537	0.868	1.345	1.761	2.145	2.624	2.977	3.787
15	0.128	0.258	0.536	0.866	1.341	1.753	2.131	2.602	2.947	3.733
16	0.128	0.258	0.535	0.865	1.337	1.746	2.120	2.583	2.921	3.686
17	0.128	0.257	0.534	0.863	1.333	1.740	2.110	2.567	2.898	3.646
18	0.127	0.257	0.534	0.862	1.330	1.734	2.101	2.552	2.878	3.610
19	0.127	0.257	0.533	0.861	1.328	1.729	2.093	2.539	2.861	3.579
20	0.127	0.257	0.533	0.860	1.325	1.725	2.086	2.528	2.845	3.552
21	0.127	0.257	0.532	0.859	1.323	1.721	2.080	2.518	2.831	3.527
22	0.127	0.256	0.532	0.858	1.321	1.717	2.074	2.508	2.819	3.505
23	0.127	0.256	0.532	0.858	1.319	1.714	2.069	2.500	2.807	3.485
24	0.127	0.256	0.531	0.857	1.318	1.711	2.064	2.492	2.797	3.467
25	0.127	0.256	0.531	0.856	1.316	1.708	2.060	2.485	2.787	3.450
26	0.127	0.256	0.531	0.856	1.315	1.706	2.056	2.479	2.779	3.435
27	0.127	0.256	0.531	0.855	1.314	1.703	2.052	2.473	2.771	3.421
28	0.127	0.256	0.530	0.855	1.313	1.701	2.048	2.467	2.763	3.408
29	0.127	0.256	0.530	0.854	1.311	1.699	2.045	2.462	2.756	3.396
30	0.127	0.256	0.530	0.854	1.310	1.697	2.042	2.457	2.750	3.385
40	0.126	0.255	0.529	0.851	1.303	1.684	2.021	2.423	2.704	3.307
60	0.126	0.254	0.527	0.848	1.296	1.671	2.000	2.390	2.660	3.232
120	0.126	0.254	0.526	0.845	1.289	1.658	1.980	2.358	2.617	3.160
∞	0.126	0.253	0.524	0.842	1.282	1.645	1.960	2.326	2.576	3.090

(After Fisher and Yates [13].)

Table 8-3 χ^2 values for the chi-square distribution

$$\int_0^{\chi^2} g(\chi'^2, f)d\chi'^2 = P_{\chi^2}(\%)$$

f	$P_{\chi^2}(\%)$ 0.5	1.0	2.5	5.0	10	20	25	30	50
1	0.000	0.000	0.001	0.004	0.016	0.064	0.102	0.148	0.455
2	0.010	0.020	0.051	0.103	0.211	0.446	0.575	0.713	1.386
3	0.072	0.115	0.216	0.352	0.584	1.005	1.213	1.424	2.366
4	0.207	0.297	0.484	0.711	1.064	1.649	1.923	2.195	3.357
5	0.412	0.554	0.831	1.145	1.610	2.343	2.675	3.000	4.351
6	0.676	0.872	1.237	1.635	2.204	3.070	3.455	3.828	5.348
7	0.989	1.239	1.690	2.167	2.833	3.822	4.255	4.671	6.346
8	1.344	1.646	2.180	2.733	3.490	4.594	5.071	5.527	7.344
9	1.735	2.088	2.700	3.325	4.168	5.380	5.899	6.393	8.343
10	2.156	2.558	3.247	3.940	4.865	6.179	6.737	7.267	9.342
11	2.603	3.053	3.816	4.575	5.578	6.989	7.584	8.148	10.341
12	3.074	3.571	4.404	5.226	6.304	7.807	8.438	9.034	11.340
13	3.565	4.107	5.009	5.892	7.042	8.634	9.299	9.926	12.340
14	4.075	4.660	5.629	6.571	7.790	9.467	10.165	10.821	13.339
15	4.601	5.229	6.262	7.261	8.547	10.307	11.036	11.721	14.339
16	5.142	5.812	6.908	7.962	9.312	11.152	11.912	12.624	15.338
17	5.697	6.408	7.564	8.672	10.085	12.002	12.792	13.531	16.338
18	6.265	7.015	8.231	9.390	10.865	12.857	13.675	14.440	17.338
19	6.844	7.633	8.907	10.117	11.651	13.716	14.562	15.352	18.338
20	7.434	8.260	9.591	10.851	12.443	14.578	15.452	16.266	19.337
21	8.034	8.897	10.283	11.591	13.240	15.445	16.344	17.182	20.337
22	8.643	9.542	10.982	12.338	14.041	16.314	17.240	18.101	21.337
23	9.260	10.196	11.688	13.091	14.848	17.187	18.137	19.021	22.337
24	9.886	10.856	12.401	13.848	15.659	18.062	19.037	19.943	23.337
25	10.520	11.524	13.120	14.611	16.473	18.940	19.939	20.867	24.337
26	11.160	12.198	13.844	15.379	17.292	19.820	20.843	21.792	25.336
27	11.808	12.879	14.573	16.151	18.114	20.703	21.749	22.719	26.336
28	12.461	13.565	15.308	16.928	18.939	21.588	22.657	23.647	27.336
29	13.121	14.256	16.047	17.708	19.768	22.475	23.567	24.577	28.336
30	13.787	14.953	16.791	18.493	20.599	23.364	24.478	25.508	29.336

where g is the probability density function for χ^2, f is the degrees of freedom, and P_{χ^2} is the probability. As an example of the use of Table 8-3, the value of χ^2 corresponding to a probability P_{χ^2} of 0.05 or 5 percent for 11 degrees of freedom is 4.575.

Example 8-4 Interval estimate for the population standard deviation

PROBLEM Derive the 95 percent upper confidence interval (single-sided) estimate for the parent population standard deviation σ in U-235 pellet enrichment from the sample given in Example 8-1.

$P_{x^2}(\%)$ f	70	75	80	90	95	97.5	99	99.5	99.9
1	1.074	1.323	1.642	2.706	3.841	5.024	6.635	7.879	10.827
2	2.408	2.773	3.219	4.605	5.991	7.378	9.210	10.597	13.815
3	3.665	4.108	4.642	6.251	7.815	9.348	11.345	12.838	16.268
4	4.878	5.385	5.989	7.779	9.488	11.143	13.277	14.860	18.465
5	6.064	6.626	7.289	9.236	11.070	12.832	15.086	16.750	20.517
6	7.231	7.841	8.558	10.645	12.592	14.449	16.812	18.548	22.457
7	8.383	9.037	9.803	12.017	14.067	16.013	18.475	20.278	24.322
8	9.524	10.219	11.030	13.362	15.507	17.535	20.090	21.955	26.125
9	10.656	11.389	12.242	14.684	16.919	19.023	21.666	23.589	27.877
10	11.781	12.549	13.442	15.987	18.307	20.483	23.209	25.188	29.588
11	12.899	13.701	14.631	17.275	19.675	21.920	24.725	26.757	31.264
12	14.011	14.845	15.812	18.549	21.026	23.337	26.217	28.300	32.909
13	15.119	15.984	16.985	19.812	22.362	24.736	27.688	29.819	34.528
14	16.222	17.117	18.151	21.064	23.685	26.119	29.141	31.319	36.123
15	17.322	18.245	19.311	22.307	24.996	27.488	30.578	32.801	37.697
16	18.418	19.369	20.465	23.542	26.296	28.845	32.000	34.267	39.252
17	19.511	20.489	21.615	24.769	27.587	30.191	33.409	35.718	40.790
18	20.601	21.605	22.760	25.989	28.869	31.526	34.805	37.156	42.312
19	21.689	22.718	23.900	27.204	30.144	32.852	36.191	38.582	43.820
20	22.775	23.828	25.038	28.412	31.410	34.170	37.566	39.997	45.315
21	23.858	24.935	26.171	29.615	32.671	35.479	38.932	41.401	46.797
22	24.937	26.039	27.301	30.813	33.924	36.781	40.289	42.796	48.268
23	26.018	27.141	28.429	32.007	35.172	38.076	41.638	44.181	49.728
24	27.096	28.241	29.553	33.196	36.415	39.364	42.980	45.558	51.179
25	28.172	29.339	30.675	34.382	37.652	40.646	44.314	46.928	52.620
26	29.246	30.434	31.795	35.563	38.885	41.923	45.642	48.290	54.052
27	30.319	31.528	32.912	36.741	40.113	43.194	46.963	49.645	55.476
28	31.391	32.620	34.027	37.916	41.437	44.461	48.278	50.993	56.893
29	32.461	33.711	35.139	39.087	42.557	45.722	49.588	52.336	58.302
30	33.530	34.800	36.250	40.256	43.773	46.979	50.892	53.672	59.703

SOLUTION From Eq. 8-15, obtain σ as:

$$\sigma = \hat{\sigma} \left[\frac{n - 1}{\chi^2} \right]^{1/2}$$

where $\hat{\sigma} = 0.00642$ from Example 8-1.

The value of χ^2 for 11 degrees of freedom and for a lower confidence limit of 95 percent, i.e., P_{χ^2}, is found in Table 8-3 as 19.675. Hence:

$$\sigma_1 = 0.00642 \left[\frac{11}{19.675} \right]^{1/2} = 0.00480$$

i.e., this value of σ_1 should be lower than the true population value in 95 percent of all estimates (note that $\sigma_2 = \infty$).

If alternatively the double-ended symmetrical 90 percent confidence limits on the population standard deviation are desired, then:

χ_2^2 represents the 5 percent confidence limit
χ_1^2 represents the 95 percent confidence limit.

From Table 8-3 for 11 degrees of freedom:

$$\chi_1^2 = 19.675$$

$$\chi_2^2 = 4.575$$

Hence, there is a (100–95) percent probability (confidence level) that $\sigma < \sigma_1$ where:

$$\sigma_1 = 0.00642 \left[\frac{11}{19.675} \right]^{1/2} = 0.00480$$

and a (5–0) percent probability (confidence level) that $\sigma \geq \sigma_2$ where:

$$\sigma_2 = 0.00642 \left[\frac{11}{4.575} \right]^{1/2} = 0.00995$$

Since there is a 10 percent probability that σ lies outside the range of σ_1, σ_2 (i.e., $\sigma \geq \sigma_2$ or $\sigma < \sigma_1$), then there is a 90 percent probability that σ lies within the range σ_1, σ_2 (i.e., $\sigma \leq \sigma_2$ and $\sigma > \sigma_1$). Hence $0.0048 < \sigma \leq 0.00995$, with a 90 percent confidence level.

III HOT SPOTS AND SUBFACTORS

In developing the thermal design of a nuclear reactor, it is customary to consider first the nominal performance of the reactor with each of the primary design variables at a completely specified nominal value and then to evaluate the effect on reactor performance of possible variations in each of the primary design variables from its nominal value. Hot spot and hot channel factors are used to express the extent to which actual reactor performance may depart from its nominal performance owing to the cumulative effect of variations of all primary design variables from their nominal values.

A few examples of hot spot and hot channel factors introduced in different design approaches will make the general notion clearer. In the design of the early pressurized water reactors, a heat flux hot spot factor F_Q was used. This factor was defined as the ratio of the highest heat flux that would possibly occur anywhere in the core, to the average heat flux. Similarly, a film temperature drop hot spot factor F_θ was defined as the ratio of the maximum temperature difference between cladding and coolant which could possibly occur anywhere in the core, to the average of this temperature difference. Finally, either a coolant temperature rise hot channel factor $F_{\Delta T}$ or a coolant enthalpy rise hot channel factor $F_{\Delta h}$ was used. The coolant temperature factor $F_{\Delta T}$ was defined as the ratio of the maximum coolant temperature rise that could possibly occur in any fuel assembly of the reactor, to the core average temperature rise. The coolant enthalpy rise hot channel factor $F_{\Delta h}$ was defined as the ratio of the maximum specific enthalpy increase of the coolant which could possibly occur in any fuel assembly, to the average specific enthalpy increase. Note that when the coolant heat capacity is independent of temperature, $F_{\Delta T}$ and $F_{\Delta h}$ are numerically equal.

The hot spot factors described above were defined for applications in LWRs primarily because they relate to the onset of the critical heat flux condition. In the LMRs, however, critical heat flux is not a limiting condition. For example, at one atmosphere sodium has a boiling point much higher than the clad temperature limit, a low vapor pressure, and a large superheat for nucleation. All these conditions are quite different from those found in LWRs and dictate the use of hot channel factors different from those defined for LWRs.

In the LMR case, the principal limits are:

1. The cladding temperature should not exceed a critical value which results in unacceptable creep rates, which in turn depend on the material, applied stresses, and irradiation-induced swelling.
2. The fuel temperature should be lower than the fuel melting point.

The temperature of interest M in an LMR core can be written in the form:

$$T_M = T_{in} + \sum_{m=1}^{M} F_m \cdot \Delta T_{m,nom} \qquad (8\text{-}17)$$

where:

$$F_m = \frac{T_M - T_{M-1}}{\Delta T_{m,nom}} = \frac{\Delta T_m}{\Delta T_{m,nom}} \qquad (8\text{-}18)$$

T_{in} = coolant inlet temperature

nom = nominal value

The subscripts M and m represent temperature values and temperature differences, respectively. Specifically:

$M = 5$ − the fuel centerline temperature
$m = 5$ − the radial temperature difference between fuel surface and fuel centerline

$M = 4$ − the fuel surface temperature
$m = 4$ − the radial temperature difference across the clad–fuel gap

$M = 3$ − the clad inside temperature
$m = 3$ − the radial temperature difference across the clad

$M = 2$ − the clad outside temperature
$m = 2$ − the radial temperature difference across the coolant film

$M = 1$ − the coolant mixed mean temperature
$m = 1$ − the axial coolant temperature rise from the inlet to an axial plane of interest

$M = 0$ − coolant inlet temperature
$m = 0$ − not defined or used.

We have now identified the properties of interest in both LWRs and LMRs. For each we are interested in finding the associated hot spot factors. For example, in LWRs the properties of interest are

- Heat flux
- Film temperature rise (θ)
- Enthalpy (Δh) or coolant temperature rise (ΔT)

whereas in LMRs the properties of interest are:

- Temperature rises (T_M where $M = 1, 2, \ldots 5$).

Since these quantities all have different nomenclature for historical reasons, let us introduce the parameter y as the property whose hot spot factor is of interest. Since this property is a function of many variables x_j write:

$$y = f(x_1, \ldots x_j, \ldots x_n)$$

where x_j ($j = 1, \ldots n$) are individual parameters that affect the property y.

For examples of x_j variables, see Table 8-4 which presents an extensive list of both x_j variables and the properties y that they affect.

Let x_j^0 be the nominal value of x_j. If Δx_j is the deviation of x_j from its nominal value x_j^0, then we can write:

$$x_j = x_j^0 + \Delta x_j$$

Further defining the nominal value y^0 of y as:

$$y^0 = y(x_1^0, \ldots, x_j^0, \ldots x_n^0) \tag{8-19}$$

We can also express the property y for the case where all x_j are not at their nominal values as:

$$y = y(\Delta x_1, \ldots, \Delta x_j, \ldots \Delta x_n) \tag{8-20}$$

where for simplicity only the Δx_j rather than x_j^0 and Δx_j are shown. There are two ways of defining the subfactor that accounts for the influence of a single variable x_j on the property y:

1. The first definition of subfactors relates deviations in x_j to the nominal value x_j^0. The specific definitions depend on the way deviations in x_j affect y and are formulated so that the subfactors are greater than unity when the deviation in x_j acts to increase y. Thus if an increase in x_j increases y, the subfactor of x_j is defined as:

$$f_j = \frac{x_j}{x_j^0} = 1 + \frac{\Delta x_j}{x_j^0} \tag{8-21}$$

On the other hand, if a decrease in x_j increases y, then the subfactor of x_j is defined as:

$$f_j = \frac{x_j^0}{x_j} = \frac{x_j^0}{x_j^0 - \Delta x_j} = \frac{1}{1 - \dfrac{\Delta x_j}{x_j^0}} \simeq 1 + \frac{\Delta x_j}{x_j^0} \tag{8-22}$$

$$(\Delta x_j \ll x_j^0)$$

2. The second way of defining subfactors (and the most used nowadays) depends on the functional relationship between x_j and the property y. This way of defining subfactors relates deviations in the property y, which are due to deviations in a single parameter x_j, to the nominal value of y. We will adopt the notation $f_{j,y}$ meaning the subfactor relative to parameter x_j affecting the property y. It is given by

$$f_{j,y} = \frac{y(0, \ldots, \Delta x_j, \ldots, 0)}{y(0, \ldots, 0, \ldots, 0)} = \frac{y_{\Delta x_j}}{y^0} \tag{8-23}$$

Example 8-5 Computation of hot spot factors from first principles

PROBLEM Assume that the pellet diameters in a core have been determined to be normally distributed with a mean ($\mu_j \equiv x_j^0$) of 0.23 inches and a standard deviation, σ_j, of 0.0115. Compute the hot spot factor, f_j, for the pellet diameter, x_j, and the hot spot factor $f_{j,y}$, for the temperature difference between the centerline and surface temperatures, ($y = T_{max} - T_{fo}$) of a one-zone cylindrical fuel pellet due to variations in pellet outside diameter. Compute these subfactors for a confidence level of 95.45 percent. Use a symmetric confidence interval.

SOLUTION Eq. 8-21 is used here because an increase in pellet diameter causes an increase in ($T_{max} - T_{fo}$) so:

$$f_j = 1 + \frac{\Delta x_j}{x_j^0} \tag{8-21}$$

We require Δx_j to be the variation in x_j^0 such that there is a 95.45 percent probability that the value of x_j is within the interval $x_j^0 + \Delta x_j$. We can

Table 8-4 Variables affecting hot spot temperatures

Uncertainty	Coolant temperature rise	Film temperature drop	Cladding temperature drop	Gap temperature drop	Fuel temperature drop
Power measurement	X	X	X	X	X
Flux distribution					
Due to calculation	X	X	X	X	X
Due to control rod position	X	X	X	X	X
Inlet flow distribution between assemblies	X	X			
Flow maldistribution within the assembly	X	X			
Orificing uncertainty	X	X			
Unit cell flow area					
Cold bow	X	X			
Hot bow	X	X			
Cladding swelling	X	X			
Pitch variation	X	X			
Variation in cladding outside diameter	X	X			
Corrosion of cladding	X	X			
Effect of cross-flow on the hot channel	X				

Parameter					
Effect of mixing on the hot channel					X
Effect of obstructions (wire wrap, grids)				X	
Circumferential temperature variation around a fuel rod				X	
Specific heat of the coolant					X
Density of the coolant					X
Heat transfer coefficient					
Film				X	
Gap		X			
Thermal conductivity					
Coolant				X	
Fuel				X	
Cladding	X				
Fissile content per pellet (pellet diameter, fuel density, enrichment)	X	X	X		
Active fuel length per rod	X	X	X		
Coolant temperature at reactor inlet			X	X	
Cladding thickness					
Manufacturing tolerances			X		
Corrosion			X		
Swelling			X		

express Δx_j as a number, z, of standard deviations, σ_j. Hence:

$$f_j = \frac{x_j^0 + z\sigma_j}{x_j^0} \text{ since } \Delta x_j \equiv z\sigma_j \tag{8-22}$$

$$f_j = 1 + z(0.05) \text{ since } \frac{\sigma_j}{x_j^0} = \frac{0.0115}{0.23} = 0.05$$

per the problem statement. A symmetrical 95.45 percent confidence interval corresponds to $\pm 2\sigma$ as seen in Table 8-1, i.e., z takes the value which corresponds to:

$$P_z = \frac{1 - 0.9545}{2} = 0.02275$$

which per Table 8-1 is $z = 2.0$.

Hence:

$$f_j = 1 + 2.0(0.05) = 1.1$$

Similarly for $z = -2.0$, $f_j = 0.9$.

From the definition of $f_{j,y}$ we have:

$$f_{j,y} = \frac{y_{\Delta x_j}}{y^0} = \frac{y_{z\sigma_j}}{y^0} \tag{8-23}$$

In this case $y^0 \equiv (T_{max} - T_{fo})^0$, and from the solution of the general energy equation for the one-zone pellet of thermal conductivity k and diameter D_f:

$$T_{max} - T_{fo} = \frac{q''' D_f^2}{16k}$$

Hence:

$$(T_{max} - T_{fo})^0 = \frac{q'''(x_j^0)^2}{16k}$$

and

$$(T_{max} - T_{fo})_{z\sigma_j} = \frac{q'''(x_j^0 + z\sigma_j)^2}{16k}$$

Substituting these results and the value of $z = \pm 2.0$ into Eq. 8-23 yields:

$$f_{j,y} = \frac{q'''(x_j^0 \pm z\sigma_j)^2/16k}{q'''(x_j^0)^2/16k} = \frac{[0.23 \pm 2.0(0.05)(0.23)]^2}{(0.23)^2}$$

$$= 1.21; \ 0.81$$

This result tells us that $0.81 < (T_{max} - T_{fo})/(T_{max} - T_{fo})^0 \le 1.21$ with a 95.45 percent probability when all other parameters affecting this quantity are at their nominal values.

A relationship between an f_j affecting y and $f_{j,y}$ can be derived in the following manner. The differential of the property y is given by:

$$dy = \frac{\partial y}{\partial x_1} dx_1 + \cdots + \frac{\partial y}{\partial x_j} dx_j + \cdots + \frac{\partial y}{\partial x_n} dx_n$$

Taking $dx_i = 0$ for $i \neq j$ (since we are only interested in the effect of x_j), we have:

$$dy_j = \frac{\partial y}{\partial x_j} dx_j$$

or for finite changes:

$$\Delta y_j = \frac{\partial y}{\partial x_j} \Delta x_j$$

Since Δy_j is defined as $y_{\Delta x_j} - y^0$, the following relationship exists:

$$y_{\Delta x_j} - y^0 = \left[\frac{\partial y}{\partial x_j} \right] (x_{j,\text{extreme}} - x_j^0)$$

The expression above can be put in the form:

$$\frac{y_{\Delta x_j}}{y^0} - 1 = \frac{x_j^0}{y^0} \left[\frac{\partial y}{\partial x_j} \right] \left[\frac{x_{j,\text{extreme}} - x_j^0}{x_j^0} \right]$$

Noting that:

$$\frac{y_{\Delta x_j}}{y^0} \equiv f_{j,y}$$

and

$$\frac{x_j^0}{y^0} \left[\frac{\partial y}{\partial x_j} \right] = \left[\frac{\partial \ell n\, y}{\partial \ell n\, x_j} \right] \quad \begin{array}{l} \text{(for small deviations of } x \text{ and } y \\ \text{about their nominal values)} \end{array}$$

and

$$\left| \frac{x_{j,\text{extreme}} - x_j^0}{x_j^0} \right| \equiv f_j - 1$$

we get:

$$(f_{j,y} - 1) = \left[\frac{\partial \ell n\, y}{\partial \ell n\, x_j} \right] (f_j - 1) \qquad (8\text{-}24)$$

In the cases where $(f_{j,y} - 1) \ll 1$ and $(f_j - 1) \ll 1$, which happens most of the time, we can write:

$$f_{j,y} - 1 \simeq \ell n[1 + (f_{j,y} - 1)] = \ell n\, f_{j,y}$$

$$f_j - 1 \simeq \ell n[1 + (f_j - 1)] = \ell n\, f_j$$

So, Eq. 8-24 can be rewritten as:

$$\ell n \, f_{j,y} = \left[\frac{\partial \, \ell n \, y}{\partial \, \ell n \, x_j} \right] \ell n \, f_j$$

$$f_{j,y} = f_j^{\left[\frac{\partial \ln y}{\partial \ln x_j} \right]} \tag{8-25}$$

It can be easily proven that this expression is exact when functionally y is the product of the parameters x_j, raised to any exponent, i.e.:

$$y = A x_1^a x_2^b \ldots x_j^n \ldots x_n^t$$

For other forms of the property y, the error involved in evaluating $f_{j,y}$ by Eq. 8-25 may be significant enough to dictate that it should be evaluated directly by Eq. 8-23.

Example 8-6 Computation of hot spot factor $f_{j,y}$ from factors f_j

PROBLEM Use Eq. 8-25 to compute the hot spot factor $f_{j,y}$ for the conditions of Example 8-5.

SOLUTION In this case:

$$x_j = x_j^0 + z \sigma_j = D_f$$

and

$$y = T_{max} - T_{fo}$$

From the solution of the energy equation for the one-zone pellet of thermal conductivity k and diameter D_f:

$$T_{max} - T_{fo} = \frac{q''' D_f^2}{16k}$$

Hence:

$$\ell n (T_{max} - T_{fo}) = 2 \, \ell n \, D_f + \ell n \, q''' - \ell n \, 16 - \ell n \, k$$

$$\frac{\partial \, \ell n (T_{max} - T_{fo})}{\partial \, \ell n \, D_f} = 2$$

From Example 8-5:

$$f_j = 1.1$$

Substituting these results into Eq. 8-25 yields:

$$f_{j,y} = f_j^{\left[\frac{\partial \ln y}{\partial \ln x_j} \right]}$$

$$= (1.1)^2$$

$$= 1.21 \text{ completing Example 8-6}$$

IV COMBINATIONAL METHODS: SINGLE HOT SPOT IN CORE

The overall hot spot factor F is determined from contributions of all of the $f_{j,y}$ that affect y. The procedure of combining these subfactors depends on the nature of the individual variables. Several methods of combination of the subfactors have been suggested, and they will be critically reviewed in this section.

The differences in properties and limits central to the thermal design of LWRs and LMRs were identified in the previous section. We will not present the various reactor thermal design approaches in use since they are in continual evolution. It is sufficient to say that for LWRs the dominant approach is to derive a critical heat flux or critical power limit that includes uncertainties in the correlation itself and in the parameters that enter the prediction of critical heat flux or power. In LMRs and gas-cooled reactors attention is focused on predicting coolant, clad or coating and fuel temperatures including relevant uncertainties.

In the remainder of this chapter attention will be focused on two common key elements in a reactor thermal design procedure:

1. A single hot pin analysis—The methods that can be used to combine the effects of the uncertainties on a parameter y from variables x_j to obtain the probability of a single hot pin exceeding defined limits
2. A multiple hot pin or core-wide analysis—The summing of the probabilities of each pin exceeding the defined limit which leads to a core-wide reliability analysis.

Returning now to the focus of this section on combinational methods, the predominant current approach is the semistatistical method. Originally proposed by Chelemer and Tong [10] this approach was the logical culmination of early work which treated all the uncertainties as either deterministic or statistical in nature. The deterministic approach takes each parameter that affects the property of interest as having the most unfavorable value. In this way worst values are taken to occur at the same location at the same time. This method is very conservative and pessimistic. Within the approximations and the confidence level of the hot spot or channel subfactors it assures that the limits will not be exceeded at any location in the reactor core at any time.

Since the probability of the most unfavorable value of all uncertainties occurring at the same position at the same time is very small, the statistical method for combining uncertainties was introduced. In this method the uncertainties are combined statistically, and the resulting hot spot (or channel) factors are a function of a chosen confidence level. This confidence level depends upon the selected safety margin assigned to the reactor:

$$F_y = F_y(z\sigma) \tag{8-26}$$

The statistical procedure of combining the hot spot (or channel) subfactors depends on the assumed statistical distributions of these subfactors. The most commonly used distributions are the normal and the rectangular distributions, although there are nonparametric procedures, e.g., method of moments, available for application to arbitrary distributions.

Table 8-5 summarizes the formulas that are developed for each of these combinational methods in the remainder of the section.

Example 8-7 Deterministic method of developing the heat flux hot spot factor

PROBLEM Find the heat flux hot spot factor $F_y \equiv F_Q$ using the deterministic approach if the nominal heat flux $y \equiv Q \equiv (q/A)^0$ is given functionally by:

$$\left(\frac{q}{A}\right)^0 = \frac{\phi^0 \sigma_f \rho^0 A_v r^0 Q_f \pi (D_f^0)^2/4}{A_f \pi D_c^0} \tag{8-27}$$

where the x_j variables are:

ϕ^0—the average neutron flux
σ_f—the U-235 fission cross-section
ρ^0—the average uranium density
r^0—the average weight fraction of uranium 235
Q_f—the energy released by fission which is transferred within the fuel as heat
D_f^0—the average diameter of fuel
D_c^0—the average outside diameter of cladding
A_f—the fuel atomic weight
A_v—Avogadro's number

since:

$$\left(\frac{q}{A}\right) = \frac{(\phi \Sigma V) Q_f}{A_{\text{fuel surface}}} = \frac{\phi \left(\sigma_f \dfrac{\rho A_v}{A_f} r\right)\left(\dfrac{\pi}{4} D_f^2 L\right) Q_f}{\pi D_c L} \tag{8-28}$$

SOLUTION The maximum possible value the heat flux can have at any point in the reactor $(q/A)_{\text{max}}$ occurs when the factors in the numerator have their maximum possible values and the factor in the denominator has its minimum possible value:

$$\left(\frac{q}{A}\right)_{\text{max}} = \frac{\phi_{\text{max}} \sigma_f \rho_{\text{max}} r_{\text{max}} Q_f \pi (D_f)^2_{\text{max}}/4}{\pi (D_c)_{\text{min}} A_f} \tag{8-29}$$

The heat flux hot spot factor F_Q is defined mathematically as:

$$F_Q = (q/A)_{\text{max}}/(q/A)^0 \tag{8-30}$$

Table 8-5 Formula for combinational methods of hot spot analysis

	Hot spot factor		Hot spot temperature	
Cumulative				
Product method	$F_m = \prod_{j=1}^{n}(f_{j,m})$	(8-34)†	$T_M = T_{in} + \sum_{m=1}^{M} F_m \cdot \Delta T_{m,nom}$	(8-52)
Sum method	$F_m = 1 + \sum_{j=1}^{n}(f_{j,m} - 1)$	(8-40)†	$T_M = T_{in} + \sum_{m=1}^{M} F_m \cdot \Delta T_{m,nom}$	(8-52)
Total statistical (sum approach)				
Vertical approach	$F_m = 1 + \sqrt{\sum_{j=1}^{n}(f_{j,m} - 1)^2}$	(8-50)†	$T_M = T_{in} + \sum_{m=1}^{M} F_m \cdot \Delta T_{m,nom}$	(8-52)
Horizontal approach	$F_m = \dfrac{T_M - T_{M-1}}{\Delta T_{m,nom}}$	(8-57)	$T_M = T_{in} + \sum_{m=1}^{M} \Delta T_{m,nom} + \sqrt{\sum_{j=1}^{n}\left[\sum_{m=1}^{M}(f_{j,m} - 1)\cdot \Delta T_{m,nom}\right]^2}$	(8-56)
Semistatistical				
Vertical approach	$F_m = F_m^D F_m^S = F_m^D \cdot \left(1 + \sqrt{\displaystyle\sum_{\substack{j=1 \\ \text{stat}}}^{n}(f_{j,m} - 1)^2}\right)$	(8-58)†	$T_M = T_{in} + \sum_{m=1}^{M} F_m \cdot \Delta T_{m,nom}$	(8-52)
Horizontal approach	$F_m = \dfrac{T_M - T_{M-1}}{\Delta T_{m,nom}}$	(8-57)	$T_M = T_{in} + \sum_{m=1}^{M} F_m^D \Delta T_{m,nom} + \sqrt{\displaystyle\sum_{\substack{j=1 \\ \text{stat}}}^{n}\left[\sum_{m=1}^{M}(f_{j,m} - 1)F_m^D \Delta T_{m,nom}\right]^2}$	(8-59)

† These formulas in the text are presented for the general reactor case, therefore, $y = m$.

379

Using the first definition of subfactor f_j in Eq. 8-21, F_Q can be put in the form:

$$F_Q = f_\phi f_\rho f_r (f_{D_f})^2 f_{D_c} \tag{8-31}$$

where the subfactors are:

flux, f_ϕ — ϕ_{max}/ϕ^0
fuel density, f_ρ — ρ_{max}/ρ^0
enrichment, f_r — r_{max}/r^0
fuel diameter, f_{D_f} — $(D_f)_{max}/D_f^0$

and

cladding diameter, f_{D_c} — $D_c^0/(D_c)_{min}$ \tag{8-32}

Some of these subfactors, in turn, may be broken down into partial subfactors. For example, the flux subfactor is often represented as:

$$f_\phi = f_R f_Z f_{Local}$$

where f_R is the ratio of peak flux to flux averaged with respect to radial distance; f_Z is the ratio of peak flux to flux averaged with respect to axial distance; and f_{Local} is a subfactor to represent local variations in flux due to control rods, water gaps, and other lattice irregularities. These factors are usually treated separately.

In some treatments of hot spot and hot channel factors, the factors f_R and f_Z representing gross variations in neutron flux are incorporated into ϕ^0 and left out of the expression for the hot spot factor. In that case ϕ^0 is the nominal maximum flux in the reactor instead of the average flux.

A Deterministic Method Formulations

In Example 8-7 the property F_Q is a function of the product of factors where each factor is a function of a single variable x_j. This is not the case for all properties. However, if a property y can be expressed as a function that is mainly the product of factors, where each factor is a function of a single variable x_j, then a good approximation for the hot spot factor F_y in terms of subfactors f_j defined to represent the maximum possible fractional variation of the variables x_j is:

$$\ell n \, F_y = \sum_{j=1}^n \left| \frac{\partial \, \ell n \, y}{\partial \, \ell n \, x_j} \right| \ell n \, f_j$$

or

$$F_y = \prod_{j=1}^n f_j^{\left[\frac{\partial \ln y}{\partial \ln x_j} \right]} \tag{8-33}$$

The advantage of this approximation procedure is that it expresses the hot channel factor as a product of individual subfactors, each raised to a power that can be evaluated from a general knowledge of how the property of interest depends on the variables which may deviate from the nominal values used in design.

The preceding development was done using the definition of subfactors given by Eqs. 8-21 and 8-22. Substituting Eq. 8-25 into Eq. 8-33 treats cases in which the subfactors are given according to the definition in Eq. 8-23 yielding the result:

$$F_y = \prod_{j=1}^{n} f_{j,y} \tag{8-34}$$

Eqs. 8-33 and 8-34 are the mathematical formulation for the approach usually called the *product deterministic* (or *cumulative*) *approach*. An application of these equations is illustrated in Table 8-6 for the hot spot factor for the heat flux F_Q. The end result F_Q, calculated by these two approaches, is identical. The subfactors f_j and $f_{j,y}$ are calculated by Eqs. 8-21 and 8-22, and by Eq. 8-23, respectively. For all variables except fuel diameter these subfactors are equal since fuel diameter is the only factor among those listed upon which the heat flux is not linearly dependent. This is illustrated by Eq. 8-29, from which we can observe that:

$$\frac{\partial \ell n \ (q/A)_{max}}{\partial \ell n \ D_f} = 2$$

Table 8-6 Subfactors affecting (q/A) for a typical early LWR reactor

	Subfactors	
Contributors	f_j defined by Eqs. 8-21 and 8-22	$f_{j,y}$ defined by Eq. 8-23
1. Radial variation of power density, f_R	2.058	2.058
2. Axial variation of power density, f_Z	1.484	1.484
3. Local variation of power density, f_{Local}	1.050	1.050
4. Fuel density, f_p	1.044	1.044
5. Fuel enrichment, f_r	1.016	1.016
6. Fuel diameter, f_{D_f}	1.003	1.006
7. Cladding diameter, f_{D_c}	1.003	1.003

F_Q (deterministic method by Eqs. 8-33 and 8-34) = 3.43.
F_Q (semistatistical method† by Eq. 8-58) = 3.36.

$$\frac{q}{A} = \frac{\phi \sigma_f \rho A_v r Q_f \pi (D_f^2)/4}{A_f \pi D_c} \tag{8-28}$$

† Taking items 1, 2, and 3 as direct contributors and 4, 5, 6, and 7 as statistical contributors. See Section IVC.

Hence the subfactors $f_{j,y} \equiv f_{(q/A)_{max},D_f}$ and $f_j = f_{D_f}$ are different and related in accordance with Eq. 8-25 as:

$$f_{(q/A)_{max},D_f} = (f_{D_f})^2$$

which is the relationship Table 8-5 reflects.

The mathematical formulation of the *sum deterministic* (or *cumulative*) approach will now be developed.

When y is a linear function of individual variables x_j such as:

$$y = \sum_{j=1}^{n} a_j x_j \tag{8-35}$$

the maximum value y can have owing to extreme variations in each of the x_j values from nominal values x_j^0 is:

$$y_{max} = y^0 + \sum_{j=1}^{n} \left[\frac{\partial y}{\partial x_j}\right] [x_{j,extreme} - x_j^0] \tag{8-36}$$

The hot spot or hot channel factor for y is defined as:

$$F_y \equiv y_{max}/y^0 \tag{8-37}$$

Further define subfactors f_j as follows:

$$f_j = 1 + \left|\frac{x_{j,extreme} - x_j^0}{x_j^0}\right| \tag{8-21}$$

Dividing Eq. 8-36 by y^0 and multiplying the second term on the right side by x_j^0/x_j^0 obtain:

$$\frac{y_{max}}{y^0} = 1 + \sum_{j=1}^{n} \left[\frac{\partial y}{\partial x_j}\frac{x_j^0}{y^0}\right] \left|\frac{x_{j,extreme} - x_j^0}{x_j^0}\right| \tag{8-38}$$

and noting Eqs. 8-37 and 8-21 this becomes:

$$F_y = 1 + \sum_{j=1}^{n} \left[\frac{\partial y}{y^0}\frac{x_j^0}{\partial x_j}\right] (f_j - 1)$$

and since $\partial (\ell n\ y) \cong \partial y/y^0$ and $\partial (\ell n\ x_j) \cong \partial x_j/x_j^0$:

$$F_y = 1 + \sum_{j=1}^{n} \left[\frac{\partial \ell n\ y}{\partial \ell n\ x_j}\right] (f_j - 1) \tag{8-39}$$

This result can be expressed in terms of the subfactors defined by Eq. 8-23 by substituting Eq. 8-24 into Eq. 8-39 yielding:

$$F_y = 1 + \sum_{j=1}^{n} (f_{j,y} - 1) \tag{8-40}$$

This approach is called the *sum deterministic* (or *cumulative*) *approach.*

Eqs. 8-39 and 8-33 become equivalent when F_y and the f_j values are close to unity. The sum form, Eqs. 8-39 and 8-40, gives more accurate results the closer y is to being a linear function of the x_j values, while the product form, Eqs. 8-33 and 8-34, is preferable the closer y is to being a product of factors that are each a function of only a single x_j. For the cases of the hot spot factors for heat flux F_Q (Example 8-7) and film drop F_θ (Problem 8-1), the product form is clearly preferable.

B Statistical Method Formulations

The two procedures that were developed for combining factors by this method parallel those introduced in the preceding section and are discussed next.

1 Product statistical method In this case F_y again is given by a product of the $f_{j,y}$ factors or $f_j^{[\partial \ln y / \partial \ln x_j]}$ factors, but each is treated as a statistical variable with a certain distribution function. The distribution function of a product of two or more statistical variables is generally difficult to evaluate even when each variable has a normal distribution because the product generally is not a random variable of the same type. Even for the simplest case of only two subfactors, f_1 and f_2, with independent probability density functions $p(f_1)$ and $p(f_2)$, respectively, the probability that a desired limit A is not exceeded is a complicated calculation given by:

$$P[f_1 \cdot f_2 < A] = \int_0^\infty \int_0^{A/f_2} p(f_1) \cdot p(f_2) df_1 df_2 \qquad (8\text{-}41)$$

For more than two subfactors, this calculation of the resulting distribution function has to be done step by step starting from the first two variables, then combining the distribution function of the first product with the third variable, and so on. The solution of Eq. 8-41 has to be done numerically which can be inconvenient and time consuming. However, if the subfactors are lognormally distributed, then the procedure is simplified because the product of lognormal variables is also lognormal. In this case Eq. 8-41 can be solved using a simple transformation that relates the product to the standardized normal distribution which is characterized by Table 8-1.

2 Sum statistical method This method was developed to avoid the originally required cumbersome solution of Eq. 8-41. It is based on the fact that the sum of n normal, independent random variables is also a random variable that is normally distributed and whose mean and variance are the sum of the means and variances of the n variables respectively. According to the central limit theorem, this relationship between means and variances is valid even if the random variables are not normally distributed, if n is a large number.

The definition of subfactors, for statistically distributed variables, will be a function of the desired confidence level. Since each confidence level is related to a specific value of z, the number of standard deviations about the distribution mean, we will use the terminology *the confidence level corresponding to z* in the remainder of the chapter.

Let x_j be a parameter with mean x_j^0 and standard deviation σ_j, affecting y. Then, according to the definition supplied by Eqs. 8-21 and 8-22, the subfactor relative to x_j is:

$$f_j = 1 + \frac{z\sigma_j}{x_j^0} \tag{8-42}$$

According to the subfactor definition of Eq. 8-23:

$$f_{j,y} = \frac{y(0, \ldots, z\sigma_j, \ldots, 0)}{y(0, \ldots, 0, \ldots, 0)} = \frac{y_{z\sigma_j}}{y^0} \tag{8-43}$$

The distribution function of y is given by:

$$f(y) = \frac{1}{\sqrt{2\pi}\sigma_y} \exp\left(\frac{-(y - y^0)^2}{2\sigma_y^2}\right)$$

where y^0 is the mean value of y, and σ_y is the standard deviation. The standard deviation σ_y is related to the standard deviations σ_j of the variables x_j by the following relationship:

$$\sigma_y^2 = \sum_{j=1}^{n} \left(\frac{\partial y}{\partial x_j}\right)^2 \sigma_j^2 \tag{8-44}$$

Eq. 8-44 is true for any number of variables x_j if y is a linear function of the x_j variables and if the x_j variables are independent.

We define:

$$z = \frac{y^* - y^0}{\sigma_y} \tag{8-45}$$

where y^* is the maximum value assumed by y, for the confidence level corresponding to z. Then from the definition of $F_{y,z}$:

$$F_{y,z} = \frac{y^*}{y^0} \tag{8-46}$$

Further, from Eqs. 8-44, 8-45, and 8-46, we get:

$$F_{y,z} - 1 = \frac{z}{y^0} \left[\sum_{j=1}^{n} \left(\frac{\partial y}{\partial x_j}\right)^2 \sigma_j^2\right]^{1/2} \tag{8-47}$$

Using Eq. 8-42, we obtain for σ_j^2:

$$\sigma_j^2 = \frac{(x_j^0)^2}{z^2} (f_j - 1)^2 \tag{8-48}$$

Substituting Eq. 8-48 into Eq. 8-47 where we assume that the z value corresponding to the confidence level of $F_{y,z}$ and that for f_j are numerically equal

obtain:

$$F_{y,z} - 1 = \left[\sum_{j=1}^{n} \left[\frac{x_j^0}{y^0} \left(\frac{\partial y}{\partial x_j} \right) \right]^2 (f_j - 1)^2 \right]^{1/2}$$

or

$$F_{y,z} = 1 + \left[\sum_{j=1}^{n} \left(\frac{\partial \ell n \, y}{\partial \ell n \, x_j} \right)^2 (f_j - 1)^2 \right]^{1/2} \qquad (8\text{-}49)$$

For the case in which we are using the definition given by Eq. 8-23 for the subfactor, we can write an equivalent expression for $F_{y,z}$ by substituting Eq. 8-24 into Eq. 8-49 and taking $f_{j,y}$ defined for the same confidence level:

$$F_{y,z} = 1 + \left[\sum_{j=1}^{n} (f_{j,y} - 1)^2 \right]^{1/2} \qquad (8\text{-}50)$$

This method was developed originally for use in LWRs in which the properties like q/A, θ, and Δh are analyzed separately, i.e., studying q/A we only have to take into account the variables that affect q/A to find F_Q. The method has also been used in the analysis of the LMR because it is quite convenient for engineering calculations. In this case the property under study is the temperature of the element M, where M can be the inside surface of the cladding, the fuel centerline, etc. This temperature can be put in the form:

$$T_M = T_{in} + \sum_{m=1}^{M} \Delta T_m \qquad (8\text{-}51)$$

A variation in a parameter x_j probably will affect more than one temperature drop ΔT_m as Table 8-4 illustrates. These effects must be combined statistically. Applying this approach for each ΔT_m separately yields T_M as:

$$T_M = T_{in} + \sum_{m=1}^{M} F_m \Delta T_{m,nom} \qquad (8\text{-}52)$$

where F_m is given by Eqs. 8-49 or 8-50. This method is quite convenient for engineering calculations because it allows the definition of an overall hot spot factor for each of the ΔT_m values. Table 8-7 presents typical LMR hot channel factors arranged in a matrix in which the rows are the effect of each factor on a temperature rise m of interest, and the columns are the temperature rises of interest. In this approach the factors affecting each temperature rise ΔT_m are arrayed vertically and grouped using Eq. 8-50 to yield F_m values that are then applied individually to each ΔT_m. Hence this method is called the *statistical vertical approach*. This approach corresponds only to a partial statistical combination because the overall effect of x_j is given by the sum of the effects in each one of the T_m values, and hence it is conservative relative to the horizontal method presented next.

Table 8-7 Typical LMR hot channel factors at 3σ confidence level

$f_{j,y}$	Coolant 1	Film 2	Clad 3	Gap 4	Fuel 5
A. Direct					
1. Inlet flow maldistribution	1.05	1.012			
2. Intrasubassembly flow maldistribution	1.14	1.035			
3. Interchannel coolant mixing	1.00†				
4. Power control band	1.02	1.02	1.02	1.02	1.02
5. Wire wrap peaking		2.00(1.214)‡			
Direct combination (Eq. 8-34)	1.221	2.137(1.297)‡	1.02	1.02	1.02
B. Statistical					
1. Fissile fuel maldistribution	1.035	1.035	1.035	1.035	1.035
2. Power level measurement	1.071	1.071	1.071	1.071	1.071
3. Nuclear power distribution	1.060	1.065	1.065	1.065	1.065
4. Rod diameter, pitch, and bow	1.011				
5. Film coefficient		1.14			
6. Gap coefficient				1.47§	
7. Fuel conductivity					1.10
8. Clad conductivity and thickness			1.12		
Statistical combination of items 1–8 (Eq. 8-50)	1.100	1.173	1.158	1.481	1.143
Product of direct and statistical	1.343	2.507(1.521)‡	1.181	1.511	1.166

† Worst condition.
‡ Numbers in parentheses should be used for calculating fuel temperatures.
§ Taken with h_g(nominal) = 7950 W/m² K (1,400 BTU/hr ft² F)

For more accurate and less conservative calculations, another method was developed called the *statistical horizontal approach*. This approach combines statistically all of the effects of a variation in the variable x_j on all of the temperature differences up to the temperature T_M under study. So, for a typical temperature T_M the variation of T_M due to the variable x_j is:

$$\delta T_{M,\Delta x_j} = T_{M,\Delta x_j} - T_{in} - \sum_{m=1}^{M} \Delta T_{m,nom} \qquad (8\text{-}53)$$

Taking $\Delta x_j = z\sigma_j$, this deviation is also given by:

$$\delta T_{M,z\sigma_j} = \sum_{m=1}^{M} (f_{j,m} - 1)\Delta T_{m,nom} \qquad (8\text{-}54)$$

From basic statistics $(\delta T_{M,z})^2 = \sum_{j=1}^{n} (\delta T_{M,z\sigma_j})^2$, i.e., the variance (σ^2) of a sum of independent random variables is the sum of the variances of each variable. Since the total deviation of δT_M for the confidence level corresponding to z, taking into account all variables, is given by:

$$\delta T_{M,z} = \left[\sum_{j=1}^{n} \left[\sum_{m=1}^{M} (f_{j,m} - 1)\Delta T_{m,nom} \right]^2 \right]^{1/2} \tag{8-55}$$

the temperature T_M can be put in the form:

$$T_M = T_{in} + \sum_{m=1}^{M} \Delta T_{m,nom} + \delta T_{M,z} \tag{8-56}$$

where $\Delta T_{M,z}$ is given by Eq. 8-55.

We can define an overall hot spot factor for this method in the following manner:

$$F_m = \frac{T_M - T_{M-1}}{\Delta T_{m,nom}} \tag{8-57}$$

Recall that in the case of $M = 1$, T_{M-1} represents the nominal coolant inlet temperature T_{in}.

The temperature differences $\Delta T_{m,nom}$ depend on the location of the hot spot under study.

C Semistatistical Methods

The statistical methods assume that all uncertainties are of a statistical nature and consequently are too optimistic. In fact, certain parameters that affect the thermal performance of a reactor core are of systematic character. To account for the distinct character of the uncertainties, the semistatistical method was developed. According to this method, the parameters are divided into two groups. The first group corresponds to the systematic or direct contributors, which are cumulatively treated. The second group is the group of the statistical contributors, and they are statistically combined. The resulting temperature distribution from application of this procedure is presented in Figure 8-3. The mean of this temperature distribution, $T_{M,cum}$, is displaced from the nominal value $T_{M,nom}$, due to the direct uncertainty contributors. The distribution about the mean reflects the consideration of the statistical contributors. Table 8-7 presents some typical LMR hot channel factors illustrating direct versus statistical groups.

In current reactor analysis, the semistatistical vertical and horizontal approaches are the most commonly used approaches.

1 Semistatistical vertical approach In this approach the overall hot spot factor is given by:

$$F_y = F_y^D \cdot F_y^S = F_y^D \left[1 + \left[\sum_{\substack{j=1 \\ stat}}^{n} [f_{j,y} - 1]^2 \right]^{1/2} \right] \tag{8-58}$$

Probability Distribution
for T_M

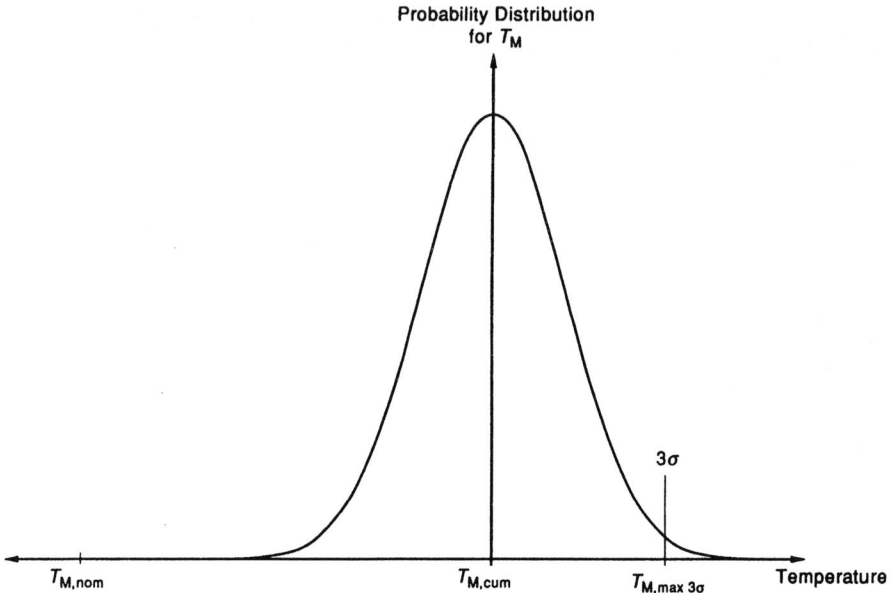

Figure 8-3 Illustration of the semistatistical method.

where:

F_y^D — Hot spot factor resulting from the deterministic contributors. It can be computed by Eqs. 8-33 and 8-34 (most cases) or Eqs. 8-39 and 8-40.

F_y^S — Hot spot factor resulting from the statistical contributors. It is calculated by Eqs. 8-49 and 8-50 taking into account only the parameters of statistical nature, i.e., $\sum\limits_{\substack{j=1 \\ \text{stat}}}^{n}$.

2 Semistatistical horizontal approach The same comments presented for the statistical horizontal approach are applicable in this case. Therefore, we only have to study the temperature of element M.

The temperature T_M, this time, is given by:

$$T_M = T_{in} + \sum_{m=1}^{M} F_m^D \Delta T_{m,nom} + \delta T_{M,z}^S \tag{8-59}$$

where:

$$\delta T_{M,z}^S = \left[\sum_{\substack{j=1 \\ \text{stat}}}^{n} \left[\sum_{m=1}^{M} (f_{j,m} - 1) F_m^D \Delta T_{m,nom} \right]^2 \right]^{1/2} \tag{8-60}$$

An overall hot spot factor for each temperature variation ΔT_m can be defined by analogy to the statistical horizontal approach.

In studies of the temperature T_M applying the methods described above, it is commonly assumed that we are computing T_M at its nominal hot spot, i.e., the point at which the function:

$$T_{M,nom}(z) = T_{in} + \sum_{m=1}^{M} \Delta T_{m,nom}(z) \qquad (8\text{-}61)$$

assumes its maximum value. This computation is done for the most critical fuel pin, and z represents the axial direction. When we include the overall hot spot factors F_m, the axial position of the maximum of the function:

$$T_M(z) = T_{in} + \sum_{m=1}^{M} F_m \Delta T_{m,nom}(z) \qquad (8\text{-}62)$$

will have a shift relative to the position of the maximum of the function of Eq. 8-61. This shift should be taken into account for more accurate calculations. Obviously the criterion for not exceeding a limit given by the methods outlined above, within the confidence level of the hot spot (or channel) subfactors, is that the maximum of the function of Eq. 8-62 is equal to the limiting temperature of the element of interest M.

Example 8-8 Computation of hot spot factors and temperatures

PROBLEM Considering the fuel hot spot, calculate the hot spot factors F_j and the hot spot temperatures T_j for each of the combinational methods of Table 8-5 using the factors of Table 8-7 and the following nominal temperature values:

T_{inlet} $= 315.6\ °C$

$\Delta T_{coolant} \equiv \Delta T_{1,nom} = 114.6\ °C$

$\Delta T_{film} \equiv \Delta T_{2,nom} = 20.4\ °C$

$\Delta T_{clad} \equiv \Delta T_{3,nom} = 44.7\ °C$

$\Delta T_{gap} \equiv \Delta T_{4,nom} = 344.9\ °C$

$\Delta T_{fuel} \equiv \Delta T_{5,nom} = 1469.8\ °C$

SOLUTION Cumulative product method:

$$F_M = \prod_{j=1}^{n} f_{j,m} \qquad (8\text{-}34)$$

$F_1 = 1.05(1.14)(1.00)(1.02)(1.035)(1.071)(1.060)(1.011) = 1.45$

$F_2 = 1.297(1.035)(1.071)(1.065)(1.14) = 1.746$

etc.

$$T_M = T_{in} + \sum_{m=1}^{M} F_m \cdot \Delta T_{m,nom} \qquad (8\text{-}52)$$

$$T_1 = 315.6 + 1.45(114.6) = 481.8 \text{ °C}$$
$$T_2 = 315.6 + 1.45(114.6) + 1.746(20.4) = 517.4 \text{ °C}$$
etc.

Cumulative sum method:

$$F_m = 1 + \sum_{j=1}^{n} (f_{j,m} - 1) \qquad (8\text{-}40)$$

$$F_1 = 1 + 0.05 + 0.14 + 0.02 + 0.035 + 0.071 + 0.060 + 0.011 = 1.387$$
etc.

$$T_M = T_{in} + \sum_{m=1}^{M} F_m \cdot \Delta T_{m,nom} \qquad (8\text{-}52)$$

$$T_1 = 315.6 + 1.387(114.6) = 474.6 \text{ °C}$$
etc.

Total statistical (sum) vertical method:

$$F_m = 1 + \left[\sum_{j=1}^{n} (f_{j,m} - 1)^2 \right]^{1/2} \qquad (8\text{-}50)$$

$$F_1 = 1 + \{(0.05)^2 + (0.14)^2 + (0.02)^2 + (0.035)^2 + (0.071)^2 + (0.060)^2 + (0.011)^2\}^{1/2} = 1.18$$

etc.

$$T_M = T_{in} + \sum_{m=1}^{M} F_m \Delta T_{m,nom} \qquad (8\text{-}52)$$

$$T_1 = 315.6 + 1.18(114.6) = 450.8 \text{ °C}$$
etc.

Total statistical (sum) horizontal method:

$$T_M = T_{in} + \sum_{m=1}^{M} \Delta T_{m,nom} + \left[\sum_{j=1}^{n} \left[\sum_{m=1}^{M} (f_{j,m} - 1) \cdot \Delta T_{m,nom} \right]^2 \right]^{1/2} \qquad (8\text{-}56)$$

$$
\begin{aligned}
T_1 &= 315.6 + 114.6 \\
&\quad + \{(0.05(114.6))^2 + (0.14(114.6))^2 + (0.02(114.6))^2 \\
&\quad + (0.035(114.6))^2 + (0.071(114.6))^2 + (0.060(114.6))^2 \\
&\quad + (0.011(114.6))^2\}^{1/2} \\
&= 315.6 + 114.6 + 0.18(114.6) = 450.8 \text{ °C} \\
&\quad \text{(same as for total statistical (sum) vertical)} \\
T_2 &= 315.6 + 114.6 + 20.4 \\
&\quad + \{(0.05(114.6) + 0.012(20.4))^2 + (0.14(114.6) \\
&\quad + 0.035(20.4))^2 + (0.02(114.6) + 0.02(20.4))^2 \\
&\quad + (0 + 0.214(20.4))^2 + (0.035(114.6) + 0.035(20.4))^2 \\
&\quad + (0.071(114.6) + 0.071(20.4))^2 + (0.060(114.6) \\
&\quad + 0.065(20.4))^2 + (0.011(114.6) + 0)^2 + (0 + 0.14(20.4))^2\}^{1/2}
\end{aligned}
$$

$$T_2 = 315.6 + 114.6 + 20.4$$
$$+ \{35.7 + 280.8 + 7.3 + 19.1 + 22.3 + 91.9 + 67.3 + 1.6 + 8.2\}^{1/2}$$
$$T_2 = 473.7 \,^{\circ}\text{C}$$

etc.

$$F_m = \frac{T_M - T_{M-1}}{\Delta T_{m,\text{nom}}} \tag{8-57}$$

$$F_1 = \frac{T_1 - T_{in}}{\Delta T_{1,\text{nom}}} = \frac{450.8 - 315.6}{114.6} = 1.18$$

$$F_2 = \frac{T_2 - T_1}{\Delta T_{2,\text{nom}}} = \frac{473.7 - 450.8}{20.4} = 1.123$$

(where T_1 is the result computed by this procedure).

Semistatistical vertical:

$$F_m = F_m^D \left(1 + \left[\sum_{\substack{j=1 \\ \text{stat}}}^{n} (f_{j,m} - 1)^2 \right]^{1/2} \right) \tag{8-58}$$

$$F_1 = 1.221(1 + \{(0.035)^2 + (0.071)^2 + (0.06)^2 + (0.011)^2\}^{1/2})$$
$$= 1.221(1.1) = 1.343$$

etc.

$$T_M = T_{in} + \sum_{m=1}^{M} F_m \cdot \Delta T_{m,\text{nom}} \tag{8-52}$$

$$T_1 = 315.6 + 1.343(114.6) = 469.5 \,^{\circ}\text{C}$$

etc.

Semistatistical horizontal:

$$T_M = T_{in} + \sum_{m=1}^{M} F_m^D \Delta T_{m,\text{nom}} + \delta T_{M,z}^S \tag{8-59}$$

$$T_1 = 469.5 \,^{\circ}\text{C} \text{ (same as semistatistical vertical for } m = 1)$$
$$T_2 = 315.6 + 114.6(1.221) + 20.4(1.297)$$
$$+ \{(0.035(1.221)(114.6) + 0.035(1.297)(20.4))^2$$
$$+ (0.071(1.221)(114.6) + 0.071(1.297)(20.4))^2$$
$$+ (0.060(1.221)(114.6) + 0.065(1.297)(20.4))^2$$
$$+ (0.011(1.221)(114.6)) + (0.14)(1.297)(20.4))^2\}^{1/2}$$
$$= 315.6 + 139.9 + 26.5$$
$$+ \{33.9 + 139.6 + 102.3 + 2.4 + 13.7\}^{1/2}$$

$$T_2 = 499.1 \,^{\circ}\text{C}$$

etc.

$$F_m = \frac{T_M - T_{M-1}}{\Delta T_{m,\text{nom}}} \tag{8-57}$$

Table 8-8 Summary of results for Example 8-8

	Coolant	Film	Clad	Gap	Fuel
	$\Delta T_{nom} = 114.6\ ^\circ C$ $m = 1$ $M = 1$	$\Delta T_{nom} = 20.4\ ^\circ C$ $m = 2$† $M = 2$	$\Delta T_{nom} = 44.7\ ^\circ C$ $m = 3$ $M = 3$	$\Delta T_{nom} = 345.0\ ^\circ C$ $m = 4$ $M = 4$	$\Delta T_{nom} = 1{,}469.8\ ^\circ C$ $m = 5$ $M = 5$
Deterministic					
Product, F_m	1.450	1.746	1.349	1.770	1.325
$T_M\ (^\circ C)$	481.8	517.4	577.7	1188.2	3135.6
Sum, F_m	1.387	1.592	1.311	1.661	1.291
$T_M\ (^\circ C)$	474.6	507.0	565.6	1138.5	3036.0
Total statistical					
Vertical, F_m	1.180	1.279	1.159	1.481	1.145
$T_M\ (^\circ C)$	450.8	476.9	528.7	1039.5	2722.4
Horizontal, F_m	1.180	1.123	1.081	1.422	1.089
$T_M\ (^\circ C)$	450.8	473.7	522.0	1012.3	2612.4
Semistatistical					
Vertical, F_m	1.343	1.522	1.181	1.511	1.166
$T_M\ (^\circ C)$	469.5	500.5	553.3	1074.5	2788.3
Horizontal, F_m	1.343	1.451	1.136	1.463	1.110
$T_M\ (^\circ C)$	469.5	499.1	549.9	1054.5	2685.7

† The clad temperatures above ($M = 2$) are those used to calculate the fuel hot spot factors.

$$F_1 = \frac{469.5 - 315.6}{114.6} = 1.343$$

$$F_2 = \frac{499.1 - 469.5}{20.4} = 1.451$$

Table 8-8 summarizes the results for F_m and T_m by all methods. Note that the deterministic results are most conservative, and the total statistical results are the least conservative. For each method individually, the product and the vertical approaches, are the most conservative.

V EXTENSION TO MORE THAN ONE HOT SPOT

In reality, a reactor core can have more than one limiting hot spot. Assume that these hot spots are independent, are of number n, and have a probability p of exceeding the temperature limit independently. The binomial distribution gives the probability $p(m)$ that exactly m of the n hot spots will exceed the temperature limit, i.e.:

$$p(m) = \frac{n!}{m!(n - m)!} p^m (1 - p)^{n-m} \qquad (8\text{-}61)$$

In the literature p is called the probability of success and $1 - p \equiv q$ the probability of failure. Care must be taken in applying these definitions to the reactor case because physically a reactor engineer considers exceeding the temperature limit as potentially leading to a pin failure. However in the statistical sense, the event is the occurrence of exceeding the temperature limit, and its occurrence is a success, not a failure. Hence, we refrain from using the word failure and rather refer to whether or not the temperature limit is exceeded.

For $m = 0$, Eq. 8-61 reduces to:

$$p(m = 0) = (1 - p)^n \qquad (8\text{-}62)$$

and $p(m = 0)$ represents the probability that the event (that of exceeding the temperature limit) will not occur. This is consistent with the result obtained by further specialization of Eq. 8-62 for the case $n = 1$. Then

$$p(m = 0) = 1 - p \equiv q \qquad (8\text{-}63)$$

where $p(m = 0)$ and q are the probabilities that an individual hot spot will not exceed the temperature limit. The use of Eqs. 8-62 and 8-63 is illustrated in Example 8-9.

As can be seen in Eq. 8-62, $p(m = 0)$ decreases as the number of limiting hot spots increases. This fact has an important ramification for reactor operation. If the reactor power distribution is flattened, the number of limiting hot spots will increase. But the probability of not exceeding the temperature limit within the core should not be allowed to decrease because of safety considerations, so we have to decrease the individual hot spot values p by decreasing the reactor power. This action mitigates the power increase derived from the flattening of the power distribution.

VI OVERALL CORE RELIABILITY

The results reached in the preceding section about the implications of power distribution flattening arose because of an implicit assumption in the combinational methods introduced in Section IV. That assumption is that if $T_{M,lim}$ is reached, it will happen only at the hot spots. In other words, the probability of not reaching $T_{M,lim}$ is assumed to be $(1 - p)$ for the hot spots and 1 for the remainder of the core.

However, in reality, the remainder of the core can contribute to the probability of exceeding the temperature limit. A simple illustration, given by Fenech and Gueron [12], clarifies this point. Assume the probability of exceeding the temperature limit of the nominal hot spot A is p. Let two other core locations, B and C, have individual probabilities of exceeding the temperature limit $p' < p$. The probability of exceeding the temperature limit at B or C (or both) is obtained by applying Eq. 8-61:

Probability of exceeding the temperature limit at either B or C $(n = 2, m = 1)$

$$= \frac{2!}{1!1!} (p')^1 (1 - p')^1 = 2p' - 2(p')^2$$

Probability of exceeding the temperature limit at B and C $(n = 2, m = 2)$

$$= \frac{2!}{2!0!} (p')^2 (1 - p')^0 = (p')^2$$

Hence the total probability of exceeding the temperature limit at B and/or at C $= 2p' - 2(p')^2 + (p')^2 = 2p' - (p')^2$

Certainly if it happens to be true that:

$$2p' - (p')^2 > p$$

then the probability of exceeding the temperature limit at B and C must be taken into account. However, even if the inequality is not achieved, these two spots B and C may contribute to the core-wide probability of exceeding the temperature, and their contributions to this probability should be considered.

The remainder of this chapter deals with methods that have been introduced to account for the nonzero probability of exceeding the limits at spots other than the nominal hot spots. These methods differ in the degree to which they have relaxed other constraints of importance. Figure 8-4 illustrates the evolution of these methods with respect to their capabilities to relax the specific constraints of:

1. Ability to distinguish between the character of variables, i.e., does a variable affect the whole core, an assembly only, a pin only, or a spot only?

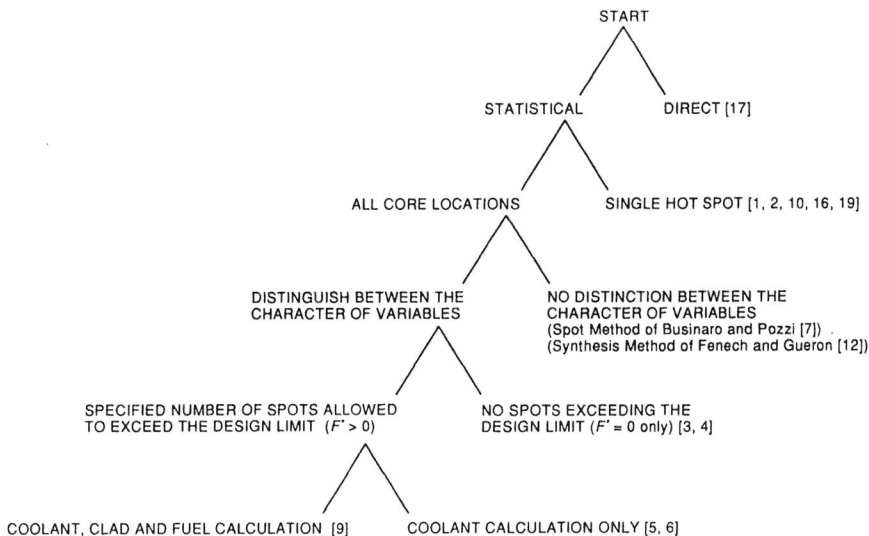

Figure 8-4 Evolution of core reliability analysis methods.

2. Evaluation of the probability that no spots will exceed a desired limit.
3. Evaluation of coolant temperatures only.
4. Consideration of threshold limits only.

A Methods That Do Not Distinguish between the Character of Variables

Both the spot method and the synthesis method made the major step of including the probability of exceeding the limiting temperature at spots throughout the core.

The spot method was introduced by Businaro and Pozzi [7] and represented the first attempts to use the Monte Carlo method in the treatment of uncertainties in reactor design.

Basically, the Monte Carlo method has the following advantages:

1. The method can be very accurate, depending on the sample size.
2. The method is conceptually simple.
3. The variables may act in the opposite sense.

For the application of the Monte Carlo method, the spatial dimension of interest of the fuel pin (or channel) is divided into a set of characteristic lengths or "spots" at which the design parameters are assumed uniform. The distribution functions of the design parameters must be known over the length of interest, and random values for the parameters are drawn over their distributions. By using the functional relations of some thermal–hydraulic model, the distribution of the maximum temperature (not the temperature at the hot spot) of the element of interest is computed. In the spot method this distribution is approximated by a normal distribution, and the overall hot spot (or channel) factors for a desired confidence level are then determined.

The spot method has a number of limitations:

1. It is computationally intensive.
2. The assumption of normally distributed maximum temperature is not quite correct as demonstrated by Fenech and Gueron [12].
3. The number of characteristic lengths selected to compose the channel has an influence on the result.

The synthesis method of Fenech and Gueron [12] presents a method for hand calculations which takes into account the fact that the limiting temperature of the element of interest M can be reached away from the hot spot position. As in the spot method, the fuel pin is divided into a number of segments of characteristic length. The deterministic method is applied to determine those spots at which the probability of exceeding the specified limit is significant. Then the statistical method is used to find the probability of exceeding the limit of each one of those spots. Assuming that these individual probabilities $p_i(T_{M,i} > T_{M,lim})$ are independent, the overall probability of exceeding the limit for a fuel pin is given by:

$$p(T_M > T_{M,lim}) = 1 - \prod_i [1 - p_i(T_{M,i} > T_{M,lim})] \qquad (8\text{-}64)$$

where i represents a segment of characteristic length.

This method allows a prediction of the expected number of locations at which the limiting temperature is exceeded. This can be seen by the following example. Let α and β characterize the overall and local shape, respectively, of the power output of the fuel pin j. The overall probability of exceeding the temperature limit of that pin is $p_j(\alpha, \beta)$ for a given $T_{M,lim}$. Assuming that $N(\alpha, \beta)$ is the number of fuel pins operating in that region, the expected total number of fuel pins in the core which exceed the temperature limit is, per Fenech and Gueron [12]:

$$\langle N \rangle = \int_{\alpha, \beta} N(\alpha, \beta) p_j(\alpha, \beta) d\alpha d\beta \qquad (8\text{-}65)$$

Besides the advantage of being able to be utilized by hand calculations, this method also is more realistic than the methods reviewed previously because it takes into account the possibility of exceeding the limit at locations other than the nominal hot spot and does not make any assumption about the maximum temperature distribution. However, this method does not account for the fact that the uncertainties can have different character.

B Methods That Do Distinguish between the Character of Variables

The uncertainties that affect the reactor performance do not have the same character. For example, the uncertainty in the power level measurement acts on the whole core, while the uncertainty in assembly flow orificing affects each assembly independently. The pellet enrichment has local influence. To take this variation in the character of variables into account, two methods were developed for the hot channel analysis: Amendola's method [3, 4] and the method of correlated temperatures [6].

In Amendola's method, the uncertainties are divided into three groups, namely:

1. Uncertainties affecting the individual channels in a subassembly
2. Uncertainties affecting individual subassemblies
3. Uncertainties affecting the whole core.

In this method Amendola works with more than one limiting hot channel. The number used is called an *equivalent number of hot channels*—equivalent in the sense of representing the net effect of all channels in a real core having a nonzero probability of exceeding the limit. In addition this method assumes a normal distribution for the probability of exceeding the limit of the limiting hot channels and only allows the evaluation of the probability that none of the hot channels will exceed the temperature limit.

This last condition is very restrictive to the designer who might otherwise realistically tolerate a certain number c of occurrences in excess of the temperature limit. Therefore, before presenting the method of correlated temperatures, the potential for allowing c to be nonzero will be demonstrated by a simplified approach.

1 The effects of allowing a nonzero number of locations to exceed the specified design limit To obtain the probability that m, the number of pins exceeding the temperature limit, is smaller than or equal to a given number c, we sum the contributions $p(m)$ for $m = 0$ to $m \leq c$. This summation of contributions from Eq. 8-61 yields:

$$P[m \leq c] = \sum_{m=1}^{c} p(m) = \sum_{m=0}^{c} \frac{n!}{m!(n-m)!} p^m (1-p)^{n-m} \qquad (8\text{-}66)$$

which represents the cumulative binomial distribution. Eq. 8-66 is also based on the assumptions that the probability of exceeding the limit is independent for each pin, and p is constant over all hot spots. Of course, this procedure is approximate because:

1. Some variables affect more than one hot spot, while others have only a local effect (global and local uncertainties will be discussed in the next section).

2. It only takes into account a designed set of n hot spots. The probability of exceeding the limit for all other core locations is not accounted for. This contribution can be accounted for by transforming other core locations into an equivalent number of limiting hot spots and using this method. An example in which the equivalent number of equally limiting channels for

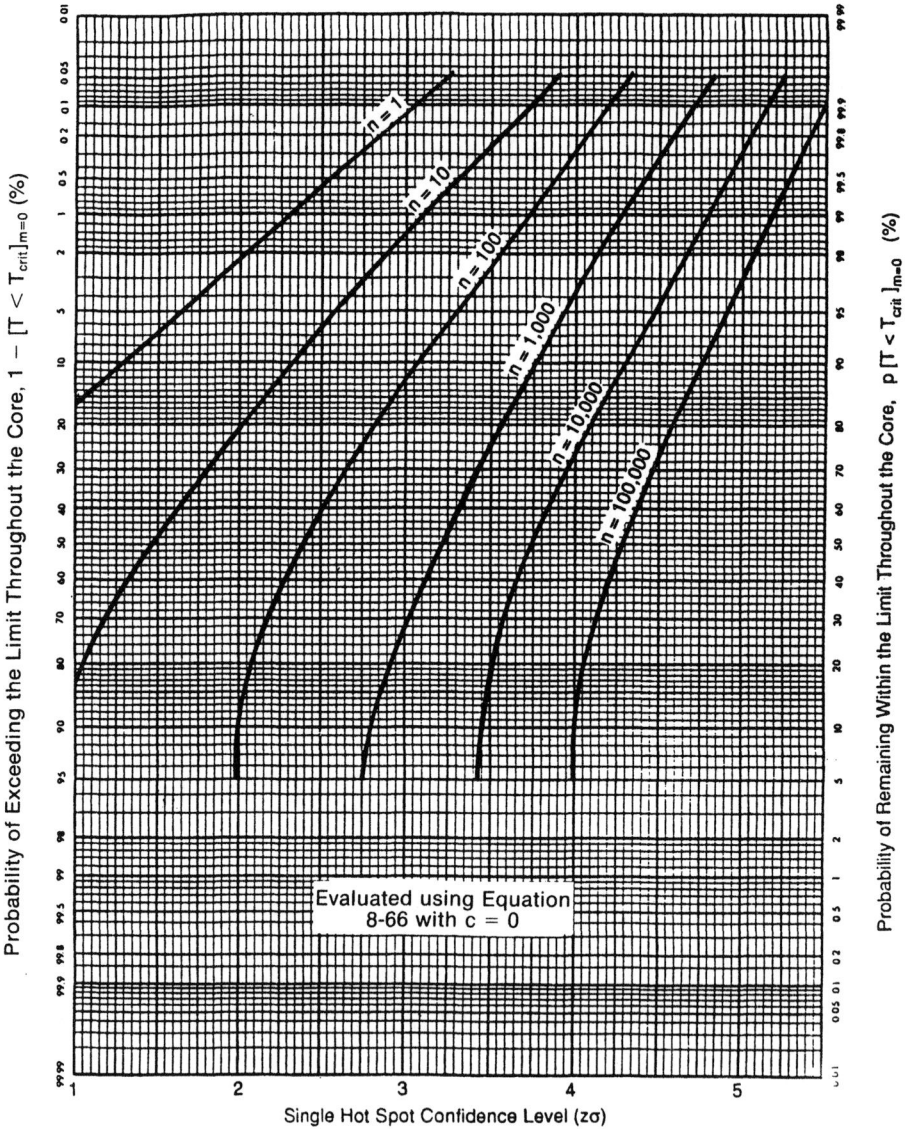

Figure 8-5 Overall core probability of not exceeding a certain temperature limit as a function of the single hot spot confidence level.

a cosine power distribution has been calculated is given by Judge and Bohl [16].

Despite these approximations, it is worthwhile to analyze this procedure to get some insight into the effect of allowing some locations to exceed the design limit. Figure 8-5 evaluated using Eq. 8-66 with $c = 0$ shows a comparison for a core of one or more equally limiting hot spots ($n = 1$ to $n = 100,000$). That figure gives the results that:

1. For a given confidence level of the single hot spot case ($n = 1$) $z\sigma$, the probability of exceeding the limit anywhere in the core increases with increasing numbers of hot spots.
2. To achieve a specific probability of remaining within the limit throughout the core, the confidence level $z\sigma$ of the single hot spot must be increased for increasing numbers of hot spots.

Allowing for a certain number c of occurrences in excess of the temperature limit, Eq. 8-66 predicts results depicted in Figure 8-6. These results are based on an overall core probability of 99.87 percent that the maximum rate of occurrences above the temperature limit will not be exceeded (3σ confidence level). The parameter ε in this figure is the number of rods exceeding the temperature limit expressed as the percent of total reactor rods.

This figure illustrates the case of a reactor with 60,000 fuel rods and 1,000 independent limiting hot spots. The excess temperature occurrence rate found

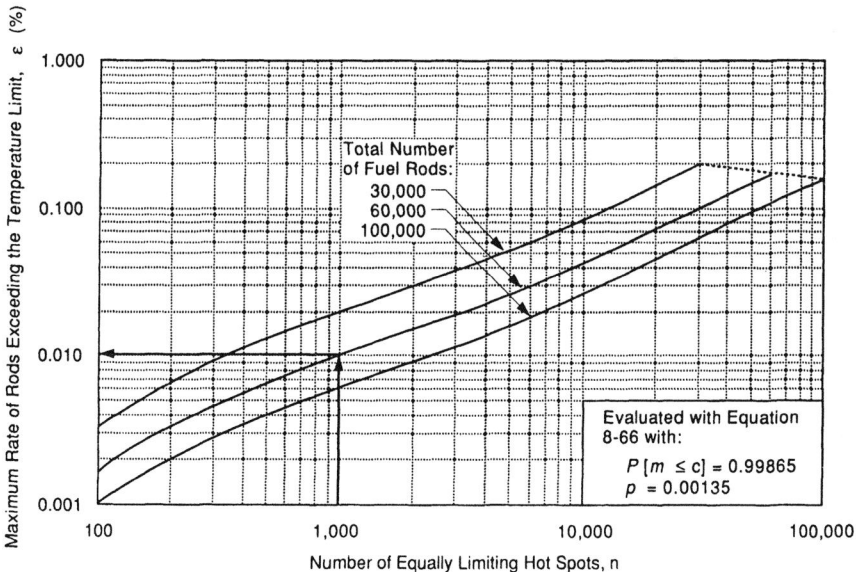

Figure 8-6 Maximum rate of rods exceeding the limiting condition in percent of total fuel rods as a function of the number of equally limiting hot spots.

is 0.01 percent (i.e., six rods). Another result shown is that in the range of 30,000 to 100,000 rods, even for the worst theoretical case of as many equally limiting hot spots as fuel rods (*dotted line* in Figure 8-6), the rate of pins exceeding the limiting temperature will never exceed 0.2 percent of the total number of rods for the 3σ confidence level.

Example 8-9 Probability computations considering varying numbers of equally limiting hot spots

PROBLEM Find the overall core probability of not exceeding a certain temperature limit for a 3σ value of the single hot spot confidence limit considering 1, 10, 100, and 1,000 equally limiting hot spots. Assume that this probability can be represented by the cumulative binomial distribution.

SOLUTION In this case, $c = 0$, and Eq. 8-62 is applicable where $p[m = 0]$ is the unknown. Recall that while $p[m]$ is the probability of exceeding the temperature limit, i.e., $P[T > T_{crit}]$, $p[m = 0]$ is the probability that this event will not occur and hence is in fact $p[T > T_{crit}]$. Further, p is the probability that a single hot spot exceeds the temperature limit corresponding to 3σ which for a normal distribution is equal to 0.001350, and $n = 1, 10, 100$, and 1000. Hence from Eq. 8-62 we obtain:

$$p[T < T_{crit}](\text{for } m = 0) = (1 - p)^n$$

$$= (1 - 0.001350)^n$$

which for the various n values yields:

n	$p(T < T_{crit})$ in percent
1,000	25.90
100	87.36
10	98.658
1	99.865

These results are illustrated in Figure 8-5.

Example 8-10 Computations of number of rods exceeding the design limit considering 1,000 equally limiting hot spots

PROBLEM Compute the number of rods exceeding the design limit in a core with 1,000 equally limiting hot spots for an overall core confidence level of 99.865 percent and an individual rod hot spot confidence level of 99.865 percent. Assume that the cumulative binomial distribution can represent the probability that the number of rods that might exceed the temperature limit is smaller than c.

SOLUTION In this case Eq. 8-66 is applicable where c is to be calculated:

$n = 1000$

$p = 0.00135$ (probability of an individual rod exceeding the hot spot limit)

and $P[m \leq c] = 0.99865$ (probability that the number of rods in the whole core might exceed the temperature limit is smaller than c).

The procedure is to evaluate the right side of Eq. 8-66 term by term to find the value of m such that the right side does not exceed the value $P[m \leq c] = 0.99865$. Hence we evaluate the terms of Eq. 8-66 as follows (note that a multiplication and division of like terms is introduced to make use of the previous term value to reduce the computation process):

$m = 0$: term 1 $= (1 - 0.001350)^{1,000} = 0.25901242$

$$m = 1: \text{term } 2 = \frac{1,000!}{1!999!}(0.001350)^1(1 - 0.001350)^{999}\frac{(1 - 0.001350)}{(1 - 0.001350)}$$

$$= \frac{1,000(0.001350)(\text{term } 1)}{(1 - 0.001350)}$$

$$= 0.35013946$$

$$m = 2: \text{term } 3 = \frac{1,000!}{2!998!}(0.001350)^2(1 - 0.001350)^{998}$$

$$= \frac{999}{2}\frac{(0.001350)}{(1 - 0.001350)}(\text{term } 2)$$

$$= 0.23642696$$

$$m = 3: \text{term } 4 = \frac{1,000!}{3!997!}(0.001350)^3(1 - 0.001350)^{997}$$

$$= \frac{998}{3}\frac{(0.001350)}{(1 - 0.001350)}(\text{term } 3)$$

$$= 0.10632288$$

$$m = 4: \text{term } 5 = \frac{1,000!}{4!996!}(0.001350)^4(1 - 0.001350)^{996}$$

$$= \frac{997}{4}\frac{(0.001350)}{(1 - 0.001350)}(\text{term } 4)$$

Completing this procedure we obtain:

m	Value of term	Sum of terms
0	0.25901242	0.25901242
1	0.35013946	0.60915188
2	0.23642696	0.84557884
3	0.10632288	0.95190172
4	0.03582468	0.98772640
5	0.00964700	0.99737340
6	0.00216264	0.99953600

Therefore, for $P[m < c] = 0.99865$, the above tabulation shows that the maximum value of m is between 5 and 6. To guarantee that the probability constraint is met, c should be an integer, i.e., 6. For a core of 60,000 rods, the maximum rate of exceeding the temperature limit is then:

$$\frac{6}{60,000} \times 100\% = 10^{-2}\%$$

as illustrated in Figure 8-6.

2 Method of correlated temperatures of Arnsberger and Mazumdar [5, 6] This method allows for a nonzero number of locations exceeding the limiting temperature and recognizes the character of the variables. It uses a combination of analytical and computational procedures.

Let $T(i,j)$ be the exit coolant temperature relative to channel j in assembly i. This temperature can be described by:

$$T(i,j) - T_{in} \equiv \Delta T(i,j) = [\Delta T_{cum}(i,j)] \cdot \alpha \cdot \beta(i) \cdot \delta(k,j) \qquad (8\text{-}67)$$

with:

$$\Delta T_{cum}(i,j) = \Delta T_{nom}(i,j) \cdot N(i,j) \qquad (8\text{-}68)$$

$$\alpha = \prod_l S_l \qquad (8\text{-}69)$$

$$\beta(i) = \prod_n S_n(i) \qquad (8\text{-}70)$$

$$\delta(i,j) = \prod_m S_m(i,j) \qquad (8\text{-}71)$$

where:

$N(i,j)$ = product of the maximum value of all nonstatistical contributors to uncertainties

S_l = statistical contributors to the whole core-wide uncertainties

$S_n(i)$ = statistical contributors to assembly-wide uncertainties

$S_m(i,j)$ = statistical contributors to channel-wide uncertainties.

Taking the natural log of both sides of Eq. 8-67:

$$\ell n[\Delta T(i,j)] - \ell n[\Delta T_{cum}(i,j)] = \ell n\ \alpha + \ell n\ \beta(i) + \ell n\ \delta(i,j)$$

$$\equiv U(i,j)$$

and defining:

$$\ell n\ \alpha\quad \equiv X \qquad (8\text{-}72)$$

$$\ell n\ \beta(i)\quad \equiv Y(i) \qquad (8\text{-}73)$$

$$\ell n\ \delta(i,j) \equiv Z(i,j) \qquad (8\text{-}74)$$

we obtain:

$$U(i,j) = X + Y(i) + Z(i,j) \tag{8-75}$$

Defining:

$$\ell n[\Delta T_{\text{lim}}] - \ell n[\Delta T_{\text{cum}}(i,j)] = U_{\text{lim}}(i,j) \tag{8-76}$$

where:

$$\Delta T_{\text{lim}} = T_{\text{lim}} - T_{\text{in}}$$

We must have for the condition of not exceeding the temperature limit:

$$U_{\text{lim}}(i,j) \geq X + Y(i) + Z(i,j) \tag{8-77}$$

The $U(i,j)$ values are not independent since they have the global random variables X and $Y(i)$ in common. However, note that for a given core and assembly, i.e., fixed X and $Y(i)$, the subchannel temperatures and hence, the $U(i,j)$ values do become independent.

We now proceed to the computation of the overall probability of not exceeding the temperature limit in a number F^* of pins in a core. Let us call this probability $P[F \leq F^*]$. The procedure is based on recognizing that for a given core and assembly, the individual channel uncertainty contributors are independent. Therefore the probability of not exceeding the temperature limit in a number F^* of pins is computed for a given core sample n. This conditional probability $P\{F \leq F | \text{sample}\}$ is well approximated by the following cumulative distribution function which is based on the Poisson distribution [14]:

$$P\{F \leq F^* | \text{sample}\} = \sum_{b=0}^{F^*} \frac{\lambda^b e^{-\lambda}}{b!} \tag{8-78}$$

where $\lambda = $ the expected number of occurrences of excess temperature must satisfy $\lambda \ll (I \times J)^{1/2}$, i.e., the probability that each channel is a hot channel is small.

The value of λ for a given core sample must include consideration of all assemblies i and channels j. Hence λ is a summation over all channels and assemblies of the probability of excess temperature of a given channel in a given assembly $P_{i,j}[X,Y(i)]$ for a given set of core- and assembly-wide uncertainties X and $Y(i)$. This is equal to:

$$\lambda = \sum_{i=1}^{I} \sum_{j=1}^{J} P_{i,j}[X,Y(i)] \tag{8-79}$$

It is permissible to obtain the expected excess temperature occurrences λ by Eq. 8-79 because the various probabilities $P_{i,j}[x,y(i)]$ are independent for a given X and $Y(i)$. Therefore we must compute $P_{i,j}[X,Y(i)]$ for each core sample. Now for a given core and assembly uncertainty set $X,Y(i)$, the associated channel conditions to yield excess temperature are by analogy from Eq. 8-77:

$$Z(i,j) > U_{\text{lim}}(i,j) - X - Y(i) \tag{8-80}$$

In this expression, $U_{lim}(i,j)$ is computed using appropriate functional relations for the channel or a subchannel code to obtain $\Delta T_{nom}(i,j)$; X and $Y(i)$ are obtained by generation of random numbers drawn over their respective distributions. Hence we can write the probability of excess temperature of a given channel in a given assembly as:

$$P_{i,j}\{X,Y(i)\} \equiv P_{i,j}[Z(i,j) > U_{lim}(i,j) - X - Y(i)|X,Y(i)] \qquad (8\text{-}81)$$

The probability $P_{i,j}\{X,Y(i)\}$ is obtained from Eq. 8-81 by computing $U_{lim}(i,j) - X - Y(i)$ for each randomly generated value of X and $Y(i)$ over their assumed distributions.

Once the $P_{i,j}\{X,Y(i)\}$ are computed, the λ values are determined from Eq. 8-79, which are then used (one λ for each sample) in Eq. 8-78. The desired overall probability of not exceeding the temperature limit in a number F^* of pins $P[F \le F^*]$, can now be obtained as:

$$P[F \le F^*] = \frac{1}{N} \sum_{n=1}^{N} P\{F \le F^*|\text{sample}\} \qquad (8\text{-}82)$$

This method was initially only defined for the coolant temperature. Since the coolant temperature is not the most limiting aspect of LMR design, Carajilescov and Todreas [9] generalized this approach to the calculation of the clad and fuel temperatures.

VII CONCLUSION

This chapter summarizes the principles that are utilized in the development of methods for the treatment of uncertainties in reactor thermal design. The basic methods in current use fall into two categories: the statistical summing of bounding sensitivities [11] and Monte Carlo techniques to obtain a departure from nucleate boiling ratio DNBR standard deviation [15, 20]. These specific methods have not been presented in detail because they are under continuing evolution. It is envisioned that future development will also address the assumption of a distribution in the design limit about a nominal rather than the threshold value concept embodied in the presentation of this chapter.

REFERENCES

1. Abernathy, F. H. The statistical aspects of nuclear reactor fuel element temperature. *Nucl. Sci. Eng.* 11:290–297, 1961.
2. ADPA introduces statistical hot spot factors. *Nucleonics* 17:92, 1958.
3. Amendola, A. *A Statistical Method for Evaluation of Hot Channel Factors in Reactor Design.* KFK-843, EUR 3979.3, July 1968.
4. Amendola, A. *Advanced Statistical Hot Spot Analysis.* EURFNR-814, KFK 1134 UNCLAS, March 1970.

5. Arnsberger, P. L., and Mazumdar, M. Some aspects of overall statistical hot spot factors analysis for a LMFBR-core. Paper No. 27, *Proceedings of the International Meeting on Fast Reactor Fuel and Fuel Elements,* Karlsruhe, Germany, September 1, 1970.

6. Arnsberger, P. L., and Mazumdar, M. The hot channel factor as determined by the permissible number of hot channels in the core. *Nucl. Sci. Eng.* 47:140–149, 1972.

7. Businaro, V. L., and Pozzi, G. P. *A New Approach on Engineering Hot Channel and Hot Spot Statistical Analysis.* EUR 1302.e, EURATOM, 1964.

8. Carajilescov, P. Current Status of Methods of Thermal Design for Reactors Cores: Compilation and Critical Assessment. S.M. thesis, Department of Nuclear Engineering, M.I.T., Cambridge, MA, December 1973.

9. Carajilescov, P., and Todreas, N. E. Expansion of the method of correlated temperatures to clad and fuel design analyses. *Nucl. Eng. Des.* 30:3–19, 1974.

10. Chelemer, H., and Tong, L. S. Engineering hot channel factors for open-lattice cores. *Nucleonics* 20:68–73, 1962.

11. Chelemer, H., Bowman, L. H., and Sharp, D. R. *Improved Thermal Design Procedure.* Westinghouse Report WCAP-8568, July 1975.

12. Fenech, H., and Gueron, H. M. The synthesis method of uncertainty analysis in nuclear reactor thermal design. *Nucl. Sci. Eng.* 31:505–512, 1968.

13. Fisher, R. A., and Yates F. *Statistical Tables for Biological, Agricultural and Medical Research* (6th ed.). London: Longman Group UK Ltd., 1978. (Previously published by Oliver and Boyd Ltd, Edinburgh.)

14. Haugen, E. B. *Probabilistic Approaches to Design.* New York: John Wiley and Sons, 1986, p. 29.

15. Heller, A. S., Jones, J. H., and D. A. Farnsworth, *Statistical Core Design Applied to the Babcock-205 Core.* BAW-10145, October 1980.

16. Judge, F. D., and Bohl, L. S. Effective hot channel factors for "flat" power reactors. *Nucl. Sci. Eng.* 28:296–300, 1967.

17. LeTourneau, B. W., and Grimble, R. E. Engineering hot channel factor for reactor design. *Nucl. Sci. Eng.* 1:359–369, 1956.

18. Lingdren, B. W., and MacElrath, G. W. *Introduction to Probability and Statistics* (2nd ed.) New York: The Macmillan Company, 1966.

19. Rude, P. A., and Nelson, A. C., Jr. Statistical analysis of hot channel factors. *Nucl. Sci. Eng.* 7:156, 1960.

20. Technical Staff Combustion Engineering Co. *Statistical Combination of Uncertainties.* CEN-139(A)-NP, November 1980.

PROBLEMS

Problem 8-1 Evaluation of the hot spot factor for film temperature drop due to variations in channel width (Section III)

For turbulent one-phase flow heat transfer in a parallel plate channel, find the engineering hot spot subfactor for the film temperature drop due to variations in the channel width d, i.e., $f_{j,y}$, where j refers to channel width, and y refers to film temperature drop. Assume that the friction factor is proportional to $Re^{-0.2}$, and the heat transfer coefficient is correlated by the Colburn equation. Consider two parallel flow channels having the same pressure drop: one having nominal dimensions, and a "hot channel" of width $d_{\text{min avg}}$ along most of the length and width $d_{\text{max loc}}$ at the hot spot.

Calculate the subfactor for a materials test reactor type fuel element for which $d_{\text{nom}} = 0.117$ in, assuming $d_{\text{max loc}} = 0.120$ in and $d_{\text{min av}} = 0.115$ in.

Answer: $f_{j,y} = \dfrac{\Delta T_{f_{\text{max loc}}}}{\Delta T_{f_{\text{nom av}}}} = \left(\dfrac{d_{\text{max loc}}}{d_{\text{min av}}}\right)\left(\dfrac{d_{\text{nom}}}{d_{\text{min av}}}\right)^{1/3} = 1.05$

Table 8-9†

Experiment number	1	2	3	4	5	6	7	8	9	10	11	12	13	14	15
$q''_{crit} \times 10^{-6}$ (BTU/hr ft²)	0.80	0.53	0.63	0.33	0.65	0.30	0.74	0.66	0.44	0.54	0.70	0.56	0.61	0.49	0.57

† Assume the data are characterized by a normal distribution.

Problem 8-2 Evaluation of the hot spot factor for pellet temperature drop due to variations in pellet thermal conductivity (Section III)

Compute the hot spot subfactor for the temperature difference $(T_{max} - T_{fo})$ between the centerline and surface temperatures of a cylindrical fuel pellet due to variations in the pellet thermal conductivity, i.e., $f_{j,y}$, where j refers to pellet thermal conductivity, and y refers to pellet centerline-to-surface temperature difference. Assume that the thermal conductivities of the pellets are normally distributed about a mean x_j^0 with a standard deviation, $\sigma = 0.04 x_j^0$. A confidence limit of 99.5 percent is desired.

Answer: $f_{j,y} = 1.115$

Problem 8-3 Evaluation of the limit DNBR by alternate combinational methods (Section IV)

Suppose Table 8-9 is a set of PWR experimental data for the critical heat flux at a certain coolant operating condition. With this data, determine:

1. The nominal value of the critical heat flux.
2. The design minimum DNBR so that there will be 90 percent probability that DNB will not occur when this DNBR limit is attained?

Assume that the uncertainties of the parameters affect the operating power as given in Table 8-10. With 97.7 percent confidence level to these measured parameters, use both cumulative and statistical methods to determine

3. The limit DNBR with the DNBR determined in part 2.
4. The nominal operating heat flux under this limit DNBR.

Answers:

1. $q''_{crit} = 0.57 \times 10^6 \dfrac{Btu}{hr\ ft^2}$

2. MDNBR = 1.46

3. Cumulative limit DNBR = 1.598 (product); 1.596 (sum)

   Statistical limit DNBR = 1.559

4. Cumulative nominal heat flux

   $= 0.357 \times 10^6 \dfrac{Btu}{hr\ ft^2}$ (product); $0.357 \times 10^6 \dfrac{Btu}{hr\ ft^2}$ (sum)

   Statistical nominal heat flux

   $= 0.366 \times 10^6 \dfrac{Btu}{hr\ ft^2}$

Table 8-10

Parameter	Nominal value (in)	Standard deviation (in); σ	Sensitivity factor; $\dfrac{\partial \ln y}{\partial \ln x_j}$
Rod diameter	0.58	0.005	2
Pitch	0.64	0.010	1.8

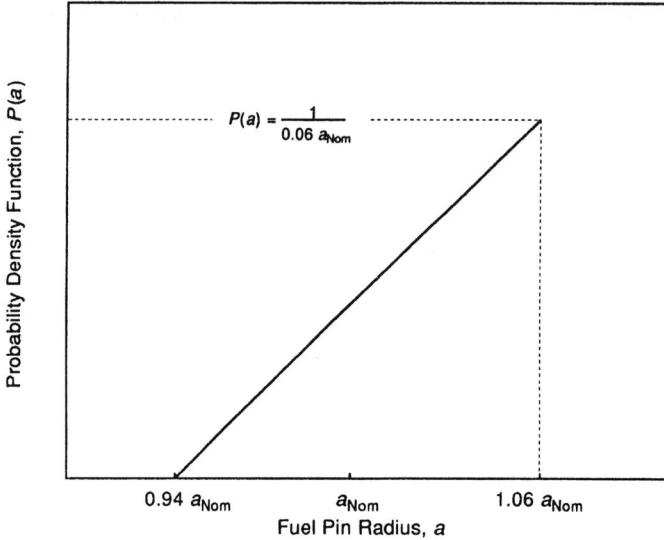

Figure 8-7 Probability density function for a, fuel pin radius.

Problem 8-4 Evaluation of subfactor for maximum of clad temperature by various combinational methods (Section IV)

Consider the hot channel subfactors for ΔT_{clad} to be used in evaluating the hot pin maximum cladding temperature. We shall take into account only variations in the fuel pellet radius and the cladding outer radius (assume that the cladding inner radius remains unchanged). The probability density functions for these variables in the observed samples of fuel pins are shown in Figures 8-7 and 8-8. Use typical PWR fuel element geometry for needed nominal dimensions. With 99 percent

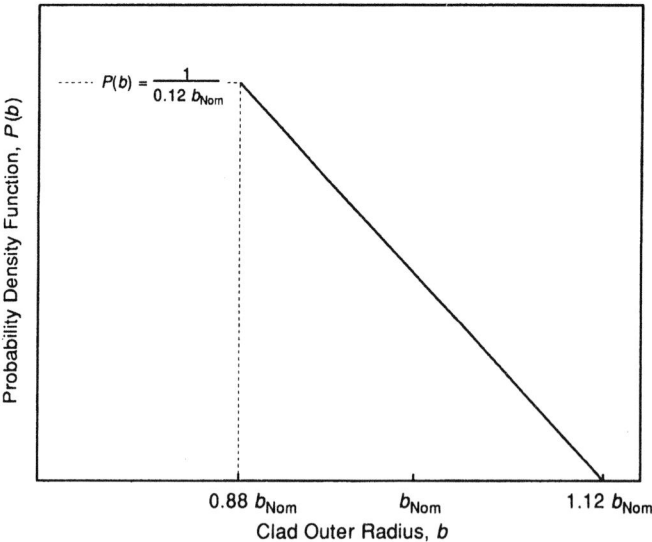

Figure 8-8 Probability density function for b, clad outer radius.

confidence limit, calculate the overall hot pin subfactor for the maximum cladding temperature using the following methods:

1. Cumulative
2. Sum statistical
3. Monte Carlo (with 100 trials); use a pair of dice as a crude random number generator. Compare the resulting sample mean and variance to the true (or analytical) mean and variance.

NOTE for part 3: if $y = y(x)$ then $p(y) = p(x) \left(\dfrac{dy}{dx}\right)^{-1}$

where $x = y^{-1}(y)$, and $p(x)$ is the probability density function for x.

Answers:

1. 1.93
2. 1.73

Problem 8-5 Semistatistical method (Section IV)

Present a general proof that the semistatistical vertical approach is more conservative than the horizontal approach, i.e., $T_{m,\text{vertical}} > T_{m,\text{horizontal}}$

Problem 8-6 Use of the semistatistical vertical method (Section IV)

It is desired to increase core power by grinding the clad to reduce the distribution of clad thickness (i.e., the nominal clad thickness is to be kept constant by starting the grinding process with a nominally thicker clad).

Using the semistatistical vertical method of combining uncertainties, find the standard deviation of the required ground clad thickness distribution such that the LMR core of Example 8-8 can achieve 1 percent increased power for the same limiting maximum inside clad temperature as the unground clad case (i.e., 553.3 °C from Table 8-8).

Assumptions:

1. All hot channel factors of Table 8-7 are maintained except that for the clad thickness.
2. The results of Table 8-8 are accurate.
3. The nominal clad thickness is 0.7 mm.
4. The clad hot spot temperature at 100% power of 550°C occurs at the core exit. At this condition and location $\Delta T_{1,\text{nom}} = 143°C$, $\Delta T_{2,\text{nom}} = 10°C$ and $\Delta T_{3,\text{nom}} = 22°C$.

Answer: $\sigma = 0.0161$

Problem 8-7 Contributions to fuel melting of all fuel spots in the core (Section VI)

For a hypothetical LMR reactor, the design procedure calls for a limited potential for fuel failure by imposing restrictions on the maximum linear power in the core. The core has 40,000 fuel pins. If each was divided into 10 spots, then the nominal linear power distribution for the operating condition is given as:

Nominal linear power (kW/ft)	No. of spots
21	100
19	1,900
17	10,000
15	19,000
13	40,000
11	50,000
9	70,000
7	99,000
5	110,000
	400,000

The core power calculation is subject to the following uncertainty factors at a 3σ level of confidence:

	3σ uncertainty factor
Direct	
Physics modeling	1.02
Control rod banking	1.02
Statistical	
Nuclear data	1.07
Criticality	1.01
Fissile fuel model distribution	1.03

The power to melt has been measured in experiments in which the power was gradually increased until melting was observed. The following results were obtained:

Measured power to melt (kW/ft)	No. of measurements
22	1
23	2
24	26
25	47
26	20
27	3
28	1
	100

Determine:

1. The probability of fuel melting at one of the hottest spots with a 3σ confidence level using the semistatistical method.
2. The probability of fuel melting for this LMR core at a 3σ level of confidence using the semistatistical method.

Answers:
1. 0.061
2. 0.998

Problem 8-8 Effect of radial power flattening on overall core reliability (Section VI)

Assume calculations of $P[F < F^*]$ have been performed for a given core for a given design limit, T^i_{lim} for $F^* = 8$ and $F^* = 16$. The results are shown in Figure 8-9. Now assume that while the total power of the core is kept constant the radial power is flattened. The $P[F < F^*]$ for $F^* = 16$ and $F^* = 8$ are reevaluated as a function of T^i_{lim}.

Will the difference between the two curves be closer together or more widely separated than in the unflattened core? Explain.

Answer: Closer

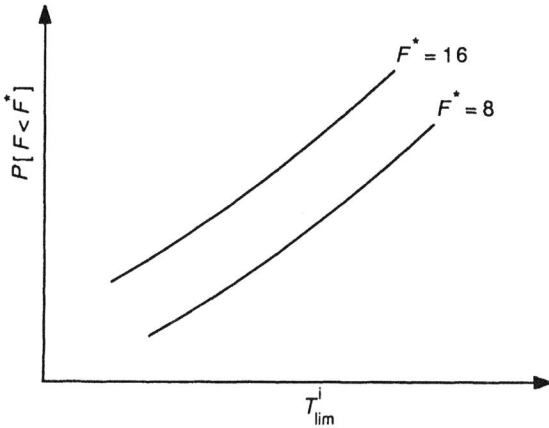

Figure 8-9

Appendices

NOMENCLATURE

A-1 COORDINATES

$\vec{i}, \vec{j}, \vec{k}$	Unit vectors along coordinate axes
$\vec{n}$	Unit normal vector directed outward from control surface
$\vec{r}$	Unit position vector
x, y, z	Cartesian
r, θ, z	Cylindrical
r, θ, ϕ	Spherical
v_x, v_y, v_z	Velocity in cartesian coordinates
v_r, v_θ, v_z	Velocity in cylindrical coordinates

A-2 EXTENSIVE AND SPECIFIC PROPERTIES

		Dimensions in two-unit systems†	
Symbol	Description	$ML\ \theta T$	FE plus $ML\ \theta T$
C, c	General property		Varies
E, e	Total energy (internal + kinetic + potential)		E, EM^{-1}
H, h	Enthalpy		E, EM^{-1}
$H°, h°$	Stagnation enthalpy		E, EM^{-1}
M, m	Total mass	M	
S, S	Entropy		E, EM^{-1}
U, u	Internal energy		E, EM^{-1}
$U°, u°$	Stagnation internal energy		E, EM^{-1}
V, v	Volume	L^3, L^3M^{-1}	

† M, mass; F, force; L, length; —, dimensionless; θ, temperature; E, energy; T, time.

A-3 DIMENSIONLESS NUMBERS

Symbol	Description	Location where symbol first appears†
Br	Brinkmann number	I, E 10-6
Ec	Eckert number	I, P 10-413
Fr	Froude number	I, E 9-36
Gr	Grashof number	I, E 10-11
Ku	Kutateladze number	I, E 11-5
L	Lorentz number	I, E 10-10
M	Mach number	I, P 10-444
M_{ij}	Mixing Stanton number	E 6-99
Nu	Nusselt number	I, E 10-11
Pe	Peclet number	I, P 10-419
Pr	Prandtl number	I, E 10-6
Re	Reynolds number	I, E 9-36
St	Stanton number	I, E 10-76

† I ≡ Volume 1; A ≡ appendix; T ≡ table; E ≡ equation; P ≡ page; F ≡ figure. For example, E10-6 = equation 10-6.

A-4 GENERAL NOTATION

Symbol	Description	Dimension in two unit systems*		Location where symbol first appears†
		$ML\,\theta T$	FE plus $ML\,\theta T$	
	General english notation			
A, A_f	General area or flow area	L^2		I, E 2-8
A	Availability function		E	I, E 4-27
	Atomic mass number	—		I, E 3-48
A_s	Projected frontal area of spacer	L^2		I, E 9-89
A_v	Avogadro's number	M^{-1}		I, E 3-4
	Unobstructed flow area in channels	L^2		I, E 9-89
A_{fb}	Sum of the area of the total fluid–solid interface and the area of the fluid	L^2		P 5-173

* M = mass; θ = temperature; E = energy; — = dimensionless; L = length; T = time; F = force.

† I ≡ Volume 1; A ≡ appendix; T ≡ table; E ≡ equation; P ≡ page; F ≡ figure. For example, E10-6 = equation 10-6.

A-4 GENERAL NOTATION (*continued*)

Symbol	Description	Dimension in two unit systems*		Location where symbol first appears†
		ML θT	*FE* plus *ML θT*	
A_{fs}	Area of the total fluid–solid interface within the volume	L^2		P 5-173
a	Atomic fraction	L^{-3}		I, E 3-14
	Half thickness of plate fuel	L		I, F 8-9
B	Buildup factor	—		I, E 3-54
C	Tracer concentration	—		E 6-58
	Constant for friction factor			I, E 9-83
C_D	Nozzle coefficient	—		I, E 9-17b
C_o	Correction factor in drift flux model for nonuniform void distribution	—		I, E 11-41
C_S	Spacer drag coefficient	—		I, E 9-89
C_v	Modified drag coefficient	—		I, E 9-90
c	Isentropic speed of sound in the fluid	LT^{-1}		I, E 11-120
c_d	Drag coefficient	—		I, E 11-24
c_p	Specific heat, constant pressure		$EM^{-1}\theta^{-1}$	I, E 4-115
c_R	Hydraulic resistance coefficient	—		E 3-35a
c_v	Specific heat, constant volume		$EM^{-1}\theta^{-1}$	I, P 4-121
D	Diameter, rod diameter	L		I, F 1-12
	Tube diameter	L		I, E 9-24
D_e	Hydraulic or wetted diameter	L		I, E 9-55
D_H	Heated diameter	L		I, E 10-12
D_s	Wire spacer diameter	L		I, F 1-14
D_V	Volumetric hydraulic diameter	L		I, P 9-391
D_{ft}	Wall-to-wall distance of hexagonal bundle	L		I, F 1-14
d	Tube diameter	L		I, F 9-4
	Diameter of liquid drop	L		I, E 11-24
E	Neutron kinetic energy		E	I, E 3-1
F	Modification factor of Chen's correlation	—		I, E 12-29

(*continued*)

A-4 GENERAL NOTATION (*continued*)

Symbol	Description	Dimension in two unit systems*		Location where symbol first appears†
		ML θT	*FE* plus *ML* θT	
	Looseness of bundle packing	—		E 4-130
	Hot spot factor	—		E 8-18
$\vec{F}, \vec{f}, F$	Force, force per unit mass		F, FM^{-1}	I, E 4-18
F_Q	Heat flux hot channel factor	—		P 8-368
F_v	Void factor	—		I, E 8-68
F_{ix}	Total drag force in the control volume i in the x direction		F	E 6-53
F_{iz}	Subchannel circumferentially averaged force for vertical flow over the solid surface in the control volume i		F	E 6-43
$F_{\Delta h}$	Coolant enthalpy rise hot channel factor	—		P 8-369
$F_{\Delta T}$	Coolant temperature rise hot channel factor	—		P 8-369
f	Moody friction factor	—		I, E 9-50
	Mass fraction of heavy atom in the fuel	—		I, E 2-13
f'	Fanning friction factor	—		I, E 9-52
$f_{c.t.}$	Friction factor in a circular tube	—		I, E 9-85
f_j	Subfactor of x_j	—		E 8-21
$f_{j,y}$	Subfactor relative to parameter x_j affecting the property y	—		E 8-23
f_{TP}	Two-phase friction factor	—		I, E 11-77
f_{tr}	Friction factor for the transverse flow	—		E 5-47a
G	Mass flux	ML^{-2}T^{-1}		I, E 5-38a
g	Distance from rod surface to array flow boundary	L		I, F 1-12
$\vec{g}$ or g	Acceleration due to gravity	LT^{-2}		I, E 4-23
H	Axial lead of wire wrap	L		I, E 9-93a
	Head (pump)	L		E 3-115

* M = mass; θ = temperature; E = energy; — = dimensionless; L = length; T = time; F = force.

† I ≡ Volume 1; A ≡ appendix; T ≡ table; E ≡ equation; P ≡ page; F ≡ figure. For example, E10-6 = equation 10-6.

A-4 GENERAL NOTATION (*continued*)

Symbol	Description	Dimension in two unit systems* $ML\,\theta T$	FE plus $ML\,\theta T$	Location where symbol first appears†
h	Wall heat-transfer coefficient		$EL^{-2}\theta^{-1}T^{-1}$	I, E 2-6
	Dimensionless pump head	—		E 3-119
h_g	Gap conductance		$EL^{-2}\theta^{-1}T^{-1}$	I, E 8-106
h_m^+	Flow mixing-cup enthalpy		EM^{-1}	I, E 5-52
I	Geometric inertia of the fluid	L^{-1}		I, E 13-38
	Energy flux		$ET^{-1}L^{-2}$	I, E 3-50
i	Irreversibility or lost work		ET^{-1}	I, E 4-47
J	Total number of neighboring subchannels	—		E 6-25
$\vec{J}$	Generalized surface source or sink for mass, momentum, and energy			I, E 4-50
$\vec{j}$ or j	Volumetric flux (superficial velocity)	LT^{-1}		I, E 5-42
j_{AX}	Flux of species A diffusing through a binary mixture of A and B due to the concentration gradient of A	$ML^{-2}T^{-1}$		P 6-256
K	Total form loss coefficient	—		I, E 9-23
K_G	Form loss coefficient in transverse direction	—		E 6-81
k	Thermal conductivity		$EL^{-1}\theta^{-1}T^{-1}$	I, E 4-114
L	Length	L		I, E 2-5
L_B	Boiling length	L		I, F 13-1
L_{NB}	Nonboiling length	L		E 3-5
ℓ	Transverse length	L		E 6-55
l	Axial dimension	L		E 3-1
ℓ_M, ℓ_H	Mixing length	L		I, E 10-66
M	Molecular mass	M		I, P 2-34
$\dot{m}$	Mass flow rate	MT^{-1}		I, E 4-30a
N	Number of subchannels	—		I, T 1-4
	Atomic density	L^{-3}		I, E 2-14

(*continued*)

A-4 GENERAL NOTATION (*continued*)

Symbol	Description	Dimension in two unit systems*		Location where symbol first appears†
		$ML\,\theta T$	FE plus $ML\,\theta T$	
N_p	Total number of rods	—		I, T 1-4
N'_p	Number of rows of rods			P 6-278
N_H, N_w, N_p	Transport coefficient of lumped subchannel	—		E 6-132a,b,c,
N_{rings}	Number of rings in a rod bundle	—		I, T 1-5
N_{rows}	Number of rows of rods	—		I, T 1-4
P	Porosity	—		I, E 8-17
	Decay power		ET^{-1}	I, E 3-68a,b
	Pitch	L		I, F 1-12
	Perimeter	L		I, E 5-64
P_H	Heated perimeter	L		I, P 10-444
P_R	Power peaking factor	—		P 6-277
P_w	Wetter perimeter	L		I, E 9-56
ΔP	Clearance on a per-pin basis	L		E 4-128
p	Pressure		FL^{-2}	I, E 2-8
$\hat{p}$	$\hat{p} = p + \rho g z$		FL^{-2}	E 7-1
Δp^+	$\dfrac{\Delta p}{\rho^* g L} - 1$	—		P 4-138
Q	Heat	ML^2T^{-2}	E	I, E 4-19
	Volumetric flow rate	L^3T^{-1}		I, E 5-45
Q'''	Power density		EL^3T^{-1}	I, P 2-22
$\dot{Q}$	Core power		ET^{-1}	I, P 2-22
	Heat flow		ET^{-1}	I, E 7-2a
	Heat-generation rate		ET^{-1}	I, E 5-124
$\dot{q}$	Rate of energy generated in a pin		ET^{-1}	I, P 2-22
q'	Linear heat-generation rate		$EL^{-1}T^{-1}$	I, P 2-22
q'''	Volumetric heat-generation rate		$EL^{-3}T^{-1}$	I, P 2-22
$\vec{q}'', q''$	Heat flux, surface heat flux		$EL^{-2}T^{-1}$	I, P 2-22
q''_{cr}	Critical heat flux		$EL^{-2}T^{-1}$	I, F 2-3
q'_o	Peak linear heat-generation rate		$EL^{-1}T^{-1}$	I, E 13-14
q'_{rb}	Linear heat generation of equivalent, dispersed heat source		$EL^{-1}T^{-1}$	E 6-34
q'''_{rb}	Equivalent dispersed heat source		$EL^{-3}T^{-1}$	E 5-49

* M = mass; θ = temperature; E = energy; — = dimensionless; L = length; T = time; F = force.

† I ≡ Volume 1; A ≡ appendix; T ≡ table; E ≡ equation; P ≡ page; F ≡ figure. For example, E10-6 = equation 10-6.

A-4 GENERAL NOTATION (*continued*)

Symbol	Description	Dimension in two unit systems*		Location where symbol first appears†
		$ML \, \theta T$	FE plus $ML \, \theta T$	
q''_w	Heat flux at the wall		$EL^{-2}T^{-1}$	I, E 5-152
R	Proportionality constant for hydraulic resistance	$L^{-4}M^n\theta^n$		E 3-35b
	Gas constant		$EM^{-1}\theta^{-1}$	I, T 6-3
	Radius	L		I, E 2-5
$\bar{R}$	Universal gas constant		$EM^{-1}\theta^{-1}$	I, AP B634
RR	Reaction rate in a unit volume	$T^{-1}L^{-3}$		I, P 3-43
$\vec{R}$	Distributed resistance		F	E 5-41
r	Enrichment	—		I, E 2-15
	Radius	L		I, T 4-7
r^*	Vapor bubble radius for nucleation	L		I, E 12-1
r_p	Pressure ratio	—		I, E 6-97
S	Surface area	L^2		I, E 2-1
	Suppression factor of Chen's correlation	—		I, E 12-32
	Slip ratio	—		I, E 5-48
S_{ij}	Minimum flow area in the transverse direction	L^2		P 6-213
$\dot{S}_{gen}$	Rate of entropy generation		$E\theta^{-1}$	I, E 4-25b
S_T	Pitch	L		I, F 9-28
s_{ij}	Gap within the transverse direction	L		P 6-213
T	Temperature	θ		I, F 2-1
	Magnitude of as-fabricated clearance or tolerance in an assembly	L		E 4-127
T_0	Reservoir temperature	θ		I, E 4-27
T_s	Surroundings temperature	θ		I, E 4-25a
t	Time	T		I, P 2-25
	Spacer thickness	L		I, F 1-12
t^*	Peripheral spacer thickness	L		I, AT J1
t_s	Time after shutdown	T		I, F 3-8
V	Mean velocity	LT^{-1}		I, E 2-8
V_m	Mean velocity	LT^{-1}		I, E 9-44
V_v	Average bundle fluid velocity	LT^{-1}		I, E 9-90
V_∞	Bubble rise velocity	LT^{-1}		I, E 11-50

(*continued*)

A-4 GENERAL NOTATION (*continued*)

Symbol	Description	Dimension in two unit systems*		Location where symbol first appears†
		ML θT	*FE* plus *ML θT*	
V_{vj}, v_{vj}	Local drift velocity of vapor	LT^{-1}		I, E 11-32
v	Velocity of a point	LT^{-1}		I, T 4-2
$\vec{v}$	Velocity vector	LT^{-1}		I, T 4-2
v_r, $\vec{v}_r$	Relative velocity of the fluid with respect to the control surface	LT^{-1}		I, E 4-15
$\vec{v}_s$	Velocity of the control volume surface	LT^{-1}		I, E 4-11
W	Work	ML^2T^{-2}	FL	I, E 4-19
	Flow rate	MT^{-1}		I, P 7-273
W_u	Useful work	ML^2T^{-2}	FL	I, E 4-29
W_{ij}	Crossflow rate per unit length of channel	$MT^{-1}L^{-1}$		P 6-213
$W_{ij}'^D$, $W_{ij}'^M$, $W_{ij}'^H$	Transverse mass flow rate per unit length associated with turbulent mass, momentum, and energy exchange	$MT^{-1}L^{-1}$		P 6-213 and P 6-214
W_{ij}^{*M}, W_{ij}^{*H}	Transverse mass flow rate associated with both molecular and turbulent momentum and energy exchange	$MT^{-1}L^{-1}$		P 6-216
X	Dimensionless radius	—		E 7-46
X^2	Lockart-Martinelli parameter	—		I, E 11-92
x	Flow quality	—		I, E 5-35
x_{cr}	Critical quality at CHF	—		I, P 12-557
x_{st}	Static quality	—		I, E 5-22
$\Delta x'$	Transverse length	L		F 6-3
Y_i	Preference for downflow in a channel			E 4-98
Z, z	Height	L		I, P 2-22
Z_B	Boiling boundary	L		I, E 13-30a
z	Axial position	L		I, E 2-2
z_c	Position of maximum temperature in clad	L		I, 13-25a
z_f	Position of maximum temperature in fuel	L		I, E 13-28

* M = mass; θ = temperature; E = energy; — = dimensionless; L = length; T = time; F = force.

† I $\equiv$ Volume 1; A $\equiv$ appendix; T $\equiv$ table; E $\equiv$ equation; P $\equiv$ page; F $\equiv$ figure. For example, E10-6 = equation 10-6.

A-4 GENERAL NOTATION (*continued*)

Symbol	Description	Dimension in two unit systems*		Location where symbol first appears†
		ML θT	*FE* plus *ML θT*	
z_{ij}^{T}	Turbulent mixing length in COBRA	L		E 6-38
z_{ij}^{L}	Effective mixing length	L		E 6-93
	General Greek symbols			
α	Confidence level	—		E 8-12
	Local void fraction	—		I, E 5-11a
	Linear thermal expansion coefficient	θ^{-1}		I, P 8-309
	Dimensionless angular speed (pump)	—		E 3-119
	Thermal diffusivity	$L^{2}T^{-1}$		I, E 10-43
α_k	Phase density function	—		I, E 5-1
β	Delayed neutron fraction	—		I, E 3-65
	Mixing parameter	—		P 6-250
	Volumetric fraction of vapor or gas	—		I, E 5-51
	Thermal volume expansion coefficient	θ^{-1}		I, E 4-117
	Ratio of flow area	—		I, T 9-1
	Dimensionless torque (pump)	—		E 3-119
	Direction angle	—		E 6-81
	Pressure loss parameter	—		I, P 6-224
Γ	Volumetric vaporization rate	$MT^{-1}L^{-3}$		I, T 5-1
γ	Fraction of total power generated in the fuel	—		I, E 3-28
	Specific heat ratio (c_p/c_v)	—		I, T 6-3
	Porosity	—		E 5-1
δ	Ratio of axial length of grid spacers to axial length of a fuel bundle	—		I, AP J681
	Gap between fuel and clad	L		I, E 8-23b
	Thickness of heat-exchanger tube	L		E 3-93
	Thickness of liquid film in annular flow	L		I, E 11-1
δ^*	Dimensionless film thickness	—		I, E 11-1

(*continued*)

A-4 GENERAL NOTATION (*continued*)

Symbol	Description	Dimension in two unit systems*		Location where symbol first appears†
		$ML\,\theta T$	FE plus $ML\,\theta T$	
δ_c	Clad thickness	L		I, F 8-27
δ_g	Gap between fuel and clad	L		I, E 2-11
δ_T	Thickness of temperature boundary layer	L		I, F 10-4
ε	$\ell^2 dw/dy$ in COBRA code	L^2T^{-1}		E 6-87
	Surface emissivity		$EL^{-2}\theta^{-4}T^{-1}$	I, E 8-107a
ε_H	Eddy diffusivity of energy	L^2T^{-1}		I, E 10-46
ε_M	Momentum diffusivity	L^2T^{-1}		I, E 9-63
ζ	Thermodynamic efficiency (or effectiveness)	—		I, E 6-31
η	Pump efficiency	—		E 3-130
η_s	Isentropic efficiency	—		I, E 6-37
η_{th}	Thermal efficiency	—		I, E 6-38
θ	Position angle	—		I, T 4-7
	Two-phase multiplier for mixing	—		E 6-126
	Film temperature drop	θ		P 8-370
θ^*	Influence coefficient	—		E 7-69
λ	Length along channel until fluid reaches saturation	L		E 2-62
λ_c	Taylor instability wavelength	L		I, E 12-51
μ	Attenuation coefficient	L^{-1}		I, E 3-53
	Dynamic viscosity	$ML^{-1}T^{-1}$	FTL^{-2}	I, E 4-84
μ'	Bulk viscosity	$ML^{-1}T^{-1}$	FTL^{-2}	I, E 4-84
$\hat{\mu}$	Estimated mean of distribution	—		E 8-3
μ_a	Absorption coefficient	L^{-1}		I, E 3-51
ν	Kinematic viscosity (μ/ρ)	L^2T^{-1}		I, P 9-361
	Time it takes fluid packet to lose its subcooling	T		E 2-57
	Dimensionless volumetric flow rate (pump)	—		E 3-119
π	Torque		FL	E 3-119
ρ	Density	ML^{-3}		I, E 2-11
ρ_m^+	Two-phase momentum density	ML^{-3}		I, E 5-66

* M = mass; θ = temperature; E = energy; — = dimensionless; L = length; T = time; F = force.

† I ≡ Volume 1; A ≡ appendix; T ≡ table; E ≡ equation; P ≡ page; F ≡ figure. For example, E10-6 = equation 10-6.

A-4 GENERAL NOTATION (*continued*)

Symbol	Description	Dimension in two unit systems*		Location where symbol first appears†
		$ML\,\theta T$	FE plus $ML\,\theta T$	
Σ	Macroscopic cross section	L^{-1}		I, E 3-3
σ	Microscopic cross section	L^2		I, E 3-3
	Normal stress component		FL^{-2}	I, F 4-8
	Standard deviation	—		I, T 3-9
	Surface tension		FL^{-1}	I, E 11-5
$\hat{\sigma}$	Estimated standard deviation of distribution	—		E 8-4
τ	Shear stress component		FL^{-2}	I, F 4-8
	Time after start-up	T		I, E 3-68a
	Time constant	T		E 3-78a
τ_s	Operational time	T		I, E 3-69a
Φ	Dissipation function		$FL^{-2}T^{-1}$	I, E 4-107
	Neutron flux	$L^{-2}T^{-1}$		I, P 3-43
ϕ	Generalized volumetric source or sink for mass, momentum, or energy			I, E 4-50
	Relative humidity	—		I, E 7-19
	Azimuthal angle	—		I, T 4-7
$\phi_{\ell o}^2$	Two-phase friction multiplier based on all-liquid-flow pressure gradient	—		I, E 11-66
ϕ_ℓ^2	Two-phase frictional multiplier based on pressure gradient of liquid flow	—		I, E 11-86
ϕ_g^2	Two-phase frictional multiplier based on pressure gradient of gas or vapor flow	—		I, E 11-86
ξ	Logarithmic energy decrement	—		I, E 3-41
χ	Energy per fission deposited in the fuel		E	I, P 3-43
ψ	Force field per unit mass of fluid		FLM^{-1}	I, E 4-21

(*continued*)

A-4 GENERAL NOTATION (*continued*)

Symbol	Description	Dimension in two unit systems*		Location where symbol first appears†
		$ML\,\theta T$	FE plus $ML\,\theta T$	
	Ratio of eddy diffusivity of heat to momentum	—		I, E 10-71
ω	Angular speed	T^{-1}		I, E 9-2b

Subscripts

Symbol	Description			Location where symbol first appears†
A	Area			E 5-4
	Annulus			E 7-50
AF	Atmospheric flow			I, E 6-68
a	Air			I, E 7-1
	Absorption			I, P 3-44
acc	Acceleration			I, E 9-22
avg	Average			I, F 2-1
B, b	Boiling			I, E 13-30a
B	Buoyancy			E 3-16
b	Bulk			I, E 2-6
C	Cold			I, F 2-1
	Condensate			I, P 7-273
CD	Condensor			I, E 6-87
CP	Compressor			I, E 6-102
$\mathcal{C}_L$	Center line			I, F 8-9
c	Capture			I, P 3-43
	Cladding			I, E 8-38
	Conduction			I, E 4-113
	Containment			I, E 7-1
	Contraction			I, F 9-35
	Coolant			I, E 6-2
	Core			P 3-74
	Critical point (thermodynamic)			I, E 9-31
ci	Clad inside			I, P 8-309
co	Clad outside			I, E 2-5
cr	Critical flow			I, E 11-113
crit	Critical			I, T 6-1
c.m.	Control mass			I, E 4-17c
c.v.	Control volume			I, E 4-30
ct	Circular tube			I, E 10-99
EA	Equivalent annulus			E 7-50
EQUIL	Equilibrium void distribution			E 6-124
e	Expansion			I, F 9-35
	Extrapolated			I, E 3-35

* M = mass; θ = temperature; E = energy; — = dimensionless; L = length; T = time; F = force.

† I ≡ Volume 1; A ≡ appendix; T ≡ table; E ≡ equation; P ≡ page; F ≡ figure. For example, E10-6 = equation 10-6.

A-4 GENERAL NOTATION (*continued*)

Symbol	Description	Dimension in two unit systems*		Location where symbol first appears†
		ML θT	*FE* plus *ML θT*	
	Equilibrium			I, E 5-53
	Eddy			P 7-303
	Electrical			I, E 10-10
eff	Effective (molecular + eddy)			I, E 5-129
ex	External			E 3-4
exit, ex, e	Indicating the position of flowing exit			E 2-84
eℓ	Elastic			I, E 3-39
FL	Flashing			I, T 7-4
f	Friction			I, E 9-91a
	Fluid			E 3-41
	Flow			I, P 10-444
	Saturated liquid			I, E 5-53
	Fuel			I, E 3-19
	Fission			I, P 3-43
fg	Difference between fluid and gas			I, T 6-3
fi	Flow without spacers			I, E 9-82
fo	Fuel outside surface			I, E 2-5
fric	Friction			I, T 9-5
fi$_s$	flow including spacers			I, AT J3
g	Gap between fuel and cladding			I, E 8-106
	Saturated vapor			I, E 5-53
H, h	Heated			I, F 2-1
HX	Heat exchanger			I, E 6-104
h, (heater)	Pressurizer heater			I, F 7-10
i	Denoting the selected subchannel control volume			P 6-213
	Index of direction			I, E 5-48
	Inner surface			E 7-69
	Index of streams into or out of the control volume			I, E 4-30a
in, IN	Indicating the position of flowing in, inlet			I, E 6-38
iso	Isothermal			I, E 13-74
iℓ	Inelastic			I, E 3-59

(*continued*)

A-4 GENERAL NOTATION (*continued*)

Symbol	Description	Dimension in two unit systems*		Location where symbol first appears†
		ML θT	*FE* plus *ML* θT	
j	Adjacent subchannel control volume of the subchannel control volume i			P 6-213
	Index of properties			E 8-1
	Index of isotopes			I, P 3-43
k	Index of phase			I, E 5-1
L	Lower			I, F 7-14
	Laminar			I, E 9-82
ℓ	Liquid phase in a two-phase flow			I, T 5-1
ℓo	Liquid only			I, E 11-66
M	Temperature of a quantity of interest			E 8-17
m	Temperature difference of a quantity of interest			E 8-17
	Mixture			I, E 5-38a
mc	Molecular conduction			E 7-104
nom	Nominal			E 8-17
n	Nuclear core			I, E 7-2d
o	Indicating initial value			I, E 3-56
	Operating			I, E 3-68
	Outer surface			E 7-69
	Reservoir conditions			I, E 4-27
out, OUT	Outlet			I, P 6-189
P	Pump			I, E 6-48
p	Primary			I, E 6-49
	Pore			I, P 8-301
R	Rated			E 3-76
	Reactor			I, E 6-53
RO	Condensation as rainout			I, E 7-128
r	Radiation			I, E 4-113
	Enrichment			E 8-31
rb	Equivalent or dispersed			E 6-34
	Rod bundle			E 5-53
ref	Reference			I, E 9-23
SC	Condensation at spray drops			I, E 7-128
SCB	Subcooled boiling			I, E 13-57
SG, S.G.	Steam generator			I, E 6-54
SP	Single phase			E 6-125
	Spray			I, E 7-135

* M = mass; θ = temperature; E = energy; — = dimensionless; L = length; T = time; F = force.

† I ≡ Volume 1; A ≡ appendix; T ≡ table; E ≡ equation; P ≡ page; F ≡ figure. For example, E10-6 = equation 10-6.

A-4 GENERAL NOTATION (*continued*)

Symbol	Description	Dimension in two unit systems*		Location where symbol first appears†
		ML θT	*FE* plus *ML θT*	
s	Scattering			I, E 3-39
	Secondary			I, E 6-49
	Sintered			I, E 8-98
	Slug flow			E 7-93
	Solid			I, P 8-301
	Spacer, wire spacer			I, F 1-14
	Surface, interface			I, E 4-11
s	Isentropic			I, E 6-10b
sat	Saturated			I, T 6-1
st	Structures			I, E 7-2a
	Static			I, E 5-22
s′	Interface between continuous vapor and falling liquid droplets			I, F 7-14
s″	Interface between the continuous vapor phase of the upper volume and the continuous liquid phase of the lower volume			I, F 7-14
s‴	Interface that separates the discontinuous phase in either the upper or lower volume from the continuous same phase in the other volume			I, F 7-14
T, t	Turbine			I, T 6-5
T	Total			I, P 7-240
TD	Theoretical density			I, E 8-17
TP	Two phase			I, E 11-67
tb	Transition boiling			I, E 12-43
th	Thermal			I, E 6-38
tr	Transverse flow			P 5-190
U	Upper			I, F 7-14
u	Useful			I, E 4-28
V	Volume			I, P 9-386
v	Cavity (void)			I, E 8-55
	Vapor or gas phase in a two-phase flow			I, T 5-1
vj	Local vapor drift			I, E 11-32

(*continued*)

A-4 GENERAL NOTATION (*continued*)

Symbol	Description	Dimension in two unit systems*		Location where symbol first appears†
		$ML\,\theta T$	FE plus $ML\,\theta T$	
vo	Vapor only			I, E 11-66
WC	Condensation at the wall			I, E 7-128
w	Wall			I, T 5-1
	Water			I, P 7-240
ϕ	Neutron flux			E 8-31
1ϕ	Single phase			I, F 13-1
2ϕ	Two phase			I, F 12-11
ρ	Fuel density			E 8-31
∞	Free stream			I, F 10-4
	Superscripts			
D	Hot spot factor resulting from direct contributors			E 8-58
j	Index of isotopes			I, P 3-43
S	Hot spot factor resulting from statistical contributors			E 8-58
t	Turbulent effect			I, E 4-132
$\vec{}$	Vector			I, P 2-22
$\overline{}$	Spatial average			I, E 3-41
', ", '''	Per unit length, surface area, volume, respectively			I, P 2-22
'	Denoting perturbation			I, E 4-125
TP	Two phase			I, E 11-77
*	Reference			P 1-10
	Denoting the velocity or enthalpy transported by the diversion crossflow			P 6-215
i	Intrinsic			E 5-12
=	Tensor			I, E 4-8
$-,\ ^-$	Averaging (time)			I, E 4-124, E 5-6
o	Stagnation			I, T 4-2
	Denoting nominal			E 8-27
	Special symbols			
Δ	Change in, denoting increment			I, E 2-8
∇	Gradient			I, E 4-2

* M = mass; θ = temperature; E = energy; — = dimensionless; L = length; T = time; F = force.

† I ≡ Volume 1; A ≡ appendix; T ≡ table; E ≡ equation; P ≡ page; F ≡ figure. For example, E10-6 = equation 10-6.

A-4 GENERAL NOTATION (*continued*)

Symbol	Description	Dimension in two unit systems*		Location where symbol first appears†
		$ML\ \theta T$	FE plus $ML\ \theta T$	
δ	Change in, denoting increment			I, T 3-3
$\langle\ \rangle$	Volumetric averaging			I, E 2-5
$\{\ \}$	Area averaging			I, E 3-20
$\equiv$	Defined as			I, P 2-22
$\simeq$, $\approx$, $\sim$	Approximately equal to			I, E 3-38

PHYSICAL AND MATHEMATICAL CONSTANTS†

Avogadro's number, A_v	0.602252×10^{24} molecules/g mol
	2.731769×10^{26} molecules/lbm mol
Barn	10^{-24} cm^2, 1.0765×10^{-27} ft^2
Boltzmann's constant, $k = \bar{R}/A_v$	1.38054×10^{-16} erg/K
	8.61747×10^{-5} eV/K
Curie	3.70×10^{10} dis/s
Electron charge	4.80298×10^{-10} esu, 1.60210×10^{-19} Coulombs
Faraday's constant	9.648×10^4 coulombs/mol
g_c conversion factor	1.0 gm cm^2/erg s^2, 32.17 lbm ft/lb$_f$ s^2
	4.17×10^8 lbm ft/lb$_f$ hr^2, 0.9648×10^{18} amu cm^2/MeV s^2
Gravitational acceleration (standard)	32.1739 ft/s^2, 980.665 cm/s^2
Joule's equivalent	778.16 ft-lb$_f$/BTU
Mass–energy conversion	1 amu = 931.478 MeV = 1.41492×10^{-13} BTU = 4.1471×10^{-17} kWhr
	1 g = 5.60984×10^{26} MeV = 2.49760×10^7 kWhr = 1.04067 MWd
	1 lbm = 2.54458×10^{32} MeV = 3.86524×10^{16} BTU
Mathematical constants	$e \equiv 2.71828$
	$\pi \equiv 3.14159$
	$\ell n\ 10 \equiv 2.30259$
Molecular volume	22413.6 cm^3/g mol, 359.0371 ft^3/lb$_m$ mol, at 1 atm and 0 °C
Neutron energy	0.0252977 eV at 2200 m/s, $\frac{1}{40}$ eV at 2187.017 m/s

† (After El-Wakil, M. M. *Nuclear Heat Transport*. Scranton, PA: International Textbook Co., 1971.

Planck's constant	6.6256×10^{-27} erg s, 4.13576×10^{-15} eV s
Rest masses:	
Electron	5.48597×10^{-4} amu, 9.10909×10^{-28} g, 2.00819×10^{-30} lbm
Neutron	1.0086654 amu, $1.6748228 \times 10^{-24}$ g, 3.692314×10^{-27} lbm
Proton	1.0072766 amu, 1.672499×10^{-24} g, 3.687192×10^{-27} lbm
Stephan-Boltzmann constant	5.67×10^{-12} W/cm^{2}K^4
Universal gas constant, $\bar{R}$	1545.08 ft-lb$_f$/lbm mol R, 1.98545 cal/g mol K, 1.98545 BTU/lbm mol R, 8.31434×10^7 erg/g mol K
Velocity of light	2.997925×10^{10} cm/s, 9.83619×10^8 ft/s

UNIT SYSTEMS

Table	Unit	Reference
C-1	Length	2
C-2	Area	2
C-3	Volume	2
C-4	Mass	2
C-5	Force	1
C-6	Density	2
C-7	Time	2
C-8	Flow	2
C-9	Pressure, momentum flux	1
C-10	Work, energy, torque	1
C-11	Power	2
C-12	Power density	2
C-13	Heat flux	2
C-14	Viscosity	1
C-15	Thermal conductivity	1
C-16	Heat transfer coefficient	1
C-17	Momentum, thermal, or molecular diffusivity	1
C-18	Surface tension	—

1. Bird, R. B., Stewart, W. E., and Lightfoot, E. N. *Transport Phenomena*. New York: John Wiley and Sons, 1960. Reprinted by permission of John Wiley and Sons, Inc.
2. El-Wakil, M. M. *Nuclear Heat Transport*. Scranton, PA: International Textbook Company, 1971.

C-1 RELEVANT SI UNITS

Quantity	Name	Symbol
SI base units		
Length	Meter	m
Time	Second	s
Mass	Kilogram	kg
Temperature	Kelvin	K
Amount of matter	Mole	mol
Electric current	Ampere	A
SI derived units		
Force	Newton	$N = kg\ m/s^2$
Energy, work, heat	Joule	$J = N\ m = kg\ m^2/s^2$
Power	Watt	$W = J/s$
Frequency	Hertz	$Hz = s^{-1}$
Electric charge	Coulomb	$C = A\ s$
Electric potential	Volt	$V = J/C$
Allowed (to be used with SI units) units		
Time	Minute	min
	Hour	h
	Day	d
Plane angle	Degree	°
	Minute	′
	Second	″
Volume	Liter	l
Mass	Tonne	$t = 1{,}000\ kg$
	Atomic mass unit	$u \approx 1.66053 \times 10^{-27}\ kg$
Fluid pressure	Bar	$bar = 10^5\ Pa$
Temperature	Degree Celsius	°C
Energy	Electron volt	$eV \approx 1.60219 \times 10^{-19}\ J$

C-2 UNIT CONVERSION TABLES

Utilization of Conversion Tables

The column and row units correspond to each other as illustrated below.

Given a quantity in units of a row, multiply by the table value to obtain the quantity in units of the corresponding column.

Example How many meters is 10 centimeters?

Answer: We desire the quantity in units of meters (the column entry) and have been given the quantity in units of centimeters (the row entry), i.e., 10 cm.

Hence meters = 0.01 cm

= 0.01 (10)

= 0.1.

Columns Rows	a Centimeters	b Meters	c	d
a = centimeters		0.01		
b = meters				
c				
d				

Table C-1 Length

	Centimeters cm	Meters† m	Inches in.	Feet‡ ft	Miles	Microns μ	Angstroms A
cm	1	0.01	0.3937	0.03281	6.214×10^{-6}	10^4	10^8
m	100	1	39.37	3.281	6.214×10^{-4}	10^6	10^{10}
in	2.540	0.0254	1	0.08333	1.578×10^{-5}	2.54×10^4	2.54×10^8
ft	30.48	0.3048	12	1	1.894×10^{-4}	0.3048×10^6	0.3048×10^{10}
Miles	1.6003×10^5	1609.3	6.336×10^4	5280	1	1.6093×10^9	1.6093×10^{13}
Microns	10^{-4}	10^{-6}	3.937×10^{-5}	3.281×10^{-6}	6.2139×10^{-10}	1	10^4
Angstroms	10^{-8}	10^{-10}	3.937×10^{-9}	3.281×10^{-10}	6.2139×10^{-14}	10^{-4}	1

Table C-2 Area

	cm²	m²†	in²	ft²‡	Mile²	Acre	Barn
cm²	1	10^{-4}	0.155	1.0764×10^{-3}	3.861×10^{-11}	2.4711×10^{-8}	10^{24}
m²	10^4	1	1550	10.764	3.861×10^{-7}	2.4711×10^{-4}	10^{28}
in²	6.4516	6.4516×10^{-4}	1	6.944×10^{-3}	2.491×10^{-10}	1.5944×10^{-7}	6.4517×10^{24}
ft²	929	0.0929	144	1	3.587×10^{-8}	2.2957×10^{-5}	9.29×10^{26}
Mile²	2.59×10^{10}	2.59×10^6	4.0144×10^{11}	2.7878×10^7	1	640	2.59×10^{34}
Acre	4.0469×10^7	4.0469×10^3	6.2726×10^6	4.356×10^4	1.5625×10^{-3}	1	4.0469×10^{31}
Barn	10^{-24}	10^{-28}	1.55×10^{-25}	1.0764×10^{-27}	3.861×10^{-35}	2.4711×10^{-32}	1

† SI units.
‡ English units.

Table C-3 Volume

	cm^3	Liters	m^3†	in^3	ft^3‡	Cubic yards	U.S. (liq.) gallons	Imperial gallons
cm^3	1	10^{-3}	10^{-6}	0.06102	3.532×10^{-5}	1.308×10^{-6}	2.642×10^{-4}	2.20×10^{-4}
Liters	10^3	1	10^{-3}	61.02	0.03532	1.308×10^{-3}	0.2642	0.220
m^3	10^6	10^3	1	6.102×10^4	35.31	1.308	264.2	220.0
in^3	16.39	0.01639	1.639×10^{-5}	1	5.787×10^{-4}	2.143×10^{-5}	4.329×10^{-3}	3.605×10^{-3}
ft^3	2.832×10^4	28.32	0.02832	1728	1	0.03704	7.481	6.229
Cubic yd	7.646×10^5	764.6	0.7646	4.666×10^4	27.0	1	202.0	168.2
U.S. gal	3785	3.785	3.785×10^{-3}	231.0	0.1337	4.951×10^{-3}	1	0.8327
Imp gal	4546	4.546	4.546×10^{-3}	277.4	0.1605	5.946×10^{-3}	1.201	1

Table C-4 Mass

	Grams g	Kilograms kg	Pounds lbm†	Tons (short)	Tons (long)	Tons (metric)	Atomic mass units, amu
g	1	0.001	2.2046×10^{-3}	11.102×10^{-6}	9.842×10^{-7}	10^{-6}	0.60225×10^{24}
kg	1000	1	2.2046	0.001102	9.842×10^{-4}	10^{-3}	6.0225×10^{26}
lbm	453.6	0.4536	1	5.0×10^{-4}	4.464×10^{-4}	4.536×10^{-4}	2.7318×10^{26}
tons (s)	9.072×10^5	907.2	2000	1	0.8929	0.9072	5.4636×10^{29}
tons (l)	1.016×10^6	1016	2240	1.12	1	1.016	6.1192×10^{29}
tons (metric)	10^6	1000	2204.7	1.1023	0.9843	1	6.0225×10^{29}
amu	1.6604×10^{-24}	1.6604×10^{-27}	3.6606×10^{-27}	1.8303×10^{-30}	1.6343×10^{-30}	1.6604×10^{-30}	1

† SI units.
‡ English units.

Table C-5 Force

	g cm s^{-2} dynes	kg m s^{-2} Newtons	lbm ft s^{-2}† poundals	lb$_f$‡
Dynes	1	10^{-5}	7.2330×10^{-5}	2.2481×10^{-6}
Newtons	10^5	1	7.2330	2.2481×10^{-1}
Poundals	1.3826×10^4	1.3826×10^{-1}	1	3.1081×10^{-2}
lb$_f$	4.4482×10^5	4.4482	32.1740	1

Table C-6 Density

	g/cm^3	kg/m^3†	lbm/in.^{3}	lbm/ft^3‡	lbm/U.S. gal	lbm/Imp gal
g/cm^3	1	10^3	0.03613	62.43	8.345	10.2
kg/m^3	10^{-3}	1	3.613×10^{-5}	0.06243	8.345×10^{-3}	0.01002
lbm/in.^{3}	27.68	2.768×10^4	1	1728	231	277.4
lbm/ft^3	0.01602	16.02	5.787×10^{-4}	1	0.1337	0.1605
lbm/U.S. gal	0.1198	119.8	4.329×10^{-3}	7.481	1	1.201
lbm/Imp gal	0.09978	99.78	4.605×10^{-3}	6.229	0.8327	1

† SI units.
‡ English units.

438

Table C-7 Time

	Microseconds μs	Seconds s	Minutes min	Hours hr	Days d	Years yr
μs	1	10^{-6}	1.667×10^{-8}	2.778×10^{-10}	1.157×10^{-11}	3.169×10^{-14}
s	10^6	1	1.667×10^{-2}	2.778×10^{-4}	1.157×10^{-5}	3.169×10^{-8}
min	6×10^7	60	1	1.667×10^{-2}	6.944×10^{-4}	1.901×10^{-6}
hr	3.6×10^9	3600	60	1	0.04167	1.141×10^{-4}
d	8.64×10^{10}	8.64×10^4	1440	24	1	2.737×10^{-3}
yr	3.1557×10^{13}	3.1557×10^7	5.259×10^5	8766	365.24	1

Table C-8 Flow

	cm^3/s	ft^3/min	U.S. gal/min	Imp gal/min
cm^3/s	1	0.002119	0.01585	0.01320
ft^3/min	472.0	1	7.481	6.229
U.S. gal/min	63.09	0.1337	1	0.8327
Imp gal/min	75.77	0.1605	1.201	1

Table C-9 Pressure, momentum flux

| | Pascal | | | | lb_f in^{-2}‡ | Atmospheres | | |
	g cm^{-1} s^{-2} dyne cm^{-2}	kg m^{-1} s^{-2} Newtons m^{-2}	lbm ft^{-1} s^{-2} poundals ft^{-2}	lb_f ft^{-2}	psia	atm	mm Hg	in. Hg
Dyne cm^{-2}	1	10^{-1}	6.7197×10^{-2}	2.0886×10^{-3}	1.4504×10^{-5}	9.8692×10^{-7}	7.5006×10^{-4}	2.9530×10^{-5}
Newtons m^{-2}	10	1	6.7197×10^{-1}	2.0886×10^{-2}	1.4504×10^{-4}	9.8692×10^{-6}	7.5006×10^{-3}	2.9530×10^{-4}
Poundals ft^{-2}	1.4882×10^1	1.4882	1	3.1081×10^{-2}	2.1584×10^{-4}	1.4687×10^{-5}	1.1162×10^{-2}	4.3945×10^{-4}
lb_f ft^{-2}	4.7880×10^2	4.7880×10^1	32.1740	1	6.9444×10^{-3}	4.7254×10^{-4}	3.5913×10^{-1}	1.4139×10^{-2}
psia	6.8947×10^4	6.8947×10^3	4.6330×10^3	144	1	6.8046×10^{-2}	5.1715×10^1	2.0360
atm	1.0133×10^6	1.0133×10^5	6.8087×10^4	2.1162×10^3	14.696	1	760	29.921
mm Hg	1.3332×10^3	1.3332×10^2	8.9588×10^1	2.7845	1.9337×10^{-2}	1.3158×10^{-3}	1	3.9370×10^{-2}
in. Hg	3.3864×10^4	3.3864×10^3	2.2756×10^3	7.0727×10^1	4.9116×10^{-1}	3.3421×10^{-2}	25.400	1

Table C-10 Work, energy, torque

	g cm^2 s^{-2} ergs	kg m^2 s^{-2}† absolute joules	lbm ft^2 s^{-2} ft-poundals	ft lb_f	cal	BTU‡	hp hr	kW hr
g cm^2 s^{-2}	1	10^{-7}	2.3730×10^{-6}	7.3756×10^{-8}	2.3901×10^{-8}	9.4783×10^{-11}	3.7251×10^{-14}	2.7778×10^{-14}
kg m^2 s^{-2}	10^7	1	2.3730×10^1	7.3756×10^{-1}	2.3901×10^{-1}	9.4783×10^{-4}	3.7251×10^{-7}	2.7778×10^{-7}
lbm ft^2 s^{-2}	4.2140×10^5	4.2140×10^{-2}	1	3.1081×10^{-2}	1.0072×10^{-2}	3.9942×10^{-5}	1.5698×10^{-8}	1.1706×10^{-8}
ft lb_f	1.3558×10^7	1.3558	32.1740	1	3.2405×10^{-1}	1.2851×10^{-3}	5.0505×10^{-7}	3.7662×10^{-7}
Thermochemical calories[a]	4.1840×10^7	4.1840	9.9287×10^1	3.0860	1	3.9657×10^{-3}	1.5586×10^{-6}	1.1622×10^{-6}
British thermal units	1.0550×10^{10}	1.0550×10^3	2.5036×10^4	778.16	2.5216×10^2	1	3.9301×10^{-4}	2.9307×10^{-4}
Horsepower hours	2.6845×10^{13}	2.6845×10^6	6.3705×10^7	1.9800×10^6	6.4162×10^5	2.5445×10^3	1	7.4570×10^{-1}
Absolute kilowatt hours	3.6000×10^{13}	3.6000×10^6	8.5429×10^7	2.6552×10^6	8.6042×10^5	3.4122×10^3	1.3410	1

[a] This unit, abbreviated cal, is used in chemical thermodynamic tables. To convert quantities expressed in International Steam Table calories (abbreviated I.T. cal) to this unit, multiply by 1.00654.

† SI units.

‡ English units.

Table C-11 Power

	Ergs/s	Joule/s watt†	kW	BTU/hr‡	hp	eV/s
Ergs/s	1	10^{-7}	10^{-10}	3.412×10^{-7}	1.341×10^{-10}	6.2421×10^{11}
Joule/s	10^7	1	10^{-3}	3.412	0.001341	6.2421×10^{18}
kW	10^{10}	10^3	1	3412	1.341	6.2421×10^{21}
BTU/hr	2.931×10^6	0.2931	2.9310×10^{-4}	1	3.93×10^{-4}	1.8294×10^{18}
hp	7.457×10^9	745.7	0.7457	2545	1	4.6548×10^{21}
eV/s	1.6021×10^{-12}	1.6021×10^{-19}	1.6021×10^{-19}	5.4664×10^{-19}	2.1483×10^{-22}	1

Table C-12 Power density

	Watt/cm³, kW/lit†	cal/s cm³‡	BTU/hr in³	BTU/hr ft³	MeV/s cm³
Watt/cm³, kW/lit	1	0.2388	55.91	9.662×10^4	6.2420×10^{12}
cal/s cm³	4.187	1	234.1	4.045×10^5	2.613×10^{13}
BTU/hr in³	0.01788	4.272×10^{-3}	1	1728	1.1164×10^{11}
BTU/hr ft³	1.035×10^{-5}	2.472×10^{-6}	5.787×10^{-4}	1	6.4610×10^7
MeV/s cm³	1.602×10^{-13}	3.826×10^{-14}	8.9568×10^{-12}	1.5477×10^{-8}	1

† SI units.
‡ English units.

Table C-13 Heat flux

	Watt/cm²†	cal/s cm²‡	BTU/hr ft²	MeV/s cm²
Watt/cm²	1	0.2388	3170.2	6.2420×10^{12}
cal/s cm²	4.187	1	1.3272×10^4	2.6134×10^{13}
BTU/hr ft²	3.155×10^{-4}	7.535×10^{-5}	1	1.9691×10^9
MeV/s cm²	1.602×10^{-13}	3.826×10^{-14}	5.0785×10^{-10}	1

Table C-14 Viscosity[a]

	g cm⁻¹ s⁻¹ poises	kg m⁻¹ s⁻¹†	lbm ft⁻¹ s⁻¹†	lbf s ft⁻²	Centipoises	lbm ft⁻¹ hr⁻¹
g cm⁻¹ s⁻¹ poises	1	10^{-1}	6.7197×10^{-2}	2.0886×10^{-3}	10^2	2.4191×10^2
kg m⁻¹ s⁻¹	10	1	6.7197×10^{-1}	2.0886×10^{-2}	10^3	2.4191×10^3
lbm ft⁻¹ s⁻¹	1.4882×10^1	1.4882	1	3.1081×10^{-2}	1.4882×10^3	3600
lbf s ft⁻²	4.7880×10^2	4.7880×10^1	32.1740	1	4.7880×10^4	1.1583×10^5
Centipoises	10^{-2}	10^{-3}	6.7197×10^{-4}	2.0886×10^{-5}	1	2.4191
lbm ft⁻¹ hr⁻¹	4.1338×10^{-3}	4.1338×10^{-4}	2.7778×10^{-4}	8.6336×10^{-6}	4.1338×10^{-1}	1

[a] When moles appear in the given and desired units, the conversion factor is the same as for the corresponding mass units.
† SI units.
‡ English units.

Table C-15 Thermal conductivity

	$g\ cm\ s^{-3}\ K^{-1}$ $ergs\ s^{-1}\ cm^{-1}\ K^{-1}$	$kg\ m\ s^{-3}\ K^{-1}$† $watts\ m^{-1}\ K^{-1}$	$lbm\ ft\ s^{-3}\ F^{-1}$	$lb_f\ s^{-1}\ F^{-1}$	$cal\ s^{-1}\ cm^{-1}\ K^{-1}$	$BTU\ hr^{-1}\ ft^{-1}\ F^{-1}$‡
$g\ cm\ s^{-3}\ K^{-1}$	1	10^{-5}	4.0183×10^{-5}	1.2489×10^{-6}	2.3901×10^{-8}	5.7780×10^{-6}
$kg\ m\ s^{-3}\ K^{-1}$	10^{5}	1	4.0183	1.2489×10^{-1}	2.3901×10^{-3}	5.7780×10^{-1}
$lbm\ ft\ s^{-3}\ F^{-1}$	2.4886×10^{4}	2.4886×10^{-1}	1	3.1081×10^{-2}	5.9479×10^{-4}	1.4379×10^{-1}
$lb_f\ s^{-1}\ F^{-1}$	8.0068×10^{5}	8.0068	3.2174×10^{1}	1	1.9137×10^{-2}	4.6263
$cal\ s^{-1}\ cm^{-1}\ K^{-1}$	4.1840×10^{7}	4.1840×10^{2}	1.6813×10^{3}	5.2256×10^{1}	1	2.4175×10^{2}
$BTU\ hr^{-1}\ ft^{-1}\ F^{-1}$	1.7307×10^{5}	1.7307	6.9546	2.1616×10^{-1}	4.1365×10^{-3}	1

Table C-16 Heat transfer coefficient

	$g\ s^{-3}\ K^{-1}$	$kg\ s^{-3}\ K^{-1}$† $Watts\ m^{-2}\ K^{-1}$	$lbm\ s^{-3}\ F^{-1}$	$lb_f\ ft^{-1}\ s^{-3}\ F^{-1}$	$cal\ cm^{-2}\ s^{-1}\ K^{-1}$	$Watts\ cm^{-2}\ K^{-1}$	$BTU\ ft^{-2}\ hr^{-1}\ F^{-1}$‡
$g\ s^{-3}\ K^{-1}$	1	10^{-3}	1.2248×10^{-3}	3.8068×10^{-5}	2.3901×10^{-8}	10^{-7}	1.7611×10^{-4}
$kg\ s^{-3}\ K^{-1}$	10^{3}	1	1.2248	3.8068×10^{-2}	2.3901×10^{-5}	10^{-4}	1.7611×10^{-1}
$lbm\ s^{-3}\ F^{-1}$	8.1647×10^{2}	8.1647×10^{-1}	1	3.1081×10^{-2}	1.9514×10^{-5}	8.1647×10^{-5}	1.4379×10^{-1}
$lb_f\ ft^{-1}\ s^{-1}\ F^{-1}$	2.6269×10^{4}	2.6269×10^{1}	32.1740	1	6.2784×10^{-4}	2.6269×10^{-3}	4.6263
$cal\ cm^{-2}\ s^{-1}\ K^{-1}$	4.1840×10^{7}	4.1840×10^{4}	5.1245×10^{4}	1.5928×10^{3}	1	4.1840	7.3686×10^{3}
$Watts\ cm^{-2}\ K^{-1}$	10^{7}	10^{4}	1.2248×10^{4}	3.8068×10^{2}	2.3901×10^{-1}	1	1.7611×10^{3}
$BTU\ ft^{-2}\ hr^{-1}\ F^{-1}$	5.6782×10^{3}	5.6782	6.9546	2.1616×10^{-1}	1.3571×10^{-4}	5.6782×10^{-4}	1

† SI units.
‡ English units.

Table C-17 Momentum, thermal, or molecular diffusivity

	cm^2 s^{-1}	m^2 s^{-1}†	ft^2 hr^{-1}‡	Centistokes
cm^2 s^{-1}	1	10^{-4}	3.8750	10^2
m^2 s^{-1}	10^4	1	3.8750×10^4	10^6
ft^2 hr^{-1}	2.5807×10^{-1}	2.5807×10^{-5}	1	2.5807×10^1
Centistokes	10^{-2}	10^{-6}	3.8750×10^{-2}	1

† SI units.
‡ English units.

Table C-18 Surface tension

	N/m†	Dyne/cm	lb$_f$/ft‡
N/m	1	1000	0.06852
Dyne/cm	0.001	1	6.852×10^{-5}
lb$_f$/ft	14.594	1.4594×10^4	1

† SI units.
‡ English units.

MATHEMATICAL TABLES

BESSEL FUNCTION*

Some useful derivatives and integrals of Bessel functions are given in Tables D-1 and D-2.

* El-Wakil, M. M. *Nuclear Heat Transport*. Scranton, PA: International Textbook Co., 1971.

Table D-1 Derivatives of Bessel functions

$$\frac{dJ_0(x)}{dx} = -J_1(x) \qquad\qquad \frac{dY_0(x)}{dx} = -Y_1(x)$$

$$\frac{dI_0(x)}{dx} = I_1(x) \qquad\qquad \frac{dK_0(x)}{dx} = -K_1(x)$$

$$\frac{dJ_v(x)}{dx} = J_{v-1}(x) - \frac{v}{x} J_v(x) \qquad\qquad \frac{dY_v(x)}{dx} = Y_{v-1}(x) - \frac{v}{x} Y_v(x)$$

$$= -J_{v+1}(x) + \frac{v}{x} J_v(x) \qquad\qquad = -Y_{v+1}(x) + \frac{v}{x} Y_v(x)$$

$$= \frac{1}{2} [J_{v-1}(x) - J_{v+1}(x)] \qquad\qquad = \frac{1}{2} [Y_{v-1}(x) - Y_{v+1}(x)]$$

$$\frac{dI_v(x)}{dx} = I_{v-1}(x) - \frac{v}{x} I_v(x) \qquad\qquad \frac{dK_v(x)}{dx} = -K_{v-1}(x) - \frac{v}{x} K_v(x)$$

$$= I_{v+1}(x) + \frac{v}{x} I_v(x) \qquad\qquad = -K_{v+1}(x) + \frac{v}{x} K_v(x)$$

$$= \frac{1}{2} [I_{v-1}(x) + I_{v+1}(x)] \qquad\qquad = -\frac{1}{2} [K_{v-1}(x) + K_{v+1}(x)]$$

$$\frac{dx^v J_v(x)}{dx} = x^v J_{v-1}(x) \qquad\qquad \frac{dx^{-v} J_v(x)}{dx} = -x^{-v} J_{v+1}(x)$$

$$\frac{dx^v Y_v(x)}{dx} = x^v Y_{v-1}(x) \qquad\qquad \frac{dx^{-v} Y_v(x)}{dx} = -x^{-v} Y_{v+1}(x)$$

$$\frac{dx^v I_v(x)}{dx} = x^v I_{v-1}(x) \qquad\qquad \frac{dx^{-v} I_v(x)}{dx} = +x^{-v} I_{v+1}(x)$$

$$\frac{dx^v K_v(x)}{dx} = -x^v K_{v-1}(x) \qquad\qquad \frac{dx^{-v} K_v(x)}{dx} = -x^{-v} K_{v+1}(x)$$

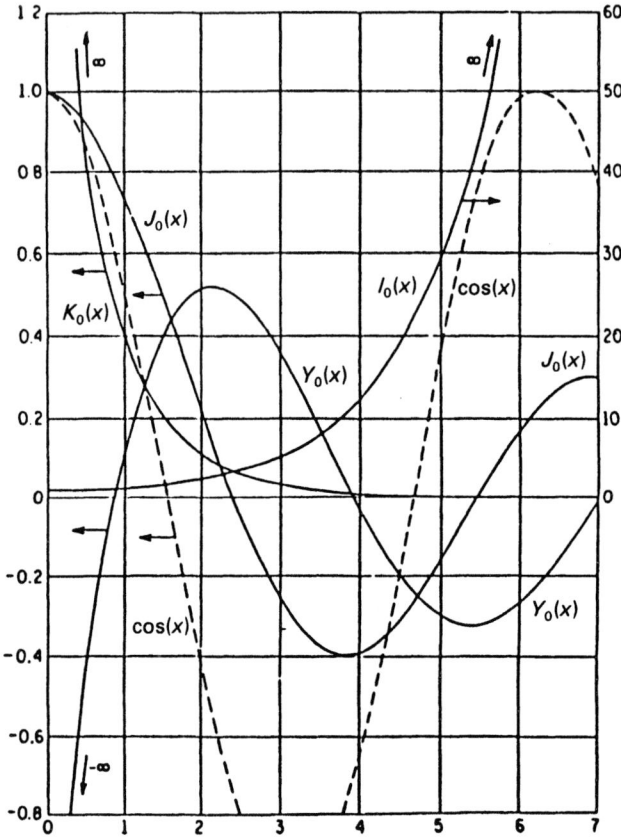

Figure D-1 The four Bessel functions of zero order.

Table D-2 Integrals of Bessel functions

$$\int J_1(x)\,dx = -J_0(x) + C \qquad \int Y_1(x)\,dx = -Y_0(x) + C$$

$$\int I_1(x)\,dx = I_0(x) + C \qquad \int K_1(x)\,dx = -K_0(x) + C$$

$$\int x^v J_{v-1}(x)\,dx = x^v J_v(x) + C \qquad \int x^{-v} J_{v+1}(x)\,dx = -x^{-v} J_v(x) + C$$

The Bessel and modified Bessel functions of zero and first order are tabulated in Table D-3 for positive values of x up to $x = 4.0$. Some roots are given below:

Roots of $J_0(x)$: $x = 2.4048, 5.5201, 8.6537, 11.7915, \ldots$
Roots of $J_1(x)$: $x = 3.8317, 7.0156, 10.1735, 13.3237, \ldots$
Roots of $Y_0(x)$: $x = 0.8936, 3.9577, 7.0861, 10.2223, \ldots$
Roots of $Y_1(x)$: $x = 2.1971, 5.4297, 8.5960, 11.7492, \ldots$

Table D-3 Some Bessel functions

x	$J_0(x)$	$J_1(x)$	$Y_0(x)$	$Y_1(x)$	$I_0(x)$	$I_1(x)$	$K_0(x)$	$K_1(x)$
0	1.0000	0.0000	$-\infty$	$-\infty$	1.000	0.0000	∞	∞
0.05	0.9994	0.0250	-1.979	-12.79	1.001	0.0250	3.114	19.91
0.10	0.9975	0.0499	-1.534	-6.459	1.003	0.0501	2.427	9.854
0.15	0.9944	0.0748	-1.271	-4.364	1.006	0.0752	2.030	6.477
0.20	0.9900	0.0995	-1.081	-3.324	1.010	0.1005	1.753	4.776
0.25	0.9844	0.1240	-0.9316	-2.704	1.016	0.1260	1.542	3.747
0.30	0.9776	0.1483	-0.8073	-2.293	1.023	0.1517	1.372	3.056
0.35	0.9696	0.1723	-0.7003	-2.000	1.031	0.1777	1.233	2.559
0.40	0.9604	0.1960	-0.6060	-1.781	1.040	0.2040	1.115	2.184
0.45	0.9500	0.2194	-0.5214	-1.610	1.051	0.2307	1.013	1.892
0.50	0.9385	0.2423	-0.4445	-1.471	1.063	0.2579	0.9244	1.656
0.55	0.9258	0.2647	-0.3739	-1.357	1.077	0.2855	0.8466	1.464
0.60	0.9120	0.2867	-0.3085	-1.260	1.092	0.3137	0.7775	1.303
0.65	0.8971	0.3081	-0.2476	-1.177	1.108	0.3425	0.7159	1.167
0.70	0.8812	0.3290	-0.1907	-1.103	1.126	0.3719	0.6605	1.050
0.75	0.8642	0.3492	-0.1372	-1.038	1.146	0.4020	0.6106	0.9496
0.80	0.8463	0.3688	-0.0868	-0.9781	1.167	0.4329	0.5653	0.8618
0.85	0.8274	0.3878	-0.0393	-0.9236	1.189	0.4646	0.5242	0.7847
0.90	0.8075	0.4059	-0.0056	-0.8731	1.213	0.4971	0.4867	0.7165
0.95	0.7868	0.4234	0.0481	-0.8258	1.239	0.5306	0.4524	0.6560
1.0	0.7652	0.4401	0.0883	-0.7812	1.266	0.5652	0.4210	0.6019
1.1	0.6957	0.4850	0.1622	-0.6981	1.326	0.6375	0.3656	0.5098
1.2	0.6711	0.4983	0.2281	-0.6211	1.394	0.7147	0.3185	0.4346
1.3	0.5937	0.5325	0.2865	-0.5485	1.469	0.7973	0.2782	0.3725
1.4	0.5669	0.5419	0.3379	-0.4791	1.553	0.8861	0.2437	0.3208
1.5	0.4838	0.5644	0.3824	-0.4123	1.647	0.9817	0.2138	0.2774
1.6	0.4554	0.5699	0.4204	-0.3476	1.750	1.085	0.1880	0.2406
1.7	0.3690	0.5802	0.4520	-0.2847	1.864	1.196	0.1655	0.2094
1.8	0.3400	0.5815	0.4774	-0.2237	1.990	1.317	0.1459	0.1826
1.9	0.2528	0.5794	0.4968	-0.1644	2.128	1.448	0.1288	0.1597
2.0	0.2239	0.5767	0.5104	-0.1070	2.280	1.591	0.1139	0.1399
2.1	0.1383	0.5626	0.5183	-0.0517	2.446	1.745	0.1008	0.1227
2.2	0.1104	0.5560	0.5208	-0.0015	2.629	1.914	0.0893	0.1079
2.3	0.0288	0.5305	0.5181	0.0523	2.830	2.098	0.0791	0.0950
2.4	0.0025	0.5202	0.5104	0.1005	3.049	2.298	0.0702	0.0837
2.5	0.0729	0.4843	0.4981	0.1459	3.290	2.517	0.0623	0.0739
2.6	-0.0968	0.4708	0.4813	0.1884	3.553	2.755	0.0554	0.0653
2.7	-0.1641	0.4260	0.4605	0.2276	3.842	3.016	0.0493	0.0577
2.8	-0.1850	0.4097	0.4359	0.2635	4.157	3.301	0.0438	0.0511
2.9	-0.2426	0.3575	0.4079	0.2959	4.503	3.613	0.0390	0.0453
3.0	-0.2601	0.3391	0.3769	0.3247	4.881	3.953	0.0347	0.0402
3.2	-0.3202	0.2613	0.3071	0.3707	5.747	4.734	0.0276	0.0316
3.4	-0.3643	0.1792	0.2296	0.4010	6.785	5.670	0.0220	0.0250
3.6	-0.3918	0.0955	0.1477	0.4154	8.028	6.793	0.6175	0.0198
3.8	-0.4026	0.0128	0.0645	0.4141	9.517	8.140	0.0140	0.0157
4.0	-0.3971	-0.0660	-0.0169	0.3979	11.302	9.759	0.0112	0.0125

DIFFERENTIAL OPERATORS*

* Bird, R. B., Stewart, W. E., and Lightfoot, E. N. *Transport Phenomena*. New York: Wiley, 1960.

Table D-4 Summary of differential operations involving the ∇-operator in rectangular coordinates* (x, y, z)

$$(\nabla \cdot \vec{v}) = \frac{\partial v_x}{\partial x} + \frac{\partial v_y}{\partial y} + \frac{\partial v_z}{\partial z} \tag{A}$$

$$(\nabla^2 s) = \frac{\partial^2 s}{\partial x^2} + \frac{\partial^2 s}{\partial y^2} + \frac{\partial^2 s}{\partial z^2} \tag{B}$$

$$(\bar{\bar{\tau}} : \nabla \vec{v}) = \tau_{xx}\left(\frac{\partial v_x}{\partial x}\right) + \tau_{yy}\left(\frac{\partial v_y}{\partial y}\right) + \tau_{zz}\left(\frac{\partial v_z}{\partial z}\right) + \tau_{xy}\left(\frac{\partial v_x}{\partial y} + \frac{\partial v_y}{\partial x}\right)$$
$$+ \tau_{yz}\left(\frac{\partial v_y}{\partial z} + \frac{\partial v_z}{\partial y}\right) + \tau_{zx}\left(\frac{\partial v_z}{\partial x} + \frac{\partial v_x}{\partial z}\right) \tag{C}$$

$$[\nabla s]_x = \frac{\partial s}{\partial x} \quad (D) \qquad [\nabla \times \vec{v}]_x = \frac{\partial v_z}{\partial y} - \frac{\partial v_y}{\partial z} \tag{G}$$

$$[\nabla s]_y = \frac{\partial s}{\partial y} \quad (E) \qquad [\nabla \times \vec{v}]_y = \frac{\partial v_x}{\partial z} - \frac{\partial v_z}{\partial x} \tag{H}$$

$$[\nabla s]_z = \frac{\partial s}{\partial z} \quad (F) \qquad [\nabla \times \vec{v}]_z = \frac{\partial v_y}{\partial x} - \frac{\partial v_x}{\partial y} \tag{I}$$

$$[\nabla \cdot \bar{\bar{\tau}}]_x = \frac{\partial \tau_{xx}}{\partial x} + \frac{\partial \tau_{xy}}{\partial y} + \frac{\partial \tau_{xz}}{\partial z} \tag{J}$$

$$[\nabla \cdot \bar{\bar{\tau}}]_y = \frac{\partial \tau_{xy}}{\partial x} + \frac{\partial \tau_{yy}}{\partial y} + \frac{\partial \tau_{yz}}{\partial z} \tag{K}$$

$$[\nabla \cdot \bar{\bar{\tau}}]_z = \frac{\partial \tau_{xz}}{\partial x} + \frac{\partial \tau_{yz}}{\partial y} + \frac{\partial \tau_{zz}}{\partial z} \tag{L}$$

$$[\nabla^2 \vec{v}]_x = \frac{\partial^2 v_x}{\partial x^2} + \frac{\partial^2 v_x}{\partial y^2} + \frac{\partial^2 v_x}{\partial z^2} \tag{M}$$

$$[\nabla^2 \vec{v}]_y = \frac{\partial^2 v_y}{\partial x^2} + \frac{\partial^2 v_y}{\partial y^2} + \frac{\partial^2 v_y}{\partial z^2} \tag{N}$$

$$[\nabla^2 \vec{v}]_z = \frac{\partial^2 v_z}{\partial x^2} + \frac{\partial^2 v_z}{\partial y^2} + \frac{\partial^2 v_z}{\partial z^2} \tag{O}$$

$$[\vec{v} \cdot \nabla \vec{v}]_x = v_x \frac{\partial v_x}{\partial x} + v_y \frac{\partial v_x}{\partial y} + v_z \frac{\partial v_x}{\partial z} \tag{P}$$

$$[\vec{v} \cdot \nabla \vec{v}]_y = v_x \frac{\partial v_y}{\partial x} + v_y \frac{\partial v_y}{\partial y} + v_z \frac{\partial v_y}{\partial z} \tag{Q}$$

$$[\vec{v} \cdot \nabla \vec{v}]_z = v_x \frac{\partial v_z}{\partial x} + v_y \frac{\partial v_z}{\partial y} + v_z \frac{\partial v_z}{\partial z} \tag{R}$$

* Operations involving the tensor τ are given for symmetrical τ only.

Table D-5 Summary of differential operations involving the ∇-operator in cylindrical coordinates* (r, θ, z)

$$(\nabla \cdot \vec{v}) = \frac{1}{r}\frac{\partial}{\partial r}(rv_r) + \frac{1}{r}\frac{\partial v_\theta}{\partial \theta} + \frac{\partial v_z}{\partial z} \tag{A}$$

$$(\nabla^2 s) = \frac{1}{r}\frac{\partial}{\partial r}\left(r\frac{\partial s}{\partial r}\right) + \frac{1}{r^2}\frac{\partial^2 s}{\partial \theta^2} + \frac{\partial^2 s}{\partial z^2} \tag{B}$$

$$(\bar{\bar{\tau}} : \nabla\vec{v}) = \tau_{rr}\left(\frac{\partial v_r}{\partial r}\right) + \tau_{\theta\theta}\left(\frac{1}{r}\frac{\partial v_\theta}{\partial \theta} + \frac{v_r}{r}\right) + \tau_{zz}\left(\frac{\partial v_z}{\partial z}\right) + \tau_{r\theta}\left[r\frac{\partial}{\partial r}\left(\frac{v_\theta}{r}\right) + \frac{1}{r}\frac{\partial v_r}{\partial \theta}\right]$$
$$+ \tau_{\theta z}\left(\frac{1}{r}\frac{\partial v_z}{\partial \theta} + \frac{\partial v_\theta}{\partial z}\right) + \tau_{\theta z}\left(\frac{\partial v_z}{\partial r} + \frac{\partial v_r}{\partial z}\right) \tag{C}$$

$$[\nabla s]_r = \frac{\partial s}{\partial r} \quad (D) \qquad\qquad [\nabla \times \vec{v}]_r = \frac{1}{r}\frac{\partial v_z}{\partial \theta} - \frac{\partial v_\theta}{\partial z} \tag{G}$$

$$[\nabla s]_\theta = \frac{1}{r}\frac{\partial s}{\partial \theta} \quad (E) \qquad\qquad [\nabla \times \vec{v}]_\theta = \frac{\partial v_r}{\partial z} - \frac{\partial v_z}{\partial r} \tag{H}$$

$$[\nabla s]_z = \frac{\partial s}{\partial z} \quad (F) \qquad\qquad [\nabla \times \vec{v}]_z = \frac{1}{r}\frac{\partial}{\partial r}(rv_\theta) - \frac{1}{r}\frac{\partial v_r}{\partial \theta} \tag{I}$$

$$[\nabla \cdot \bar{\bar{\tau}}]_r = \frac{1}{r}\frac{\partial}{\partial r}(r\tau_{rr}) + \frac{1}{r}\frac{\partial}{\partial \theta}\tau_{r\theta} - \frac{1}{r}\tau_{\theta\theta} + \frac{\partial \tau_{rz}}{\partial z} \tag{J}$$

$$[\nabla \cdot \bar{\bar{\tau}}]_\theta = \frac{1}{r}\frac{\partial \tau_{\theta\theta}}{\partial \theta} + \frac{\partial \tau_{r\theta}}{\partial r} + \frac{2}{r}\tau_{r\theta} + \frac{\partial \tau_{\theta z}}{\partial z} \tag{K}$$

$$[\nabla \cdot \bar{\bar{\tau}}]_z = \frac{1}{r}\frac{\partial}{\partial r}(r\tau_{rz}) + \frac{1}{r}\frac{\partial \tau_{\theta z}}{\partial \theta} + \frac{\partial \tau_{zz}}{\partial z} \tag{L}$$

$$[\nabla^2\vec{v}]_r = \frac{\partial}{\partial r}\left(\frac{1}{r}\frac{\partial}{\partial r}(rv_r)\right) + \frac{1}{r^2}\frac{\partial^2 v_r}{\partial \theta^2} - \frac{2}{r^2}\frac{\partial v_\theta}{\partial \theta} + \frac{\partial^2 v_r}{\partial z^2} \tag{M}$$

$$[\nabla^2\vec{v}]_\theta = \frac{\partial}{\partial r}\left(\frac{1}{r}\frac{\partial}{\partial r}(rv_\theta)\right) + \frac{1}{r^2}\frac{\partial^2 v_\theta}{\partial \theta^2} + \frac{2}{r^2}\frac{\partial v_r}{\partial \theta} + \frac{\partial^2 v_\theta}{\partial z^2} \tag{N}$$

$$[\nabla^2\vec{v}]_z = \frac{1}{r}\frac{\partial}{\partial r}\left(r\frac{\partial v_z}{\partial r}\right) + \frac{1}{r^2}\frac{\partial^2 v_z}{\partial \theta^2} + \frac{\partial^2 v_z}{\partial z^2} \tag{O}$$

$$[\vec{v} \cdot \nabla\vec{v}]_r = v_r\frac{\partial v_r}{\partial r} + \frac{v_\theta}{r}\frac{\partial v_r}{\partial \theta} - \frac{v_\theta^2}{r} + v_z\frac{\partial v_r}{\partial z} \tag{P}$$

$$[\vec{v} \cdot \nabla\vec{v}]_\theta = v_r\frac{\partial v_\theta}{\partial r} + \frac{v_\theta}{r}\frac{\partial v_\theta}{\partial \theta} + \frac{v_r v_\theta}{r} + v_z\frac{\partial v_\theta}{\partial z} \tag{Q}$$

$$[\vec{v} \cdot \nabla\vec{v}]_z = v_r\frac{\partial v_z}{\partial r} + \frac{v_\theta}{r}\frac{\partial v_z}{\partial \theta} + v_z\frac{\partial v_z}{\partial z} \tag{R}$$

* Operations involving the tensor τ are given for symmetrical τ only.

Table D-6 Summary of differential operations involving the ∇-operator in spherical coordinates* (r, θ, ϕ)

$$(\nabla \cdot \vec{v}) = \frac{1}{r^2}\frac{\partial}{\partial r}(r^2 v_r) + \frac{1}{r \sin\theta}\frac{\partial}{\partial \theta}(v_\theta \sin\theta) + \frac{1}{r \sin\theta}\frac{\partial v_\phi}{\partial \phi} \tag{A}$$

$$(\nabla^2 s) = \frac{1}{r^2}\frac{\partial}{\partial r}\left(r^2 \frac{\partial s}{\partial r}\right) + \frac{1}{r^2 \sin\theta}\frac{\partial}{\partial \theta}\left(\sin\theta \frac{\partial s}{\partial \theta}\right) + \frac{1}{r^2 \sin^2\theta}\frac{\partial^2 s}{\partial \phi^2} \tag{B}$$

$$(\bar{\bar{\tau}} : \nabla\vec{v}) = \tau_{rr}\left(\frac{\partial v_r}{\partial r}\right) + \tau_{\theta\theta}\left(\frac{1}{r}\frac{\partial v_\theta}{\partial \theta} + \frac{v_r}{r}\right) + \tau_{\phi\phi}\left(\frac{1}{r \sin\theta}\frac{\partial v_\phi}{\partial \phi} + \frac{v_r}{r} + \frac{v_\theta \cot\theta}{r}\right)$$

$$+ \tau_{r\theta}\left(\frac{\partial v_\theta}{\partial r} + \frac{1}{r}\frac{\partial v_r}{\partial \theta} - \frac{v_\theta}{r}\right) + \tau_{r\phi}\left(\frac{\partial v_\phi}{\partial r} + \frac{1}{r \sin\theta}\frac{\partial v_r}{\partial \phi} - \frac{v_\phi}{r}\right) \tag{C}$$

$$+ \tau_{\theta\phi}\left(\frac{1}{r}\frac{\partial v_\phi}{\partial \theta} + \frac{1}{r \sin\theta}\frac{\partial v_\theta}{\partial \phi} - \frac{\cot\theta}{r}v_\phi\right)$$

$$[\nabla s]_r = \frac{\partial s}{\partial r} \tag{D}$$

$$[\nabla s]_\theta = \frac{1}{r}\frac{\partial s}{\partial \theta} \tag{E}$$

$$[\nabla s]_\phi = \frac{1}{r \sin\theta}\frac{\partial s}{\partial \phi} \tag{F}$$

$$[\nabla \times \vec{v}]_r = \frac{1}{r \sin\theta}\frac{\partial}{\partial \theta}(v_\phi \sin\theta) - \frac{1}{r \sin\theta}\frac{\partial v_\theta}{\partial \phi} \tag{G}$$

$$[\nabla \times \vec{v}]_\theta = \frac{1}{r \sin\theta}\frac{\partial v_r}{\partial \phi} - \frac{1}{r}\frac{\partial}{\partial r}(r v_\phi) \tag{H}$$

$$[\nabla \times \vec{v}]_\phi = \frac{1}{r}\frac{\partial}{\partial r}(r v_\theta) - \frac{1}{r}\frac{\partial v_r}{\partial \theta} \tag{I}$$

$$[\nabla \cdot \bar{\bar{\tau}}]_r = \frac{1}{r^2}\frac{\partial}{\partial r}(r^2 \tau_{rr}) + \frac{1}{r \sin\theta}\frac{\partial}{\partial \theta}(\tau_{r\theta} \sin\theta) + \frac{1}{r \sin\theta}\frac{\partial \tau_{r\phi}}{\partial \phi} - \frac{\tau_{\theta\theta} + \tau_{\phi\phi}}{r} \tag{J}$$

$$[\nabla \cdot \bar{\bar{\tau}}]_\theta = \frac{1}{r^2}\frac{\partial}{\partial r}(r^2 \tau_{r\theta}) + \frac{1}{r \sin\theta}\frac{\partial}{\partial \theta}(\tau_{\theta\theta} \sin\theta) + \frac{1}{r \sin\theta}\frac{\partial \tau_{\theta\phi}}{\partial \phi} + \frac{\tau_{r\theta}}{r} - \frac{\cot\theta}{r}\tau_{\phi\phi} \tag{K}$$

$$[\nabla \cdot \bar{\bar{\tau}}]_\phi = \frac{1}{r^2}\frac{\partial}{\partial r}(r^2 \tau_{r\phi}) + \frac{1}{r}\frac{\partial \tau_{\theta\phi}}{\partial \theta} + \frac{1}{r \sin\theta}\frac{\partial \tau_{\phi\phi}}{\partial \phi} + \frac{\tau_{r\phi}}{r} + \frac{2 \cot\theta}{r}\tau_{\theta\phi} \tag{L}$$

$$[\nabla^2\vec{v}]_r = \nabla^2 v_r - \frac{2 v_r}{r^2} - \frac{2}{r^2}\frac{\partial v_\theta}{\partial \theta} - \frac{2 v_\theta \cot\theta}{r^2} - \frac{2}{r^2 \sin\theta}\frac{\partial v_\phi}{\partial \phi} \tag{M}$$

$$[\nabla^2\vec{v}]_\theta = \nabla^2 v_\theta + \frac{2}{r^2}\frac{\partial v_r}{\partial \theta} - \frac{r_\theta}{r^2 \sin^2\theta} - \frac{2 \cos\theta}{r^2 \sin^2\theta}\frac{\partial v_\phi}{\partial \phi} \tag{N}$$

$$[\nabla^2\vec{v}]_\phi = \nabla^2 v_\phi - \frac{v_\phi}{r^2 \sin^2\theta} + \frac{2}{r^2 \sin\theta}\frac{\partial v_r}{\partial \phi} + \frac{2 \cos\theta}{r^2 \sin^2\theta}\frac{\partial v_\theta}{\partial \phi} \tag{O}$$

$$[\vec{v} \cdot \nabla\vec{v}]_r = v_r\frac{\partial v_r}{\partial r} + \frac{v_\theta}{r}\frac{\partial v_r}{\partial \theta} + \frac{v_\phi}{r \sin\theta}\frac{\partial v_r}{\partial \phi} - \frac{v_\theta^2 + v_\phi^2}{r} \tag{P}$$

$$[\vec{v} \cdot \nabla\vec{v}]_\theta = v_r\frac{\partial v_\theta}{\partial r} + \frac{v_\theta}{r}\frac{\partial v_\theta}{\partial \theta} + \frac{v_\phi}{r \sin\theta}\frac{\partial v_\theta}{\partial \phi} + \frac{v_r v_\theta}{r} - \frac{v_\phi^2 \cot\theta}{r} \tag{Q}$$

$$[\vec{v} \cdot \nabla\vec{v}]_\phi = v_r\frac{\partial v_\phi}{\partial r} + \frac{v_\theta}{r}\frac{\partial v_\phi}{\partial \theta} + \frac{v_\phi}{r \sin\theta}\frac{\partial v_\phi}{\partial \phi} + \frac{v_\phi v_r}{r} + \frac{v_\theta v_\phi \cot\theta}{r} \tag{R}$$

* Operations involving the tensor τ are given for symmetrical τ only.

THERMODYNAMIC PROPERTIES

Table E-1 Saturation state properties of steam and water

Temperature °C	Pressure bar†	Specific volume m/³kg		Specific enthalpy kJ/kg	
		Water	Steam	Water	Steam
0 01	0·006 12	$1·000\ 2 \times 10^{-3}$	206·146	0·000 611	2501
10	0·012 271	$1·000\ 4 \times 10^{-3}$	106·422	41·99	2519
20	0·023 368	$1·001\ 8 \times 10^{-3}$	57·836	83·86	2538
30	0·042 418	$1·004\ 4 \times 10^{-3}$	32·929	125·66	2556
40	0·073 750	$1·007\ 9 \times 10^{-3}$	19·546	167·47	2574
50	0·123 35	$1·012\ 1 \times 10^{-3}$	12·045	209·3	2592
60	0·199 19	$1·017\ 1 \times 10^{-3}$	7·677 6	251·1	2609
70	0·311 61	$1·022\ 8 \times 10^{-3}$	5·045 3	293·0	2626
80	0·473 58	$1·029\ 0 \times 10^{-3}$	3·408 3	334·9	2643
90	0·701 90	$1·035\ 9 \times 10^{-3}$	2·360 9	376·9	2660
100	1·013 25	$1·043\ 5 \times 10^{-3}$	1·673 0	419·1	2676
110	1·432 7	$1·051\ 5 \times 10^{-3}$	1·210 1	461·3	2691
120	1·985 4	$1·060\ 3 \times 10^{-3}$	0·891 71	503·7	2706
130	2·701 1	$1·069\ 7 \times 10^{-3}$	0·668 32	546·3	2720
140	3·613 6	$1·079\ 8 \times 10^{-3}$	0·508 66	589·1	2734
150	4·759 7	$1·090\ 6 \times 10^{-3}$	0·392 57	632·2	2747
160	6·180 4	$1·102\ 1 \times 10^{-3}$	0·306 85	675·5	2758
170	7·920 2	$1·114\ 4 \times 10^{-3}$	0·242 62	719·1	2769
180	10·027	$1·127\ 5 \times 10^{-3}$	0·193 85	763·1	2778
190	12·553	$1·141\ 5 \times 10^{-3}$	0·156 35	807·5	2786
200	15·550	$1·156\ 5 \times 10^{-3}$	0·127 19	852·4	2793
210	19·080	$1·172\ 6 \times 10^{-3}$	0·104 265	897·7	2798
220	23·202	$1·190\ 0 \times 10^{-3}$	0·086 062	943·7	2802
230	27·979	$1·208\ 7 \times 10^{-3}$	0·071 472	990·3	2803
240	33·480	$1·229\ 1 \times 10^{-3}$	0·059 674	1037·6	2803
250	39·776	$1·251\ 2 \times 10^{-3}$	0·050 056	1085·8	2801
260	46·941	$1·275\ 5 \times 10^{-3}$	0·042 149	1135·0	2796
270	55·052	$1·302\ 3 \times 10^{-3}$	0·035 599	1185·2	2790
280	64·191	$1·332\ 1 \times 10^{-3}$	0·030 133	1236·8	2780
290	74·449	$1·365\ 5 \times 10^{-3}$	0·025 537	1290	2766
300	85·917	$1·403\ 6 \times 10^{-3}$	0·021 643	1345	2749
310	98·694	$1·447\ 5 \times 10^{-3}$	0·018 316	1402	2727
320	112·89	$1·499\ 2 \times 10^{-3}$	0·015 451	1462	2700
330	128·64	$1·562 \times 10^{-3}$	0·012 967	1526	2666
340	146·08	$1·639 \times 10^{-3}$	0·010 779	1596	2623
350	165·37	$1·741 \times 10^{-3}$	0·008 805	1672	2565
360	186·74	$1·894 \times 10^{-3}$	0·006 943	1762	2481
370	210·53	$2·22 \times 10^{-3}$	0·004 93	1892	2331
374·15	221·2	$3·17 \times 10^{-3}$	0·003 17	2095	2095

From *U.K. Steam Tables in S.I. Units*. London: Edward Arnold Pub., Ltd., 1970.)
† 1 bar = 10^5 N/m².

	Water						Steam				
c_{pf} kJ/kg K	$\sigma \times 10^3$ N/m	$\mu_f \times 10^6$ N s/m²	$\nu_f \times 10^6$ m²/s	k_f W/mK	$(Pr)_f$	c_{pg} kJ/kg K	$\mu_g \times 10^6$ N s/m²	$\nu_g \times 10^6$ m²/s	$k_g \times 10^3$ W/mK	$(Pr)_g$	Temperature °C
4·218	75·60	1786	1·786	0·569	13·2	1·863	8·105	1672	17·6	0·858	0 01
4·194	74·24	1304	1·305	0·587	9·32	1·870	8·504	905	18·2	0·873	10
4·182	72·78	1002	1·004	0·603	6·95	1·880	8·903	515	18·8	0·888	20
4·179	71·23	798·3	0·802	0·618	5·40	1·890	9·305	306	19·5	0·901	30
4·179	69·61	653·9	0·659	0·631	4·33	1·900	9·701	190	20·2	0·912	40
4·181	67·93	547·8	0·554	0·643	3·56	1·912	10·10	121	20·9	0·924	50
4·185	66·19	467·3	0·473	0·653	2·99	1·924	10·50	80·6	21·6	0·934	60
4·191	64·40	404·8	0·414	0·662	2·56	1·946	10·89	54·9	22·4	0·946	70
4·198	62·57	355·4	0·366	0·670	2·23	1·970	11·29	38·5	23·2	0·959	80
4·207	60·69	315·6	0·327	0·676	1·96	1·999	11·67	27·6	24·0	0·973	90
4·218	58·78	283·1	0·295	0·681	1·75	2·034	12·06	20·2	24·9	0·987	100
4·230	56·83	254·8	0·268	0·684	1·58	2·076	12·45	15·1	25·8	1·00	110
4·244	54·85	231·0	0·245	0·687	1·43	2·125	12·83	11·4	26·7	1·02	120
4·262	52·83	210·9	0·226	0·688	1·31	2·180	13·20	8·82	27·8	1·03	130
4·282	50·79	194·1	0·210	0·688	1·21	2·245	13·57	6·90	28·9	1·05	140
4·306	48·70	179·8	0·196	0·687	1·13	2·320	13·94	5·47	30·0	1·08	150
4·334	46·59	167·7	0·185	0·684	1·06	2·406	14·30	4·39	31·3	1·10	160
4·366	44·44	157·4	0·175	0·681	1·01	2·504	14·66	3·55	32·6	1·13	170
4·403	42·26	148·5	0·167	0·677	0·967	2·615	15·02	2·91	34·1	1·15	180
4·446	40·05	140·7	0·161	0·671	0·932	2·741	15·37	2·40	35·7	1·18	190
4·494	37·81	133·9	0·155	0·664	0·906	2·883	15·72	2·00	37·4	1·21	200
4·550	35·53	127·9	0·150	0·657	0·886	3·043	16·07	1·68	39·4	1·24	210
4·613	33·23	122·4	0·146	0·648	0·871	3·223	16·42	1·41	41·5	1·28	220
4·685	30·90	117·5	0·142	0·639	0·861	3·426	16·78	1·20	43·9	1·31	230
4·769	28·56	112·9	0·139	0·628	0·850	3·656	17·14	1·02	46·5	1·35	240
4·866	26·19	108·7	0·136	0·616	0·859	3·918	17·51	0·876	49·5	1·39	250
4·985	23·82	104·8	0·134	0·603	0·866	4·221	17·90	0·755	52·8	1·43	260
5·134	21·44	101·1	0·132	0·589	0·882	4·575	18·31	0·652	56·6	1·48	270
5·307	19·07	97·5	0·130	0·574	0·902	4·996	18·74	0·565	60·9	1·54	280
5·520	16·71	94·1	0·128	0·558	0·932	5·509	19·21	0·491	66·0	1·61	290
5·794	14·39	90·7	0·127	0·541	0·970	6·148	19·73	0·427	71·9	1·69	300
6·143	12·11	87·2	0·126	0·523	1·024	6·968	20·30	0·372	79·1	1·79	310
6·604	9·89	83·5	0·125	0·503	1·11	8·060	20·95	0·324	87·8	1·92	320
7·241	7·75	79·5	0·124	0·482	1·20	9·580	21·70	0·281	99·0	2·10	330
8·225	5·71	75·4	0·123	0·460	1·35	11·87	22·70	0·245	114	2·36	340
10·07	3·79	69·4	0·121	0·434	1·61	15·8	24·15	0·213	134	2·84	350
15·0	2·03	62·1	0·118	0·397	2·34	27·0	26·45	0·184	162	4·40	360
55	0·47	51·8	0·116	0·340	8·37	107	30·6	0·150	199	16·4	370
∞	0	41·4	0·131	0·240		∞	41·4	0·131	240		374·15

Table E-2 Thermodynamic properties of dry saturated steam, pressure

Abs. press., psia	Temperature F	Specific volume		Enthalpy			Entropy		
		Sat. liquid	Sat. vapor	Sat. liquid	Evap.	Sat. vapor	Sat. liquid	Evap.	Sat. vapor
p	T (F)	v_f (ft³/lbm)	v_g (ft³/lbm)	h_f (BTU/lbm)	h_{fg} (BTU/lbm)	h_g (BTU/lbm)	s_f (BTU/lbm R)	s_{fg} (BTU/lbm R)	s_g (BTU/lbm R)
1.0	101.74	0.01614	333.6	69.70	1036.3	1106.0	0.1326	1.8456	1.9782
2.0	126.08	0.01623	173.73	93.99	1022.2	1116.2	0.1749	1.7451	1.9200
3.0	141.48	0.01630	118.71	109.37	1013.2	1122.6	0.2008	1.6855	1.8863
4.0	152.97	0.01636	90.63	120.86	1006.4	1127.3	0.2198	1.6427	1.8625
5.0	162.24	0.01640	73.52	130.13	1001.0	1131.1	0.2347	2.6094	1.8441
6.0	170.06	0.01645	61.98	137.96	996.2	1134.2	0.2472	1.5820	1.8292
7.0	176.85	0.01649	53.64	144.76	992.1	1136.9	0.2581	1.5586	1.8167
8.0	182.86	0.01653	47.34	150.79	988.5	1139.3	0.2674	1.5383	1.8057
9.0	188.28	0.01656	42.40	156.22	985.2	1141.4	0.2759	1.5203	1.7962
10	193.21	0.01659	38.42	161.17	982.1	1143.3	0.2835	1.5041	1.7876
14.696	212.00	0.01672	26.80	180.07	970.3	1150.4	0.3120	1.4446	1.7566
15	213.03	0.01672	26.29	181.11	969.7	1150.8	0.3135	1.4415	1.7549
20	227.96	0.01683	20.089	196.16	960.1	1156.3	0.3356	1.3962	1.7319
25	240.07	0.01692	16.303	208.42	952.1	1160.6	0.3533	1.3606	1.7139
30	250.33	0.01701	13.746	218.82	945.3	1164.1	0.3680	1.3313	1.6993
35	259.28	0.01708	11.898	227.91	939.2	1167.1	0.3807	1.3063	1.6870
40	267.25	0.01715	10.498	236.03	933.7	1169.7	0.3919	1.2844	1.6763
45	274.44	0.01721	9.401	243.36	928.6	1172.0	0.4019	1.2650	1.6669
50	281.01	0.01727	8.515	250.09	924.0	1174.1	0.4110	1.2474	1.6585
55	287.07	0.01732	7.787	256.30	919.6	1175.9	0.4193	1.2316	1.6509

60	292.71	0.01738	7.175	262.09	915.5	1177.6	0.4270	1.2168	1.6438
65	297.97	0.01743	6.655	267.50	911.6	1179.1	0.4342	1.2032	1.6374
70	302.92	0.01748	6.206	272.61	907.9	1180.6	0.4409	1.1906	1.6315
75	307.60	0.01753	5.816	277.43	904.5	1181.9	0.4472	1.1787	1.6259
80	312.03	0.01757	5.472	282.02	901.1	1183.1	0.4531	1.1676	1.6207
85	316.25	0.01761	5.168	286.39	897.8	1184.2	0.4587	1.1571	1.6158
90	320.27	0.01766	4.896	290.56	894.7	1185.3	0.4641	1.1471	1.6112
95	324.12	0.01770	4.652	294.56	891.7	1186.2	0.4692	1.1376	1.6068
100	327.81	0.01774	4.432	298.40	888.8	1187.2	0.4740	1.1286	1.6026
110	334.77	0.01782	4.049	305.66	883.2	1188.9	0.4832	1.1117	1.5948
120	341.25	0.01789	3.728	312.44	877.9	1190.4	0.4916	1.0962	1.5878
130	347.32	0.01796	3.455	318.81	872.9	1191.7	0.4995	1.0817	1.5812
140	353.02	0.01802	3.220	324.82	868.2	1193.0	0.5069	1.0682	1.5751
150	358.42	0.01809	3.015	330.51	863.6	1194.1	0.5138	1.0556	1.5694
160	363.53	0.01815	2.834	335.93	859.2	1195.1	0.5204	1.0436	1.5640
170	368.41	0.01822	2.675	341.09	854.9	1196.0	0.5266	1.0324	1.5590
180	373.06	0.01827	2.532	346.03	850.8	1196.9	0.5325	1.0217	1.5542
190	377.51	0.01833	2.404	350.79	846.8	1197.6	0.5381	1.0116	1.5497
200	381.79	0.01839	2.288	355.36	843.0	1198.4	0.5435	1.0018	1.5453
250	400.95	0.01865	1.8438	376.00	825.1	1201.1	0.5675	0.9588	1.5263
300	417.33	0.01890	1.5433	393.84	809.0	1202.8	0.5879	0.9225	1.5104
350	431.72	0.01913	1.3260	409.69	794.2	1203.9	0.6056	0.8910	1.4966
400	444.59	0.0193	1.1613	424.0	780.5	1204.5	0.6214	0.8630	1.4844
450	456.28	0.0195	1.0320	437.2	767.4	1204.6	0.6356	0.8378	1.4734
500	467.01	0.0197	0.9278	449.4	755.0	1204.4	0.6487	0.8147	1.4634
550	476.94	0.0199	0.8424	460.8	743.1	1203.9	0.6608	0.7934	1.4542
600	486.21	0.0201	0.7698	471.6	731.6	1203.2	0.6720	0.7734	1.4454
650	494.90	0.0203	0.7083	481.8	720.5	1202.3	0.6826	0.7548	1.4374
700	503.10	0.0205	0.6554	491.5	709.7	1201.2	0.6925	0.7371	1.4296
750	510.86	0.0207	0.6092	500.8	699.2	1200.0	0.7019	0.7204	1.4223

(continued)

Table E-2 Thermodynamic properties of dry saturated steam, pressure (*continued*)

Abs. press., psia	Temperature F	Specific volume		Enthalpy			Entropy		
		Sat. liquid	Sat. vapor	Sat. liquid	Evap.	Sat. vapor	Sat. liquid	Evap.	Sat. vapor
p	T (F)	v_f (ft³/lbm)	v_g (ft³/lbm)	h_f (BTU/lbm)	h_{fg} (BTU/lbm)	h_g (BTU/lbm)	s_f (BTU/lbm R)	s_{fg} (BTU/lbm R)	s_g (BTU/lbm R)
800	518.23	0.0209	0.5687	509.7	688.9	1198.6	0.7108	0.7045	1.4153
850	525.26	0.0210	0.5327	518.3	678.8	1197.1	0.7194	0.6891	1.4085
900	531.98	0.0212	0.5006	526.6	668.8	1195.4	0.7275	0.6744	1.4020
950	538.43	0.0214	0.4717	534.6	659.1	1193.7	0.7355	0.6602	1.3957
1000	544.61	0.0216	0.4456	542.4	649.4	1191.8	0.7430	0.6467	1.3897
1100	556.31	0.0220	0.4001	557.4	630.4	1187.7	0.7575	0.6205	1.3780
1200	567.22	0.0223	0.3619	571.7	611.7	1183.4	0.7711	0.5956	1.3667
1300	577.46	0.0227	0.3293	585.4	593.2	1178.6	0.7840	0.5719	1.3559
1400	587.10	0.0231	0.3012	598.7	574.7	1173.4	0.7963	0.5491	1.3454
1500	596.23	0.0235	0.2765	611.6	556.3	1167.9	0.8082	0.5269	1.3351
2000	635.82	0.0257	0.1878	671.7	463.4	1135.1	0.8619	0.4230	1.2849
2500	668.13	0.0287	0.1307	730.6	360.5	1091.1	0.9126	0.3197	1.2322
3000	695.36	0.0346	0.0858	802.5	217.8	1020.3	0.9731	0.1885	1.1615
3206.2	705.40	0.0503	0.0503	902.7	0	902.7	1.0580	0	1.0580

(From El-Wakil, M. 1971. Wherein abridged from Keenan, J. H. and Keyes, F. G. *Thermodynamic Properties of Steam.* New York: John Wiley and Sons, Inc., 1937.)

Table E-3 Thermodynamic properties of dry saturated steam, temperature

| Temperature F | Abs. press., psia | Specific volume | | Enthalpy | | | Entropy | | |
| | | Sat. liquid | Sat. vapor | Sat. liquid | Evap. | Sat. vapor | Sat. liquid | Evap. | Sat. vapor |
T (F)	p	v_f (ft³/lbm)	v_g (ft³/lbm)	h_f (BTU/lbm)	h_{fg} (BTU/lbm)	h_g (BTU/lbm)	s_f (BTU/lbm R)	s_{fg} (BTU/lbm R)	s_g (BTU/lbm R)
32	0.08854	0.01602	3306	0.00	1075.8	1075.8	0.0000	2.1877	2.1877
35	0.09995	0.01602	2947	3.02	1074.1	1077.1	0.0061	2.1709	2.1770
40	0.12170	0.01602	2444	8.05	1071.3	1079.3	0.0162	2.1435	2.1597
45	0.14752	0.01602	2036.4	13.06	1068.4	1081.5	0.0262	2.1167	2.1429
50	0.17811	0.01603	1703.2	18.07	1065.6	1083.7	0.0361	2.0903	2.1264
60	0.2563	0.01604	1206.7	28.06	1059.9	1088.0	0.0555	2.0393	2.0948
70	0.3631	0.01606	867.9	38.04	1054.3	1092.3	0.0745	1.9902	2.0647
80	0.5069	0.01608	633.1	48.02	1048.6	1096.6	0.0932	1.9428	2.0360
90	0.6982	0.01610	468.0	57.99	1042.9	1100.9	0.1115	1.8972	2.0087
100	0.9492	0.01613	350.4	67.97	1037.2	1105.2	0.1295	1.8531	1.9826
110	1.2748	0.01617	265.4	77.94	1031.6	1109.5	0.1417	1.8106	1.9577
120	1.6924	0.01620	203.27	87.92	1025.8	1113.7	0.1645	1.7694	1.9339
130	2.2225	0.01625	157.34	97.90	1020.0	1117.9	0.1816	1.7296	1.9112
140	2.8886	0.01629	123.01	107.89	1014.1	1122.0	0.1984	1.6910	1.8894
150	3.718	0.01634	97.07	117.89	1008.2	1126.1	0.2149	1.6537	1.8685
160	4.741	0.01639	77.29	127.89	1002.3	1130.2	0.2311	1.6174	1.8485
170	5.992	0.01645	62.06	137.90	996.3	1134.2	0.2472	1.5822	1.8293
180	7.510	0.01651	50.23	147.92	990.2	1138.1	0.2630	1.5480	1.8109
190	9.339	0.01657	40.96	157.95	984.1	1142.0	0.2785	1.5147	1.7932
200	11.526	0.01663	33.64	167.99	977.9	1145.9	0.2938	1.4824	1.7762

(*continued*)

457

Table E-3 Thermodynamic properties of dry saturated steam, temperature (*continued*)

Temperature F	Abs. press., psia	Specific volume		Enthalpy			Entropy		
T (F)	p	Sat. liquid v_f (ft³/lbm)	Sat. vapor v_g (ft³/lbm)	Sat. liquid h_f (BTU/lbm)	Evap. h_{fg} (BTU/lbm)	Sat. vapor h_g (BTU/lbm)	Sat. liquid s_f (BTU/lbm R)	Evap. s_{fg} (BTU/lbm R)	Sat. vapor s_g (BTU/lbm R)
210	14.123	0.01670	27.82	178.05	971.6	1149.7	0.3090	1.4508	1.7598
212	14.696	0.01672	26.80	180.07	970.3	1150.4	0.3120	1.4446	1.7566
220	17.186	0.01677	23.15	188.13	965.2	1153.4	0.3239	1.4201	1.7440
230	20.780	0.01684	19.382	198.23	958.8	1157.0	0.3387	1.3901	1.7288
240	24.969	0.01692	16.323	208.34	952.2	1160.5	0.3531	1.3609	1.7140
250	29.825	0.01700	13.821	216.48	945.5	1164.0	0.3675	1.3323	1.6998
260	35.429	0.01709	11.763	228.64	938.7	1167.3	0.3817	1.3043	1.6860
270	41.858	0.01717	10.061	238.84	931.8	1170.6	0.3958	1.2769	1.6727
280	49.203	0.01726	8.645	249.06	924.7	1173.8	0.4096	1.2501	1.6597
290	57.556	0.01735	7.461	259.31	917.5	1176.8	0.4234	1.2238	1.6472
300	67.013	0.01745	6.466	269.59	910.1	1179.7	0.4369	1.1980	1.6350
310	77.68	0.01755	5.626	279.92	902.6	1182.5	0.4504	1.1727	1.6231
320	89.66	0.01765	4.914	290.28	894.9	1185.2	0.4637	1.1478	1.6115
330	103.06	0.01776	4.307	300.68	887.0	1187.7	0.4769	1.1233	1.6002
340	118.01	0.01787	3.788	311.13	879.0	1190.1	0.4900	1.0992	1.5891
350	134.63	0.01799	3.342	321.63	870.7	1192.3	0.5029	1.0754	1.5783
360	153.04	0.01811	2.957	332.18	852.2	1194.4	0.5158	1.0519	1.5677
370	173.37	0.01823	2.625	342.79	853.5	1196.3	0.5286	1.0287	1.5573

380	195.77	0.01836	2.335	353.45	844.6	1198.1	0.5413	1.0059	1.5471
390	220.37	0.01850	2.0836	364.17	835.4	1199.6	0.5539	0.9832	1.5371
400	247.31	0.01864	1.8633	374.97	826.0	1201.0	0.5664	0.9608	1.5272
410	276.75	0.01878	1.6700	385.83	816.3	1202.1	0.5788	0.9386	1.5174
420	308.83	0.01894	1.5000	396.77	806.3	1203.1	0.5912	0.9166	1.5078
430	343.72	0.01910	1.3499	407.79	796.0	1203.8	0.6035	0.8947	1.4982
440	381.59	0.01926	1.2171	418.90	785.4	1204.3	0.6158	0.8730	1.4887
450	422.6	0.0194	1.0993	430.1	774.5	1204.6	0.6280	0.8513	1.4793
460	466.9	0.0196	0.9944	441.4	763.2	1204.6	0.6402	0.8298	1.4700
470	514.7	0.0198	0.9009	452.8	751.5	1204.3	0.6523	0.8083	1.4606
480	566.1	0.0200	0.8172	464.4	739.4	1203.7	0.6645	0.7868	1.4513
490	621.4	0.0202	0.7423	476.0	726.8	1202.8	0.6766	0.7653	1.4419
500	680.8	0.0204	0.6749	487.8	713.9	1201.7	0.6887	0.7438	1.4325
520	812.4	0.0209	0.5594	511.9	686.4	1198.2	0.7130	0.7006	1.4136
540	962.5	0.0215	0.4649	536.6	656.6	1193.2	0.7374	0.6568	1.3942
560	1133.1	0.0221	0.3868	562.2	624.2	1186.4	0.7621	0.6121	1.3742
580	1325.8	0.0228	0.3217	588.9	588.4	1177.3	0.7872	0.5659	1.3532
600	1542.9	0.0236	0.2668	610.0	548.5	1165.5	0.8131	0.5176	1.3307
620	1786.6	0.0247	0.2201	646.7	503.6	1150.3	0.8398	0.4664	1.3062
640	2059.7	0.0260	0.1798	678.6	452.0	1130.5	0.8679	0.4110	1.2789
660	2365.4	0.0278	0.1442	714.2	390.2	1104.4	0.8987	0.3485	1.2472
680	2708.1	0.0305	0.1115	757.3	309.9	1067.2	0.9351	0.2719	1.2071
700	3093.1	0.0369	0.0761	823.3	172.1	995.4	0.9905	0.1484	1.1389
705.4	3206.2	0.0503	0.0503	902.7	0	902.7	1.0580	0	1.0580

(From El-Wakil, M. 1971. Wherein abridged from Keenan, J. H. and Keyes, F. G. *Thermodynamic Properties of Steam*. New York: John Wiley and Sons, Inc., 1937.)

Table E-4 Thermodynamic properties of sodium

Temperature, R (Sat. press., psia)		Sat. liquid	Sat. vapor	Temperature of superheated vapor, R				
				800	900	1000	1100	1200
700 (8.7472×10^{-9})	v...... h...... s......	1.7232×10^{-2} 219.7 0.6854	$>10^{10}$ 2180.5 3.4866	...... 2203.1 3.5169	.... 2224.7 3.5424	.... 2246.4 3.5652	.... 2268.0 3.5857	.... 2289.5 3.6043
800 (5.0100×10^{-7})	v...... h...... s......	1.7548×10^{-2} 252.3 0.7290	7.4375×10^{8} 2200.1 3.1637	....	8.3835×10^{8} 2224.4 3.1925	9.3168×10^{8} 2246.3 3.2155	1.0249×10^{9} 2267.9 3.2360	1.1180×10^{9} 2289.5 3.2546
900 (1.1480×10^{-5})	v...... h...... s......	1.7864×10^{-2} 284.3 0.7667	3.6411×10^{7} 2217.6 2.9148	....	....	4.0267×10^{7} 2245.2 2.9440	4.4718×10^{7} 2267.7 2.9653	4.8789×10^{7} 2289.4 2.9811
1000 (1.3909×10^{-4})	v...... h...... s......	1.8180×10^{-2} 325.9 0.7999	3.323×10^{6} 2232.7 2.7168	....	....	....	3.6834×10^{6} 2264.8 2.7474	4.0254×10^{6} 2288.6 2.7680
1100 (1.0616×10^{-3})	v...... h...... s......	1.8496×10^{-2} 347.0 0.8296	4.7592×10^{5} 2245.1 2.5551	....	....	....	....	5.2512×10^{5} 2282.7 2.5878
1200 (5.7398×10^{-3})	v...... h...... s......	1.8812×10^{-2} 377.7 0.8563	9.5235×10^{4} 2254.9 2.4207	....	....	....	....	....
1300 (2.3916×10^{-2})	v...... h...... s......	1.9128×10^{-2} 408.2 0.8807	2.4520×10^{4} 2262.8 2.3073	....	....	....	....	....
1400 (8.1347×10^{-2})	v...... h...... s......	1.9444×10^{-2} 438.4 0.9031	7.6798×10^{3} 2269.3 2.2109	....	....	....	....	....
1500 (2.3351×10^{-1})	v...... h...... s......	1.9760×10^{-2} 468.5 0.9239	2.8334×10^{3} 2274.9 2.1282	....	....	....	....	....
1600 (5.8425×10^{-1})	v...... h...... s......	2.0076×10^{-2} 498.5 0.9433	1.1935×10^{3} 2280.0 2.0567	....	....	....	....	....
1700 (1.3170)	v...... h...... s......	2.0329×10^{-2} 528.5 0.9615	5.5585×10^{2} 2285.3 1.9948	....	....	....	....	....
1800 (2.7164)	v...... h...... s......	2.0708×10^{-2} 558.6 0.9786	2.8200×10^{2} 2291.1 2.9411	....	....	....	....	....
1900 (5.1529)	v...... h...... s......	2.1024×10^{-2} 588.8 0.9949	1.5512×10^{2} 2297.2 1.8941	....	....	....	....	....
2000 (9.1533)	v...... h...... s......	2.1340×10^{-2} 619.1 1.0105	90.914 2304.1 1.8530	....	....	....	....	....
2100 (15.392)	v...... h...... s......	2.1656×10^{-2} 649.7 1.0255	56.185 2312.1 1.8171	....	....	....	....	....
2200 (24.692)	v...... h...... s......	2.1972×10^{-2} 680.7 1.0399	36.338 2321.0 1.7885	....	....	....	....	....
2300 (38.013)	v...... h...... s......	2.2288×10^{-2} 712.0 1.0538	24.454 2330.7 1.7576	....	....	....	....	....

Temperature of superheated vapor, R							
1400	1600	1800	2000	2200	2400	2600	2700
....	....	....	....	....	....	....	....
2332.7	2375.9	2419.1	2462.3	2505.4	2548.6	2591.8	2613.4
3.6381	3.6665	3.6924	3.7148	3.7354	3.7545	3.7713	3.7796
1.3044×10^9	1.4907×10^9	1.6771×10^9	1.8634×10^9	2.0498×10^9	2.2361×10^9	2.4224×10^9	2.5156×10^9
2332.7	2375.9	2419.1	2462.3	2505.4	2548.6	2591.8	2613.4
3.2884	3.3169	3.3428	3.3652	3.3858	3.4048	3.5217	3.4299
5.6924×10^7	6.5056×10^7	7.3188×10^7	8.1320×10^7	8.9452×10^7	9.7584×10^7	1.0572×10^8	1.0978×10^8
2332.7	2375.9	2419.1	2462.3	2505.4	2548.6	2591.8	2613.4
3.0179	3.0464	3.0723	3.0947	3.1153	3.1343	3.1511	3.1594
4.6978×10^6	5.3693×10^6	6.0406×10^6	6.7118×10^6	7.383×10^6	8.0541×10^6	8.7253×10^6	9.0609×10^6
2332.6	2375.9	2419.1	2462.3	2505.4	2548.6	2591.8	2613.4
2.8024	2.8309	2.8568	2.8792	2.8998	2.9188	2.9357	2.9439
6.1515×10^5	7.0339×10^5	7.9139×10^5	8.7935×10^5	9.6729×10^5	1.0552×10^6	1.1432×10^6	1.1871×10^6
2331.7	2375.7	2419.0	2462.3	2505.4	2548.6	2591.8	2613.4
2.6263	2.6552	2.6812	2.7036	2.7243	2.7433	2.7601	2.7684
1.1345×10^5	1.3001×10^5	1.4635×10^5	1.6263×10^5	1.7891×10^5	1.9517×10^5	2.1144×10^5	2.1957×10^5
2327.6	2374.7	2418.7	2462.2	2505.4	2548.6	2591.8	2613.4
2.4778	2.5089	2.5353	2.5578	2.5785	2.5975	2.6143	2.6226
2.6936×10^4	3.1124×10^4	3.5095×10^4	3.9019×10^4	4.2931×10^4	4.6838×10^4	5.0743×10^4	5.2695×10^4
2312.2	2371.1	2417.6	2461.7	2505.1	2548.5	2591.8	2613.4
2.3445	2.3836	2.4115	2.4343	2.4551	2.4742	2.4911	2.4993
....	9.0793×10^3	1.0292×10^4	1.1460×10^4	1.2616×10^4	1.3767×10^4	1.4916×10^4	1.5491×10^4
....	2359.9	2414.0	2460.3	2504.5	2548.1	2591.6	2613.2
....	2.2715	2.3040	2.3280	2.3491	2.3683	2.3853	2.3935
....	3.1025×10^3	3.5625×10^3	3.9820×10^3	4.3896×10^3	4.7929×10^3	5.1944×10^3	5.3948×10^2
....	2332.6	2404.7	2456.5	2502.7	2547.2	2591.0	2612.8
....	2.1651	2.2083	2.2352	2.2573	2.2769	2.2940	2.3023
....	....	1.4040×10^3	1.5823×10^3	1.7496×10^3	1.9128×10^3	2.0743×10^3	2.1548×10^3
....	....	2384.7	2448.0	2498.7	2545.1	2589.7	2611.8
....	....	2.1192	2.1523	2.1765	2.1969	2.2144	2.2228
....	....	6.0659×10^2	6.9378×10^2	7.7180×10^2	8.4601×10^2	9.1858×10^2	9.5458×10^2
....	....	2347.7	2431.3	2490.5	2540.7	2587.1	2609.8
....	....	2.0309	2.0747	2.1031	2.1252	2.1433	2.1519
....	....	....	3.2952×10^2	3.7033×10^2	4.0785×10^2	4.4385×10^2	4.6158×10^2
....	....	....	2402.3	2475.6	2532.5	2582.2	2605.9
....	....	....	1.9996	2.0347	2.0597	2.0792	2.0882
....	....	....	1.6838×10^2	1.9197×10^2	2.1298×10^2	2.3265×10^2	2.4224×10^2
....	....	....	2359.5	2451.8	2518.8	2573.9	2599.3
....	....	....	1.9259	1.9701	1.9996	2.0212	2.0209
....	....	....	....	1.0543×10^2	1.1816×10^2	1.2980×10^2	1.3539×10^2
....	....	....	....	2417.4	2498.0	2560.9	2588.9
....	....	....	....	1.9072	1.9426	1.9673	1.9780
....	....	....	....	60.665	68.825	76.167	79.656
....	....	....	....	2372.9	2469.0	2451.9	2573.5
....	....	....	....	1.8455	1.8876	1.9164	1.9284
....	....	....	....	....	41.754	46.622	48.920
....	....	....	....	....	2431.8	2516.2	2552.3
....	....	....	....	....	1.8340	1.8674	1.8811
....	....	....	....	....	26.244	29.585	31.163
....	....	....	....	....	2388.2	2484.0	2525.0
....	....	....	....	....	1.7820	1.8201	1.8356

(continued)

Table E-4 Thermodynamic properties of sodium (*continued*)

Temperature, R (Sat. press., psia)		Sat. liquid	Sat. vapor	Temperature of superheated vapor, R				
				800	900	1000	1100	1200
2400 (56.212)	v......	2.2604×10^{-2}	17.109	......	....	....	....	....
	h......	743.8	2341.2	......	....	....	....	....
	s......	1.0673	1.7329	......	....	....	....	....
2500 (80.236)	v......	2.2920×10^{-2}	12.388	......	....	....	....	....
	h......	776.2	2352.6	......	....	....	....	....
	s......	1.0805	1.7111	......	....	....	....	....
2600 (1.1116×10^2)	v......	2.3236×10^{-2}	9.2328	......	....	....	....	....
	h......	809.1	2365.1	......	....	....	....	....
	s......	1.0934	1.6919	......	....	....	....	....
2700 (1.052×10^2)	v......	2.3552×10^{-2}	7.0380					
	h......	842.7	2378.8					
	s......	1.1061	1.6751					

(From Meisl, C. J., and Shapiro, A. *Thermodynamic Properties of Alkali Metal Vapors and Mercury* (2nd ed.). Gen. Elec. Flight Propulsion Lab. Rept., R60FPD358-A, November 1960. Reprinted by permission of General Electric Company, 1988.

Units: v in ft^3/lbm, h in BTU/lbm, s in BTU/lbm R.

Temperature of superheated vapor, R							
1400	1600	1800	2000	2200	2400	2600	2700
. . . .	. . . .	. . . .	. . . .	. . . .	. . . .	19.460	20.580
. . . .	. . . .	. . . .	. . . .	. . . .	. . . .	2446.8	2492.6
. . . .	. . . .	. . . .	. . . .	. . . .	. . . .	1.7748	1.7922
. . . .	. . . .	. . . .	. . . .	. . . .	. . . .	13.219	14.032
. . . .	. . . .	. . . .	. . . .	. . . .	. . . .	2406.5	2456.5
. . . .	. . . .	. . . .	. . . .	. . . .	. . . .	1.7321	1.7501
. . . .	. . . .	. . . .	. . . .	. . . .	. . . .	. . . .	9.8326
. . . .	. . . .	. . . .	. . . .	. . . .	. . . .	. . . .	2418.1
. . . .	. . . .	. . . .	. . . .	. . . .	. . . .	. . . .	1.7120

Table E-5 Thermodynamic properties of helium

Pressure (psia)	Temperature (F)					
	100	200	300	400	500	600
14.696						
v	102.23	120.487	138.743	157.00	175.258	193.515
ρ	0.0097820	0.0082997	0.0072076	0.0063694	0.0057059	0.0051676
h	707.73	827.56	952.38	1077.20	1202.02	1326.83
s	6.8421	7.0472	7.2233	7.3776	7.5149	7.6386
50						
v	30.085	35.451	40.817	46.183	51.549	56.915
ρ	0.033239	0.028208	0.024500	0.021653	0.019399	0.017570
h	703.08	827.90	952.72	1077.54	1202.36	1327.18
s	6.2342	6.4393	6.6153	6.7697	6.9070	7.0307
150						
v	10.063	11.8522	13.6407	15.4293	17.2183	19.008
ρ	0.099372	0.084372	0.073310	0.064812	0.058078	0.052610
h	704.08	828.91	953.73	1078.55	1203.37	1328.19
s	5.6886	5.8937	6.0698	6.2241	6.3614	6.4852
400						
v	3.8062	4.4775	5.1487	5.8197	6.4905	7.1616
ρ	0.26273	0.22334	0.194225	0.171831	0.154072	0.139633
h	706.58	831.42	956.24	1081.06	1205.88	1330.70
s	5.2013	5.4065	5.5827	5.7371	5.8744	5.9981
600						
v	2.5546	3.0023	3.44995	3.8973	4.3449	4.7923
ρ	0.39146	0.33308	0.28986	0.25658	0.23016	0.20867
h	708.49	833.33	958.15	1082.97	1207.79	1332.61
s	4.9998	5.2050	5.3813	5.5357	5.6730	5.7968
900						
v	1.7200	2.0187	2.3173	2.6157	2.91399	3.2124
ρ	0.58139	0.49537	0.43154	0.38230	0.34317	0.31129
h	710.29	835.38	960.40	1085.42	1210.42	1335.36
s	4.7981	5.0035	5.1797	5.3342	5.4715	5.5953
1500						
v	1.05192	1.2314	1.4108	1.58994	1.7690	1.9483
ρ	0.95064	0.81207	0.70880	0.62897	0.56528	0.51328
h	715.54	840.77	965.88	1090.92	1215.93	1340.97
s	4.5437	4.7475	4.9257	5.0801	5.2176	5.3414
2500						
v	0.65044	0.75847	0.86635	0.97410	1.08176	1.18947
ρ	1.53741	1.31845	1.15427	1.02659	0.92442	0.84071
h	724.37	849.73	974.95	1100.10	1225.22	1350.29
s	4.2887	4.4928	4.6712	4.8258	4.9634	5.0873
4000						
v	0.42377	0.49161	0.55932	0.62694	0.69444	0.76191
ρ	2.3598	2.0341	1.78789	1.59503	1.44000	1.31248
h	736.48	862.24	987.70	1113.12	1238.46	1363.73
s	4.0531	4.2576	4.4363	4.5912	4.7287	4.8530

Units: v in ft³/lbm, h in BTU/lbm, and s in BTU/lbm °R

From Fabric Filter Systems Study. In: *Handbook of Fabric Filter Technology* (Vol. 1). PB200-648, APTD-0690, National Technical Information Service, December 1970; wherein reprinted from El-Wakil, M. *PowerPlant Technology*. New York: McGraw-Hill, 1984.

Table E-6 Thermodynamic properties of CO$_2$

Pressure, psia	Property†	−75 F	−50	0	50	100	150	200	300	400	600	800	1000	1200	1400	1600	1800 F
1.00	v	93.90	100.0	112.2	124.4	136.0	148.8	161.0	185.4	209.7	258.5	307.2	356.0	404.8	453.6	502.3	551.0
	h	283.2	288.0	297.8	307.7	318.0	328.4	339.1	361.4	384.7	434.4	487.1	542.4	599.6	658.6	718.8	780.0
	s	1.4772	1.4892	1.5112	1.5316	1.5506	1.5684	1.5852	1.6165	1.6451	1.6969	1.7423	1.7829	1.8197	1.8533	1.8838	1.9123
10.0	v	9.280	9.902	11.15	12.38	13.61	14.84	16.06	18.51	20.96	25.85	30.73	35.61	40.49	45.36	50.24	55.11
	h	282.6	287.5	297.3	307.3	317.7	328.2	339.0	361.3	384.6	434.4	487.1	542.4	599.6	658.6	718.8	780.0
	s	1.3733	1.3853	1.4073	1.4277	1.4467	1.4645	1.4813	1.5126	1.5412	1.5930	1.6384	1.6790	1.7158	1.7494	1.7799	1.8084
20.0	v	4.586	4.904	5.542	6.119	6.778	7.407	8.016	9.247	10.47	12.92	15.36	17.80	20.24	22.68	25.11	27.55
	h	281.9	287.0	296.8	306.8	317.3	327.9	338.8	361.1	384.5	434.3	487.1	542.4	599.6	658.6	718.8	780.0
	s	1.3417	1.3538	1.3759	1.3964	1.4154	1.4332	1.4500	1.4813	1.5099	1.5617	1.6071	1.6477	1.6845	1.7181	1.7486	1.7771
40.0	v	2.239	2.404	2.738	3.053	3.363	3.688	3.993	4.615	5.230	6.458	7.688	8.901	10.12	11.37	12.56	13.78
	h	280.6	285.9	295.8	305.9	316.5	327.4	338.4	360.9	384.3	434.2	487.0	542.4	599.6	658.6	718.8	780.0
	s	1.3088	1.3211	1.3435	1.3642	1.3834	1.4014	1.4184	1.4499	1.4787	1.5305	1.5759	1.6165	1.6533	1.6869	1.7174	1.7459
80.0	v	. . .	1.154	1.335	1.498	1.657	1.828	1.982	2.298	2.608	3.226	3.839	4.448	5.060	5.670	6.281	6.887
	h	. . .	283.8	293.8	304.1	315.1	326.4	337.7	360.2	383.9	434.0	486.9	542.3	590.5	658.6	718.8	780.0
	s	. . .	1.2778	1.3044	1.3284	1.3490	1.3679	1.3855	1.4177	1.4468	1.4991	1.5446	1.5852	1.6220	1.6556	1.6861	1.7146
120	v	. . .	. . .	0.8665	0.9799	1.088	1.208	1.311	1.525	1.734	2.148	2.559	2.966	3.373	3.781	4.188	4.592
	h	. . .	. . .	291.7	302.2	313.6	325.4	337.0	359.7	383.5	433.8	486.8	542.3	599.5	658.6	718.8	780.0
	s	. . .	. . .	1.2833	1.3086	1.3297	1.3488	1.3666	1.3993	1.4285	1.4808	1.5263	1.5669	1.6037	1.6373	1.6678	1.6963
160	v	. . .	. . .	0.6305	0.7207	0.8033	0.8986	0.9760	1.139	1.297	1.610	1.918	2.224	2.530	2.836	3.141	3.445
	h	. . .	. . .	289.7	300.4	312.1	324.4	336.3	359.1	383.1	433.6	486.6	542.2	599.6	658.6	718.8	780.0
	s	. . .	. . .	1.2666	1.2928	1.3154	1.3350	1.3529	1.3857	1.4151	1.4675	1.5133	1.5539	1.5907	1.6243	1.6548	1.6833
200	v	. . .	. . .	0.4891	0.5652	0.6376	0.7125	0.7748	0.9075	1.035	1.287	1.534	1.779	2.024	2.269	2.513	2.757
	h	. . .	. . .	287.7	298.6	310.6	323.4	335.6	358.5	382.7	433.4	486.5	542.2	599.5	658.5	718.8	780.0
	s	. . .	. . .	1.2519	1.2805	1.3038	1.3239	1.3421	1.3753	1.4049	1.4574	1.5033	1.5439	1.5807	1.6143	1.6448	1.6733

(*continued*)

Table E-6 Thermodynamic properties of CO₂ (*continued*)

Pressure, psia	Property†	-75 F	-50	0	50	100	150	200	300	400	600	800	1000	1200	1400	1600	1800 F
240	v	...	...	0.3948	0.4614	0.5237	0.5886	0.6407	0.7532	0.8604	1.071	1.273	1.482	1.687	1.891	2.095	2.297
	h	...	...	285.6	296.7	309.1	322.4	334.9	358.0	382.3	433.1	486.4	542.1	599.5	658.5	718.8	780.0
	s	...	...	1.2395	1.2694	1.2940	1.3145	1.3330	1.3671	1.3963	1.4490	1.4948	1.5356	1.5724	1.6060	1.6365	1.6650
300	v	...	...	...	0.3563	0.4100	0.4636	0.5065	0.5985	0.6868	0.8556	1.021	1.186	1.349	1.513	1.676	1.838
	h	...	...	...	294.0	306.9	320.9	333.9	357.1	381.6	432.8	486.2	542.0	599.4	658.5	718.7	780.0
	s	...	...	...	1.2562	1.2813	1.3029	1.3219	1.3560	1.3862	1.4389	1.4848	1.5256	1.5624	1.5960	1.6265	1.6550
360	v	...	...	...	0.2858	0.3341	0.3780	0.4171	0.4958	0.5693	0.7212	0.8502	0.9874	1.125	1.261	1.397	1.533
	h	...	...	...	291.2	304.6	319.4	332.8	356.3	381.0	432.5	486.0	541.9	599.4	658.5	718.7	779.9
	s	...	...	...	1.2436	1.2699	1.2925	1.3124	1.3475	1.3779	1.4307	1.4766	1.5174	1.5542	1.5878	1.6183	1.6468
440	v	...	...	...	0.2216	0.2652	0.3040	0.3358	0.4022	0.4633	0.5817	0.6950	0.8079	0.9201	1.032	1.142	1.255
	h	...	...	...	287.6	301.6	317.4	331.4	355.1	380.2	432.1	485.8	541.7	599.3	658.4	718.6	779.9
	s	...	...	...	1.2282	1.2559	1.2797	1.3006	1.3370	1.3681	1.4215	1.4675	1.5083	1.5451	1.5787	1.6092	1.6377
520	v	...	...	...	0.1772	0.2174	0.2513	0.2795	0.3374	0.3901	0.4912	0.5881	0.6832	0.7785	0.8733	0.9672	1.062
	h	...	...	...	283.9	298.7	315.4	330.0	354.0	379.4	431.7	485.5	541.5	599.2	658.3	718.6	779.9
	s	...	...	...	1.2148	1.2438	1.2687	1.2905	1.3281	1.3599	1.4138	1.4599	1.5007	1.5375	1.5711	1.6010	1.6301
600	v	...	...	...	0.1452	0.1823	0.2123	0.2383	0.2898	0.3363	0.4250	0.5093	0.5921	0.6747	0.7571	0.8385	0.9202
	h	...	...	...	280.3	295.7	313.4	328.6	352.8	378.6	431.1	485.3	541.4	599.0	658.2	718.6	779.8
	s	...	...	...	1.2020	1.2323	1.2583	1.2809	1.3198	1.3525	1.4071	1.4534	1.4942	1.5310	1.5646	1.5951	1.6236
800	v	...	...	...	...	0.1196	0.1483	0.1712	0.2126	0.2489	0.3173	0.3812	0.4436	0.5060	0.5680	0.6292	0.6906
	h	...	...	...	...	288.2	308.4	325.1	350.0	376.5	430.1	484.7	541.0	598.8	658.0	718.4	779.7
	s	...	...	...	...	1.2111	1.2391	1.2631	1.3041	1.3380	1.3935	1.4404	1.4812	1.5180	1.5516	1.5821	1.6106

P (psia)												
1000	v	0.1101	0.1310	0.1663	0.1966	0.2526	0.3048	0.3547	0.4049	0.4545	0.5037	0.5526
	h	303.4	321.6	347.1	374.5	429.1	484.0	540.6	598.5	657.8	718.3	779.6
	s	1.2218	1.2472	1.2903	1.3258	1.3828	1.4302	1.4712	1.5080	1.5416	1.5721	1.6006
1200	v		0.1042	0.1356	0.1621	0.2096	0.2531	0.2953	0.3374	0.3789	0.4199	0.4609
	h		318.4	344.2	372.5	428.1	483.5	540.2	598.2	657.7	718.2	779.5
	s		1.2343	1.2791	1.3158	1.3740	1.4216	1.4628	1.4996	1.5332	1.5637	1.5922
1400	v			0.1136	0.1375	0.1788	0.2160	0.2529	0.2892	0.3249	0.3601	0.3551
	h			341.4	370.4	427.0	482.9	539.8	598.0	657.5	718.0	779.5
	s			1.2703	1.3078	1.3668	1.4145	1.4558	1.4927	1.5263	1.5568	1.5853
1600	v				0.1191	0.1557	0.1898	0.2211	0.2530	0.2843	0.3153	0.3461
	h				367.6	426.0	482.3	539.5	597.7	657.2	717.9	779.4
	s				1.3002	1.3602	1.4083	1.4497	1.4867	1.5193	1.5508	1.5793
1800	v				0.1047	0.1377	0.1675	0.1964	0.2249	0.2528	0.2804	0.3079
	h				364.0	424.9	481.7	539.1	597.4	657.0	717.8	779.3
	s				1.2930	1.3539	1.4023	1.4440	1.4812	1.5148	1.5453	1.5738
2200	v					0.1120	0.1368	0.1605	0.1840	0.2070	0.2296	0.2522
	h					421.7	480.5	538.4	596.9	656.6	717.5	779.2
	s					1.3426	1.3925	1.4344	1.4720	1.5057	1.5362	1.5647
2600	v						0.1156	0.1357	0.1557	0.1752	0.1945	0.2137
	h						479.3	537.6	596.4	656.3	717.2	779.1
	s						1.3834	1.4260	1.4640	1.4981	1.5286	1.5571
3000	v						0.1001	0.1176	0.1349	0.1519	0.1687	0.1856
	h						478.1	536.8	595.8	656.0	717.0	778.8
	s						1.3752	1.4183	1.4569	1.4915	1.5220	1.5505

† v = cu ft/lb; h = BTU/lb; s = BTU/(lb)(R). To make the enthalpies of saturated CO_2 consistent with those of superheated CO_2, add 143.3 BTU/lb to h_{sat}.
(From *Reactor Handbook of Engineering*, US AEC, 1955. Wherein reprinted from Perry, J. H. *Chemical Engineer's Handbook*. (3rd ed.). New York: McGraw-Hill, 1950. Data from Sweigert, Weber, and Allen, *Ind. Eng. Chem.* 38:185, 1946.)

THERMOPHYSICAL PROPERTIES
OF SOME SUBSTANCES

This appendix is adapted from Poppendiek, H. F., and Sabin, C. M. *Some Heat Transfer Performance Criteria for High Temperature Fluid Systems*. American Society of Mechanical Engineers, 75-WA/HT-103, 1975. The section on sodium is from *Liquid Metals Handbook*. NAV EXOS P-733 Rev., June 1952, AEC.

CAUTION: The following property values are presented in a comparative manner which, however, may not yield accuracies needed for detailed assessments. In such cases recent tabulated property listings should be consulted:

Density
Specific heat
Prandtl modulus
Vapor pressure

The section on physical properties of some solids is from Collier, J. G. *Convective Boiling and Condensation* (2nd ed.). New York: McGraw-Hill, 1981.

NOTE: For dynamic and kinematic fluid viscosities, see Figures I, 9-10 and I, 9-11, respectively. For thermal conductivity of engineering materials, see Figure I, 10-1.

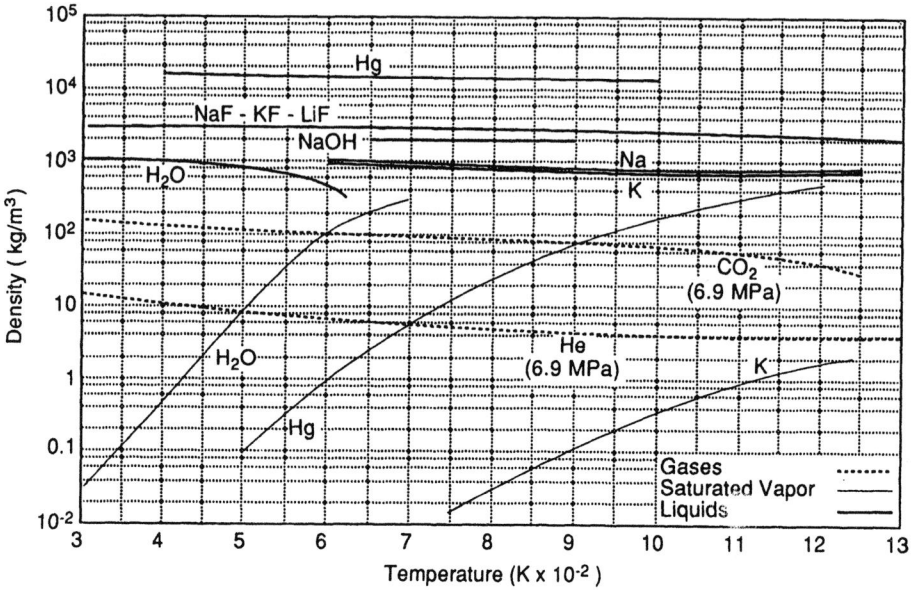

Figure F-1 Density versus temperature.

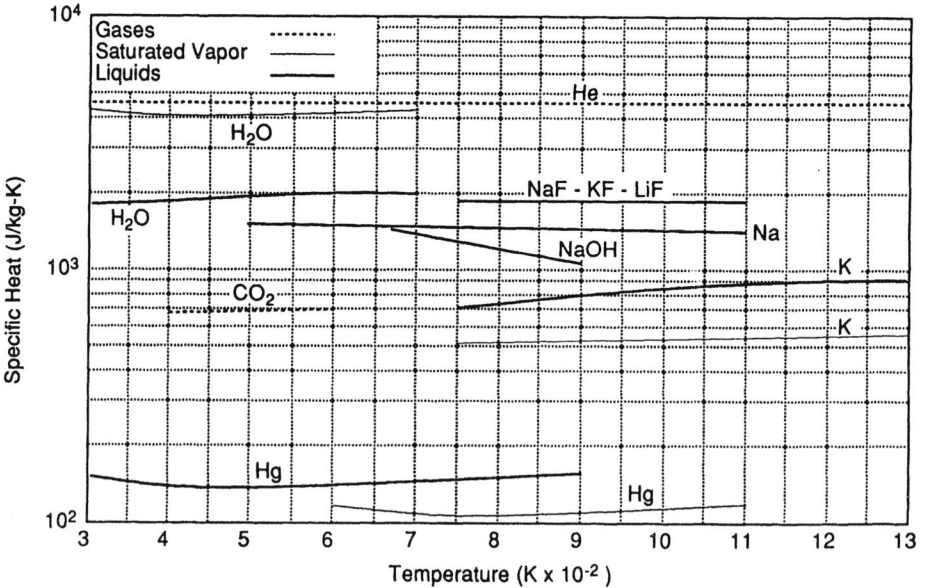

Figure F-2 Specific heat versus temperature.

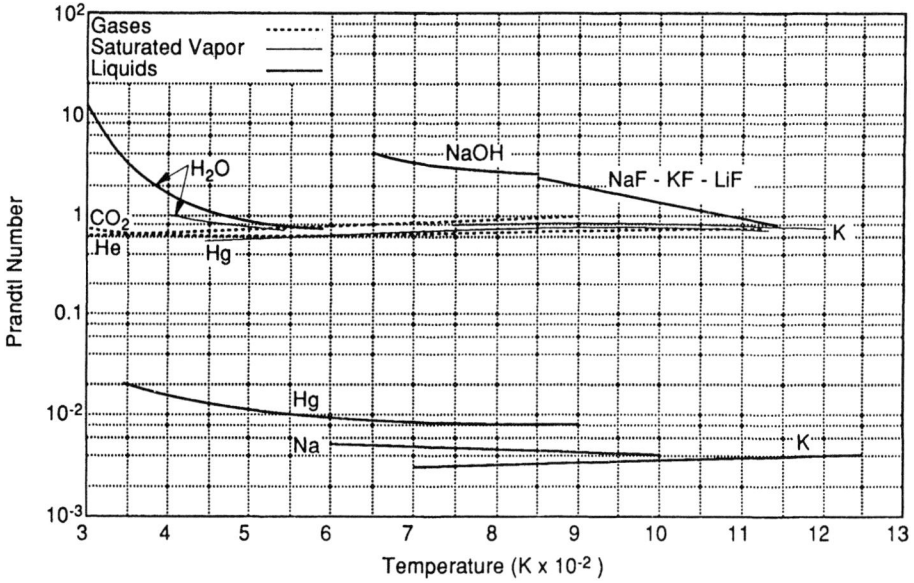

Figure F-3 Prandtl number versus temperature.

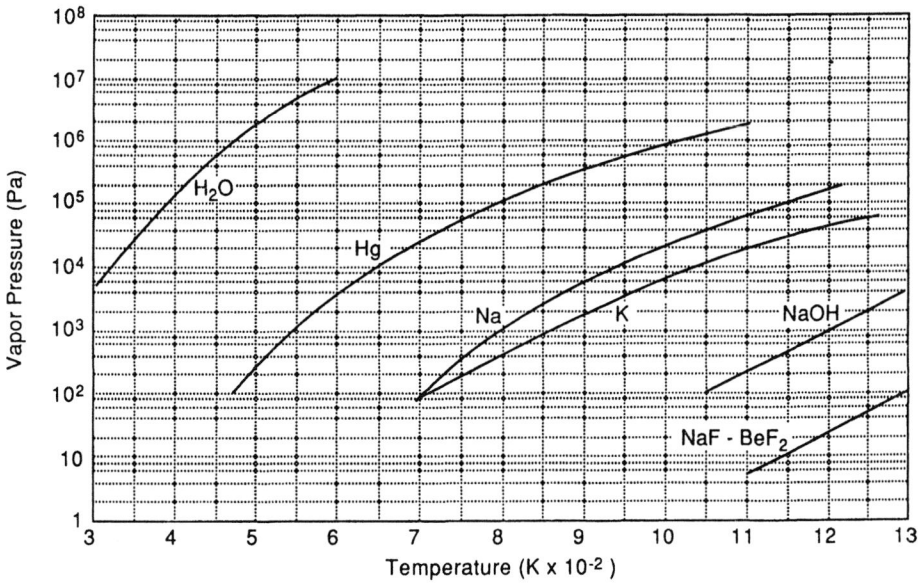

Figure F-4 Vapor pressure versus temperature.

Table F-1 Thermal properties of solids: Metals

| | Properties at 20 °C | | | | k, W/m deg C | | | | |
| | | | | | Temperature | | | | |
	ρ $\left(\dfrac{kg}{m^3}\right)$	c_p $\left(\dfrac{J}{kg\ deg\ C}\right)$	k $\left(\dfrac{W}{m\ deg\ C}\right)$	α $\left(\dfrac{m^2}{s}\right)$	100 °C	200 °C	300 °C	400 °C	600 °C
Aluminum, pure	2707	896	204	$8 \cdot 42 \times 10^{-5}$	206	215	229	249	
Duralumin, 94–96 Al, 3–5 Cu	2787	883	164	$6 \cdot 68 \times 10^{-5}$	182	194			
Lead	11,370	130	34·6	$2 \cdot 34 \times 10^{-5}$					
Iron, pure	7897	452	72·7	$2 \cdot 03 \times 10^{-5}$	67·5	62·3	55·4	48·5	39·8
Iron, wrought, C < 0·5%	7849	460	58·9	$1 \cdot 63 \times 10^{-5}$	57·1	51·9	48·5	45·0	36·4
Iron, cast, C ≈ 4%	7272	419	51·9	$1 \cdot 70 \times 10^{-5}$					
Carbon steel, C ≈ 0·5%	7833	465	53·7	$1 \cdot 47 \times 10^{-5}$	51·9	48·5	45·0	41·5	34·6
Carbon steel, C = 1·5%	7753	486	36·4	$0 \cdot 97 \times 10^{-5}$	36·3	36·3	34·6	32·9	31·2
Nickel steel, 10%	7945	460	26·0	$0 \cdot 72 \times 10^{-5}$					
Nickel steel, 30%	8073	460	12·1	$0 \cdot 33 \times 10^{-5}$					
Nickel steel, 50%	8266	460	13·8	$0 \cdot 36 \times 10^{-5}$					
Nickel steel, 70%	8506	460	26·0	$0 \cdot 67 \times 10^{-5}$					
Nickel steel, 90%	8762	460	46·7	$1 \cdot 16 \times 10^{-5}$					
Chrome steel, 1%	7865	460	60·6	$1 \cdot 67 \times 10^{-5}$	55·4	51·9	46·7	41·5	36·4

Chrome steel, 5%	7833	460	39.8	1.11×10^{-5}	38.1	36.4	36.4	32.9	29.4
Chrome steel, 10%	7785	460	31.2	0.87×10^{-5}	31.2	31.2	29.4	29.4	31.2
Cr–Ni steel, 18% Cr, 8% Ni	7817	460	16.3	0.44×10^{-5}	17.3	17.3	19.0	19.0	22.5
Ni–Cr steel, 20% Ni, 15% Cr	7865	460	14.0	0.39×10^{-5}	15.1	15.1	16.3	17.3	19.0
Manganese steel, 2%	7865	460	38.1	1.05×10^{-5}	36.4	36.4	36.4	34.6	32.9
Tungsten steel, 2%	7961	444	62.3	1.76×10^{-5}	58.9	53.7	48.5	45.0	36.4
Silicon steel, 2%	7673	460	31.2	0.89×10^{-5}					
Copper, pure	8954	383	386	11.2×10^{-5}	379	374	369	364	353
Bronze, 75 Cu, 25 Sn	8660	343	26.0	0.86×10^{-5}					
Brass, 70 Cu, 30 Zn	8522	385	111	3.41×10^{-5}	128	144	147	147	
German silver, 62 Cu 15 Ni, 22 Zn	8618	394	24.9	0.73×10^{-5}	31.2	39.8	45.0	48.5	
Constantan, 60 Cu, 40 Ni	8922	410	22.7	0.61×10^{-5}	22.2	26.0			
Magnesium, pure	1746	1013	171	9.71×10^{-5}	168	163	158		
Molybdenum	10,220	251	123	4.79×10^{-5}	118	114	111	109	106
Nickel, 99.9% pure	8906	446	90.0	2.27×10^{-5}	83.1	72.7	64.0	58.9	
Silver, 99.9% pure	10,520	234	407	16.6×10^{-5}	415	374	362	360	
Tungsten	19,350	134	163	6.27×10^{-5}	151	142	133	126	113
Zinc, pure	7144	384	112	4.11×10^{-5}	109	106	100	93.5	
Tin, pure	7304	227	64.0	3.88×10^{-5}	58.9	57.1			

(Adapted from Table A-1, Eckert, E. R. G., and Drake, R. M., Jr. *Heat and Mass Transfer*. New York: McGraw-Hill, 1959.)

DIMENSIONLESS GROUPS OF FLUID MECHANICS AND HEAT TRANSFER

Name	Notation	Formula	Interpretation in terms of ratio
Biot number	Bo	$\dfrac{hL}{k_s}$	Surface conductance ÷ internal conduction of solid
Cauchy number	Ca	$\dfrac{V^2}{B_s/\rho} = \dfrac{V^2}{a^2}$	Inertia force ÷ compressive force = (Mach number)^2
Eckert number	Ek	$\dfrac{V^2}{c_p \Delta T}$	Temperature rise due to energy conversion ÷ temperature difference
Euler number	Eu	$\dfrac{\Delta p}{\rho V^2}$	Pressure force ÷ inertia force
Fourier number	Fo	$\dfrac{kt}{\rho c_p L^2} = \dfrac{\alpha t}{L^2}$	Rate of conduction of heat ÷ rate of storage of energy
Froude number	Fr	$\dfrac{V^2}{gL}$	Inertia force ÷ gravity force
Graetz number	Gz	$\dfrac{D}{L} \cdot \dfrac{V\rho c_p D}{k}$	Re Pr ÷ (L/D); heat transfer by convection in entrance region ÷ heat transfer by conduction
Grashof number	Gr	$\dfrac{g\beta \, \Delta T L^3}{\nu^2}$	Buoyancy force ÷ viscous force
Knudsen number	Kn	$\dfrac{\lambda}{L}$	Mean free path of molecules ÷ characteristic length of an object
Lewis number	Le	$\dfrac{\alpha}{D_c}$	Thermal diffusivity ÷ molecular diffusivity
Mach number	M	$\dfrac{V}{a}$	Macroscopic velocity ÷ speed of sound

475

Name	Notation	Formula	Interpretation in terms of ratio
Nusselt number	Nu	$\dfrac{hL}{k}$	Temperature gradient at wall ÷ overall temperature difference
Péclet number	Pé	$\dfrac{V\rho c_p D}{k}$	(Re Pr); heat transfer by convection ÷ heat transfer by conduction
Prandtl number	Pr	$\dfrac{\mu c_p}{k} = \dfrac{\nu}{\alpha}$	Diffusion of momentum ÷ diffusion of heat
Reynolds number	Re	$\dfrac{\rho VL}{\mu} = \dfrac{VL}{\nu}$	Inertia force ÷ viscous force
Schmidt number	Sc	$\dfrac{\mu}{\rho D_c} = \dfrac{\nu}{D_c}$	Diffusion of momentum ÷ diffusion of mass
Sherwood number	Sh	$\dfrac{h_D L}{D_c}$	Mass diffusivity ÷ molecular diffusivity
Stanton number	St	$\dfrac{h}{V\rho c_p} = \dfrac{h}{c_p G}$	Heat transfer at wall ÷ energy transported by stream
Stokes number	Sk	$\dfrac{\Delta p L}{\mu V}$	Pressure force ÷ viscous force
Strouhal number	Sl	$\dfrac{L}{tV}$	Frequency of vibration ÷ characteristic frequency
Weber number	We	$\dfrac{\rho V^2 L}{\sigma}$	Inertia force ÷ surface tension force

MULTIPLYING PREFIXES

tera	T	10^{12}	deci	d	10^{-1}
giga	G	10^{9}	centi	c	10^{-2}
mega	M	10^{6}	milli	m	10^{-3}
kilo	k	10^{3}	micro	μ	10^{-6}
hecto	h	10^{2}	nano	n	10^{-9}
deca	da	10^{1}	pico	p	10^{-12}
(deka)			femto	f	10^{-15}
			atto	a	10^{-18}

LIST OF ELEMENTS

Atomic number	Symbol	Name	Atomic weight†	Atomic number	Symbol	Name	Atomic weight†
1	H	Hydrogen	1.00794 (7)	53	I	Iodine	126.9045
2	He	Helium	4.002602 (2)	54	Xe	Xenon	131.29 (3)
3	Li	Lithium	6.941 (2)	55	Cs	Cesium	132.9054
4	Be	Beryllium	9.01218	56	Ba	Barium	137.33
5	B	Boron	10.811 (5)	57	La	Lanthanum	138.9055 (3)
6	C	Carbon	12.011	58	Ce	Cerium	140.12
7	N	Nitrogen	14.0067	59	Pr	Praseodymium	140.9077
8	O	Oxygen	15.9994 (3)	60	Nd	Neodymium	144.24 (3)
9	F	Fluorine	18.998403	61	Pm	Promethium	[145]
10	Ne	Neon	20.179	62	Sm	Samarium	150.36 (3)
11	Na	Sodium	22.98977	63	Eu	Europium	151.96
12	Mg	Magnesium	24.305	64	Gd	Gadolinium	157.25 (3)
13	Al	Aluminum	26.98154	65	Tb	Terbium	158.9254
14	Si	Silicon	28.0855 (3)	66	Dy	Dysprosium	162.50 (3)
15	P	Phosphorus	30.97376	67	Ho	Holmium	164.9304
16	S	Sulfur	32.066 (6)	68	Er	Erbium	167.26 (3)
17	Cl	Chlorine	35.453	69	Tm	Thulium	168.9342
18	Ar	Argon	39.948	70	Yb	Ytterbium	173.04 (3)
19	K	Potassium	39.0983	71	Lu	Lutetium	174.967
20	Ca	Calcium	40.078 (4)	72	Hf	Hafnium	178.49 (3)
21	Sc	Scandium	44.95591	73	Ta	Tantalum	180.9479
22	Ti	Titanium	47.88 (3)	74	W	Tungsten	183.85 (3)
23	V	Vanadium	50.9415	75	Re	Rhenium	186.207
24	Cr	Chromium	51.9961 (6)	76	Os	Osmium	190.2
25	Mn	Manganese	54.9380	77	Ir	Iridium	192.22 (3)
26	Fe	Iron	55.847 (3)	78	Pt	Platinum	195.08 (3)
27	Co	Cobalt	58.9332	79	Au	Gold	196.9665
28	Ni	Nickel	58.69	80	Hg	Mercury	200.59 (3)

Z	Symbol	Name	Atomic weight	Z	Symbol	Name	Atomic weight
29	Cu	Copper	63.546 (3)	81	Tl	Thallium	204.383
30	Zn	Zinc	65.39 (2)	82	Pb	Lead	207.2
31	Ga	Gallium	69.723 (4)	83	Bi	Bismuth	208.9804
32	Ge	Germanium	72.59 (3)	84	Po	Polonium	[209]
33	As	Arsenic	74.9216	85	At	Astatine	[210]
34	Se	Selenium	78.96 (3)	86	Rn	Radon	[222]
35	Br	Bromine	79.904	87	Fr	Francium	[223]
36	Kr	Krypton	83.80	88	Ra	Radium	226.0254
37	Rb	Rubidium	85.4678 (3)	89	Ac	Actinium	227.0278
38	Sr	Strontium	87.62	90	Th	Thorium	232.0381
39	Y	Yttrium	88.9059	91	Pa	Protactinium	231.0359
40	Zr	Zirconium	91.224 (2)	92	U	Uranium	238.0289
41	Nb	Niobium	92.9064	93	Np	Neptunium	237.0482
42	Mo	Molybdenum	95.94	94	Pu	Plutonium	[244]
43	Tc	Technetium	[98]	95	Am	Americium	[243]
44	Ru	Ruthenium	101.07 (2)	96	Cm	Curium	[247]
45	Rh	Rhodium	102.9055	97	Bk	Berkelium	[247]
46	Pd	Palladium	106.42	98	Cf	Californium	[251]
47	Ag	Silver	107.8682 (3)	99	Es	Einsteinium	[252]
48	Cd	Cadmium	112.41	100	Fm	Fermium	[257]
49	In	Indium	114.82	101	Md	Mendelevium	[258]
50	Sn	Tin	118.710 (7)	102	No	Nobelium	[259]
51	Sb	Antimony	121.75 (3)	103	Lr	Lawrencium	[260]
52	Te	Tellurium	127.60 (3)	104	Rf	Rutherfordium‡	[261]
				105	Ha	Hahnium‡	[262]
				106	unnamed	Unnamed	[263]

† Values in parentheses are the uncertainty in the last digit of the stated atomic weights. Values without a quoted error in parentheses are considered to be reliable to ±1 in the last digit except for Ra, Ac, Pa and Np which are not given by Walker et al., 1983. Brackets indicate most stable or best known isotope.

‡ The names of these elements have not been accepted because of conflicting claims of discovery.

(From Walker, F. W., Miller, D. G., and Feiner, F. *Chart of the Nuclides* (13th ed.) Revised 1983, General Electric Co.)

SQUARE AND HEXAGONAL ROD ARRAY DIMENSIONS

J-1 LWR FUEL BUNDLES: SQUARE ARRAYS

Tables J-1 and J-2 present formulas for determining axially averaged unit subchannel and overall bundle dimensions, respectively.

The presentation of these formulas as axially averaged values is arbitrary and reflects the fact that grid-type spacers occupy a small fraction of the axial length of a fuel bundle. Therefore this fraction δ has been defined as:

$$\delta = \frac{\text{total axial length of grid spacers}}{\text{axial length of the fuel bundle}}$$

Formulas applicable at the axial grid locations or between grids are easily obtained from Tables J-1 and J-2 by taking $\delta = 1$ or $\delta = 0$, respectively.

Determination of precise dimensions for an LWR assembly would require knowledge of the specific grid configuration used by the manufacturer. Typically the grid strap thickness at the periphery is slightly enhanced. Here it is taken as thickness t^*. It should also be carefully noted that these formulas are based on the assumptions of a rectangular grid and no support tabs or fingers. The dimension g is the spacing from rod surface to the flow boundary of the assembly. In the grid plane, the segment along g open for flow is $g - t/2$.

Table J-1 Square arrays: Axially averaged unit subchannel dimensions for ductless assembly

	$i \equiv 1$ INTERIOR	$i \equiv 2$ EDGE	$i \equiv 3$ CORNER
1. A_{fi} Area for flow without spacer	$P^2 - \dfrac{\pi D^2}{4}$	$P\left(\dfrac{D}{2}+g\right) - \dfrac{\pi D^2}{8}$	$\left(\dfrac{D}{2}+g\right)^2 - \dfrac{\pi D^2}{16}$
2. A_{fis} Average area for flow with spacer	$A_{f1} - (2Pt - t^2)\delta$	$A_{f2} - \left[\left(\dfrac{D}{2}+g-t^*\right)t + \dfrac{Pt^*}{2}\right]\delta$	$A_{f3} - \left[2\left(\dfrac{D}{2}+g\right)t^* - t^{*2}\right]$
3. P_{wi} Wetted perimeter without spacer and duct	πD	$\dfrac{\pi}{2}D$	$\dfrac{\pi}{4}D$
4. P_{wis} Average wetted perimeter including spacer but without duct	$\pi D + 4(P - t)\delta$	$\dfrac{\pi}{2}D + \left[\left(\dfrac{D}{2}+g-t^*\right)2 + (P-t)\right]\delta$	$\dfrac{\pi}{4}D + \left(\dfrac{D}{2}+g-t^*\right)2\delta$

Table J-2 Square arrays: Axially averaged overall dimensions for ductless assembly (assuming grid spacer around N_p rods of thickness t and that $g = (P - D)/2$

1. Total area inside square A_T:

$$A_T = D_\ell^2$$

where D_ℓ is the length of one side of the square.

2. Total average cross-sectional area for flow A_{fT}:

$$A_{fT} = D_\ell^2 - N_p \frac{\pi}{4} D^2 - N_p \frac{t}{2} P4\delta + N_p \left(\frac{t}{2}\right)^2 4\delta$$

$$A_{fT} = D_\ell^2 - \left[N_p \frac{\pi}{4} (D^2) + 2\delta(\sqrt{N_p})(D_\ell t) - N_p t^2 \delta \right]$$

since $D_\ell = \sqrt{N_p} P$ and $t^* = t$

where:

D = rod diameter
N_p = number of rods
t = interior spacer thickness
t^* = peripheral spacer thickness.

3. Total average wetted perimeter P_{wT}:

$$P_{wT} = N_p \pi D + 4\sqrt{N_p} D_\ell \delta - 4N_p t\delta$$

4. Equivalent hydraulic diameter for overall square D_{eT}:

$$D_{eT} = \frac{4A_{fT}}{P_{wT}} = \frac{4D_\ell^2 - [N_p \pi D^2 + 8\sqrt{N_p} D_\ell t\delta - 4N_p t^2\delta]}{N_p \pi D + 4\sqrt{N_p} D_\ell \delta - 4N_p t\delta}$$

J-2 LMR FUEL BUNDLES: HEXAGONAL ARRAYS

Tables J-3 and J-4 summarize the formulas for determining axially averaged unit subchannel and overall bundle dimensions, respectively. Fuel pin spacing is illustrated as being performed by a wire wrap. In practice both grids and wires are used as spacers in LMR fuel and blanket bundles. The axially averaged dimensions in these tables are based on averaging the wires over one lead length.

Table J-3 Hexagonal array with wire wrap spacer: Axially averaged unit subchannel dimensions

	$i = 1$ INTERIOR	$i = 2$ EDGE	$i = 3$ CORNER
1. A_{fi} Area for flow without wire wrap spacers	$\dfrac{1}{2}P\left(\dfrac{\sqrt{3}}{2}P\right) - \dfrac{\pi D^2}{8} = \dfrac{\sqrt{3}}{4}P^2 - \dfrac{\pi D^2}{8}$	$P\left(\dfrac{D}{2}+g\right) - \dfrac{\pi D^2}{8}$ NOTE: $g \equiv$ spacer diameter plus wall pin clearance, if any.	$\left[\dfrac{1}{\sqrt{3}}\left(\dfrac{D}{2}+g\right)^2 - \dfrac{\pi D^2}{24}\right]$
2. A_{fis} Area for flow including wire wrap spacers	$A_{f1} - \left(\dfrac{3}{6}\right)\dfrac{\pi}{4}D_s^2 = A_{f1} - \dfrac{\pi D_s^2}{8}$ (Three spacers traverse cell per unit lead; each traverse is 60° of 360°)	$A_{f2} - \left(\dfrac{2}{4}\right)\dfrac{\pi D_s^2}{4}$ (Two spacers traverse cell per unit lead; each traverse is 90° of 360°)	$A_{f3} - \left(\dfrac{1}{6}\right)\dfrac{\pi}{4}D_s^2$ (One spacer traverse cell per unit lead; each traverse is 60° of 360°)
3. P_{wis} Wetted perimeter including wire wrap spacers	$\dfrac{\pi D}{2} + \dfrac{\pi D_s}{2}$ (Three spacers traverse cell per unit lead; each traverse is 60° of 360°)	$\dfrac{\pi D}{2} + P + \dfrac{\pi D_s}{2}$ (Two spacers traverse cell per unit lead; each traverse is 90° of 360°)	$\dfrac{\pi}{6}(D + D_s) + \dfrac{2}{\sqrt{3}}\left(\dfrac{D}{2}+g\right)$ (One spacer traverse cell per unit lead; each traverse is 60° of 360°)

Table J-4 Hexagonal array: Axially averaged overall dimensions (assuming wire wrap spacers around each rod)

1. Total area inside hexagon A_{hT}:

$$A_{hT} = D_{ft}D_\ell + 2\left(\frac{1}{2}\right)D_{ft}D_\ell \sin 30 \text{ degrees} = \frac{\sqrt{3}}{2}D_{ft}^2$$

since:

$$D_\ell = \frac{D_{ft}}{2} = \frac{1}{\cos 30 \text{ degrees}} = \frac{\sqrt{3}}{3}D_{ft}$$

where:

D_ℓ is the length of one side of the hexagon.
D_{ft} is the distance across flats of the hexagon, for a bundle considering clearances or tolerances between rods and duct.

Now:

$$D_\ell = (N_{pS} - 1)(D + D_s) + \frac{2\sqrt{3}}{3}\left(\frac{D}{2} + g\right)$$

$$D_{ft} = 2\left[\left(\frac{\sqrt{3}}{2}\right)N_{rings}(D + D_s) + \frac{D}{2} + g\right]$$

where:

D is the rod diameter
D_s is the wire wrap diameter
g is the rod-to-duct spacing
N_p is the number of rods
N_{pS} is the number of rods along a side
N_{rings} is the number of rings.

2. Total cross-sectional area for flow A_{ft}:

$$A_{ft} = A_{hT} - N_p\frac{\pi}{4}(D^2 + D_s^2)$$

3. Total wetted perimeter P_{wT}:

$$P_{wT} = 6D_\ell + N_p\pi D + N_p\pi D_s = 2\sqrt{3}D_{ft} + N_p\pi(D + D_s)$$

4. Equivalent diameter for overall hexagonal array D_{eT}:

$$D_{eT} = \frac{4A_{fT}}{P_{wT}}$$
$$= \frac{2\sqrt{3}D_{ft}^2 - N_p\pi(D^2 + D_s^2)}{2\sqrt{3}D_{ft} + N_p\pi(D + D_s)}$$

PROOF OF LOCAL VOLUME-AVERAGING THEOREMS OF CHAPTER 5

Two theorems dealing with local volume averaging were presented in Section III without proof. They were the theorem for local volume-averaging of a divergence, Eq. 5-5 and an expression for the divergence of an intrinsic local volume average, Eq. 5-6. A formal derivation has been given by Slattery [1]. A simple geometrical derivation was more recently presented by Whitaker [2]. Whitaker's derivation is presented in this Appendix in nearly directly in the manner he proposed it. Although it is performed using the scalar fluid property B, the results can be directly generalized to vector forms such as Eqs. 5-5 and 5-6.

The essential question that these theorems involve is the rule for interchanging the sequence of spatial integration with spatial differentiation for the function B associated with the fluid in a system consisting of a continuous fluid phase and a separate discontinuous phase. This discontinuous phase can be either a solid (porous or nonporous), liquid or gas. In Chapter 5 our interest was focused on a system in which the continuous fluid phase was a single-phase, liquid and the discontinuous phase was stationary solids. This question can be expressed as the determination of the second term on the right side in the following equation:

$$\langle \text{div } B \rangle = \text{div}\langle B \rangle + ? \tag{K-1}$$

Start by geometrically identifying the average volume for our system for convenience as a sphere arbitrarily placed within the fluid–solid system. Following the definitions of Chapter 5 as illustrated in Figure 5-1, this sphere has total volume V, fluid volume V_f, fluid–solid interfacial area A_{fs}, and fluid en-

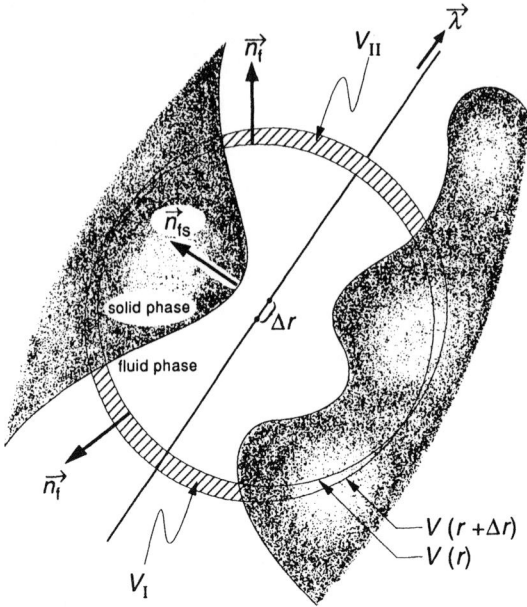

Figure K-1 A uniform translation of the averaging volume.

trance and exit area A_f. For the derivation two such averaging volumes are to be considered with centroids at positions r and $r + \Delta r$ along a straight line. The orientation of this straight line is designated by the unit vector $\vec{\lambda}$. These averaging volumes, the second of which is created by a uniform translation of the first in the $\vec{\lambda}$ direction, are illustrated in Figure K-1. The directional derivative d/dr of the function B integrated over the fluid volume $V_f(r, t)$ is written using the definition of the derivative as:

$$\frac{d}{dr} \int_{V_f(r,t)} B dV = \lim_{\Delta r \to 0} \left[\frac{\int_{V_f(r+\Delta r,t)} B dV - \int_{V_f(r,t)} B dV}{\Delta r} \right] \tag{K-2}$$

The intersection of the two integrals of Eq. K-2 cancel yielding:

$$\frac{d}{dr} \int_{V_f(r,t)} B dV = \lim_{\Delta r \to 0} \left[\frac{\int_{V_{II}(\Delta r,t)} B dV_{II} - \int_{V_I(\Delta r,t)} B dV_I}{\Delta r} \right] \tag{K-3}$$

where the fluid volumes $V_I(\Delta r, t)$ and $V_{II}(\Delta r, t)$ are illustrated in Figure K-1 as V_I and V_{II}.

Next we represent the volume integrals over fluid volumes $V_I(\Delta r, t)$ and $V_{II}(\Delta r, t)$ in terms of the fluid surface areas A_I and A_{II}, which are identified by the unit outwardly directed normal vector $\vec{n}_f$ shown in Figure K-1. Note that as $\Delta r \to 0$, these two areas will be coincident with the area of fluid entrances and exits, $A_f(t)$. The essential concept in the approach used to represent the volumes V_I and V_{II} is to visualize that the displacement of a circular element of surface area dA_I or dA_{II} by the amount Δr creates a cylinder of volume dV

whose cross-sectional area dA_{cs} depends on the angle between the surface normal $\vec{n}_f$ and the orientation $\vec{\lambda}$ of the line segment Δr. Specifically, the cylinder cross-sectional area dA_{cs} is related to the surface elements dA_I and dA_{II} by:

$$dA_{cs} = \vec{\lambda} \cdot \vec{n}_f dA_{II} \text{ over } A_{II} \qquad \text{(K-4a)}$$

and

$$dA_{cs} = -\vec{\lambda} \cdot \vec{n}_f dA_I \text{ over } A_I \qquad \text{(K-4b)}$$

Hence the volume integrals in Eq. K-3 can be expressed in terms of area integrals as:

$$dV_I = -\Delta r \vec{\lambda} \cdot \vec{n}_f dA_I \qquad \text{(K-5a)}$$

and

$$dV_{II} = +\Delta r \vec{\lambda} \cdot \vec{n}_f dA_{II} \qquad \text{(K-5b)}$$

Now these relations do not properly account for the volume elements located in the vicinity of the contact point between the surface of the averaging volume and the fluid–solid interface because the shape of the solid surface is not necessarily aligned in the $\vec{\lambda}$ direction. Figure K-2 illustrates this region. The error in volume can be approximated by assuming that Δr also represents the average width of such a typical volume element and that P is the length of the contact line between the surface of the averaging volume and the fluid–solid

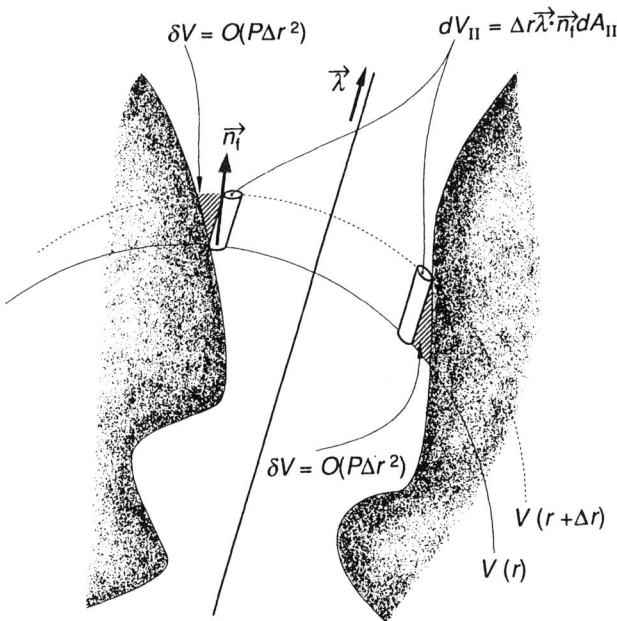

Figure K-2 Detail of volumes near the fluid–surface interface.

interface. Then the error in volume is of order $P\Delta r^2$, i.e.:

$$\delta V = 0(P\Delta r^2) \tag{K-6}$$

Substituting the results of Eqs. K-5 and K-6 into K-3 yields:

$$\frac{d}{dr}\int_{V_f(r,t)} B dV$$

$$= \lim_{\Delta r \to 0} \left[\frac{\int_{A_{II}(r,t)} B \Delta r \vec{\lambda} \cdot \vec{n}_f dA_{II} + \int_{A_I(r,t)} B \Delta r \vec{\lambda} \cdot \vec{n}_f dA_I + 0(BP\Delta r^2)}{\Delta r} \right] \tag{K-7}$$

Since Δr and $\vec{\lambda}$ are independent of position, they can be removed from the integrals. Further, since:

$$A_f(r,t) = \lim_{\Delta r \to 0} (A_I(r,t) + A_{II}(r,t)) \tag{K-8}$$

Eq. K-7 in the limit as $\Delta r \to 0$ becomes:

$$\frac{d}{dr}\int_{V_f(r,t)} B dV = \vec{\lambda} \cdot \int_{A_f(r,t)} \vec{n}_f B dA \tag{K-9}$$

Expressing the derivative with respect to r in the form:

$$\frac{d}{dr} = \vec{\lambda} \cdot \vec{\nabla} \tag{K-10}$$

leads to the desired result, an expression for the divergence of the intrinsic local volume average of B over the fluid volume only:

$$\vec{\nabla}\int_{V_f(t)} B dV = \int_{A_f(t)} \vec{n}_f B dA \tag{5-6a}$$

where $\vec{n}_f$ here is identically $\vec{n}$ in Chapter 5.

The second desired result, the local volume average of a divergence, can be obtained by utilizing Eq. 5-6a in the divergence theorem, i.e.:

$$\frac{1}{V}\int_{V_f(t)} \vec{\nabla} B dV = \frac{1}{V}\int_{A_f(t)} \vec{n}_f B dA + \frac{1}{V}\int_{A_{fs}(t)} \vec{n}_{fs} B dA$$

becomes:

$$\langle \text{div } B \rangle = \text{div}\langle B \rangle + \frac{1}{V}\int_{A_{fs}} \vec{n}_{fs} B dA$$

REFERENCES

1. Slattery, J. C. *Momentum, Energy and Mass Transfer in Continua.* New York: McGraw-Hill, 1972.
2. Whitaker, S. A simple geometrical derivation of the spatial averaging theorem. Chemical Engineering Education 19:18–21, 1985.

INDEX

Numbers in *italics* indicate figures. Page numbers followed by t indicate tables.